Protozoa and Their Role in Marine Processes

NATO ASI Series

Advanced Science Institutes Series

A series presenting the results of activities sponsored by the NATO Science Committee, which aims at the dissemination of advanced scientific and technological knowledge, with a view to strengthening links between scientific communities.

The Series is published by an international board of publishers in conjunction with the NATO Scientific Affairs Division

A Life Sciences B Physics	Plenum Publishing Corporation London and New York
C Mathematical and Physical Sciences D Behavioural and Social Sciences E Applied Sciences	Kluwer Academic Publishers Dordrecht, Boston and London
F Computer and Systems Sciences G Ecological Sciences H Cell Biology	Springer-Verlag Berlin Heidelberg New York London Paris Tokyo Hong Kong Barcelona

Series G: Ecological Sciences Vol. 25

Protozoa and Their Role in Marine Processes

Edited by

P. C. Reid
C. M. Turley
Plymouth Marine Laboratory
Citadel Hill
Plymouth, Devon PL1 2PB
United Kingdom

and

P. H. Burkill
Plymouth Marine Laboratory
Prospect Place
West Hoe
Plymouth, Devon PL1 3DH
United Kingdom

Springer-Verlag Berlin Heidelberg New York
London Paris Tokyo Hong Kong Barcelona
Published in cooperation with NATO Scientific Affairs Division

Proceedings of the NATO Advanced Study Institute on Protozoa and Their Role in Marine Processes held in Plymouth (United Kingdom), 24 July–5 August 1988.

ISBN-13: 978-3-642-73183-9 e-ISBN-13: 978-3-642-73181-5
DOI: 10.1007/978-3-642-73181-5

Softcover reprint of the hardcover 1st edition 1991

2131/3140-543210 – Printed on acid-free-paper

PREFACE

In the summer of 1988, under NATO sponsorship, approximately 80 scientists lived and worked together in Plymouth for two weeks to evaluate the ecological role of protozoa in the sea. Through the convivial surroundings, close working conditions and special facilities that had been brought together for NATO ASI 604/87 a 'melting pot' of ideas was formed, which stimulated the multidisciplinary creativity which is expressed in this book and in a second series of papers which will be published in Marine Microbial Food Webs under the title - "Protozoa and their Role in Marine Microbial Food Webs".

Discussions of the role of protozoa in the microbial food web, in the cycling of carbon and nitrogen and the extent to which this web acts as a link or sink to metazoa in the water column were major themes of the ASI. Structured sessions covering oral and poster presentations, field work, modelling, laboratory practicals and demonstrations of techniques such as image analysis and flow cytometry, formed the core of the meeting. Participants took part enthusiastically in the practical sessions developing new concepts and obtaining new insights into their work. The practicals included a 'protozoo' and some beautiful films and videos. Field excursions were made to a range of sites including a unique marine sewage farm at Looe in Cornwall, (Jones this volume). Interactive workshops allowed scientists with no modelling experience to input their results to three simulation models and a flow analysis package.

The major discussions of the ASI are encapsulated in the papers of this book which comprise both invited keynote addresses and session summaries. In the index the papers are grouped under headings representing the main sessions of the meeting, which in general, follow the order of presentation at the ASI.

The session on Taxonomy drew attention to the huge diversity of 'protists' found in marine environments, their varied nutritional modes and outlined a revised evolutionary tree. Few marine protists have so far been isolated in culture and biotechnological applications are in their infancy, although many of the results of the ASI have relevance in this area. One presentation outlined the potential for manipulating the microbial assemblage at the base of the food chain to maximise yields in commercial prawn farming. Rapid advances are being made in methodology, in particular, epifluorescence microscopy has made a very great contribution to microbial studies since the last ASI on a related subject in 1980 [Bougis P (ed) (1982) Marine Pelagic Protozoa and Microzooplankton Ecology. Ann Inst océanogr Paris 58(S): 1-352]. This microscopic technique contributed to one of the highlights of the meeting, the recognition of the importance of chloroplast retention by protists. Examples of endosymbiosis from a wide range of algal groups were described, some retaining their original tests within the host, others selectively adsorbing plastids. In the sessions on Trophic Behaviour and Community Grazing, phagotrophic protists were seen to play an important role in structuring food webs because of their fast growth rates and efficient energetics and production. Estimate of production were seen to vary by two orders of magnitude. One study reported selective grazing by heterotrophic nanoplankton on bivalve gametes. This discovery has implications for microzooplankton dynamics and may be one of the causes of year-to-year variability of intertidal settlement by gamete releasing organisms. The session on Mineral Flux demonstrated interannual variability of oceanic sedimentation from long-term sediment trap studies - a discovery which has direct relevance to studies of global climate change. The parallel session on 'Snow and Fluff', another highlight of the meeting, showed the important role that protozoa play in the vertical transport of particles in the sea and in the formation and decomposition of aggregates in the water column and on arrival on the sea-bed. Protozoa were also seen to act as 'crucial' agents in the transfer and/or recycling of nutrients a theme which was further developed in the modelling sessions. Ecological discussions stressed the extreme diversity of habitats and niches occupied by protozoa and their great temporal and spatial variation. Larry Pomeroy, in his summation of the ASI, using hierarchical theory, stressed the need to examine this complexity at similar levels of ecological development.

The success of the meeting was only possible through the hard work and enthusiasm of a dedicated team of local and international helpers - see appended list of acknowledgements. The local team included three sub-committees to cover logistics, catering and the social programme. Delegation of responsibilities to these committees, spread the workload, broadened the input of ideas and proved essential to the efficient running of the ASI. During the ASI Plymouth celebrated the 400th anniversary of the Armada providing a background and theme for a highly successful social programme with Dave Robins as master of ceremonies.

I wish, in particular, to thank Peter Burkill for his imagination and help in the planning and running of the ASI and Carol Turley for her financial wizardry and great assistance in the production of this book. The whole operation would not have been possible without Angie Smith who ran the ASI secretariat and typed, revised and concatenated this volume.

We are all indebted to Professor B L Bayne, Director, Plymouth Marine Laboratory, for his support and for the use of facilities during, prior and subsequent to the ASI. We also thank Professor L Heath, Polytechnic South West, for access to the well equipped laboratories of the Department of Biological Sciences and for the enthusiastic support of his technicians.

The memories of this Plymouth ASI have left a lasting impression on all participants which will, I believe, strongly influence the future direction of marine microbial research. It is our hope that this book will aid this process and provide a fitting record of the hard work and team spirit engendered in all who were involved.

Chris Reid
Director NATO ASI 604/87

ACKNOWLEDGEMENTS

The Director and Organising Committee of the NATO Advanced Study Institute 604/87 on 'Protozoa and their Role in Marine Processes' wish to acknowledge and thank NATO and the following organisations, companies and individuals for their support and assistance.

Commission of the European Communities
UNESCO
National Science Foundation, USA
Office of Naval Resarch, USA

Plymouth Marine Laboratory
Polytechnic South West
The Marine Biological Association of the UK
Plymouth City Council

Analytical Measuring Systems
The British Petroleum Company plc
Cell Systems
Coulter Electronics Ltd
Duncan and Associates
Ellis Horwood Ltd
Elsevier Applied Science Publishers Ltd
Nikon UK Ltd
Olympus Optical Co (UK) Ltd
St Austell Brewery Co Ltd
Sterilin Ltd
Vospers Rentals Ltd
Carl Zeiss (Oberkochen) Ltd

Main Committee Members:

Dr PC Reid (Director)
Dr PH Burkill (Secretary)
Dr CM Turley (Treasurer)

Other Committee Members:

Dr H Ducklow (USA)
Dr JC Green (UK)
Dr MB Jones (UK)
Prof F Rassoulzadegan (F)
Dr Gene Small (USA)
Prof V Smetacek (FRG)

Local Organisers:

D Robins
J Green
A Smith
E Pilling
R Leakey
A John
B Matthews
P Radford
A Taylor
J Stephens
G Tarran
G Siley
M Wilson at Buckland Abbey
National Trust Staff at Cotehele

For preparation of manuscripts and their production in camera-ready form, we thank Mrs Angie Smith, Dr Howard Bottrell and Mrs Linda Jones, and for photographic assistance, Mr David Nicholson.

CONTENTS

TAXONOMY

METHODS

POLLUTION

SYMBIOSIS

AUTECOLOGY

TROPHIC BEHAVIOUR

COMMUNITY GRAZING

ENERGETICS AND PRODUCTION

MARINE 'SNOW' AND 'FLUFF'

MINERAL FLUX

NUTRIENT CYCLING

MODELLING

TAXONOMY (46 - OR MORE - PROTISTAN PHYLA) - SESSION SUMMARY

Eugene B Small
University of Maryland
Department of Zoology
College Park Campus
Maryland 20742
USA

and

John C Green
Plymouth Marine Laboratory
Citadel Hill
Plymouth PL1 2PB
UK

INTRODUCTION

This session summarizes current thinking concerned with classification of protists and outlines a scheme for their presumed phylogenetic relationships. It is presented in three sections 1) a descriptive summary of the papers presented, 2) a discussion of current ideas concerning phylogeny and 3) a dendrogram of phylogenetic relationships among eukaryotic organisms.

As an introduction, **Gene Small** presented a ten minute, 50 projection slide and sound show that attempted to illustrate, using Normarski interference contrast photomicrography, the diversity (as well as aesthetic beauty) of the protist assemblage. Broad overviews of multiple phyletic groups were provided by the keynote lecturer, **Michael Sleigh** - (heterotrophic protists important in marine ecology, this volume), and by **John Green** (the chromophyte assemblage). **John Green** in characterizing the chromophyte assemblage of protistan phyla (those protist phyla containing chlorophyll *a* and *c* plastid pigments) illustrated the use of criteria other than pigments for understanding both the diversity and perhaps phylogenetic relatedness of this phyletic cluster. The characters that relate chromophytes to each other in addition to the chlorophyll *a* and *c* pigments are derived, *inter alia*

NATO ASI Series, Vol. G 25
Protozoa and Their Role in Marine Processes
Edited by P. C. Reid et al.

from transmission electron microscopy studies of the ultrastructural morphology of species a number of which are now clonally cultivated. Some of the more significant findings revealed by these methods are the importance of 1) the structure and arrangement of flagellar hairs, 2) the comparative ultrastructure and arrangement of flagella and flagellar root systems, and 3) the morphology of mitochondrial cristae, as indicators of affinity within chromophyte protists.

The next three papers presented new findings important to the understanding of systematic relationships within two different monophyletic protist assemblages, namely the ciliophorans and the foraminiferans. **David Montagnes**' data dealt with the results of his recent studies of the naked planktonic choreotrich ciliates,based on the use of protargol silver impregnation. As he and **Denis Lynn** demonstrated in a separate laboratory session, the protargol procedure permanently preserves and stains nuclear and ciliary systems. The morphology of these systems forms the basis of the systematics of all ciliates.

Andrew Gooday presented his recent findings of encrusting foraminiferans, the komokiaceans, and the presumably related megaloprotistan xenophyophoreans of the deep sea floor. This latter group is amongst the largest for a protist, and has been photographed with sizes up to 25 cm across on the ocean's floor by cameras mounted in the mouth of epibenthic sleds. **Jelle Bijma** presented a film, which utilized Normarski interference contrast imagery, on the life cycle and biology of the planktonic foraminiferan, *Hastigerina pelagica*. This film recorded that *H. pelagica* (1) has a voracious appetite for metazoan copepod nauplii (digestion takes place in 3-6 hours; (2) contains dinoflagellate endosymbiotes (*Paracysta robusta* and *Pyrocystis noctiluca*) which thus makes this planktonic foraminiferan a true mixotroph; and (3) has a sexual and reproductive cycle of twenty-nine days that appears to relate to the lunar cycle.

Two papers illustrated the new systematic information that is becoming available through the comparative analysis of chronometric biochemical molecules. Denis Lynn presented a paper on the phylogenetic and ecological implications derived from the recent studies by himself (Lynn and Sogin 1988) and by Gunderson and Sogin (see Gunderson *et al.* 1987 and references cited therein) on the use of 16s-like, ribosomal RNA for insights into our understanding of the time sequences for the evolution of 18 separate protistan phyla see Figure 1. As Lynn pointed out: 'A recent protist phylogeny (Gunderson *et al.* 1987) demonstrated: 1) genetic distances within some protist groups are as great as those **between** some eukaryotic kingdoms; 2) trypanosomatids and euglenids are a monophyletic group; 3) dinoflagellates and ciliates are a monophyletic group; 4) some chrysophytes and oomycetes are a monophyletic group; 5) the chlorophyte *Chlamydomonas* is associated with metaphytes; and 6) amoeboid forms evolved independently several times.

In the evolution of planktonic food webs eukaryotes may have arisen from a gram-positive bacterium, acquired mitochondria from a purple non-sulfur photosynthetic bacterium, and chloroplasts from a cyanobacterium. Four stages are envisaged: 1) a two-component 'web' with primary producers as cyanobacteria and heterotrophic bacteria as consumers: 2) a four component web with evolution of planktonic eukaryotic flagellates, including dinoflagellates, as additional primary producers and consumers; 3) a six-component web with evolution of ciliates as consumers, autotrophs, and mixotrophs; and 4) a seven-component web with evolution of metazooplankton, initially larval forms of annelids and molluscs followed by crustaceans'.

A word of caution is perhaps suggested by **Tony Soldo**'s report in this conference session. He noted from literature that reports of ribosomal coding DNA sequences for two separate amoeboid groups (the amoeboflagellate, *Naegleria gruberi*; and *Entamoeba histolytica*, an amoebozoan are to be found on circular, nonchromosomal DNA and are thus perhaps derived from

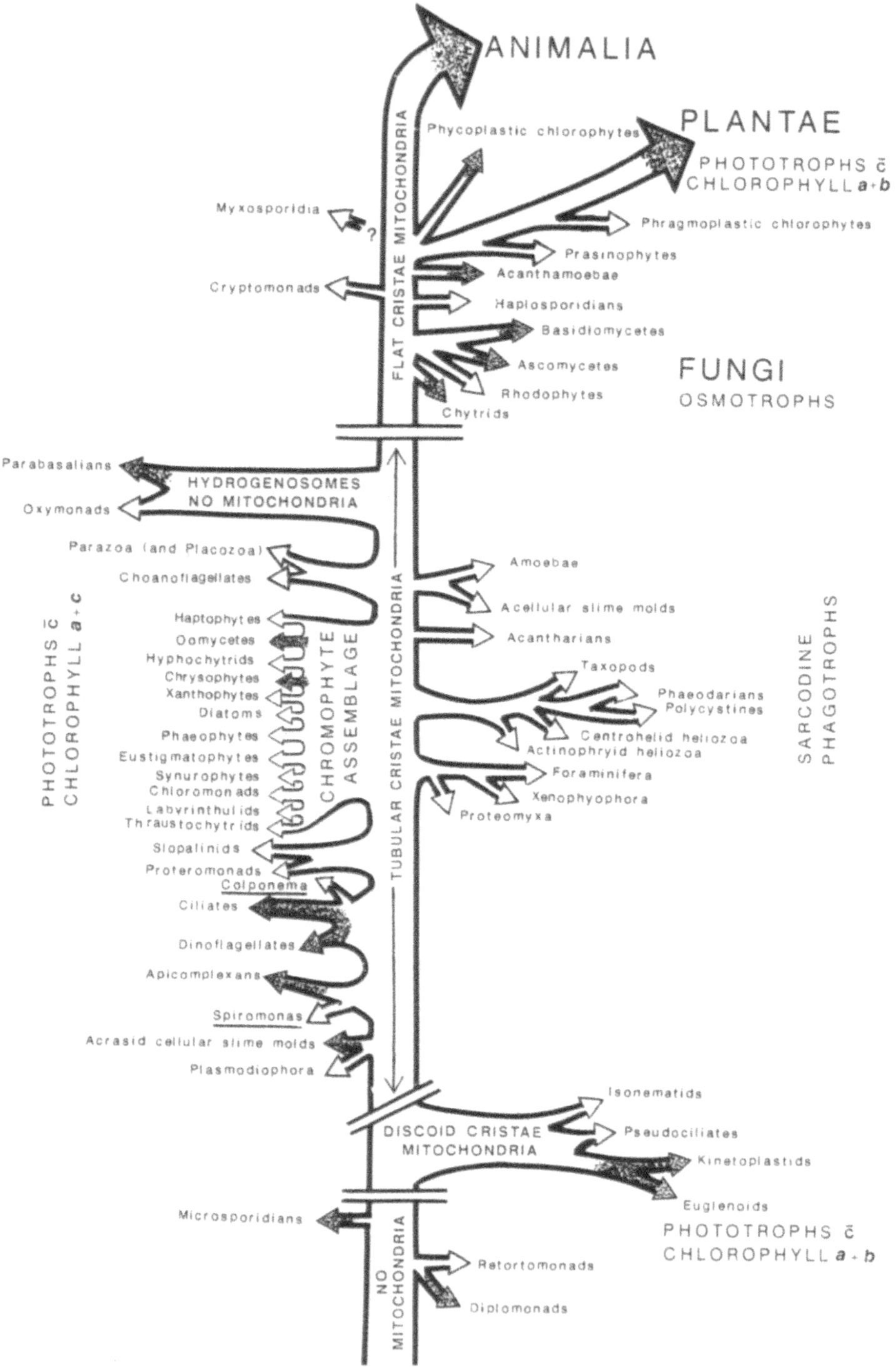

URKARYOTE ANCESTRY

Fig. 1. Dendrogram drawn by E B Small of presumed phylogenetic relationships amongst the recognized groups of eukaryotic organisms. The shaded arrows represent those groups for which their relative position to each other is based on comparative 16s or 16s-like ribosomal RNA sequence data (Sogin *et al.* 1986 Gunderson *et al.* 1987; Gunderson per. comm.). The placement of other arrows relative to each other is based on the comparative morphology of mitochondrical cristae, the comparative morphology of the flagellar' apparatus, and other more classic criteria (see Taylor 1978)

a separate plasmid ancestry. The same is true for *Paramecium* sp. **Tony Soldo** reported from his own studies that the DNA responsible for coding ribosomal RNA in some marine scuticociliates (*Parauronema, Miamiens* and *Anophryoides* spp.) is not circular but linear and extra-chromosomal (see Soldo and Brickson 1988). Such linear RNA encoding DNA molecules is also known for the ciliates *Tetrahymena, Stylonychia,* and *Oxytricha* as well as for the cellular slime mold, *Dictyostelium* sp. These data suggest a possible multiple origin for ribosomal RNA, thereby opening to question the monophyletic nature of RNA molecular sequence homologies. On the otherhand, these DNA sequences, be they circular of linear, may simply be nuclear genes present on small, hitherto unrecognized chromosomes. The comparative ease by which Soldo has isolated these linear molecules from the marine ciliates may facilitate future molecular analysis of the RNA forming DNA sequences.

SUMMARY (E B Small)

Our present understanding of the diversity of higher taxa has in the recent past undergone some major shifts in direction. As explained by **Michael Sleigh** in this volume, terms like 'protozoa' and 'algae' albeit useful as general descriptive words, have been replaced by the older Haekelian name Protista. Even that term as a word for a single kingdom of largely unicellular organisms is suspect, when we look at the positional relationships of the component protistan phyla on a single dendrogram (Figure 1; Small, unpublished) that illustrates in juxtaposition, the phylogenetic and biotic relationships of eukaryotic organisms. Such a tree must include the kinds of evidence (some of which were presented in this session) which determine the branching patterns: 1) comparative sequences of 16s-like ribosomal RNA (see Sogin *et al.* 1986, Gunderson *et al.* 1987, Gunderson pers. comm.) 2) the comparative ultrastructural morphology of mitochondrial cristae (for those groups of organisms, possessing mitochondria); 3) the comparative ultrastructural organization of the motility organelles or organellar systems (e.g. the flagellar apparatus of 'chromophyte' related protists, or its complete absence in the 'akonts'; 4) the ultrastructural organization of various types of pseudopods in sarcodinids).

Based on this evidence (see review by Taylor 1978) and especially molecular studies on RNA sequence data, some ideas are emerging concerning the 'Protista' (Gunderson *et al.* 1987). Protist phyla are more heterogeneous than are the plant and multicellular animal kingdoms and the true fungi, another multicellular assemblage of living organisms, which many phylogenetists also recognize at a kingdom level (e.g. Margulis 1981, and refs. cited therein). These kingdoms, all with flattened mitochondrial cristae, sit as the 3 major upper-most branches on top of the main tree trunk. Conversely, the protistan phyletic groups are found as unicellular components of multicellular kingdom branches (e.g. the chloroplastic chlorophytes) as well as the kingdom sized protistan phyletic clusters themselves (e.g. the chromophyte assemblage, the euglenoid - kinetoplastid assemblage; the ciliate - dinoflagellate cluster). Thus, it appears that a single unified **kingdom** protista is unreal. Rather, the protista represent several structural phyletic grades in the evolution of organisms (for instance, see Leedale 1974). In a similar sense there is no 16s-like ribosomal RNA evidence to support the unity of the chlorophyll pigmented 'phytoflagellates', since the chlorophyll *a* + *b* chlorophytes, the chlorophyll *a* + *c* chromophytes, and the chlorophyll *a* + *b* euglenoids appear as three separate branches on the main tree trunk. Apparently chloroplast symbioses occured quite independently and separately of one another (Gunderson *et al.* 1987). Other chlorophyll-bearing phyla (cryptomonads, dinoflagellates, rhodophytes) are similarly, separately placed. Protists whose motility has been described as pseudopodial are probably equally unrelated, since those sarcodinid forms for which 16s-like ribosomal RNA data is available (the acanthamoebae and the cellular acrasid slime molds) are at opposite ends of the dendrogram while the other sarcodine phagotrophs with tubular cristae mitochondria sit somewhere, presumably on the centre of the trunk.

There is, therefore, obvious tree-like configuration of phylogenetic relationships, a single trunk with the protistan groups as relatively short branches. The Y-shaped configuration of Taylor (1978) with chlorophyte and euglenoid (chlorophyll *a* + *b*) branch to one side and a chromophyte (chlorophyll *a* + *c*) branch to the other side appears untenable since the euglenoids rest as a short branch close to the base, the chromophytes are in the middle and the chlorophytes and plantae close to the top along with the fungi and animalia to either side.

REFERENCES

Gunderson JH, Elwood H, Ingold A, Kindle K, Sogin M (1987) Phylogenetic relationships between chlorophytes, chrysophytes, and oomycetes. Proc Natl Acad Sci USA 84:5823-5827

Leedale GF (1974) How many are the kingdoms of organisms? Taxon 23:261-270

Lynn DH, Sogin ML (1988) Assessment of phylogenetic relationships among ciliated protists using partial ribosomal RNA sequences derived from reverse transcripts. BioSystems 21:249-254

Margulis L (1981) Symgiosis in Cell Evolution: Life and its Environment on the Early Earth. WH Freeman, San Francisco, p 419

Sogin ML, Elwood HJ, Gunderson JH (1986) Evolutionary diversity of eukaryotic small-subunit rRNA genes. Proc Natl Acad Sci USA 83: 1383-1387

Soldo AT, Brickson SA (1988) Extrachromosomal DNA in marine ciliates. Soc Protozool Abstracts No 96:47

Taylor FJR, (1978) Problems in the development of an explicit hypothetical phylogeny of the lower eukaryotes. BioSystems 10:67-89

A TAXONOMIC REVIEW OF HETEROTROPHIC PROTISTS IMPORTANT IN MARINE ECOLOGY

Michael A Sleigh
Department of Biology
University of Southampton
Southampton SO9 3TU
United Kingdom

INTRODUCTION

The easy distinction between higher animals and higher plants led earlier biologists to extend the animal and plant kingdoms down to the unicellular level and separate algae taxonomically from protozoa. There-after one group was studied by botanists and the other by zoologists, and the separation was perpetuated in spite of the difficulty of defining a boundary between algae and protozoa. Closer study raised many questions about the taxonomy of these 'lower' organisms. The current view of cellular evolution (e.g. Woese 1987; Sleigh 1989) is that among the first cells (progenotes) were prokaryotes of different types which evolved to give the diversity of bacteria; other progenotes gave rise to nucleated, and probably amoeboid, phagotrophic eukaryotes with cytoskeletal systems, and at an early stage flagella and golgi systems evolved. Most eukaryotes developed symbiotic associations with aerobic bacteria to form mitochondria, different groups having different forms of internal cristae. In some cases symbiotic associations developed with photosynthetic cells to form plastids, different groups having different combinations of photosynthetic pigments and different arrangements of enclosing membranes. Evolutionary offshoots at various stages of this radiation of unicellular eukaryotes produced a great variety of organisms fitted for different environments and modes of life, including anaerobes, aerobes, phagotrophs, saprotrophs, autotrophs and parasites, comprising the assemblage (actually a grade of organization) that is now regarded as the Kingdom Protista or Protoctista. This kingdom contains all algal and protozoan groups as well as the flagellate fungi, so that the names Algae and Protozoa no longer have taxonomic meaning, but

NATO ASI Series, Vol. G 25
Protozoa and Their Role in Marine Processes
Edited by P. C. Reid et al.

(without capital letters) the terms algae and protozoa still have descriptive value.

The Kingdom Protista includes many branches of the evolutionary tree of life, some large, some small; the fundamental differences that separate the groups represented by these branches have led to suggestions that they should each be regarded as separate phyla (e.g. Corliss 1984), and 50 or more such protistan phyla have been recognised. If better understanding of relationships can be obtained in the future from rRNA sequencing and other studies, it may be possible to cluster some of these groups and recognise larger evolutionary branches (e.g. Euglenozoa, comprising Euglenophyta and Kinetoplasta) that will make the taxonomy of protists more manageable. In many groups of protists colonial forms have developed, leading to the evolution of macroscopic life-forms; some of these gained only limited complexity and are still regarded as protists, e.g. green, brown and red algae, but others adapted structurally and physiologically to a greater range of environments and comprise the separate plant, animal and fungal kingdoms.

Nutritionally, few of the photosynthetic protists that have been adequately studied are strict autotrophs, for autotrophs may show auxotrophy (a requirement for organic compounds, e.g. vitamins like cyanocobalamin, from the environment) to a greater or lesser degree. Many eukaryotes that contain chloroplasts also practise heterotrophy (i.e. are mixotrophs), gaining organic compounds by active uptake through their surface membranes (saprotrophy or osmotrophy) or by phagotrophy. Heterotrophic eukaryotes that depend upon osmotrophy for their organic carbon needs are restricted to habitats rich in appropriate compounds, e.g. as parasites or in areas of active decomposition, since in dilute media they compete poorly with bacteria whose surface area/volume ratio is usually much greater. Phagotrophic organisms require specific means of food collection; whether they be bacterivores, herbivores, carnivores or detritivores, they need to recognise, isolate from the environment, engulf and digest the food, so they are specialised for particular types of food uptake and restricted to sites where their food occurs. Particulate food may be deposited on or attached to a surface, or may be suspended in water; in either case some protists may use raptorial techniques to capture individual bacteria, plant cells, motile protists or even small metazoa, while other protists may sweep surfaces or filter water more indiscriminately to concentrate and separate food particles from the environment. In some cases heterotrophic protists may not engulf the prey, but suck out its fluid contents (myzocytosis).

The marine environment includes many different habitats, conditions of life for protists and food sources. In the open sea the surface film, photic zone and deeper aphotic regions provide different planktonic environments, and the oligotrophic ocean centres differ from eutrophic coastal and upwelling regions, with neritic zones often favouring auxotrophs because of vitamins derived from terrestrial sources. Among benthic habitats, substratum type, depth and illumination, oxygenation and water movement create many differences in physical and chemical conditions for the movement, attachment and respiration of protists and particularly for the types of food available to them. In both planktonic and benthic situations local conditions may lead to formation of specialised small-scale communities of organisms, bacteria and protists particularly, with mutual dependence upon one another. These may depend upon, for example, aggregates of organic matter sinking as marine snow or faecal pellets, or the bodies of dead plants or animals on the sea bed, or comprise perhaps the symbiotic associations of protists like foraminiferans or radiolarians with endosymbiotic algae and/or bacteria; certain protists are characteristic of such communities, while others are more evenly dispersed. In this discussion about protists of ecological importance, emphasis will be placed upon the relationship between taxonomic features and the ecology of the organisms concerned, but space will not allow more detailed discrimination of habitats than noted in this paragraph. Although parasitic protists may have substantial ecological impact, they are beyond the scope of this discussion; the systematic positions of some groups will only be mentioned in passing.

FEATURES OF HETEROTROPHIC PROTISTS

The purpose of this contribution is to point to the salient taxonomic features (especially those visible with the light microscope) of the more ecologically important heterotrophic protists in marine habitats and to refer to sources for more detailed information. Important technical terms used to describe protists will also be introduced. With the recognition of 50 or more phyla of protists, a taxonomic survey could become a simple catalogue, but fortunately many of the groups are small and of limited ecological importance, although their evolutionary and phylogenetic interest may be substantial. To make the diversity easier to handle, the groups will be divided into four categories on the basis of shared characters. These are the ciliates (Ciliophora), the flagellate and allied protists, the amoeboid protists and the spore-forming parasitic groups. Only the first of these is probably monophyletic, each of the others is believed to contain several or many branches of the protistan evolutionary tree, so their shared features should not be taken to imply close relationships. Parasites of the spore-forming groups (notably the Sporozoa, Microspora and Myxospora), as well as chytrids and others among the flagellate fungi may cause considerable damage, even epidemics, to populations of marine organisms, as may parasitic flagellates, ciliates and amoebae, but none of these will be discussed here.

FLAGELLATES AND ALLIED PROTISTS

This assemblage is taken to include the algal and fungal groups with flagellated gametes and/or zoospores, and the protistan groups that appear closely related to these, as well as the

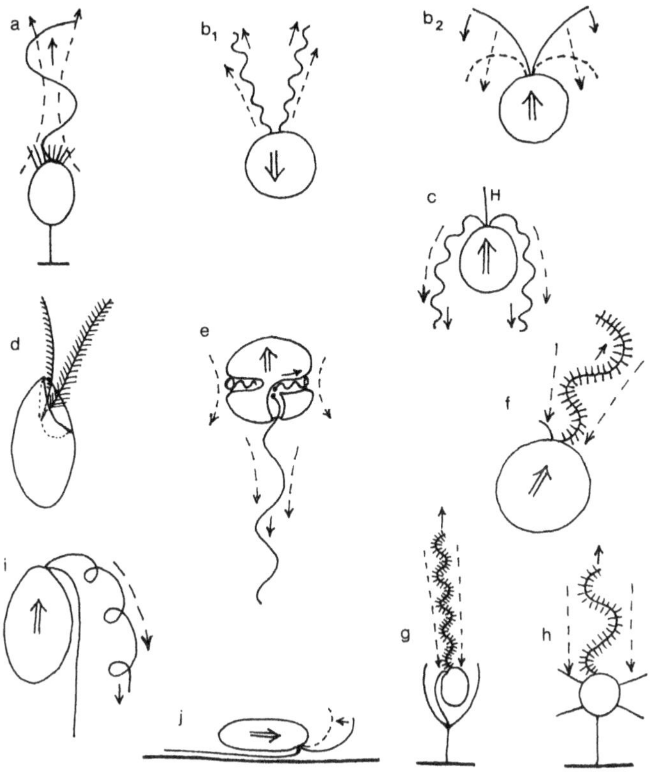

Fig. 1. Typical ways in which flagella are used by members of various flagellate groups. The motion of waves along the flagella is shown by a simple arrow, dotted arrows show the pattern of water flow and double-shafted arrows show the direction in which the cell is propelled (unless it is sedentary). a. A choanoflagellate drawing a feeding current through the collar; b. a chlorophyte being propelled in reverse by undulating flagella (b_1) and forward by a 'ciliary breast-stroke' (b_2); c. a haptophyte propelled with the haptonema (H) in front; d. a cryptophyte whose flagellar action is unknown; e. a dinoflagellate whose cell is rotated by the transverse flagellum in the equatorial groove (cingulum) and propelled by both this flagellum and the longitudinal flagellum that extends from the sulcus groove; f. a chrysophyte propelled by a long mastigoneme-bearing flagellum, while the shorter naked flagellum may help to trap the food; g. a loricate bicoecid and h. a helioflagellate both use mastigoneme-bearing flagella to create a feeding current; i. a euglenid drawn along by helical undulations of one flagellum while the other trails, often along a surface; j. a bodonid glides on one (usually longer) flagellum while the other may assist in drawing the cell along (figure modified from Sleigh 1989)

purely flagellate groups. Only the more important groups with marine representatives are discussed here, although a more complete list of the larger groups is given in the Appendix. Many of the groups contain autotrophic members, although few if any are entirely autotrophic, but several are entirely heterotrophic. Groups with chlorophyll-containing members possessing rigid walls or other continuous periplasts may contain osmotrophic mixotrophs or osmotrophic heterotrophs, whilst other chlorophyll-containing groups in which the body surface is naked or the periplast is interrupted at a cell mouth may include phagotrophic mixotrophs or phagotrophic heterotrophs. Those groups that are entirely heterotrophic are normally phagotrophic, except for osmotrophs among parasitic and necrophagous forms.

Cell size, shape, colour, stored food products and features of surface coats may be used in identification (Table 1), but a primary distinction between the groups of this assemblage is provided by the number of flagella, their appendages and the way in which the flagella are used in propulsion (Figure 1). Uniflagellate cells in which the flagella push the organism along (as in animal sperm) are found in chytrid gametes and in collar flagellates. Except when a flagellum has been reduced or lost, other flagellate protists have two or more flagella. These flagella may move in a similar manner, homodynamically, in chlorophytes and some haptophytes, or in dissimilar ways, heterodynamically, in other haptophytes, dinoflagellates, cryptophytes, heterokonts, euglenids and bodonids. Each of these groups will be considered briefly. Heterotrophic marine flagellates are in general not well known, and investigators are likely to encounter members of smaller and more obscure groups, as well as undescribed species.

Choanoflagellates are commonly found attached to all kinds of solid surfaces including quite small floating particles and the surfaces of other organisms, and are also abundant as free-swimming cells in the plankton; in both situations they are important bacterivores. The small (about 5 µm) colourless

Table 1. FEATURES OF FLAGELLATE GROUPS

Group	No: flagella	Hetero-/ Homo-dynamic	Basal body orientation	Shaft features	Colour	Body Surface
Choanoflagellata	1	-	-	(h)	t	na,ot,sl
Chlorophyta	2(4,+)	ho,he	or	na,sc	gr,t	(na)cw
Haptophyta	2	ho,he	or	na	gb,t	os,cs
Cryptophyta	2	he	pa	ma(2)	var,t	pp
Dinophyta	2	he	or	h,pr	br,t	na,tp
Chrysophyceae	2	he	or	ma(1)	gb,t	na,ss
Bicosoecida	2	he	or	ma(1)	t	ol
Helioflagellata	1(2)	(he)	pa/or	ma(1)	t	os,na,ss
Euglenophyta	2	he	pa	h,pr	gr,t	pl,ol
Bodonida	2	he	pa	h,pr	t	na

Group	Usual size range µm	Nutrition	Feeding mechanism	Food	Habit	Habitats
Choanoflagellata	5-15	ph	f	b	n,s,cl	o,c,w,xs
Chlorophyta	3-50	au,om	-	-	n,y,cl	o?c,xs
Haptophyta	3-15	au,pm	i	b,a,p	n,y	o,c,xs
Cryptophyta	2-40	au,oh	-	-	n,y	o,c,xs
Dinophyta	10-2000	au-ph	i,g	b,a,p	n,y,cl	o,c,xs
Chrysophyceae	3-20	au-ph	i	b,p,a	n,s,cl	o,c,w,xs
Bicosoecida	5-15	ph	i	b	s,cl	o,c,w,xs
Helioflagellata	5-20	pm,ph	f,i	b,a,p	n,s	o,c,w,xs
Euglenophyta	5-300	au-ph	i,g	b,p,a,d	n,v,s	w,as,xs
Bodonida	5-15	ph	i,g	b,d	n,v	o,c,w,xs,as

Key to abbreviations. Paired flagella may be homodynamic (ho) or heterodynamic (he), their bases may be oriented parallel (pa) or orthogonally (or) to one another; flagellar shafts may be naked (na), bear slender hairs (h), scales (sc), or tripartite mastigonemes (ma) on one or two flagella, the flagellar shaft may contain a paraxial rod (pr); the cells may be colourless (t), green (gr), golden-brown (gb), brown (br) or varied (var); the body surface may be naked (na), be enclosed in an organic test (ot) or a lorica containing silica (sl) or organic material (ol), a cell wall (cw), thecal plates (tp), a thin periplast (pp) or a pellicle (pl) or bear scales of organic (os), calcareous (cs) or siliceous nature (ss); their nutrition may be phagotrophic heterotrophy (ph), autotrophy (au), osmotrophic mixotrophy (om), osmotrophic heterotrophy (oh), phagotrophic mixotrophy (pm), or a combination of these; feeding may involve filtration (f), interception (i) or picking items from a surface (g); the food may be bacteria (b), algae (a), protozoa (p) or detritus (d); different forms swim (n) or creep over (v) or are sessile (s) upon surfaces, may live as symbionts (y) or form colonies (cl); they are found mainly in oceanic (o) or coastal (c), planktonic habitats, attached to suspended particles (w) or in oxic (xs) or anoxic (as) sediments.

cells carry a single flagellum which propels water away from the cell (Figure 1a), drawing it through a collar of fine pseudopodia (filopods) that surround the flagellar base. Particles as small as 0.2 µm are filtered from the water and carried down the filopods to be taken into food vacuoles near the base of the collar. Leadbeater (1985) divides collar flagellates into three families according to the presence or absence of a theca or of an open basket-like lorica composed of silica strips; members of all three groups may be attached or free-swimming, solitary or colonial. Because of the structure and movement of the flagellum, swimming cells are pushed along by a posterior flagellum.

Among the chlorophyte groups, typically containing chlorophylls a and b and storing starch, green marine flagellate unicells occur in the scale-less Chlorophyceae (Leedale and Hibberd 1985) and in the scale-bearing Prasinophyceae (Leedale 1985). All of the latter and most of the former are autotrophic, although the extent of osmotrophic mixotrophy is not known. There are no known phagotrophs in the group, and the known chlorophycean heterotrophs are freshwater rather than marine forms. Where there are four or more flagella they move homodynamically, as do the paired flagella in Chlorophyceae (Figure 1b), but in Prasinophyceae with two flagella they tend to be used heterodynamically, and cells with a single flagellum occur.

Members of the Haptophyta (Prymnesiophyta) are best known as small (about 5 µm) cells (Hibberd 1980; Hibberd and Leedale 1985a), mostly from marine plankton, where they are often abundant, although their ecological importance as heterotrophs is unknown. Typically the cells possess two equal hairless flagella which move together homodynamically (Figure 1c), but in some forms the flagella are unequal and heterodynamic, including forms where one flagellum trails and the other is undulated laterally. The group gains its name from the haptonema, a third flagellum-like appendage borne between the flagella. The haptonema may be longer or shorter than the flagella (or occasionally absent), and held in front of the cell or trailed behind it; it is thinner than a flagellum and moves differently, long ones often coiling; the name haptonema (adhesive thread) was given because it often attaches to other bodies by the tip or other parts. The cells normally contain two golden brown chloroplasts with chlorophylls a, c_1 and c_2 and with fucoxanthin as the dominant xanthophyll. Around the cell body are one or more layers of golgi-derived unmineralised scales, further surrounded in the coccolithophorids by scales bearing crystalline calcium carbonate. Some species have been shown to have complex life histories, sometimes with benthic stages. Most species are photosynthetic, storing leucosin, but many are phagotrophic mixotrophs, ingesting bacteria and small eukaryote cells at the non-flagellate pole of the cell.

Two flagella of almost the same length emerge laterally or antero-laterally from a vestibular slit on cryptophyte cells (Figure 1d). They are often held in front and are believed to be moved heterodynamically during bursts of swimming, but are not as easy to see as in many groups. The two flagellar bases lie parallel to each other and long mastigonemes (tripartite flagellar hairs) with a single terminal filament are borne in two rows on the longer flagellum and shorter mastigonemes with two terminal filaments in a single row on the shorter flagellum. The cells are usually between 2 and 20 µm long and are covered with a very delicate periplast of rectangular or hexagonal plates. Large ejectisomes occur in the vestibular walls and smaller ones under the general body surface. Autotrophic members contain chlorophylls a and c_2 as well as phycobiliproteins in unique chloroplasts that suggest an origin from eukaryote symbionts. No cryptophyte is proven to be phagotrophic, but apparent food vacuoles have been reported several times from electron micrographs: osmotrophic mixotrophy is reported in some cases and obligate osmotrophic heterotrophs occur in polluted situations. The group is not well known, and Santore and Leedale (1985) suggest that characters used in keys to the group may be unreliable.

Dinoflagellates are usually between 10 µm and 2 mm in diameter and most exist as flagellate unicells (Dodge 1982), though colonial forms, including filamentous stages, and many parasites and important photosynthetic symbionts are found in invertebrate animals and amoeboid protists (Taylor 1987). The cell bodies of dinoflagellates are often complex with a large nucleus whose chromosomes are permanently condensed, and in photosynthetic forms there are usually many brown plastids with chlorophylls a and c_2 and peridinin as a prominent xanthophyll. The swimming cells are enclosed in a theca of multiple (usually 3) membranes, between two of which cellulosic plates are laid down in many groups; These plates give the bodies rigidity and characteristic shapes. The two flagella of dinoflagellates are dissimilar (Figure 1e); flagellate cells of this group have one ribbon-like transverse flagellum which contains a striated paraxial strand, bears a single row of fine hairs and usually lies in a cingular groove, while the second, posterior, flagellum emerges from a longitudinal sulcus groove, bears simple hairs and ends in a fine point. In *Prorocentrum* and related forms both flagella emerge at the anterior end. Many autotrophic dinoflagellates are phagotrophic mixotrophs and about half of all dinoflagellate species are obligate heterotrophs. While photosynthetic species are common in neritic waters, Gaines and Elbrachter (1987) observed that "A majority of open ocean dinoflagellates are colourless heterotrophic forms, yet they remain almost totally unknown", and Lessard and Swift (1987) reported clearance rates for oceanic heterotrophic dinoflagellates feeding on phytoplankton of 1-28 µl $individual^{-1}$ h^{-1}, which is equivalent to that of ciliates in the same region. Some species are associated with surface sediments (Larsen 1985). In general the food of phagotrophic dinoflagellates seems to be the larger phytoplankton and ciliates, captured by 'pseudopodial veils' emerging from the sulcus region, and digested outside the theca, but smaller forms like *Oxyrrhis* may phagocytose smaller prey into internal food vacuoles; *Noctiluca* uses trailing mucus to capture prey. The phagotrophic marine ebriids with internal silica skeletons, e.g. *Hermesinum*, are also (doubtfully) considered by some to belong here (Lee 1985); better information is urgently needed. Heterotrophic dinoflagellates deserve more attention from protistologists, and it was good to see that several papers about them were given at this meeting.

The heterokont series of groups have flagellate cells in which one flagellum bears two rows of tripartite mastigonemes with two (or 3?) terminal filaments, and often a second more or less hairless flagellum, which usually emerges perpendicular to the first. Many members of these groups have a typical helical strand in the transition zone of the axoneme. The presence of mastigonemes on the hairy (tinsel) flagellum causes a reversal of the water flow around the beating flagellum, so that when waves pass along the flagellum from base to tip, the surrounding water is drawn towards the flagellar base (Figure 1f). The cells therefore swim with the (usually slightly curved) flagellum held in front of the cell, and in both swimming and attached cells the water flow created by the flagellum draws particles towards the cell surface where they can be intercepted and phagocytosed. The naked flagellum (when present) usually trails backwards and may be used for anchorage. This pattern of flagellation is found on the zoospores and/or the gametes of brown algae (Phaeophyceae), oomycete fungi, labyrinthulids and xanthophytes (Tribophyceae), on the cells of raphidophytes (chloromonads), and a single tinsel flagellum is found on silicoflagellates (*Dictyocha*), on the zoospores of eustigmatophytes and hyphochytrids and on the male gametes of some diatoms (Bacillariophyceae) (references in Sleigh 1989). None of these groups concern us here, although many oomycetes and labyrinthulids are saprophytes or parasites on marine plants, and there is a tendency towards osmotrophic mixotrophy in some other groups, with a few obligate heterotrophs among diatoms that live on the surface of seaweed (Li and Volcani 1987), the other groups are almost entirely photosynthetic.

Three remaining groups of heterokonts contain common marine phagotrophic species. The Chrysophyceae are a predominantly photosynthetic group of small (about 5-10 µm) cells with typical heterokont flagellation (Figure 1f), although the group contains a few coccoid, filamentous or palmelloid forms (Hibberd and Leedale 1985b). Typically one or two golden brown chloroplasts are present containing chlorophylls a, c_1 and c_2 and fucoxanthin, and leucosin is stored in the cytoplasm. A number of photosynthetic as well as colourless species phagocytose bacteria, and similar particles, brought to the cell surface by flagellar action and taken into food vacuoles near the flagellar base. Flagellate cells may be attached or free-swimming, and may form colonies. The group is best known from fresh water, but they are common bacterivores in the sea, and their true abundance and importance there is yet to be determined. The bicosoecids (often written bicoecids) have often been grouped with Chrysophyceae, but are always phagotrophic heterotrophs (Moestrup and Thomsen 1976). They are heterokonts anchored within a sessile or stalked cup-shaped lorica by the posterior flagellum (Figure 1g), which can retract to draw the cell to the base of the lorica. The anterior tinsel flagellum draws a narrow feeding current to the cell surface where there is a prominent lip at one side of the flagellar base. Bicosoecids are commonly found attached to particles like marine snow or larger surfaces in any marine habitat. Similar marine sites are occupied by diverse helioflagellates (Figure 1h), which have a single anterior tinsel flagellum that is used like the chrysophycean flagellum to draw particles towards the cell, where particles over a wide size range may be intercepted and held by slender, stiff, but contractile, axopods that radiate out from the body. These are colourless heterotrophs, but *Pedinella* is a photosynthetic form with short axopods that appears intermediate between chrysophyceans and some other helioflagellates (Patterson and Fençhel 1985). Although most of these forms are stalked, some are found free-swimming.

A final pair of flagellate groups, the euglenids and bodonid kinetoplastids are considered to be fairly closely related, and can be considered together. Both are basically biflagellate with one flagellum that beats to the anterior or the side and another that trails behind, often in contact with the substratum as a sort of 'ski' (Figures 1i, j). Either or both flagella may have slender hairs and other attached material and both usually contain a striated paraxial rod. The small bodonids are usually easily recognised from their movement, as well as from the presence of a large concentration of mitochondrial DNA (the kinetoplast) near to the almost parallel flagellar bases (Vickerman 1976). Bodonids have a tubular cytostome through which they endocytose single bacteria or similar particles from the surface over which they move or from the water current created by the anterior flagellum. The most familiar euglenids are the large green free-swimming forms with chlorophylls a and b commonly found in fresh water and occasionally in marine sediments, but there are numerous, usually smaller, colourless species living osmotrophically or phagotrophically in marine sediments (Larsen 1987). Phagotrophic euglenids often possess a prominent ingestion apparatus associated with a cytostome near to the flagellar pocket; the food is commonly diatoms or other flagellates. The body is reinforced by a pellicle, usually of strips of material lying just beneath the cell membrane. While euglenids are most commonly associated with sediments, bodonids are rather ubiquitous and occur in sediments and on small floating particles as well as swimming in more open waters.

THE AMOEBOID PROTISTS

The twelve or more major taxa of amoeboid protists are grouped into two assemblages (see Appendix) on the basis of the form of the pseudopodia; this is convenient and so it will be followed, even though it is recognised that the main taxa within an assemblage may not be at all closely related and may be regarded as independent phyla by some authorities. In the rhizopod assemblage the pseudopodia are flexible and constantly changing shape, whilst the axopodia of actinopods are needle-like, supported by axial groups of microtubules, and more permanent, though they can change in length. Rhizopodia are further subdivided into blunt lobopodia, slender hyaline and often branched filopodia and the anastomosing networks of reticulopodia. A recent guide to the classification of rhizopods has been written by Page (1987). Some features of amoeboid groups are summarised in Table 2, and a classification of the assemblage is given in the Appendix.

The most commonly found naked marine forms with rhizopodia are likely to be small amoebae in the 5-25 µm size range. They are generally bacterivorous forms that collect bacteria from the surfaces of small and large floating particles, sediment particles, animals or plants or the water/air interface, as well as being found (dislodged?) in the plankton. Small marine amoebae occur in at least three of the main taxa (regarded by Page (1987) as classes, but sometimes given higher rank): the Heterolobosea (Figure 2a), which are generally monopodial and move eruptively, have discoidal mitochondrial cristae and often a flagellate stage, suggesting a link with flagellate protists; the Lobosea (Figure 2b), whose more broadly lobed and smoothly-moving pseudopods may carry sub-pseudopodia, and with tubular mitochondrial cristae but no flagellate stage; and the Filosea (Figure 2c), with filopodia that do not arise from lobopods, have either discoidal or vesicular mitochondrial cristae and no flagellate stage. Further details of the forms of the pseudopodia, the movement of the amoebae, the form of the nuclei and the nature of any surface specializations are used to identify families of amoebae defined by Page (1987); they are notoriously difficult and little studied. Larger naked amoebae (mainly Lobosea) may be found in marine sediments, as may some testate (shelled) amoebae with lobed (Testacealobosea) or more usually filose (Testaceafilosea) pseudopods.

The reticulate pseudopodial networks of foraminiferids and their relatives (Figure 2d) appear studded with moving granules (hence the name of the class Granuloreticulosea). These foraminiferids have tests with one to many chambers; some tests are simply composed of mucopolysaccharide (e.g. *Allogromia*), but in others arenaceous or similar fragments are incorporated into the test, and in the best-known forms the organic layer is reinforced with calcite, aragonite, or occasionally silica spicules, the common calcite forms being either porcellanous shells formed from small crystals or hyaline shells formed from

Table 2. FEATURES OF AMOEBOID GROUPS

Group	Surface structure	Food	Habit	Habitats	Special Features	Usual size range μm
Rhizopoda:						
Lobosea	na,ot,at	b,a,p,m,d	v(n)	o,c,w,xs,as		5-500
Filosea	na,ot,ss	b,a,d	v	xs,as		10-200
Granulo-reticulosea	ot,ct,at,st	b,a,p,m,d	n,v	o,c,xs	(symbionts)	20-20,000
Xenophyophorea	at	b,d+?	s	do,xs	Barite	20,000
Actinopoda:						
Heliozoea	na,ss	b,a,p,m	n,s	xs		10-400
Phaeodarea	sh	b,a,p,m	n	o,do	Phaeodium	100-500
Polycystinea	sk	b,a,p,m	n,cl	o,pz	symbionts	50-2000
Acantharea	sr	b,a,p,m	n	o,pz	symbionts	100-1000

Key to abbreviations. The body surface may be naked (na), or enclosed in a test of organic (ot), agglutinated (at), calcareous (ct) or siliceous (st) material, bear siliceous scales (ss) or be supported by an internal skeleton of hollow siliceous spicules (sh), solid siliceous spicules (sk) or spicules of strontium sulphate (sr); other abbreviations as in Table 1, except that some amoeboid forms eat metazoa (m), are typical of the deep ocean (do) or photic zone (pz).

larger radial crystals. Foraminiferid life cycles vary, they may have amoeboid or flagellate gametes, and some have a haplo-diploid alternation of generations, with slight differences in test structure in the two generations. The reticulopods emerge from one or more openings of the test and move constantly as they extend widely over the substratum or in the water as a sweeping or fishing net, trapping and engulfing any small organisms like bacteria and diatoms, or, in the case of planktonic forms, even large active prey like copepods, small medusae or *Sagitta*. There is a small number of genera of large (500 μm) and more bulbous planktonic foraminiferids, e.g. *Globigerina* with radiating calcareous spines and dinoflagellate symbionts, but most of the members of this very large class are flattened or irregular benthic forms found on rocks, weeds and

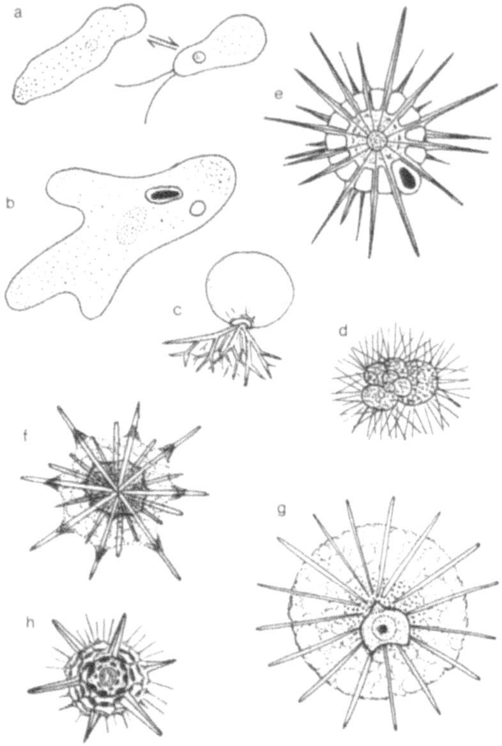

Fig. 2. Body forms in amoeboid protozoa. The small heterolobosean amoeba (a) with an eruptive pseudopodium may reversibly transform to a flagellate stage; the (usually) larger lobose amoeba (b) moves more smoothly; the slender, branching, pseudopods of filose amoebae are hyaline (c - this form has a test); foraminifera (d) often have calcareous tests, in this case with spines, and fine anastomosing pseudopods in an extensive reticular system; among actinopods the heliozoa (e) normally have no skeleton except the microtubular axes of axopodia, while acantharia (f) have spicules of strontium sulphate, phaeodarea (g) have hollow silica spines and polycystinea (h) have skeletons of solid silica elements (redrawn from Sleigh 1989)

sediments at all depths, those nearer the surface occasionally containing symbiotic diatoms or other algae. They are classified according to the nature of the test, and the shape, number, arrangement and extent of perforation and decoration of the chambers of which the test is composed. Bock, Hay and Lee (1985) give a key to families and Loeblich and Tappan (1984) give a recent classification of the group.

The Xenophyophorea are a class of presumed amoeboid forms with a branched organic test several cm long, found in deep-sea sediments. The tests carry agglutinated material and the living strands within contain barium sulphate crystals (Gooday and Nott 1982). Details of their structure, physiology and taxonomic relationships are awaited with interest. The numbers of xenophyophores and foraminifera on sea-floor sediments suggest that such amoebae have a substantial role in cycling benthic materials, and perhaps in sequestering barium.

Actinopods tend to be spherical planktonic forms, often with symbiotic algae, but probably represent a life form evolved several times. In all cases the axopodia are supported by bundles of microtubules in arrays that characterise each group: the axopodia carry extrusomes whose contents may be extruded during the capture of prey that contact their surface: the axopodia may then retract, often with great rapidity, to draw their prey to the cell surface, or a cytoplasmic veil may flow around the trapped prey to engulf it for digestion. Four classes are generally recognised, the Heliozoea lack internal skeletons, but the Acantharea, Polycystinea and Phaeodarea possess internal skeletons; the last two classes are subdivisions of the former radiolaria, and all of the last three groups are sometimes loosely (though incorrectly) referred to as radiolaria.

The class Heliozoea has been claimed to contain up to six groups whose ancestry is different, and forms in this group are seldom planktonic, since most are attached to some surface by stalks or axopods (Febvre-Chevalier 1985). In the familiar actinophryid heliozoon *Actinophrys* (Figure 2e), found in marine and freshwater habitats, the microtubular axes of the axopods radiate from the surface of a central nucleus, and the cell surface is naked. The axopodial axes of centrohelids (e.g. *Heterophrys*) radiate from a central centroplast structure and the nucleus is eccentric, the axopods are slender with prominent extrusome granules and highly contractile, and the cell surface bears organic or siliceous scales or spines between the axopods. Marine centrohelids, and similar forms (e.g. *Actinocoryne*) with strongly contractile stalks, are found attached to bryozoa, algae and similar surfaces. The desmothoracid heliozoa with an organic extracellular capsule are freshwater forms, but the curious bilaterally symmetrical taxopodid *Sticholonche* is a marine pelagic form which carries lateral axopodia that can be moved like oars around a basal articulation, as well as fixed dorsal axopodia and clusters of siliceous spicules. The heliozoa probably have less ecological significance than helioflagellates, but their local abundance suggests some importance.

Although the skeletons of radiolaria are well known from the fossil record, their biology and that of acantharea is poorly known because they are planktonic forms of the open sea that are not easily studied in the laboratory. In the members of all three classes the cytoplasm is divided by a capsule membrane into inner and outer zones, but the capsule and the cytoplasmic and skeletal features of the three groups are different. In acantharea (Figure 2f) 20 needle-shaped spicules of strontium sulphate radiate from the cell centre, and a number of axopodia also radiate from a central granule. Both spicules and axopodia emerge through a delicate capsule, which encloses an endoplasmic zone containing crystals, oil drops, symbiotic dinoflagellate, haptophyte or occasionally cryptophyte(?) algae and one (when young) to many (when adult) small nuclei. The outer cortical region contains a reticulopodial network enclosed within a filamentous meshwork which is attached by contractile myonemes to the emergent spicules; axopodia and slender filopodia emerge from the surface and capture diverse microplanktonic prey. These are quite large protists and are important members of the plankton in the surface zones of warmer seas. Cachon and Cachon (1985) follow a classification based on the way in which the spicules meet at the cell centre, but suggest this is imperfect; the arrangement of myonemes may also be used.

In both of the radiolarian classes internal siliceous skeletons are present in most species and provide a basis for most taxonomic schemes (Anderson 1983). In the Phaeodarea (Figure 2g) the skeleton is typically formed of hollow silica spines linked by organic material to each other or to a more continuous internal shell. Members of this group are mostly of intermediate (100-500 µm) size, with a capsule perforated by one large and (usually) two smaller pores, without symbionts, but usually with a prominent brown granular mass, the phaeodium, outside the main pore of the capsule. Large bundles of axopodia radiate from an axoplast granule at each smaller pore, and the large pore is said to act as a cytostome. The cytoplasm within the capsule contains oil droplets and vacuoles and a single large nucleus (except in reproducing cells). The capsule is surrounded by a vacuolated calymma region. These radiolaria are mostly members of the deep ocean plankton; commonly at depths of several thousand metres.

Most members of the Polycystinea (Figure 2h) have solid siliceous spines or (sometimes concentric) internal shells in a great variety of shapes. Axopodia radiate from one or more intra-capsular axoplast granules, passing through the thick capsule by a collar-like fusule. Sometimes the axoplast is at the centre of the large central nucleus, or beside the nucleus, and in other cases there are many axoplasts on the nuclear surface; the form and position of the axoplast(s) may be used in classification (Cachon and Cachon, 1985). The capsule is perforated by many small scattered pores in the Spumellarida or by a cluster of pores at one pole in the Nassellarida; this arrangement corresponds with the more spherical organisation of the skeleton and body of spumellarids and the pyramidal arrangement of structures in nassellarids. The cytoplasm within the capsule contains oil droplets and vacuoles and communicates through these pores in the capsule with the highly vacuolate calymma region which contains food vacuoles and symbiotic algae (usually dinoflagellates, but occasionally prasinophytes in spumellarids). Axopods and filopods extend outwards from the calymma and from the projecting spicules to form a feeding net capable in larger spumellarids of capturing copepods and similar small metazoans as well as larger protists, though the smaller nassellarids (50-100 μm) are said to feed on bacteria and smaller protists. Some solitary spumellarids reach diameters of several mm, and colonial species containing hundreds or thousands of cells may form gelatinous masses several cm in diameter or strands a metre or more long. These radiolarians are important in the oceanic plankton, where, for example, colonial species with symbionts may be the dominant macrozooplankton of tropical surface waters.

THE CILIATES

Ciliates are normally easily distinguished from all other organisms. Typically they possess nuclei of two types and a characteristic cortical organization of cilia with underlying infraciliature of basal bodies and associated fibres. Each cell normally contains one or more larger and usually polyploid macronuclei which are the source of RNAs for protein synthesis, while the one or more smaller diploid micronuclei provide genetic continuity through sexual reproduction but do not directly control cell metabolism. The cilia of the general body surface (somatic ciliature) are used primarily for locomotion and the oral ciliature in and around the cytostome area for food collection; patterns of ciliation have long been used in classification and the details of the infraciliature have assumed increased importance in more recent systematics. Most ciliates have one or more contractile vacuoles, many possess extrusomes of characteristic types, some develop tentacles or loricas and sessile forms may have stalks; all of these features aid identification.

The taxonomy of the phylum Ciliophora has been rearranged many times, including several major revisions in the last 15 years, and is still unstable. Among widely available classifications one outlined by Corliss (1979) and Levine *et al.* (1980) is primarily based on oral morphology, while another by Small and

Lynn (1985) is based largely on somatic and oral ultrastructural characters. These classifications are compared in the Appendix and examples of the important marine taxa are shown in Figure 3; typical features of these groups are

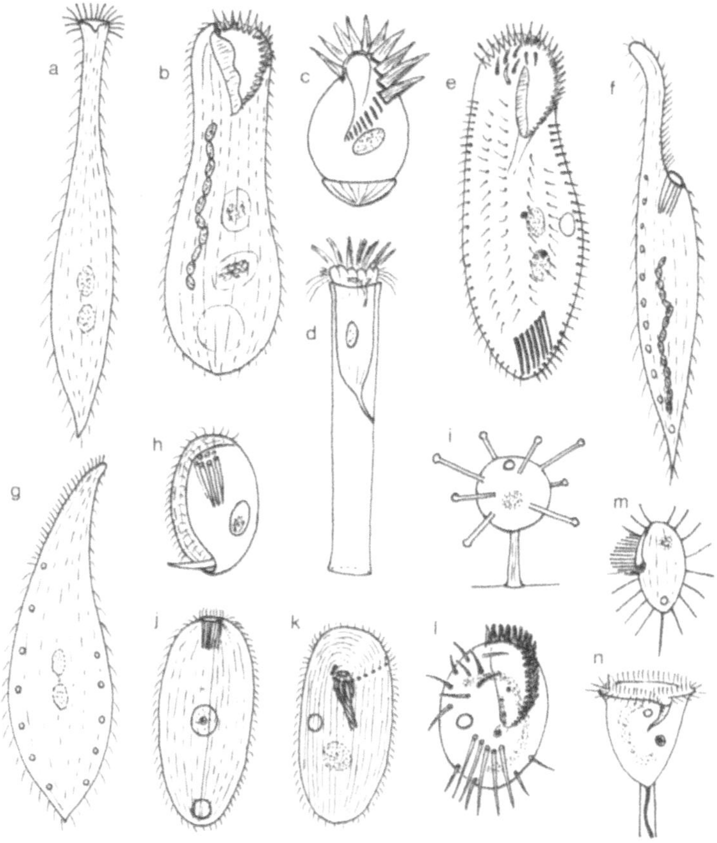

Fig. 3. Body forms in ciliated protozoa representing groups commonly found in marine habitats. a: *Trachelocerca* (Karyorelictea) 200 µm, b. *Condylostoma* (Heterotrichida) 1mm, c. *Strombidium* (Oligotrichida) 50 µm, d. *Eutintinnus* (Tintinnida) 300 µm, e. *Urostyla* (Stichotrichida) 100 µm, f. *Dileptus* (Haptorida) 500 µm, g. *Amphileptus* (Pleurostomatida) 250 µm, h. *Dysteria* (Cyrtophorida) 50 µm, i. *Podophrya* (Suctorida) 50 µm, j. *Prorodon* (Prostomatida) 200 µm, k. *Nassula* (Nassulida) 150 µm, 1. *Euplotes* (Euplotida) 100 µm, m. *Cyclidium* (Scuticociliatida) 20 µm, n. *Vorticella* (Peritrichida) 50 µm. Typical lengths are indicated. See also text and Table 3 (figures mostly redrawn from Sleigh 1989)

outlined in Table 3. Keys to place many ciliates within families and sometimes within genera have been produced by Small and Lynn (1985), and family characters and lists of genera are given by Corliss (1979), but species identification requires the use of Kahl's (1930-35) descriptions and often the use of more recent detailed accounts. Groups that are predominantly parasitic or endocommensal, or are uncommon in marine habitats, will not be mentioned further.

The common and important free-swimming ciliates of the open sea are the oligotrichids and tintinnids (the largest ciliate order), the former are naked and most of the latter secrete loricas into which foreign material may be incorporated. In both of these orders most of the cilia present form a loop of membranelles leading to the mouth; by means of these membranelles they swim and also feed on nanoplankton and even picoplankton, being apparently the main predators on microflagellates in many situations. Members of both orders, especially oligotrichs, may also be found in shallow water and benthic habitats. Oligotrich ciliates frequently contain chloroplast symbionts derived from their prey.

Members of several other groups are found in the plankton, being either neritic or are forms that are associated for most of the time with surfaces, e.g. scuticociliates, peritrichs and hypotrichs; these may move over or anchor to particles of floating detritus (such as marine snow), to floating animals and plants or to any benthic surface; all of these groups feed largely on bacteria or small algae, collected by filtration using membranelles and/or paroral membranes, although larger particles may also be eaten by some species. In scuticociliates there is usually a prominent paroral membrane (often in 3 parts) in an extended shallow buccal cavity and accompanied by smaller membranelles, and in peritrichs the paroral membrane and an elongated single membranelle extend out of a deep infundibular buccal cavity to run one or more times around the apical end of the body. Scuticociliates have a

Table 3. FEATURES OF CILIATE GROUPS

Subphylum Class Order	Body Ciliature	Mouth Ciliature	Cytopharynx	Feeding Mechanism	Food	Habit	Habitat	Usual size range µm
Postciliodesmatophora								
Karyorelictea	extensive/full	varies	simple	i,g	b,p,a,d,m	v,n,th	xs,as	20-2000
Spirotrichea								
Heterotrichida	full/reduced	azm (pm)	simple	f(g)	b,a,p,m,d	n,s,	xs,as	30-1500
Oligotrichida	reduced	azm(pm/em)	simple	f(g)	b,a,p,(d)	n	o,c,xs,as	20-100
Tintinnida	reduced	azm	simple	f(g)	b,a,p,(d)	n	o,c	40-200
Stichotrichida	ventral cirri	azm,em/pm	simple	f	b,a,p,d,	v,th,(n)	xs,as	40-250
Rhabdophora								
Litostomatea								
Haptorida	extensive	simple	rhabdos	i,g	p	v,n	o,c,xs	30-1000
Pleurostomatida	extensive	simple	rhabdos	i,g	p	v,n,th	xs	50-1000
Cyrtophora								
Phyllopharyngea								
Cyrtophorida	ventral cilia	simple or none	cyrtos	i,g	b,a,d	v,th	xs,ez,as	10-50
Suctorida	'larva' only	none	none	i	p	s	o,c,w,ez,xs	25-200
Prostomatea,								
Prostomatida	full	simple	cyrtos	i,g	p,d	v,n,s	c,xs	20-200
Nassophorea								
Nassulida	full	mm (pm)	cyrtos	i,g	b,a,d	v,th	xs	50-300
Euplotida	ventral cirri	azm, pm	simple	f	b,a,p,d	v,th,(n)	o,c,w,xs,as	25-200
Oligohymenophorea								
Scuticociliatida	full/reduced	mm,pm	simple	f	a,b	n,th	o,c,w,xs,as	15-250
Peritrichida	very reduced or absent	pm, (mm)	simple	f	b	s(v,th),cl	o,c,w,ez,xs,as	25-200

Key: Abbreviations as in Tables 1 and 2 except that some ciliates are typically thigmotactic (th) or commonly epizoic (ez), and have mouth ciliature composed of a few membranelles (mm), or a long adoral zone of membranelles (azm), and a paroral membrane (pm) and/or an endoral membrane (em).

sparse body ciliature with often a few long thigmotactic cilia. Peritrichs lack somatic cilia on the sedentary (often stalked) forms, but with an aboral girdle on migratory stages or motile species. Scuticociliates are abundant in sediments and common as commensal species, while peritrichs attach to many substrata, including the surfaces of planktonic animals, particularly crustaceans. Hypotrichs are flattened forms with a long row of membranelles, small paroral membrane and motile somatic cilia located in clusters as ventral cirri: the shorter disc-shaped species of the order Euplotida (e.g. *Euplotes, Uronychia, Diophrys, Aspidisca*) may often be found on suspended particles, or swimming among them, as well as in the sediments where the larger, elongate, and more omnivorous species of the order Stichotrichida (e.g. *Holosticha, Keronopsis, Oxytricha*) are found. These two orders of 'hypotrichs' are separated into different classes by Small and Lynn (1985) (see Appendix). The odontostomes are a small group of bacterivores found especially in anaerobic sediments.

Other ciliates that abound in the sediments are the omnivorous karyorelictids, heterotrichs and prostomatids, carnivorous haptorids and pleurostomatids and algivorous nassulids and cyrtophorids. The karyorelictids are generally elongate and flattened interstitial forms, mostly with a superficial, terminal or ventral slit-like cytostome with often (but not always) simple ciliation and a simple cytopharynx (but occasionally with a large ventral ingesting area), and with diploid macronuclei that never divide; large numbers of species from genera like *Remanella, Geleia* and *Tracheloraphis* occur in marine sediments, including anaerobic ones. The pleurostomatids also tend to be large, laterally flattened forms with a ventral slit-like cytostome and simple surrounding cilia, but have a flattened rhabdos (in which rod-like bundles of microtubules reinforce the walls of a straight cytopharynx) and polyploid macronuclei, and often groups of trichocysts are present; many species of genera like *Loxophyllum* are again numerous in marine interstitial habitats. A more obvious and more complex rhabdos leads from an apical or sub-apical cytostome in haptorids, which deploy toxicysts in the cytostome region for food capture, and are often large e.g. *Dileptus* and extensible e.g. *Lacrymaria*; while most of these forms are found in or near sediments, *Mesodinium* and *Didinium* are found in neritic marine plankton. *Mesodinium rubrum* contains a cryptophyte symbiont and is an important primary producer in many coastal situations (Lindholm 1985).

In nassulids and cyrtophorids a cyrtos (a curved tube of cytopharyngeal rods) extends into the cytoplasm from a ventral cytostome; in cyrtophorids the body is dorso-ventrally flattened with a reduced (or nearly absent) dorsal ciliature. In the nassulids there are usually some prominent hypostomal compound oral cilia (polykinetids) posterior and to the left of the mouth. An adhesive 'podite' occurs near the posterior end of many cyrtophorids e.g. *Dysteria*. Members of both groups e.g. *Nassula, Chlamydodon,* ingest filamentous cyanobacteria or diatoms, whilst smaller forms of both groups collect bacteria and small cells from surfaces, occasionally living as parasites. The sub-apical (often apparently apical) cytostome of prostomatids, e.g. *Holophrya, Prorodon,* leads to a simple cytopharynx (now thought to be a cyrtos) which handles an omnivorous diet. The heterotrichs are generally large ciliates with an extensive somatic ciliature as well as a long row of oral membranelles (polykinetids) and sometimes a large undulating (paroral) membrane, e.g. *Blepharisma, Condylostoma.* The folliculinids are an interesting group of sessile loricate marine heterotrichs whose body extends into two wing-like lobes bearing a fringe of membranelles. Many of these large heterotrichs create a powerful feeding current, able to draw in even large ciliates, flagellates, diatoms and small metazoans as well as smaller food particles and sometimes detritus. Several large genera of heterotrichs occur commonly in sediments and such genera as *Metopus* are typical of anaerobic habitats.

The chonotrichs are epibionts on crustaceans, and suctorians are generally predators on ciliates and are usually sedentary on particles or the surfaces of plants or animals. Their ecological roles are unlikely to be substantial. Although both groups have ciliated larval stages, adults appear unciliated, the former have a few cilia in an apical funnel and the latter loose the cilia when they develop the suctorial tentacles.

CONCLUSIONS

The eukaryote Protista are diverse in structure, physiology and ecology. They evolved early in the history of life, alongside the bacteria, and among the 50 or more phyla now recognised they display a great diversity of life-styles. They occupy aerobic and anaerobic habitats in which they live as autotrophs, phagotrophs and free-living or parasitic saprotrophs. This paper was only concerned with protists that practice heterotrophy, many of whom are mixotrophs also capable of photosynthesis. These forms were subdivided into three categories, flagellated protists, amoeboid protists and ciliates; both of the former assemblages are certainly polyphyletic in origin, while the ciliates appear to form a more compact natural group.

In planktonic environments the most numerous ciliates are tintinnids and naked oligotrichs, eating bacteria, flagellates and coccoid algae, though haptorids like *Mesodinium* may be locally abundant and some attached or creeping forms (largely bacterivorous hypotrichs, scuticociliates and peritrichs) may be associated with the surfaces of drifting or sinking detritus. Floating omnivorous amoeboid forms include acantharea and certain foraminifera as well as polycystine radiolaria, while smaller lobose amoebae are associated with detritus and the air-water interface. The most important bacterivorous flagellates of open water are the swimming and attached choanoflagellates and chrysophytes, probably followed by the haptophytes and dinoflagellates among swimming forms while helioflagellates, bodonids and bicosoecids are commonly associated with detritus; most of these forms may also consume larger prey, with dinoflagellates covering the widest size range.

The benthic ciliates are much more diverse than planktonic ones, covering a wide range of bacterivorous, algivorous, carnivorous and omnivorous species in a size range from 10 to 200 µm; euplotids, scuticociliates and various spirotrich groups in particular are likely to be found in almost any habitat. Rhizopod amoebae of all groups are encountered in sediments, with the foraminifera being usually the most prominent and probably the most ecologically important of the surface-dwelling forms; their diets are usually varied, including algae, protozoa, detritus and (among small forms particularly) bacteria. Many of the flagellate groups found in plankton also abound here, notably bodonids, chrysophytes and dinoflagellates, but also many phagotrophic and saprotrophic euglenids and saprotrophic cryptophytes. Most of these flagellates are small or very small and eat bacteria or small algae, but the euglenids, and particularly the dinoflagellates, take larger food items.

Although a large number of heterotrophic groups have been mentioned here, it is clear that many of the groups, particularly among the flagellate phyla, are imperfectly known, and many species, and possibly even phyla, remain to be described.

REFERENCES

Anderson OR (1983) Radiolaria. Springer, New York

Bock W, Hay W, Lee JJ (1985) Order Foraminiferida. In: Lee JJ, Hutner SH, Bovee EC (eds) An illustrated guide to the Protozoa. Society of Protozoologists, Lawrence Kansas, p 252

Cachon J, Cachon M (1985) Classes Acantharea, Polycystinea and Phaeodarea. In: Lee JJ, Hutner SH, Bovee EC (eds) An illustrated guide to the Protozoa. Society of Protozoologists, Lawrence Kansas, p 274

Corliss JO (1979) The ciliated protozoa, characterization, classification and guide to the literature, 2nd edn. Pergamon, Oxford

Corliss JO (1984) The kingdom Protista and its 45 phyla. BioSystems 17:87-126

Dodge JD (1982) Marine dinoflagellates of the British Isles. HM Stationery Office, London

Febvre-Chevalier C (1985) Class Heliozoea. In: Lee JJ, Hutner SH, Bovee EC (eds) An illustrated guide to the Protozoa. Society of Protozoologists, Lawrence Kansas, p 302

Gaines G, Elbrachter M (1987) Heterotrophic nutrition. In: Taylor FJR (ed) The biology of dinoflagellates. Blackwell, Oxford, p 224

Gooday AJ, Nott JA (1982) Intracellular barite crystals in two xenophyophores, *Aschemonella ramuliformis* and *Galatheammina* sp (Protozoa: Rhizopoda) with comments on the taxonomy of *A. ramuliformis*. J Mar Biol Ass UK 62:595-605

Hibberd DJ (1980) Prymnesiophytes (= Haptophytes). In: Cox ER (ed) Phytoflagellates. Elsevier North Holland, New York, p 273

Hibberd DJ, Leedale GF (1985a) Order Prymnesiida. In: Lee JJ, Hutner SH, Bovee EC (eds) An illustrated guide to the Protozoa. Society of Protozoologists, Lawrence Kansas, p 74

Hibberd DJ, Leedale GF (1985b) Order Chrysomonadida. In: Lee JJ, Hutner SH, Bovee EC (eds) An illustrated guide to the Protozoa. Society of Protozoologists, Lawrence Kansas, p 54

Kahl A (1930-35) Urtiere oder Protozoa 1: Wimpertiere oder Ciliata (Infusoria). In: Dahl F (ed) Die Tierwelt Deutschlands. Fischer, Jena, Tiel 18, 21, 25, 30 p 1

Larsen J (1985) Algal studies of the Danish Wadden Sea II. A taxonomic study of psammobious dinoflagellates. Opera Bot Soc Bot Lund 79:14-39

Larsen J (1987) Algal studies of the Danish Wadden Sea V. A Taxonomic study of bethic-interstitial euglenoid flagellates. Nord J Bot 7: 589-607

Leadbeater BSC (1985) Order Choanoflagellida. In: Lee JJ, Hutner SH, Bovee EC (eds) An illustrated guide to the Protozoa. Society of Protozoologists, Lawrence Kansas, p 106

Lee JJ (1985) Order Ebriida. In: Lee JJ, Hutner SH, Bovee EC (eds) An illustrated guide to the Protozoa. Society of Protozoologists, Lawrence Kansas, p 140

Leedale GF (1985) Order Prasinomonadida. In: Lee JJ, Hutner SH, Bovee EC (eds) An illustrated guide to the Protozoa. Society of Protozoologists, Lawrence Kansas, p 97

Leedale GF, Hibberd DJ (1985) Order Volvocida. In: Lee JJ, Hutner SH, Bovee EC (eds) An illustrated guide to the Protozoa. Society of Protozoologists, Lawrence Kansas, p 88

Lessard EJ, Swift E (1987) Species-specific grazing rates of heterotrophic dinoflagellates in oceanic waters, measured with a dual-label radiosotope technique. Mar Biol 87: 289-296

Levine ND *et al.* (1980) A newly revised classification of the Protozoa. J Protozool 27:37-58

Li GW, Volcani BE (1987) Four new apochlorotic diatoms. Br Phycol J. 22:375-382

Lindholm T (1985) *Mesodinium rubrum* a unique photosynthetic ciliate. Adv Aquat Microbiol 3:1-48

Loeblich AR, Tappan H (1984) A new suprageneric classification of the Foraminifera. Micropaleontology 30:1-70

Moestrup Ø, Thomsen HA (1976) Fine structural studies on the flagellate genus *Bicoeca* i. *Bicoeca maris* with particular emphasis on the flagellar apparatus. Protistologica 12:101-120

Page FC (1987) The classification of 'naked' amoebae (Phylum Rhizopoda). Arch Protistenk 133:199-217

Patterson DJ, Fenchel T (1985) Insights into the evolution of heliozoa (Protozoa, Sarcodina) as provided by ultrastructural studies on a new species of flagellate from the genus *Pteridomonas*. Biol J Linn Soc 34:381-403

Santore UJ, Leedale GF (1985) Order Cryptomonadida. In: Lee JJ, Hutner SH, Bovee Ec (eds) An illustrated guide to the Protozoa. Society of Protozoologists, Lawrence Kansas, p 19

Sleigh MA (1989) Protozoa and other protists. Edward Arnold, London

Small EB, Lynn DH (1985) Phylum Ciliophora. In: Lee JJ, Hutner SH, Bovee EC (eds) An illustrated guide to the Protozoa. Society of Protozoologists, Lawrence Kansas, p 393

Taylor FJR (ed) (1987) The biology of dinoflagellates. Blackwell, Oxford

Vickerman K (1976) The diversity of the kinetoplastid flagellates. In: Lumsden WHR, Evans DA (eds) Biology of the Kinetoplastida, Vol 1. Academic Press, London New York p 1

Woese CR (1987) Bacterial evolution. Microbiol Rev 51:221-271

APPENDIX

List of groups in the different assemblages of protists (see also Sleigh 1989)

FLAGELLATE AND ALLIED PROTISTS

Phylum Dinophyta: the dinoflagellates (+ Ebriida ?)
- Parabasalia: trichomonads and hypermastigids (parasites)
- Metamonada: retortamonads, diplomonads and oxymonads (parasites)
- Kinetoplasta: bodonids and trypanosomatids (many parasites)
- Euglenophyta: the euglenids (*Stephanopogon* and *Hemimastix* belong near here?)
- Cryptophyta
- Opalinata: opalinids (parasites) (Proteromonads belong near here?)
- Heterokonta: includes Chrysophyceae, Synurophyceae, Bacillariophyceae, Phaeophyceae, Xanthophyceae (=Tribophyceae), Eustigmatophyceae, Oomycotea, Hyphochytridiomycotea, Raphidophyceae, Labyrinthulea, Thraustochytridea and perhaps Bicosoecida and some helioflagellates (actinomonads)
- Haptophyta (Prymnesiophyta): includes coccolithophorids
- Chlorophyta: includes Prasinophyceae, Chlorophyceae, Oedogoniophyceae, Ulvaphyceae, Conjugatophyceae and Charophyceae
- Chytridiomycota: chytrids (parasites, often in algae)
- Choanoflagellata: collar flagellates
- Rhodophyta: red algae

AMOEBOID PROTISTS

Phylum Rhizopoda

Class Karyoblastea (Pelobiontea): *Pelomyxa*
- Heterolobosea: schizopyrenids and acrasids (some amoeboflagellates)
- Lobosea: gymnamoebae and testacealobosea
- Eumycetozoea: cellular and syncytial slime moulds
- Plasmodiophorea: intracellular parasites in plants
- Filosea: aconchilinids and testaceafilosea
- Granuloreticulosea: foraminifera etc.
- Xenophyophorea

Phylum Actinopoda

Class Heliozoea: actinophryids, centrohelids, taxopodids, desmothoracids
- Polycystinea: spumellarids and nassellarids
- Phaeodarea
- Acantharea

THE CILIATES

Classification by Corliss (1979) and Levine *et al.* (1980)

Phylum Cilipohora
- Class Kinetofragminophorea
  - Subclass Gymnostomata: karyorelictids, prostomatids, haptorids and pleurostomatids
  - Vestibulifera: trichostomatids, entodiniomorphids and colpodids
  - Hypostomata: synhymeniids, nassulids, cyrtophorids, chonotrichs, rhynchodids and apostomes
  - Suctoria: suctoria
- Class Oligohymenophorea
  - Subclass Hymenostomata: hymenostomes, peniculines, scuticociliates and astomes
  - Peritricha: peritrichs
- Class Polyhymenophorea
  - Subclass Spirotricha: heterotrichs, odontostomes, oligotrichs, tintinnids, stichotrichs and hypotrichs

Classification by Small and Lynn (1985)

Phylum Ciliophora
- Subphylum Postciliodesmatophora
  - Class Karyorelictea: karyorelictids
  - Spirotrichea: heterotrichs, oligotrichs, tintinnids and stichotrichs
- Subphylum Rhabdophora
  - Class Litostomatea: haptorids, pleurostomatids, trichostomes and entodiniomorphids
- Subphylum Cyrtophora
  - Class Phyllopharyngea: cyrtophorids, rhynchodids, chonotrichs and suctoria
  - Prostomatea: prostomatids (Small (pers. comm.) suggests this position)
  - Nassophorea: nassulids, peniculines and hypotrichs
  - Oligohymenophorea: hymenostomes, scuticociliates, peritrichs, astomes, apostomes and plagiopylids
  - Colpodea: colpodids

METHODS FOR THE STUDY OF MARINE MICROZOOPLANKTON - SESSION SUMMARY

Fereidoun Rassoulzadegan
Station Zoologique
B P 28
06230 - Villefranche-sur-Mer
France

INTRODUCTION

As protozoan and microbial ecologists our 'raison d'être'rests with the study of the structure function of ocean systems. To progress this we need field and experimental sampling, observation, and rate measurement techniques. Since the work of Pomeroy (1974), a widespread interest in, and a need for, an understanding of the dynamics of aquatic microbial communities has stimulated a search for appropriate methods to describe pathways and rates of energetic interactions within marine microbial food webs. A significant part of this session was therefore devoted to experimental methods.

Madhu Paranjape (1990) in his Keynote talk, presented new information on protozoan sampling in specific environments, such as the surface microlayer, sea-ice and ice-water interfaces. He stressed the diversity of the sampling methods needed to accommodate the wide size and abundance ranges of marine protozoa (almost 6 orders of magnitude).

Farooq Azam discussed new developments in the study of the mesopelagic microbial community and of the vertical flux of organic matter in the ocean. Although most of our knowledge is concerned with the role of microbes in the euphotic zone of the water column, the importance of the protozoa in both mesopelagial and ocean bottom communities is beginning to be appreciated (Cho and Azam 1988, Karl *et al.* 1988, Gooday 1988a b, Lochte 1990, Lochte and Turley 1988, Turley *et al.* 1988, Suess 1988). He questioned whether free-living bacteria in the deep ocean can effectively utilize sinking particles, or whether most of the sinking particulate organic matter is taken by mesopelagic particle eaters (zooplankton).

NATO ASI Series, Vol. G 25
Protozoa and Their Role in Marine Processes
Edited by P.C. Reid et al.

After describing the role of microbes in the production and utilisation of dissolved organic matter (DOM) in the mesopelagic, he outlined recent work (Cho and Azam 1988) that indicated that bacteria utilize most of the sinking organic matter. Since most mesopelagic bacteria are free living and, since all bacteria are osmotrophic, this suggests that large scale particle solution occurs, and this is presumably caused by attached bacteria. Azam proposed a hypothesis that particle-attached bacteria, by way of their exoenzyme activities, rapidly convert the particle to soluble material and liberate both hydrolysate and progeny into the surrounding water. In consequence, a large number of small particles (free-living bacteria 0.3 - 0.6 µm) are produced at the expense of a small number of large, rapidly sinking, particles. This causes a decrease in the total sinking rate and an enhancement of the total particulate adsorbing surface (represented mostly by small bacteria which account for ~ 90% of the particulate surface area). Recent data has shown that bacterial carbon represents about 50% of the total particulate organic carbon (POC) in the mesopelagic zone.

The process of the production of bacteria from large sinking particles may be an important pathway for the fate of radionuclides and surface-reactive matter in the ocean's interior. The protozoan contribution to the turnover of non-sinking particles is considered to be one of the possible pathways for reaggregation of bacteria and their subsequent downward transport in the ocean. To study this important microbial community appropriate observation methods must be employed. Separation of these microbial communities according to size requires careful techniques as between 30 to 50% of the small bacteria can go through GFC and GFF filters. These bacteria occur at densities around 0.1-0.2 million cells ml^{-1} in the mesopelagic zone, and are as small as those usually found in the euphotic zone. This is an important point to consider when studying total POC in the aphotic zone as GFC or GFF filters are often used and these would pass a significant fraction of the bacterial carbon. The bacterial heterotrophic production in the mesopelagic zone becomes an important parameter to study. Water column integration of bacterial production (plus respiration) may also provide a minimum estimate of the vertical flux, with the thymidine incorporation method (Fuhrman and Azam 1982) as a useful tool for studying depth-dissipation of sinking flux.

Evelyn and Barry Sherr presented a review of methods used to count marine protozoa by epifluorescence microscopy (EPIM). Unfortunately, there is no standard method because of problems associated with preservation and staining, and because of subjectivity when counting with the microscope. The problems of flagellate body loss, decrease of chlorophyll *a* fluorescence, and release of food vacuole contents during preservation were discussed (Murphy and Haugen 1985, Bloem *et al.* 1986, Børsheim and Bratbak 1987, Sieracki *et al.* 1987, Booth 1988). Two preservation methods were recommended:
1) Lugol Discolouration Method (LDM). This is a new method that I recently developed with E and B Sherr. Seawater samples are instantly fixed with alkaline Lugol solution to a final volume of 0.5%. Borate buffered formalin (BBF) is then immediately added (to a final volume of 3%) in order to decolourise the stain. After this treatment items in the food vacuoles of microprotozoa can easily be seen by EPIM.
2) Glutaraldehyde Method. In this the sample is mixed with ice-cold glutaraldehyde to a final volume of 4%. Both methods minimise cell shrinkage, but this still may occur with certain kinds of protozoa.

The major problem with the fluorescent staining of colourless protists is the overlap of the induced fluorescence emission spectra with that of the autofluorescence (chlorophyll), leading to an underestimation of chlorophyll, mainly in the picoplankton (Caron 1983). This depends on the fluorochrome used e.g. acridine orange, (Wood 1962), proflavin (Hass 1982), and fluorescein isothiocyanate (Sherr and Sherr 1983). The stains : 4',6-diamidino-2-phenylindole, DAPI (Porter and Feig 1980), and primulin (Caron 1983) were also recommended.

As an alternative to tedious, time-consuming, and probably subjective microscopic enumeration the use of image analysis and microphotometry techniques was suggested, especially for processing large numbers of samples. Finally, the role of EPIM was described in relation to trophodynamic studies within the microbial food webs, such as protozoan bacterivory (McManus and Fuhrman 1986, Sherr and Sherr 1987), ingestion of algae (Rassoulzadegan *et al.* 1988), and uptake of high molecular weight DOM by protozoa (Sherr 1988).

Eugene Small discussed the use of the protargol staining procedure (Lee *et al.* 1985). Following an historical review of the method, which goes back to the 1940's when Kostlow first used 'bleaching stuff', he presented some variants of the method. The use of a modified Bouin fixative was suggested to overcome problems of rapid changes of salinity in unstable environments. This modified Bouin's fixative, is made from picric acid saturated in concentrated calcium carbonate buffered formaldehyde containing 5% (v/v) acetic acid. To fix the sample the fixative is diluted with sample in ratios (sample : fixative) varying from 10:1 for seawater to 19:1 for brackish water (Coats and Heinbokel 1982, Montagnes and Lynn 1987). Picric fixation is superior as it enhances the ability of cells to pick up silver stain. The problems of pH regulation were also discussed. The protargol solution procedure involves a shift in pH from 8.6 (pH of solution) to 6.8-7. Better results are obtained if the initial pH of the solution is adjusted to 8.8. To prevent loss of Bouin-protargol treated material from preparations the use of parlodion, dissolved in absolute methanol, was recommended. This covers the whole preparation as a plastic film. See Lee *et al.* (1985).

David Montagnes and **Denis Lynn** presented a description of the use of quantitative protargol stain (QPS) for the observation and counting of marine protozoa (Montagnes and Lynn 1987). The advantages of QPS are two-fold; 1) it facilitates specific identification of organisms, and 2) it permits quantitative ennumeration. A further advantage of the method is that the various preparation steps involved lead naturally to a random distribution of organisms on the coverslips, which is a necessary condition for the estimation of the total number of the organisms in water samples (e.g. coverslip total count).

It was pointed out that for the study of protozoan standing stocks, the use of the QPS is as accurate as the Utermöhl method, providing that fragile species are accounted for. With the QPS method, the density of the protozoa in a given volume of water is determined by concentrating cells on cellulose filters where they are embedded in agar prior to protargol (or Chatton-Lwoff) silver staining.

The additional advantages of QPS are:

- it stains the ciliate natural assemblages for species identification,
- it provides a permanent record,
- it reveals ecologically important structures and ontogenetic stages, and
- it can be used in growth rate experiments

Chris Reid outlined sampling and observation of protozoan resting cysts in sediment and water. Cysts are widely distributed in the water column but need to be concentrated before they can be adequately studied. Three aspects were discussed : 1) sampling methods, 2) cleaning and concentration, and 3) quantitative and qualitative microscopy.

Conventional plankton sampling techniques are inadequate since they do not account for encystment in planktonic protists which takes place over short periods; sediment traps may be used to integrate and concentrate cysts over longer periods of time (Reid 1987). The value of the Continous Plankton Recorder was also mentioned especially for the study of the seasonal and geographic patterns of distribution (Reid and John 1983). Unfixed cysts may be incubated for species identification. The use of the polarizing microscope as a tool for the identification of dinoflagellate cysts (Reid and Boalch 1987) and to distinguish the different materials used for construction of cyst walls in protists was recommended.

Johan Wikner and Åke Hagström gave a lecture on the use of the genetically marked minicell technique (GMMT) to measure predation on bacteria in aquatic environments. Wikner showed how genetic labelling can be applied to the measurement of the absolute rate and regulation of bacterivory in protozoa. The GMMT (Wikner *et al.* 1986) involves osmotically stable, non-growing minicells which are used as a probe. Their size (0.2 - 0.6 µm) and surface characteristics are similar to bacteria, and they have a specific marker to distinguish them from other organisms (e.g. minicells produced by *Escherichia coli* M2141, Adler *et al.* 1967). The *E. coli* minicell system is a standard technique in microbiology to study in vivo protein synthesis. Minicells comparable in size to bacteria are obtained by a sucrose-gradient centrifugation of the minicell preparation. These lack chromosomal DNA but may contain plasmid DNA. By introducing the plasmid pACYC184 carrying the cat gene, one obtains high synthesis of chloramphenicol acetyltransferase within the minicells, which can be radio-labelled, and serves as a marker protein. He also mentioned the use of the b-lactamase protein, originating from the plasmid pBR322, as a marker protein in another minicell strain, used as an internal standard during the recapture of the minicells.

The use of marker proteins permits quantitative estimates of minicells to be made in aqueous samples. The minicell and its marked protein are unaffected by incubation in sterile sea water for at least 50 hours. By mixing minicells and natural bacteria in ratios of about 1:5 and following the disappearance of the minicells, it is possible to measure, in a few hours, predation rates of between $1x10^4$ and $20x10^4$ bacteria $ml^{-1}h^{-1}$ (Hagstrøm *et al.* 1988).

The GMMT may be used potentially for the study of diel patterns of *in situ* bacterivory (Wikner *et al.* 1990). Data were presented to show that the maximum rate of disappearance of minicell label from the bacteria fraction coincided with the appearance of the label in the 1-3 µm size fraction. The results of the size fractionation experiments show that the predation capacity increases from 0.6 µm to larger sizes, and reaches a peak around 3-5 µm. They also demonstrate the existence of at least three trophic steps between bacteria and the higher levels. This means that there is less of the bacterial production potentially available to higher trophic levels. However, these results do not necessarily support the sink hypothesis since there is an input of autotrophic carbon fixation at each of these levels by autotrophy or mixotrophy. In addition, the microbial food web , even if it does not contribute in terms of carbon transfer to the higher trophic levels, might be crucial in supplying inorganic phosphorous and nitrogen for phytoplankton (Wiadnyana and Rassoulzadegan 1989).

Evelyn Lessard gave a presentation on the use of radioisotopes to estimate grazing in protozoa. She stressed the need for intercalibration as different methods often do not give comparable results. A major problem with rate measurements is the long incubation time required by the old techniques, which were essentially adapted from earlier copepod feeding studies. She suggested that new techniques, such as Wikner's GMMT, Capriulo's fluorescence technique (Capriulo 1988), and her dual-label radioisotope technique (DLRT) (Lessard and Swift 1985) would allow analysis at the level of individual cells with short incubation time scales (minutes to hours), and with minimum manipulation of the populations.

The experimental protocol for her technique was outlined. This permits examination of community grazing by microplankton populations and simultaneous determination of species-specific grazing rates. The problem of isotope recycling is minimised by the use of high specific activity isotopes and short incubation times. With the DLRT one can also make simultaneous measurements of herbivory (labelling algal prey with ^{14}C) and bacterivory in protozoa (labelling bacterial prey with ^3H-thymidine, Fuhrman and Azam 1982, Hollibaugh *et al.* 1980). A limitation of the DLRT is that it is difficult to estimate herbivory by nanozooplankton because consumers and producers may not differ greatly in size.

Raymond Leakey, Peter Burkill and **Michael Sleigh** compared fixatives used in the quantification of pelagic ciliate populations. The data obtained by counting and sizing individual cells from fixed samples allowed the authors to compare the effects of formaldehyde, glutaraldehyde, Lugol and Bouin solutions on the abundance and cell volume of both cultured and natural ciliate populations. The superiority of Lugol and Bouin as fixatives for the aloricate species was noted, and data for *Uronema* sp. cultures showed that Lugol and formaldehyde fixation resulted in insignificant changes in cell volume, whereas glutaraldehyde increased and Bouin decreased cell volume by 10% (see also Revelante and Gilmartin 1983).

Andrew Gooday, outlined sampling, fixation and sorting of benthic foraminifera. Because of their relatively large size (45-1000 µm), they may be collected in the same way as meiofaunal and macrofaunal metazoans. They are taken from the upper centimetre of sediment by scooping up or by inserting a hand-held coring tube. For deep water sediments a coring device (Phleger, box, or multiple corer) must be employed in order to minimize disturbance of the sediment-water interface. For rich environments, towed gears, epibenthic sledges, Agassiz trawls, and anchor dredges can be used. Attached or epiphytic communities can be separated by vigorous shaking in a container and susequent sieving.

For fixation, BBF is used at final v/v concentration of 2-4%. This is also used for storage, rather than alcohol, as alcohol causes shrinkage of soft shelled forms. Phosphate or cacodylate buffered glutaraldehyde was recommended for cytological purposed. After washing through a series of sieves (e.g. 500, 150, 106, 63, 45 µm) the samples are stained with Rose Bengal.

This permits the recognition of 'live' specimens. Dry, hard shelled specimens are stored on micropaleontological slides; but the soft-shelled forms must be gradually transferred to anhydrous glycerine, mounted under a coverglass supported by glass beads and sealed with a suitable agent. For very delicate allogromiids he recommended mounting in 4% formaldehyde sealing the coverglass in two stages with 'Clearseal' and 'Bioseal' (Arnold 1974, Gooday 1988).

DISCUSSION AND SYNTHESIS

An important source of the controversy concerning problems such as the 'sink' or 'link' status of the microbial loop or food web (Azam *et al.* 1983, Ducklow *et al.* 1986, Sherr *et al.* 1987a, Sherr and Sherr 1988) is the lack of suitable techniques for sampling sea microbes and for making rate measurements on their populations. Pomeroy and Wiebe (1988) quoted a variance of 521, for laboratory measurements of growth yield in Protozoa. This high variance could partly be due to the different physiological stages of the species studied, but could be mainly caused by disagreement between the results from a diversity of techniques.

The methodological approach to protozoan handling and observation should take into consideration the fact that these organisms exist as communities rather than as individual groups. The microbial food web covers a large range of microorganisms from bacteria (0.3-0.6 µm) to large protozoa (several 100 µm). These are of different structural consistancy and composition. The phagotrophic protozoa can have functional chloroplasts (Stoecker *et al.* 1987, Jonsson 1987, Laval-Peuto and Rassoulzadegan 1988), whereas the nanosized autotrophs can consume particles (Estep *et al.* 1986, Porter 1988). There can be no one sampling or observation technique to cover all types.

The size and structural composition of the organisms determines the sampling methods and the observation techniques to be used. Some groups of large individuals possess solid structures (e.g.

lorica in tintinnids, shell in forminifera, cell walls in thecate dinoflagellates) and can be sampled with fine mesh nets (≤ 20 µm), but bottles are normally used to sample water-column protozoa. Standard Niskin bottles may be toxic to protozoa and it has been suggested that ultraclean techniques or GOFLO bottles should be used instead (Fitzwater *et al.* 1982, Chavez and Barber 1987). For benthic protozoans, especially ciliates, the salinity or temperature gradient technique (Ühlig 1964, Fenchel 1967) is still frequently used. In the case of most marine protozoa, good sampling methods alone cannot resolve quantitative problems if they are not followed up with a suitable fixation and observation technique.

The Lugol-Utermöhl method (see Sournia 1978) is an efficient tool, although it cannot be used for organisms ≤ 5 µm because of the long settling time, nor can it be used for mixotrophs which require EPIM (Wood 1962, Daley and Hobbie 1975, Hobbie *et al.* 1977, Porter and Feig 1980, Caron 1983). The use of EPIM permits the analysis of the trophic status of protozoa (Davis and Sieburth 1982, Haas 1982, Sherr and Sherr 1983, 1987, Børsheim 1984, Fuhrman and McManus 1984, Cynar and Sieburth 1986, Estep *et al.* 1986, McManus and Fuhrman 1986, Sherr *et al.* 1987b, Sieracki *et al.* 1987, Porter 1988, Rassoulzadegan *et al.* 1988). Image analysis coupled with EPIM can increase accuracy by permitting rapid counting and sizing of a statistically significant number of particles, while at the same time determining their trophic status. With counts on filters it may be difficult to identify organisms. This may be resolved by using the filter-transfer-freeze (FTF) technique of Hewes and Holm-Hansen (1983). The use of any of these techniques can result in some cell breakage, due to the use of formalin as fixative, and/or of vacuum filtration (Sorokin 1981). A recent technique described by Grumpton (1987) uses neutralized glutaraldehyde as fixative at a final concentration of 0.5-1% to prepare permanent mounts of freshwater phytoplankton. A gentle vacuum filtered concentrate is mounted in methacrylic resin. This seems to be suitable for delicate naked flagellates but there is some doubt as to whether the technique can be applied to seawater.

The live microscope counting method (Zukov 1973, Pavlovskaya 1976, Sorokin 1977, Dale and Burkill 1982), cannot be easily applied to quantitative studies because of the need to examine a large number of samples in a short period of time. Other techniques for live counting can employ electronic particle counters. The method has the advantages that living populations are studied and the counts are statistically significant ($\geq 10^5$ counts per size class). A disadvantage is the difficulty of identification of trophic groups, but the data can be appropriately supplemented by microscope observations (Sheldon 1978). Another disadvantage of Coulter counting is that for it to be effective considerable experience in the optimal use of the instrument is required. The recognition and elimination of electrical noise is of paramount importance otherwise the counts at small sizes can be meaningless (Sheldon 1978). In the past it may have been this factor more than any other that has led to the rather mixed reputation that the method has held among the majority of us.

Finally, to conclude the comments on plankton counts, I should mention the potential of flow cytometry for particle characterisation (Trask *et al.* 1982, Olson *et al.* 1983, Yentsch *et al.* 1983, Wood *et al.* 1985). Some studies of protozoan grazing rates have been made on discrete laboratory prey which have given encouraging results (Cucci *et al.* 1985, Stoecker *et al.* 1986, Gerritsen *et al.* 1987, Burkill 1987, Burkill *et al.* 1988). If these methods become more widely applicable they could overcome many of the difficulties discussed above.

With respect to rate measurements, an intensive effort has been made to improve the methodology during the last decade but there has not been much success. The Fuhrman and Azam (1982) ^3H-thymidine method is frequently used for production measurments of heterotrophic bacteria, and there is a sister method that measures bacterial production through protein synthesis using leucine incorporation (Kirchman *et al.* 1985).

The advantage of these methods is that it may be possible to measure community growth and production at the same time without any manipulation of the populations. An older method, the frequency of dividing cells (FDC) of Hagstrøm *et al.* (1979) has also been successfully employed (Newell and Christian 1981, Davis and Sieburth 1984).

Techniques for measuring grazing rates are more numerous. The major problem of measuring bacterivory in protozoa (Sieburth 1984, Fenchel 1982,1984,1986a)is the stimulation of additional bacterial growth by the protozoa, due to feed-back exchanges. This makes the use of any traditional control (growth of the prey in absence of the predators) difficult. To overcome this, three short-time incubation methods have been developed. Hollibaugh *et al.* (1980) used ^{3}H-thymidine as marker. This measures pre-labelled bacteria after ingestion by protozoa. The method was refined by Lessard and Swift (1985), who developed the DLRT using ^{14}C-bicarbonate and ^{3}H-Thymidine. This permits the measurement of protozoan grazing on both hetero- and autotrophic prey in the same community. The third technique (GMMT) uses genetically marked minicells as a probe (GMMT) for the measurement of predation on bacteria (Wikner *et al.* 1986, Hagstrøm *et al.* 1988). These methods have the advantage of measuring community grazing on living prey. However, they have some disadvantages, such as the limitation in the measurement of herbivory by nanoflagellates through size fractionation. The size fractionation itself may cause problems, where incorporation of the label in the protozoans retained on a filter must be measured. Also, we do not know if some of them are destoyed before they can be measured either by vacuum filtration or by air-drying the filter.

Another approach to the measurement of community growth and grazing is the simultaneous measurement of size screened populations by Coulter counter and EPIM during incubation (Rassoulzadegan and Sheldon 1986, Sheldon and Rassoulzadegan 1987). This permits the visualization of real population interactions that are impossible with microscopy. However, there is a significant disadvantage, which is the pre-incubation screening, which disrupts community feed-back.

Other studies of grazing rates have employed inert artificial particles or natural, heat killed bacteria as prey. Fenchel (1986b) studied the rate and pathways of *in vitro* uptake of inert particles by protozoa using videomicroscopy (Inouè 1986) and stroboscopic light; and other techniques were concerned with the observation of food vacuole packaging using fluorescent beads (Børsheim 1984, McManus and Fuhrman 1986, Cynar and Sieburth 1986). Sherr *et al.* (1987b) developed an alternative method using the heat-killed DTAF fluorescently labelled bacteria (FLB) as food particles. McManus and Pace (1988) recently suggested the use of fluorescently-labelled minicells (FLM), combining the Sherr *et al.* (1987b) FLB technique with that of Wikner *et al.* (1986). The advantages of these approches are : 1) use of non-growing bacterial-sized particles; 2) *in vitro* observations of food intake pathways and rates (videomicroscopy); 3) Virtually instantaneous *in situ* community grazing rate measurements using 'natural' particles (FLB or FLM). The disadvantage is mainly with the use of inert particles. If chemoreceptivity in the predators can be demonstrated (Lapidus and Levandowsky 1981, Van Houten *et al.* 1981, Landry *et al.* 1987, Fenchel and Jonsson 1988) then this could cause problems. Another difficulty when dealing with EPIM of the food vacuole package is the shrinkage and ejection of vacuole contents of protozoa by certain fixatives, which otherwise are desirable for EPIM observations (Sieracki *et al.* 1987). However, the lugol discolourisation method (LDM), may provide a reasonable alternative.

In parallel with the above approaches, protozoan bacterivory has been assessed by blocking either prokaryotic or eukaryotic growth with metabolic inhibitors (Newell *et al.* 1983, Fuhrman and McManus 1984, Sherr *et al.* 1986, Sanders and Porter 1986). This method has the disadvantage of needing long incubation times (~ 12 h) since it is the change in bacterial standing stock that is assessed.

For herbivory in large protozoa, some new methods have been developed since an earlier review (Rassoulzadegan 1982, including the dilution method of Landry and Hassett (1982), which permits indirect estimation of grazing rate (Paranjape 1987), Video Tape Recording (VTR), to measure the food intake rate (Buskey and Stoecker, 1988, Taniguchi and Takeda 1988), and Capriulo's *in situ* technique for measuring food vacuole packaging (Capriulo 1988) by *in vivo* microphotometric quantification of chlorophyll autofluorescence (Piller 1981, Iturriaga *et al.* 1988). There is also another new technique, which uses fluorescently labelled algae (Rublee and Gallegos 1989). This may be useful for specialized studies, since algae are usually 'naturally labelled' for EPIM. It has been demonstrated that it is possible to make grazing rate measurements on autotrophic pico- and nanoplankton by observing the *in situ* autofluorescing vacuole contents (Rassoulzadegan *et al.* 1988). For this technique protozoa are fixed with cacodylate buffered glutaraldehyde (CBG). This is organised so that the final samples contain 2% v/v glutaraldehyde in a 0.2 M cacodylate buffer, and may be used between 4°C and 20°C (Laval-Peuto and Rassoulzadegan 1988). The disadvantage of the technique is that the digestion rate has to be estimated, either directly or from data available in the literature, in order to extrapolate the food vacuole content to the grazing rates. This problem applies to Capriulo's (1988) method as well.

None of the available methods can satisfactorily describe the role of marine microbial communities in the total ocean flux of particulate matter and in the context of 'New Science' (Pearce 1988). It has been suggested by Paranjape (1990), that an updated UNESCO technical manual on marine protozoa should be produced. In the interim, increased ship-board collaboration using complementary techniques is needed.

ACKNOWLEDGEMENTS

I thank Ray Sheldon and two anonymous revewiers for their comments on, and corrections to the manuscript.

REFERENCES

Adler HI, Fischer WD, Cohen A, Hardigree AA (1967) Miniature *E. coli* cells deficient in DNA. Proc Natl Acad Sci USA 57:321-326.
Arnold ZM (1974) Field and laboratory techniques for the study of living Foraminifera. In: Hedley RH Adams CG (eds) Foraminifera Vol 1 Academic Press, London p 153
Azam F, Fenchel T, Field JG, Gray JS, Meyer-Reil LA, Thingstad F (1983). The ecological role of water column microbes in the sea. Mar Ecol Prog Ser 10:257-263
Balch WM, Reid PC, Surrey-Gent SC (1983) Spatial and temporal variability of dinoflagellate cyst abundance in a tidal estuary. Can J Fish Aquat Sci 40:244-261
Bloem *et al*. (1986) Fixation, counting, and manipulation of heterotrophic nanoflagellates. Appl Environ Microbiol 52:1266-1272
Booth BC (1988) Autofluorescence for analyzing oceanic phytoplankton communities. Bot Mar 30:101-108
Børsheim KY (1984) Clearance rates of bacteria-sized particles by freshwater ciliates, measured with mono-disperse fluorescent latex beads. Oecologia 63:286-288
Børsheim KY, Bratback G (1987) Cell volume to cell carbon conversion factors for a bacterivorous *Monas* sp. enriched from seawater. Mar Ecol Prog Ser 36:171-175
Burkill PH (1987) Analytical flow cytometry and its application to marine microbial ecology. In: Sleigh MA (ed) Microbes in the sea. Ellis Horwood, Chichester p 139
Burkill PH, Robins DB, Tarran G (1988) Rapid ingestion of algae by the heterotrophic dinoflagellates, *Oxyrrhis marina*: A demonstration using the Coulter Epics 741 flow cytometer. NATO ASI 88 [Abstracts] Plymouth
Buskey EJ, Stoecker DK (1988) Locomotory patterns of the planktonic ciliate *Favella* sp: adaptation for locating food patches and food particles. Bull Mar Sci 43:783-796
Capriulo GM (1988) The use of quantitative microscope epifluorescence to measure the ingestion rates of ciliates feeding on algae. NATO ASI 88 [Abstracts] Plymouth
Caron DA (1983) Technique for the enumeration of heterotrophic and phototrophic nanoplankton, using epifluorescence microscopy, and comparison with other procedures. Appl Environ Microbiol 46:491-498
Chavez FP, Barber RT (1987) An estimate of the new production in the equatorial Pacific. Deep-sea Res 34:1229-1244
Cho BC, Azam F (1988) Major role of bacteria in biogeochemical fluxes in the ocean's interior. Nature 332:441-443
Coats DW Heinbokel JF (1982) A study of reproduction and other life cycle phenomena in planktonic protists using an acridine orange fluorescence technique. Mar Biol 67:71-79
Conover RJ, Herman AW, Prinsenberg SJ, Harris LR (1986) Distribution of and feeding by copepod *Pseudocalanus* under fast ice during the arctic spring. Science 232:1245-1247
Cucci TL *et al*. (1985) Flow cytometry: a new method for characterization of differential ingestion, digestion and egestion by suspension feeders. Mar Ecol Prog Ser 24:201-204
Cynar FJ, Sieburth JMcN (1986) Unambiguous detection and improved quantification of phagotrophy in apochlorotic nano-flagellates using fluorescent microspheres and concomitant phase contrast and epifluorescence microscopy. Mar Ecol Prog Ser 32:61-70
Dale T, Burkill PH (1982) Live counting - A quick and simple technique for enumerating pelagic ciliates. Ann Inst Oceanogr Paris 58:267-276
Daley RJ, Hobbie JE (1975) Direct counts of aquatic bacteria by a modified epifluorescence technique. Limnol Oceanogr 20:875-882

Davis PG Sieburth JMcN Sieburth (1982) Differentiation of phototrophic and heterotrophic nanoplankton populations in marine waters by epifluorescence microscopy. Ann Inst Oceanogr Paris 58:249-260

Davis PG, Sieburth JMcN (1984) Estuarine and oceanic microflagellate predation of actively growing bacteria: estimation by frequency of dividing bacteria. Mar Ecol Prog Ser 19:237-246

Ducklow HW, Purdie DA, Williams PJLeB, Davies JM (1986) Bacterioplankton: a sink for carbon in coastal marine plankton community. Science 232:865-867

Estep KW, Davis PG, Keller MD, Sieburth JMcN (1986) How important are oceanic algal nanoflagellates in bacterivory? Oceanogr 31:646-65

Fenchel T (1967) The quantitative importance of ciliates as compared with metazoans in various types of sediments. Ophelia 4:121-137

Fenchel T (1982) Ecology of heterotrophic microflagellates. Quantitative occurence and importance as bacterial consumers. Mar Ecol Prog Ser 9:35-42

Fenchel T (1984) Suspended marine bacteria as food source. In: M J Fasham (ed) Energy and Materials in marine Ecosystems Plenum Press NY and London p 301

Fenchel T (1986a) The ecology of heterotrophic microflagellates. Adv Microb Ecol 9:57-97

Fenchel T (1986b) Protozoan Filter Feeding Progress in Protistology 1:65-113

Fenchel T, Jonsson P (1988) The functional biology of *Strobidium sulcatum*, a marine oligotrich ciliate (Ciliophora Oligotrichina). Mar Ecol Prog Ser 48:1-15

Fizwater SE, Knauer GA, Martin JH (1982) Metal contamination and its effect on primary production measurments Limnol Oceanogr 27:544-551

Fuhrman JA, Azam F (1982) Thymidine incorporation as a measure of heterotrophic bacterioplankton production in marine surface waters: evaluation and field results. Mar Biol 66:109-120

Fuhrman JA, McManus GB (1984) Do bacteria-sized marine eukaryotes consume significant bacterial production? Science 22:1257-1260

Gerritsen J, Sanders RW, Bradley SW, Porter KG (1987) Individual feeding variability of protozoan and crustacean zooplankton analyzed with flow cytometry Limnol Oceanogr 32:691-699

Gooday AJ (1988a) A response by benthic Foraminifera to the deposition of phytodetritus in the deep-sea Nature 332:70-73

Gooday AJ (1988b) The response of deep-sea benthic Foraminiferal communities to phytodetritus ('Fluff') NATO ASI 88 [Abstracts] Plymouth

Gooday (1988) Sarcomastigophora In : Higgins RP, Theil H (eds) An introduction to the study of meiofauna

Grumpton WG (1987) A simple and reliable method for making permanent mounts of phytoplankton for light and fluorescence microscopy. Limnol Oceanogr 32:1154-1159

Haas LW (1982) Improved epifluorescence microscopy for observing planktonic micro-organisms. Ann Inst Oceanogr Paris 58:261-266

Hagstrøm A, Larsson U, Hörstedt HP, Normark S (1979) Frequency of dividing cells a new approach to the determination of bacterial growth rates in aquatic environments. Appl Environ Microbiol 37:805-812

Hagstrøm A, Azam F, Andersson A, Wikner J, Rassoulzadegan F (1988) Microbial loop in an oligotrophic pelagic marine ecosystem: possible roles of cyanobacteria and nanoflagellates in the organic fluxes. Mar Ecol Prog Ser 49:171-178

Hewes CD, Holm-Hansen O (1983) A method for recovering nannoplankton from filters for identification with microscope: the filter-transfer-freeze (FTF) technique. Limnol Oceanogr 28:389-394

Hobbie JE, Daley RJ, Jasper S (1977) Use of Nuclepore filters for counting bacteria by epifluorescence microscopy. Appl Environ Microbiol 33:1225-1228

Hollibaugh JT, Fuhrman JA, Azam F (1980) Radioactive labelling of natural assemblages of bacterioplankton for use in trophic studies. Limnol Oceanogr 25:172-181

Inouè S (1986) Video microscopy Plenum Press N Y
Iturriaga RB, Mitchell BG, Kiefer DA (1988) Microphotometric analysis of individual particle absorption spectra. Limnol Oceanogr 33:128-135
Jonsson PR (1987) Photosynthetic assimilation of inorganic carbon in marine oligotrich ciliates (Ciliophora Oligotrichina) Mar Microb Food Webs 2:55-68
Karl DM, Knauer GA, Martin JH (1988) Downward flux of particulate organic matter in the ocean: a particle decomposition paradox Nature 332:438-441
Kirchman DE, K'nees E, Hodson R (1985) Leucine incorporation and its potential as a measure of protein synthesis by bacteria in natural aquatic systems. Appl Environ Microbiol 49:599-607
Landry MR, Hasset RP (1982) Estimating the grazing impact of microzooplankton Mar Biol 67:283-288
Landry MR, Lehner-Fournier JH, Sundstrom JA, Selph KE (1987) Discriminate feeding of marine Protozoa on living versus heat-killed bacteria [Abstract] EOS Trans Amer Geophys Union 68:1782
Lapidus R, Levandowsky M (1981) Mathematical models of behavioral responses to sensory stimuli by protozoa In: Levandowsky M, Hutner SH (eds) Biochemistry and Physiology of Protozoa. Vol 4, Academic Press p 235
Laval-Peuto M, Rassoulzadegan F (1988) Autofluorescence of marine planktonic Oligotrichina and other ciliates. Hydrobiologia 159:99-110
Lee JJ, Small EB, Lynn DH, Bovee EC (1985) Some techniques for collecting cultivating and observing protozoa In: Lee JJ, Hunter SH, Bovee EC (eds) An illustrated guide to the protozoa. Soc Protozool Special Publ, Allen Press Lawrence Kansas p 1
Legner M Phytoplankton quantity assessment by means of flow cytometry. In: Mar Microb Food Webs Burkill PH, Turley CM, Reid PC (eds) (in press)
Lessard EJ, Swift E (1985) Species-specific grazing rates of heterotrophic dinoflagellates in oceanic waters measured with a dual-label radioisotope technique. Mar Biol 87:289-296
Lochte K, Turley CM (1988) Bacteria and cyanobacteria associated with phytodetritus in the deep-sea Nature 333:67-69
Lochte K (1990) Protozoa as makers and breakers of marine aggregates. In: Reid PC, Turley CM, Burkill PH (eds) Protozoa and their role in marine processes. Springer, Berlin Heidelberg, New York
McManus GB, Fuhrman JA (1986) Bacterivory in seawater studied with the use of inert fluorescent particles. Limnol Oceanogr 31:420-426
McManus GB, Pace ML (1988) Measurement of protozoan bacterivory using fluorescently-labelled minicells NATO ASI 88 [Abstracts] Plymouth
Montagnes DJS, Lynn DH (1987) A quantitative Protargol Stain (QPS) for ciliates: method description and test of its quantitative nature. Mar Microb Food Webs Paris 2:83-93
Murphy LS, Haugen EM (1985) The distribution and abundance of phototrophic ultraplankton in the North Atlantic. Limnol Oceanogr 30:47-58
Newell SY, Christian RR (1981) Frequency of dividing cells as an estimator of bacterial productivity. Appl Environ Microbiol 42:23-31
Newell SY, Sherr BF, Sherr EB, Fallon RD (1983) Bacterial response to presence of eukaryotic inhibitors in water from a coastal marine environment. Mar Environ Res 10:147-157
Olson RJ, Frankel SL, Chisholm SW, Shapiro HM (1983) An inexpensive flow cytometer for the analysis of fluorescence signals in phytoplankton: Chlorophyll and DNA distributions J Exp Mar Biol Ecol 68:129-144
Paranjape MA (1987) Grazing by microzooplankton in the eastern canadian arctic in summer 1983. Mar Ecol Prog Ser 40:239-246
Paranjape MA (1990) Quantitative sampling of field populations of protozooplankton. In: Reid PC, Turley CM, Burkill PH (eds) Protozoa and their role in marine processes. Springer, Berlin Heidelberg, New York

Pavlovskaya TV (1976) Distribution of microzooplankton in the coastal waters of the Black Sea In: Biological Studies in the Black Sea. Naukova Dumka kiev p 75 (in Russian)
Pearce F (1988) The science of the thin green smear New Scientist 1639:24-26
Piller H (1977) Microscope photometry Springer
Pomeroy LR (1974) Oceanic productivity - a changing paradigm BioScience 24:499-50
Pomeroy LR, Wiebe WJ (1988) Energetics of microbial food webs. Hydrobiologia 159:7-18
Porter KG (1988) Phagotrophic phytoflagellates in microbial food webs. Hydrobiologia 159:89-97
Porter KG, Feig YS (1980) The use of DAPI for identiying and counting aquatic microflora. Limnol Oceanogr 25:943-948
Rassoulzadegan F (1982) Feeding in marine planktonic protozoa. Ann Inst Oceanogr Paris 58:191-206
Rassoulzadegan F, Sheldon RW (1986) Predator-prey interactions of nanozoo- plankton and bacteria in an oligotrophic marine environment. Limnol Oceanogr 31:1010-1021
Rassoulzadegan FM, Laval-Peuto M, Sheldon RW (1988) Partitioning of the food ration of marine ciliates between pico- and nanoplankton. Hydrobiologia 159:75-88
Revelante N, Gilmartin M (1983) Microzooplankton distribution in the Northern Adriatic Sea with emphasis on the relative abundance of ciliated protozoa. Oceanologica Acta 6:407-415
Reid PC (1987) Mass encystment of a planktonic oligotrich ciliate. Mar Biol 95:221-230
Reid PC, Boalch GT (1987) A new method for the identification of dinoflagellate cysts. J Plank Res 9:249-253
Rublee PA, Gallegos CL (1989) Use of fluorescently labelled algae (FLA) to estimate microzooplankton grazing. Mar Ecol Prog Ser 51:221-227
Sanders RW, Porter KG (1986) Use of metabolic inhibitors to estimate protozooplankton grazing and bacterial production in a monomictic eutrophic lake with an anaerobic hypolimnion. Appl Environ Microbiol 52:101-107
Sheldon RW (1978) Electronic counting: Sensing-zone counters in the laboratory In: Sournia A (ed) Phytoplankton manual. Unesco Monographs on Oceanographic methodology 6 London p 202
Sheldon RW, Parsons TR (1967) A practical manual on the use of Coulter counter in marine science. Coulter Electronics Toronto: p 66
Sheldon RW, Rassoulzadegan F (1987) A method for measuring plankton production by particle counting. Mar Microb Food Webs 2:29-44
Sherr EB (1988) Direct use of high molecular weight polysaccharide by heterotrophic flagellates Nature 335:348-351
Sherr EB, Sherr BF (1983) Double-staining epifluorescence technique to assess frequency of dividing cells and bacterivory in natural populations of heterotrophic microprotozoa. Appl Environ Microbiol 46:1388-1393
Sherr EB, Sherr BF (1987) High rates of consumption of bacteria by pelagic ciliates. Nature 325:710-711
Sherr E, Sherr B (1988) Role of microbes in pelagic food webs: a revised concept. Limnol Oceanogr 33:1225-1227
Sherr BF, Sherr EB, Andrew TL, Fallon RD, Newell SY (1986) Trophic interactions between heterotrophic Protozoa and bacterioplankton in estuarine water analyzed with selective metabolic inhibitors. Mar Ecol Prog Ser 32:169-179
Sherr EB, Sherr BF, Albright LJ (1987a) Bacteria: link or sink? Science 235:88-89
Sherr BF, Sherr EB, Fallon RD (1987b) Use of monodispersed fluorescently labeled bacteria to estimate *in situ* protozoan bacterivory. Appl Environ Microbiol 53:958-965
Sieburth JMcN (1984) Protozoan bacterivory in pelagic marine waters. In: Hobbie JE, Williams PJLeB (eds) Heterotrophic activity in the sea. Plenum Publishing Corp NY p 405
Sieracki ME, Johnson PW, Sieburth JMcN (1985) Detection

enumeration and sizing of planktonic bacteria by image-analyzed epifluorescence microscopy. Appl Environ Microbiol 49:799-810
Sieracki ME, Hass LW, Caron DA, Lessard EJ (1987) The effect of fixation on particle retention by microflagellates: understimation of grazing rates. Mar Ecol Progr Ser 38:251-258
Smetaceck V, Scharek R, Farhrbach E, Rohardt G, Zaucker F, Weppernig R (1987) A simple method for obtaining high-resolution profiles from the water layer immediately underlying sea ice. EOS Trans Amer Geophys Union 68:1768
Sorokin Yu (1977) The heterotrophic phase of plankton succession in the Sea of Japan. Mar Biol 41:107-117
Sorokin YuI (1981) Microheterotrophic organisms in marine ecosystems In: Longhurst AR (ed) Analysis of marine ecosystems. Academic Press NY p 293
Sournia A (1978) Phytoplankton manual. Unesco. Monographs on Oceanographic methodology. London p 337
Steedman HF (1976) Zooplankton fixation and preservation. Unesco. Monographs on Oceanographic methodology. United Kingdom, Paris p 350
Stoecker DK, Cucci TL, Hulburt EM, Yentsch CM (1986) Selective feeding by *Balanion* sp (Ciliata: Balanionidae) on phytoplankton that best support its growth. J Exp Mar Biol Ecol 95:113-130
Stoecker DK, Michaels AE, Davis LH (1987) Large proportion of marine planktonic ciliates found to contain functional chloroplasts. Nature 326:790-792
Suess E (1988) Oceanography Effects of microbe activity Nature 333:17-18
Taniguchi A, Takeda Y (1988) Feeding rate and behavior of the Tintinnid Ciliate *Favella taraikaensis* observed with a high speed VTR system. Mar Microb Food Webs 3:21-34
Trask BJ, Van Den Engh GJ, Elgershuizen JH (1982) Analysis of phytoplankton by flow cytometry. Cytometry 2:258-264
Turley CM, Lochte L, Patterson DJ (1988) A barophilic flagellate isolated from 4500 m in the mid-North Atlantic. Deep-Sea Research 35:1079-1092
Uhlig G (1964) Eine einfache Methode zur Extraction der valigen microfauna aus marinen sedimenten. Helgol Wiss Meeresunters 11:178-185
Van Houten J, Hauser DCR, Levandowsky M (1981) Chemosensory behavior of protozoa. In: Levandowsky M, Hunter SH (eds) Biochemistry and Physiology of Protozoa. Vol 4, Academic Press p 67
Wiadnyana NN, Rassoulzadegan F (1989) Selective feeding of Acartia clausi and Centropages typicus on microzooplankton. Mar Ecol Prog Ser 53:37-45
Wikner J, Andersson A, Normark S, Hagstrøm A (1986) Use of genetically marked minicells as a probe in measurement of predation on bacteria in aquatic environments. Appl Environ Microbiol 52:4-8
Wikner J, Rassoulzadegan F, Hagstrøm (1990) A Periodic bacterivore activity counterbalances bacterial growth in the marine environment. Limnol Oceanogr (in press)
Wood AM *et al* (1985) Discrimination between types of pigments in marine *synechococcus* sp by scanning spectroscopy epifluorescence microscopy and flow cytometry. Limnol Oceanogr 30:1303-1315
Wood EJF (1962) A method for phytoplankton study. Limnol Oceanogr 7:32-35
Yentsch CM *et al* (1983) Flow cytometry and cell sorting: a technique for analysis of aquatic particles. Limnol Oceanogr 28:1275-1280
Zukov BF (1973) Colourless flagellates. Hydrobiol J (Kiev) 9:28-31 (in Russian)

QUANTITATIVE SAMPLING OF FIELD POPULATIONS OF PROTOZOOPLANKTON

Madhu A Paranjape
Northwest Atlantic Fisheries Centre
Science Branch
Department of Fisheries and Oceans
PO Box 5667
St John's
Newfoundland A1C 5X1
Canada

INTRODUCTION

Microzooplankton is defined as the phagotrophic animal forms which will pass through a 200 µm mesh netting (Beers and Stewart 1967). This is a size grouping exhibiting a common trophic capacity. However, since the lower threshold in size is not identified in this definition, the community includes organisms in the size range of 2-20 µm which may have different names according to some classification schemes (Dussart 1965, Sieburth *et al.* 1978). The upper size limit is also somewhat ambiguous as there are colonial protozoa that can be several times larger than the 200 µm cut-off, but are still considered to be part of the microzooplankton. As a result of better methods of preservation and observation, it is becoming apparent that many forms of flagellates, considered to be autotrophs because of their pigments, are as capable of utilising particulate food sources as other heterotrophic flagellates that are taxonomically and structurally related (Estep *et al.* 1986, Bird and Kalff 1987, Porter 1988). Similarly many ciliates are now known to harbour functional plastids and apparently can utilise their photosynthetic products while still being phagotrophic (Laval-Peuto *et al.* 1986, McManus and Fuhrman 1986, Stoecker *et al.* 1987, Laval-Peuto and Rassoulzadegan 1988). The existence of such mixotrophy makes the conventional trophic classification ambiguous. Sieburth and Estep (1985) proposed a new, more precise terminology which emphasised evolutionary relationships and trophic status. Terminology based on size-grouping will remain in use however, as studies based on size-classes

NATO ASI Series, Vol. G 25
Protozoa and Their Role in Marine Processes
Edited by P. C. Reid et al.

provide insights into metabolic and growth rates, and predator-prey relationships. They are also easily measured by automatic non-destructive optical and electronic sizing devices (Platt 1985).

Protozooplankton is a taxonomically and morphologically related group of organisms and may encompass several size- and trophic categories. They are, however a ubiquitous and abundant component of the microzooplankton community in marine and freshwater habitats. A majority of the protists will fall into the size categories of nanoplankton (2-20 µm) or microplankton (20-200 µm), however, some members may exist in the picoplankton (0.2-2 µm) size-class as well (Furhman and McManus 1984, Cynar *et al.* 1985). At the other end of the size scale, colonial radiolarians of the genus *Collozoum* can be as large as 2-3 m in length (Swanberg and Harbison 1980, Swanberg and Anderson 1981). Thus, from the known size distributions, protozoa can range over almost 6 orders of magnitude in size.

The spectrum of abundance of protozoa also has a wide range. The heterotrophic and mixotrophic nanoflagellates which include chrysomonads, cryptomonads, bodonids, bicoecids, dinoflagellates, and choanoflagellates are present in virtually all marine and freshwater environments in the range of 10^5 to 10^7 cells per litre (Sorokin 1981, Sieburth and Davis 1982, Sherr and Sherr 1984, Pick and Caron 1987). Ciliate (mostly oligotrichs and tintinnids) abundance generally ranges from 10^2 to 10^4 per litre in most habitats, including highly productive estuarine zones (Sorokin 1981, Porter *et al.* 1985, Taylor and Heynen 1987). Sarcodine protozoa are generally a minor component of the total microzooplankton. These include naked amoebae, foraminifera, radiolaria, acantharia and heliozoa. Naked amoebae are found in concentrations of 10^1 to 10^3 per litre in open ocean water (Davies *et al.* 1978, Paranjape *et al.* 1985) and may be concentrated in the surface microlayer in the range of 10^3 to 10^5 per litre (Davies *et al.* 1978). Other sarcodine taxa, foraminifera, radiolaria and acantharia, can be transiently abundant but generally are present in concentrations of about 10^1 or less per litre (Bé 1977, Anderson 1983). In this context, it should be remembered that the potential prey organisms of these protists, bacteria and algae, occur in the range of 10^9 and 10^6 per litre respectively (Azam and Ammerman 1984).

Because of the wide range in the size of protozoa and the similarly wide range in their abundance, the problems of field sampling are similar for both phytoplankton and mesozooplankton. The broad spectrum in size and abundance requires a different approach in selecting a sample size. While bacteria could be counted in a few ml, flagellates in 10's of ml and ciliates in 100's of ml, sarcodines require several thousands of ml to obtain reliable estimates of abundance.

Most workers concentrate their research efforts on a specific taxon which may require more than one method of collection, handling and preservation. For example, physiological studies on large foraminifera or radiolaria are seldom done on specimens collected from routine net tows. Hand collection by SCUBA diving is still the most preferred method. However, estimates of quantitative abundance for these sarcodines are routinely made by the traditional methods used for mesozooplankton studies. Therefore, no one method can be recommended for the study of protozoa. The methods to be used are mostly based on the taxa to be studied, the nature of the study and perhaps the choice of the individual scientist.

Despite the diverse requirements for the study of protozoa, there have been attempts to standardise sampling methodologies and procedure. Under the sponsorship of UNESCO and SCOR, several monographs on oceanographic methodologies have been published. Three of these monographs (Zooplankton sampling, UNESCO 1978, Zooplankton fixation and preservation, Steedman 1976, Phytoplankton manual, Sournia 1978 have sections that deal directly with the study of protozoa. Other sections, such as sampling design and strategies (Cassie 1968, Venrick 1978,

Margalef 1978), although not written explicitly for protozoa, are directly applicable to the study of protozoa.

These monographs need to be updated. The Phytoplankton Manual was published in 1978. It has been 10 years since then and 20 years since Zooplankton Sampling was published. I would like to make a plea to the sponsoring organisations (UNESCO, SCOR, IABO, etc.) to initiate a new round of Working Groups to revise and update these monographs to take into account the recent refinements in methodologies.

Has there been any dramatic progress in methodology? Regrettably, the answer is no; although there have been several refinements in existing sampling equipment and observational techniques to give more precise and accurate information on protozooplankton field populations. The most commonly used methods are described fully in the UNESCO monographs. Other new methods have been developed to sample specific sites of biological activity, such as the air-water interface and the sea-ice-water interface. A brief review of methods used in the water column sampling and at these latter sites is presented.

WATER-COLUMN SAMPLING

Free-living protozoan populations in the water column show a high degree of diversity in terms of size and abundance. As mentioned earlier, sizes may vary from >0.2 µm to >200 µm and abundance may vary by several orders of magnitude. Since the majority of the protozooplankton overlaps phytoplankton in size and habitat, the typical sampling methods used to study phytoplankton are also used extensively to study protozoa. It is no surprise that some of the early studies on pelagic protozoa were conducted by phytoplankton specialists. Similarly, for the field studies on larger protozoa such as foraminifera, radiolaria, and acantharia, the traditional sampling methods used for mesozooplankton are used. Juvenile stages of many of these sarcodines and metazoans are smaller and can be sampled by techniques used for microzooplankton study.

To sample the majority of the protozoans some method of capturing water from various depths is necessary. What is to be done with this water and the amount of water required will depend on the nature of the study and the type of taxa to be studied.

WATER BOTTLES

To obtain discrete samples from various depths, water bottles are essential. Numerous designs of commercially available water bottles of varying capacity are summarised by Venrick (1978). In biomass-rich inshore waters small capacity water bottles are adequate, but in the oceanic oligotrophic waters large capacity bottles are necessary to sample the sparse populations. These large bottles unfortunately lack a mixing mechanism before the removal of subsamples. As a result, serious errors can be introduced in the abundance estimates of particulates due to incomplete extraction of the rapidly settling particles. Gardner (1977) noted significant losses of foraminifera, acantharia, dinoflagellates, faecal pellets and organic aggregates from 30-litre Niskin bottles, as they settled below the sampling port. The particulate concentration of the settled material was 1.1 to 1.7 times higher than the initial water sample. Similar incomplete recovery was also reported by Calvert and McCartney (1979).

Water samplers to study vertical or horizontal microdistribution have been designed and used by Blaker (1979) and Owen (1981). Both of these instruments are capable of taking several samples simultaneously at 10 to 20 cm intervals. The volume of water captured is relatively small if one needs to study a variety of biological and chemical parameters, but if used in an appropriate environment, such as a chlorophyll maximum or a density discontinuity, these samplers should be able to collect sufficient number of cells to give reasonable data. Another water sampler of simple construction for gently collecting protozoa is described by Graham *et al.* (1976).

PUMP SAMPLING

The main advantage of pump sampling is the potential ability to resolve spatial patterns of distribution of plankton at a much finer scale than can be achieved by discrete sampling by bottles or towed nets. Pumps can be used to sample at discrete depths or over integrated vertical, oblique or horizontal distances. A servo-controller developed for the pumping system at the Bedford Institute of Oceanography (Canada) can maintain the pump-intake at predetermined depths (±0.5 m) even when the ship is drifting or rolling (A Herman, per. comm). The use of continuous-flow sensors to measure other physical and chemical variables, such as temperature, salinity, fluorescence, dissolved nutrients simultaneously with plankton concentration makes it possible to sample the environment at biologically relevant time- and space-scales.

Pump sampler designs vary with requirements and applications. Miller and Judkins (1981) have reviewed the general engineering considerations and hydraulic theory in the design and operation of pump systems (see also Beers 1978). These authors also discuss the advantages and disadvantages of pumps relative to other means of sampling. A simple empirical method to determine the fine-scale resolution of a continuous-flow sampling system is described by Mackas and Owen (1982).

Pumps can sample almost the entire spectrum of size-classes of planktonic organisms in the water from the same source, although some degree of damage to delicate, soft-bodied organisms can be expected. Recently described large-volume pump systems ($>0.5 m^3/min$) were specifically designed to sample ichthyoplankton and their potential food simultaneously (Taggart and Leggett 1984, Harris *et al.* 1986). The damage to certain species of larval fish was recognised in these studies, but comparison with net sampling showed that the pumps were as effective or, in some cases, more effective in capturing mobile organisms than plankton nets, towed either horizontally or vertically. There were differences in efficiency of sampling different size-classes of fish larvae, however (see also, Leithiser *et al.* 1979, Cada and Loar 1982, Solemdal and Ellertsen 1984). Such comparisons with the low-volume zooplankton sampling pumps and nets were made by Mullin and Brooks (1976) and Pillar (1984). Since pump systems vary in design and operation, each system should be evaluated for the specific sampling it is designed to achieve.

Generally, a portion of water collected by pumps is preserved as a 'whole water' sample for enumeration of organisms in the lower end of the size spectrum. The larger forms (usually less numerous) are concentrated on screens of desired porosity. Sorokin (1977, 1981) has questioned the use of pumps in collection of water and subsequent procedures of concentration and preservation. He further argues that even the most gentle methods of concentrating water samples destroy ciliates and flagellates and the chemical fixatives used would also further distort and destroy these fragile organisms. He recommends whenever possible, 'live-counting' of these organisms in specially-designed counting chambers, using water-immersion objectives (Sorokin and Kogelschatz 1979, Sorokin 1980, Sorokin *et al.* 1985).

Clearly, this is a moot question that has been ignored. There have been few attempts to compare the sampling efficiency and the damage inflicted by pumps on microplankton. Herman *et al.* (1984) compared rates of primary production and enzyme uptake of pumped water with that obtained by conventional, water bottles and found no significant difference. Similarly, I have compared the microzooplankton community from the Nova Scotian Shelf collected by the two sampling methods and the species composition and abundance of ciliates was found to be similar (Paranjape unpublished data). Effects of concentration were not tested, however. 'Live-counting' may be an ideal method, but it is difficult to use on routine biological surveys onboard ships. Besides, the concentration of organisms in the whole water sample will have to be sufficiently large (1 to 10 per ml) to give precision to the count (Dale and Burkill 1982, Bak and Nieuwland 1987). Until these methods have been tested rigorously, there appears to be no substitute for preserved whole water or concentrated/preserved samples to estimate species abundance and biomass of protozoa in routine biological surveys.

NET SAMPLING

Net samples have mostly been used for qualitative taxonomic surveys and have rarely been used to sample protozooplankton quantitatively. Even now small-mesh nets are used only for those organisms with rigid or hard walls (tintinnids and sarcodines). There is increasing evidence however, that plankton nets of even small-mesh (25 µm) underestimate the abundance of a variety of organisms with rigid walls. Michaels' study (1988) showed the abundance of net collected acantharia, radiolaria, foraminifera, dinoflagellates and diatoms were lower by 12 to 50 times than those estimated from 5-1 water bottles.

SURFACE MICROLAYER

The surface layer of the ocean is a distinct and separate environment with its own collection of organisms. Zaitsev's (1971) monograph on marine neuston gave momentum to the study of surface layers as an important specialised marine environment. These layers also serve as a concentration site for many substances of anthropogenic origin such as heavy metals, hydrocarbons and pesticides (Lion and Leckie 1982, Hardy 1982). As in many instances, the surface film or layer is defined operationally by the methods used to collect it. Parker (1978) defined neuston as all biota in microlayers between <1000 µm depth and the air-water interface, while Zaitsev (1971) included all organisms in the upper 5 cm in this community.

More than 20 different techniques have been described; each

method samples a different depth of the layer which makes comparison of sampling efficiency and results somewhat difficult (Hardy 1982). To collect large-volume samples of the microlayer, three types of samplers are commonly used. The screen sampler collects a surface film of 100 to 200 μm thickness, while the plate-sampler collects a surface film of 60 to 100 μm thickness (Parker 1978). The rotating drum method is becoming increasingly popular. The collecting surface or drum on these samplers can be ceramic (Harvey 1966) or PVC (Brockmann *et al.* 1976), glass (Carlson *et al.* 1988), hydrophilic Teflon (Garrett and Barger 1974) or hydrophobic Teflon (Hardy *et al.* 1988). For small samples sufficient for microbiological and other specialised studies membrane filters, Teflon pads, discs and perforated plates and glass slides have been used (Crow *et al.* 1975, Van Vleet and Williams 1980, Larsson *et al.* 1974, Sieburth 1982) to lift the surface film off the underlying water. To sample larger organisms skimmer nets of various designs are used (Zaitsev 1971, Hempel and Weikert 1972). In order to avoid contamination from sampling platforms, most of the samplers are best deployed from small boats away from the research ship or upwind from a slowly-moving ship. Consequently, the methods are sea-state dependent. The major methods of sampling are reviewed and illustrated by Sieburth (1979) and useful guidelines for selection of an appropriate technique are summarised by Garrett and Duce (1980) and Lion and Leckie (1981).

Whichever method is used to collect the samples of the surface microlayer, often a subsurface water sample is collected to compare the measured biological and chemical parameters. The major findings of such comparisons indicate enrichment of the microlayers by particulate organic matter, dissolved organics, chlorophyll *a* and ATP, often by orders of magnitude (Nishizawa and Nakajima 1971, Sieburth *et al.* 1976, Davis *et al.* 1978, Freedman *et al.* 1982, Carlucci *et al.* 1985). In addition, a diverse and abundant heterotrophic community of bacteria, flagellates, ciliates and amoebae in concentrations much higher than in the underlying water has been reported (Harvey and

Burzell 1972, Sieburth *et al.* 1976, Sieburth 1982, Hardy 1982, Hardy *et al.* 1988). Zaitsev (1971) has summarised abundance data on bacteria, tintinnids, and radiolaria from the Black Sea. However, the study of Davis *et al.* (1978) on distribution, abundance and taxonomy of marine amoebae from oceanic water is the only quantitative study of a protozoan taxon from the surface microlayer that I am aware of.

SEA-ICE AND ICE-WATER INTERFACE

Another site of intense biological activity that has received much attention lately is sea-ice. The microbial community associated with sea-ice is an important feature of the productivity of the polar oceans, as it effectively extends the length of the otherwise short growing season. The development and growth of sea-ice and the microbial communities associated with it in the Arctic and Antarctic has been summarised by Grainger *et al.* (1985) and Garrison *et al.* (1986).

High concentrations of algal populations dominated by pennate diatoms in sea-ice have been known for a long time (see Horner 1976, for review). The presence of protozoa was also reported in the early part of the century (Nansen 1906). It is only recently, however, that the diversity of protozoan taxa in sea-ice has been recognised. In the Arctic, a variety of autotrophic and heterotrophic nanoflagellates, heliozoa, amoebae, ciliates, and small metazoa have been recovered from sea-ice (Horner and Alexander 1972, Grainger *et al.* 1985). Similar communities have also been observed in Antarctic sea-ice (Fenchel and Lee 1972, Garrison *et al.* 1984). The organisms are mostly associated with the bottom of the ice, where thick algal mats grow. In the Arctic, this bottom assemblage may be up to 10 cm thick while in the Antarctic, it is reported to be as much as 4 m thick (Horner *et al.* 1988). Microbial communities can also be found on the surface of the ice, scattered in the interior of ice, in discreet bands or layers and in brine channels. The different mechanisms involved in the formation of sea-ice presumably cause these distributional patterns.

Sea-ice biota are usually sampled by coring. Cores can be obtained by using an ice-coring auger from the surface of the ice or from the bottom of the ice by SCUBA divers. Whichever method is used, the sample is allowed to melt before being analysed. Garrison and Buck (1986) pointed out that when ice samples are allowed to melt, the microorganisms living within the brine channels are subjected to rapid and extreme changes in salinity. This osmotic shock results in substantial losses of flagellates and ciliates, thereby, over-representing organisms with rigid cell walls. Such losses can be minimised if ice-samples are allowed to melt in sterile filtered seawater, which acts as an osmotic buffer.

The majority of protozoan taxa described from sea-ice are not truly pelagic. For example, commonly observed species of the ciliate genera, *Euplotes* and *Stylonychia* require solid surfaces. The large 'internal surface' of the slush ice, such as large ice crystals, brine channels, diatom nets on the underside of the ice, is more comparable with the interstitial space of a sandy bottom. Therefore, many of the methods used to sample meiofauna (Higgins and Thiel 1988) may also be applicable to the study of sea-ice fauna.

Other equipment and instruments designed to sample sea-ice fauna are described in internal technical reports or documents for limited circulation (e.g.. Grainger and Hsiao 1982, Brooks 1982). A number of prototypes have also been constructed in various laboratories engaged in polar research. For example, the Metrology Division in Bedford Institute of Oceanography (Canada) has developed a variety of instruments to study under-ice communities (A Herman and D Knox, pers. comm.).

Our knowledge of the sea-ice microbial communities is still rudimentary. We need to know their origin and also their fate when the ice melts in Spring, their abundance and species composition, their food and predators, and their relationship with the fauna in the underlying water column.

The water column immediately below the sea-ice can also function as a unique habitat under certain conditions, particularly when the ice is ready to break up and the algal growth in the ice is most conspicuous. A viscous non-turbulent layer in the top few millimetres of water, presumably due to the ice-algal exudates and strong density stratification due to freshening of the surface layer appears as the warmer season approaches (Dr T Chriss, Dalhousie University, Halifax, pers. comm.). Such stable layers receive substantial input of fauna and flora from the underside of the ice. This is where the mixing of the two distinct communities occurs. These stable layers can be sites of intense biological activity. Several systems have been described recently to profile these layers. Conover *et al.* (1986) described an under-ice pumping system that could be lowered through a 25 cm hole but could sample the ice-water interface away from the physical and hydraulic disturbance of the hole. The pumping system is capable of profiling the upper 2.5 m of the water column at a resolution of about 20 cm. At that time of the year (May) in the Arctic, Conover *et al.* (1986) did not find stratification of biological parameters. However, adults and juvenile stages of *Pseudocalanus* sp. were aggregated in concentrations of 10^3 l^{-1} at 10 cm depth below the ice and were feeding intensively on algae presumably eroded off the ice. A similar pumping system has been described and used in the Antarctic by Smetacek *et al.* (1987).

OTHER METHODS

Other more specialised and non-routine methods used to collect particulate matter may be used to quantify certain taxa of protozooplankton. Some examples of these include: Collection of individual sarcodines by SCUBA for physiological studies (Bé *et al.* 1977. 1981, Anderson *et al.* 1986, Swanberg and Anderson 1981, Michaels 1988), collection of macroaggregates and associated protozoa by SCUBA or submersibles (Silver *et al.*

1978, Silver and Alldredge 1981, Caron *et al.* 1982). Large-volume *in situ* filtration systems (Bishop and Edmonds 1976, Johnson *et al.* 1987), non-destructive *in situ* electronic counters (Scura 1982, Herman 1988).

Methods of sample observation have improved enormously in parallel with refinements of traditional sampling techniques. In the past 15 years, reflected fluorescence microscopy coupled with a variety of fluorochromes has greatly increased our ability to observe and discriminate between autotrophic and heterotrophic components of the microzooplankton community with ease and precision. The labour-intensive methods used to count and measure size, shape, volume and surface area of large numbers of organisms and the subjectivity associated with such measurements has been largely overcome with computer-assisted image analysis systems (Sieracki *et al.* 1985, Bjørnson 1986, Estep *et al.* 1986) and flow cytometry (see review by Yentsch and Pomponi, 1986).

Recent novel advances in quantitative optical microscopy such as Video Enhanced Light Microscopy (VELM) and Confocal Light Microscopy (CLM) are closing the gap between light microscopy and scanning electron microscopy. The VELM technology in particular has the greatest potential for solving many of the problems associated with epifluorescence microscopy. A highly sensitive low-light level video-camera attached to a conventional fluorescence microscope amplifies the signals obtained from specimens allowing visualisation of faint fluorescence that cannot be seen by eye or recorded by photo-micrography. It also allows one to observe cells at such low excitation intensities that fluorochrome photo-bleaching is no longer a problem. In a conventional transmitted-light mode, the system allows enhancement of extremely low contrast images generated by minute cellular structure so these can be clearly seen and recorded. In the biomedical field such systems have been used to observe small objects (e.g. microtubules) with dimensions that are of an order of magnitude smaller than the resolution limit of the light microscope. The VELM field is advancing so rapidly that the recent excellent reviews by Inoué (1986) and Shotton (1988) may become outdated before long. These exciting new tools will soon find applications in biological oceanography. Perhaps old fashioned microscopy but with a new twist, will be fashionable once again as was the case in the 1880's and early 1900's.

ACKNOWLEDGEMENTS

I am grateful to John R Beers for his constructive suggestions.

REFERENCES

Anderson OR (1983) Radiolaria. Springer, New York p 355
Anderson OR, Swanberg NR, Lindset JL, Bennett P (1986) Functional morphology and species characteristics of a large solitary radiolarian *Physematium muelleri*. Biol Bull 171:175-187
Azam F, Ammerman JW (1984) Cycling of organic matter by bacterioplankton in pelagic marine ecosystems: micro-environmental considerations. In: Fasham MJR (ed) Flows of energy and materials in marine ecosystems: theory and practice.Plenum Press, New York p 345
Bak RPM, Nieuwland G (1987) Densities of protozoan nanoplankton populations in intertidal mesocosms: influence of oil pollution and a self-igniting clearing agent. Neth J Sea Res 21:303-315
Bé AWH (1977) An ecological, zoogeographic and taxonomic review of recent planktonic foraminifera. In: Oceanic micropaleontology, vol I. Acad Press, London, p 1
Bé AWH, Caron DA, Anderson OR (1981) Effects of feeding frequency on life processes of the planktonic foraminifer *Globigerinoides sacculifer* in laboratory culture. J Mar Biol Ass UK 61:257-277
Bé AWH, Hemleben C, Anderson OR, Spindler, M, Hacunda J, Tuntivate-Choy S (1977) Laboratory and field observations of living planktonic foraminifera. Micropaleo 23:155-179
Beers JR (1978) Sampling techniques. Pump sampling. In: Sournia A (ed) Phytoplankton manual. Monogr Oceanogr Method 6:41-49
Beers JR, Stewart GL (1967) Microzooplankton in the euphotic zone at five locations across the California current. J Fish Res Board Can 24:2053-2068
Bird DF, Kalff J (1984) Algal phagotrophy: regulating factors and importance relative to photosynthesis in *Dinobryon* (Chrysophycae). Limnol Oceanogr 32:277-284
Bishop JKB, Edmond JM (1976) A new large volume filtration system for the sampling of oceanic particulate matter. J Mar Res 34:181-198
Bjørnson PK (1986) Automated determination of bacterioplankton biomass by image analysis. Appl Environ Microbiol 51:1199-1204
Blaker IA (1979) A close-interval water sampler with minimal disturbance properties. Limnol Oceanogr 24:983-988
Brockmann UH, Kattner G, Hentzschel G, Wandschneider K, Junge GH, Huhnerfuss H (1976) Naturliche oberflachenfilme in Seegebeit vor Sylt. Mar Biol 36 135-146
Brooks DJ (1982) Arctic through-the-ice zooplankton sampling nets. Can Fish Oceans Technical Note Ser 82-3, Unpub MS, p 21
Cada GF, Loar JM (1982) Relative effectiveness of two ichthyoplankton sampling techniques. Can J Fish Aquat Sci 39:811-814
Calvert SE, McCartney MJ (1979) The effect of incomplete

recovery of large particles from water samplers on the chemical composition of oceanic particulate matter. Limnol Oceanogr 24:532-536
Carlson DJ, Canty J, Cullen JJ (1988) Description of and results from a new surface microlayer sampling device. Deep Sea Res 35:1205-1213
Caron DA, Davis PG, Madin LP, Seiburth JMcN (1982) Heterotrophic bacteria and bacterivorous protozoa in oceanic macroaggregates. Science 218:795-797
Carlucci AF, Gravin DB, Henrichs SM (1985) Surface-film microheterotrophs: amino acid metabolism and solar radiation effects on their activities. Mar Biol 85:13-22
Cassie RM (1968) Sample design. In: Anon (ed) Zooplankton sampling. Monogr Oceanogr Method 2:105
Conover RJ, Herman AW, Prinsenberg SJ, Harris LR (1986) Distribution of and feeding by the copepod *Pseudocalanus* under fast ice during the arctic spring. Science 232:1245-1247
Crow SA, Ahern DG, Cook WL, Bourquin AW (1975) Densities of bacteria and fungi in coastal surface films as determined by a membrane absorption procedure. Limnol Oceanogr 20:644-646
Cynar FJ, Estep KW, Sieburth JMcN (1985) The detection and characterization of bacteria-sized protists in 'protist-free' filterates and their potential impact on experimental marine ecology. Microb Ecol 11:281-288
Dale T, Burkill PH (1982) 'Live counting' - a quick and simple technique for enumerating pelagic ciliates. Ann Inst Oceanogr 58(S):267-276
Davis PG, Caron DA, Sieburth JMcN (1978) Oceanic amoebae from the North Atlantic: culture, distribution, taxonomy. Trans Am Microsc Soc 96:73-88
Dussart BM (1965) Les différentes catégories de plancton. Hydrobiologia 26:72-74
Estep KW, Davis PG, Keller MD, Sieburth JMcN (1986) How important are oceanic algal nanoflagellates in bacterivory? Limnol Oceanogr 31:646-650
Estep KW, MacIntyre F, Hjörleifsson E, Sieburth, JMcN (1986) MacImage: a user-friendly image-analysis system for the accurate mensuration of marine organisms. Mar Ecol Prog Ser 33:243-253
Fenchel T, Lee CC (1972) Studies on ciliates associated with sea ice from Antarctica. 1.The nature of the fauna. Arch Protistenkd 114:231-236
Freedman ML, Hains JJ Jr, Schindler JE (1982) Diel changes in neuston biomass as measured by ATP and cell counts, Lake Louise, Georgia, USA. J Freshwater Ecol 1:373-381
Furhman JA, McManus GB (1984) Do bacteria-sized marine eukaryotes consume significant bacterial production? Science 224:1257-1260
Gardner WD (1977) Incomplete extraction of rapidly settling particles from water samplers. Limnol Oceanogr 22:764-768
Garrett WD, Barger WR (1974) Sampling and determining the concentration of film-forming organic constituents of the air-water interface. NRL Memo Rep 2852, Naval Res Lab, Washington, DC
Garrett WD, Duce RA (1980) Surface microlayer samplers. In: Dobson F, Hasse L, Davis R (eds) Air-sea interactions. Plenum Publ Corp, NY, p 471
Garrison DL, Buck KR (1986) Organism losses during ice melting: a serious bias in sea ice community studies. Polar Biol 6:237-239
Garrison DL, Buck KR, Silver MW (1984) Microheterotrophs in the ice-edge zone. Antarct J US 19:109-111
Garrison, DL, Sullivan CW, Ackley SF (1986) Sea ice microbial communities in Antarctica. Bioscience 36:243-250
Graham LB, Colburn AD, Burke JC (1976) A new, simple method for gently collecting planktonic protozoa. Limnol Oceanogr 21:336-341
Grainger EH, Hsias SIC (1982) A study of the ice biota of Frobisher Bay, Baffin Island 1979-1981. Can MS Rep Fish Aquat Sci 1647, p 128
Grainger EH, Mohammed AA, Lovrity JE (1985) The sea ice fauna of Frobisher Bay, the Arctic Canada. Arctic 38:23-30
Hardy JT (1982) The sea surface microlayer: biology, chemistry and anthropogenic enrichment. Prog Oceanogr 11:307-328

Hardy JT, Coley JA, Antrim LD, Kiesser SL (1988) A hydrophobic large-volume sampler for collecting aquatic surface microlayers: characterization and comparison with the glass plate method. Can J Fish Aquat Sci 45:822-826
Harris RP, Fortier L, Young RK (1986) A large-volume pump system for studies of the vertical distribution of fish larvae under open sea conditions. J Mar Biol Assoc UK 66:845-854
Harvey GW (1966) Microlayer collection from the sea surface. A new method and initial results. Limnol Oceanogr 11:608-613
Harvey GW, Burzell LA (1972) A simple microlayer method for small samples. Limnol Oceanogr 17:156-157
Hempel G, Weikert H (1972) The neuston of the subtropical and boreal north-eastern Atlantic ocean. A Review. Mar Biol 13:70-88
Herman AW (1988) Simultaneous measurement of zooplankton and light attenuance with a new optical plankton counter. Continent Shelf Res 8 205-221
Herman AW, Mitchell MR, Young SW (1984) A continuous pump sampler for profiling copepods and chlorophyll in the upper oceanic layers. Deep-Sea Res 31:439-450
Higgins RP, Thiel M (1988) Introduction to the study of meiofauna. Smithsonian Inst Press, Washington DC p 488
Horner RA (1976) Sea ice organisms. Oceanogr Mar Biol Ann Rev 14:167-182
Horner RA, Alexander V (1972) Algal populations in arctic sea-ice: an investigation of heterotrophy. Limnol Oceanogr 17:454-458
Horner RA, Syvertsen EE, Thomas DP, Lange C (1988) Proposed terminology and reporting units for sea ice algal assemblages. Polar Biol 8:249-253
Inoué S (1986) Video microscopy. Plenum NY p 612
Johnson BD, Wangersky PJ, Zhou XL (1987) An *in situ* pump sampler for trace materials in sea water. Mar Che 22:353-361
Larsson K, Odham G, Södergren A (1974) On lipid surface films on the sea. 1.Simple method for sampling and studies of composition. Mar Chem 2:49-57
Laval-Peuto M, Salvano P, Gayol P, Gruet C (1986) Mixotrophy in marine planktonic ciliates: ultrastructural study of *Tontonia appendiculariformis* (Ciliophora, Oligotrichina) Mar Microb Food Webs 1:81-104
Laval-Peuto M, Rassoulzadegan F (1988) Autofluorescence of marine planktonic Oligotrichina and other ciliates. Hydrobiologia 159:99-110
Leithiser RM, Ehrich KF, Thum AB (1979) Comparison of high volume pump and conventional plankton nets for collecting fish larvae entrained in power plant cooling systems. J Fish Res Board Can 36:81-84
Lion LW, Leckie JD (1981) The biogeochemistry of the air-sea interface. Ann Rev Ecol Earch Planet Sci 9:449-486
Mackas DL, Owen RW (1982) Temporal and spatial resolution of pump sampling systems. Deep-Sea Res 29:883-892
Margalef R (1978) Sampling design: some examples. In: Sournia A (ed) Phytoplankton manual. Monogr Oceanogr Methodol 6:17-31
McManus GB, Fuhrman JA (1986) Photosynthetic pigment in the ciliate *Laboea strobila* from Long Island Sound, USA. J Plankton Res 8:317-327
Michaels AFC (1988) Vertical distribution and abundance of Acantharia and their symbionts. Mar Biol 97:559-579
Miller CB, Judkins DC (1981) Design of pumping systems for sampling zooplankton, with descriptions of two high-capacity samplers for coastal studies. Biol Oceanogr 1:29-56
Mullin MM, Brooks ER (1976) Some consequences of distributional heterogeneity of phytoplankton and zooplankton. Limnol Oceanogr 21:784-796
Nansen F (1906) Protozoa on the ice-floes of the North Polar Sea. The Norwegian North Polar expedition 1883-1886. Scientific results 5 p 22
Nishizawa S, Nakajima K (1971) Concentrations of particulate organic material in the sea surface layer. Bull Plankton Soc Japan 18:12-19
Owen RW (1981) Microscale plankton patchiness in the larval anchovy environment. Rapp P-v Réun Cons int Explor Mer 178:364-368

Paranjape MA, Conover RJ, Harding GC, Prouse NJ (1985) Micro- and macrozooplankton on the Nova Scotian shelf in the prespring bloom period: a comarison of their resource utilization. Can J Fish Aquat Sci 42:1484-1492
Parker BC (1978) Neuston sampling. In: Sournia A (ed) Phytoplankton manual. Monogr Oceanogr Methodol 6:64-67
Pick FR, Caron DA (1987) Picoplankton and nanoplankton biomass in Lake Ontario: relative contribution of phototrophic and heterotrophic communities. Can J Fish Aquat Sci 44:2164-2172
Pillar SC (1984) A comparison of the performance of four zooplankton samplers. S Afr J Mar Sci 2:1-18
Platt T (1985) Structure of marine ecosystems: its allometric basis. In: Ulanowicz RE, Platt T (eds) Ecosystem theory for biological oceanography. Can Bull Fish Aquat Sci 213:55-64
Porter KD (1988) Phagotrophic phytoflagellates in microbial food webs. Hydrobiologia 159:89-97
Porter, KG, Sherr EB, Sherr BF, Pace M, Sanders RW (1985) Protozoa in planktonic food webs. J Protozool 32:409-415
Scura ED (1982) An *in situ* device for sensing and collecting microplankton CalCoFI Rep 23:205-211
Sherr BF, Sherr EB (1984) Role of heterotrophic protozoa in carbon and energy flow in aquatic ecosystems. In: Klug MJ, Reddy CA (eds) Current perspectives in microbial ecology. Am Soc Microbiol, Washington DC, p 412
Shotton DM (1988) Review: a video-enhanced light microscope and its application in cell biology. J Cell Sci 89:129-150
Sieburth JMcN (1979) Sea microbes. Oxford Univ Press, New York, p 491
Sieburth JMcN (1982) Microbiological and organic-chemical processes in the surface and mixed layers. In: Liss PS, Slinn WGN (eds) Air-sea exchange of gases and particles. D Reidel Pub Co, Dordrecht, p 121
Sieburth JMcN, Davis PG (1982) The role of heterotrophic nanoplankton in the grazing and nurturing of planktonic bacteria in the Sargasso and Caribbean sea. Ann Inst Oceanogr 58(S):285-296
Sieburth JMcN, Estep KW (1985) Precise and meaningful terminology in marine microbial ecology. Mar Microb Food Webs 1:1-15
Sieburth JMcN, Smetacek V, Lenz J (1978) Pelagic ecosystem structure: heterotrophic compartments of the plankton and their relationship to plankton size fractions. Limnol Oceanogr 23:1256-1263
Sieburth JMcN, Willis PJ, Johnson KM, Burney CM, Lavoie DM, Hinga KE, Caron DA, French FW III, Johnson PW, Davis PG (1976) Dissolved organic matter and heterotrophic microneuston in the surface microlayers of the North Atlantic. Science 194:1415-1418
Sieracki ME, Johnson PA, Sieburth JMcN (1985) Detection, enumeration, and sizing of planktonic bacteria by image-analyzed epifluorescence microscopy. Appl Environ Microbiol 49:799-810
Silver MW, Alldredge AL (1981) Bathypelagic marine snow:deep-sea algal and detrital community. J Mar Res 39:501-530
Silver MW, Shanks AL, Trent JD (1978) Marine snow: Microplankton habitat and source of small-scale patchiness in pelagic populations. Science 201:371-373
Smetacek V, Scharek R, Farhrbach E, Rohardt G, Zaucker F, Weppernig R (1987) A simple method for obtaining high-resolution profiles from the water layer immediately underlying sea ice. EOS Trans Am geophys Un 68:1768
Solemdal P, Ellersten B (1984) Sampling fish larvae with large pumps: qualitative and quantitative comparisons with traditional gear. In: Dahl D, Danielssen D, Moksness E, Solemdal P (eds) The propagation of cod (*Gadus morhus* L.). Flødevigen Rapp 1:335-363
Sorokin YuI (1977) The heterotrophic phase of plankton succession in the Sea of Japan. Mar Biol 41:107-117
Sorokin YuI (1980) A chamber for counts of protozoa and nanoplankton organisms. Hydrobiol J 16:84-86
Sorokin YuI (1981) Microheterotrophic organisms in marine ecosystems. In: Longhurst AR (ed) Analysis of marine ecosystems. Acad Press, London, p 293
Sorokin YuI, Kogelschatz J (1979) Analysis of heterotrophic microplankton in upwelling areas. Hydrobiologia 66:195-208

Sorokin YuI, Kopylov AI, Mamaeva NV (1985) Abundance and dynamics of microplankton in the central tropical Indian Ocean. Mar Ecol Prog Ser 24:27-41
Sournia A (ed) (1978) Phytoplankton manual. Monographs on oceanographic methodology. 6. UNESCO, Paris, p 337
Steedman HF (ed) (1976) Zooplankton fixation and presevation. Monographs on Oceanographic Methodology. 4. UNESCO, Paris, p 350
Stoecker DK, Michaels AE, Davis LH (1987) Large proportion of marine planktonic ciliates found to contain functional choroplasts. Nature 326:790-792
Swanberg NR, Anderson OR (1981) *Collozoum caudatum* sp. nov., a giant colonial radiolarian from equatorial and Gulf Stream waters. Deep-Sea Res 28:1033-1047
Swanberg NR, Harbison GR (1980) The ecology of *Collozoum longiforme* sp. nov. a new colonial radiolarian from the equatorial Atlantic Ocean. Deep-Sea Res 27:715-731
Taggert CT, Leggett WC (1984) Efficiency of large-volume plankton pumps, and evaluation of a design suitable for deployment from small boats. Can J Fish Aquat Sci 41:1428-1435
Taylor WD, Heynen ML (1987) Seasonal and vertical distribution of Ciliophora in Lake Ontario. Can J Fish Aquat Sci 44:2185-2191
UNESCO (1968) Zooplankton sampling. Monographs on Oceanographic Methodolgy 2. UNESCO, Paris, p 174
Van Vleet ES, Williams PM (1980) Sampling sea surface films: a laboratoy evaluation of techniques and collecting materials. Limnol Oceanogr 25:764-770
Venrick EL (1978) Sampling design: sampling strategies. In: Sournia A (ed) Phytoplankton manual. Monogr Oceanogr Methodol 6:7-16
Venrick EL (1978) Sampling techniques: Water bottles. In: Sournia A (ed) Phytoplankton manual. Monogr Oceanogr Methodol 6:33-40
Yentsch CM, Pomponi SA (1986) Automated individual cell analysis in aquatic research. Int Rev Cytol 105:183-243
Zaitsev YuP (1971) Marine neustonology (in Russian). Israel Prog Scient Transl, Jerusalem, p 207

THE APPLICATION OF IMAGE ANALYSED FLUORESCENCE MICROSCOPY FOR CHARACTERISING PLANKTONIC BACTERIA AND PROTISTS

Michael E Sieracki and L Kenneth Webb
Virginia Institute of Marine Science
College of William and Mary
Gloucester Point, Virginia 23062
USA

INTRODUCTION

Fluorescence microscopy has contributed to a major and recent shift in our understanding of plankton ecology. (Pomeroy 1974, Sieburth 1977, Azam *et al.* 1983, Ducklow 1983). This shift is illustrated by an increased awareness of the important role microorganisms play in marine carbon and nutrient cycles (Williams 1981, Goldman and Caron 1985). Examples of the application of fluorescence microscopy include: 1) the discovery, using fluorochrome-stained samples, of bacterial populations in marine waters two orders of magnitude greater in number than previously shown (Ferguson and Rublee 1976, Hobbie *et al.* 1977, Watson *et al.* 1977); 2) the discovery of ubiquitous populations of unicellular cyanobacteria, *Synechococcus* sp., in surface waters of oceans and lakes (Johnson and Sieburth 1979, Waterbury *et al.* 1979, Caron *et al.* 1985); 3) improved estimates of heterotrophic and phototrophic flagellate populations in marine waters (Davis *et al.* 1985); 4) the discovery of the importance of aloricate ciliates as grazers of bacteria (Sherr and Sherr 1987); 5) the development of techniques for measuring protozoan grazing rates (McManus and Fuhrman 1986, Sieracki *et al.* 1987); 6) the discovery of significant bacterivory by photosynthetic organisms (Bird and Kalff 1986); and 7) the discovery of the high degree of photosynthetic endosymbiosis in marine ciliates (Stoecker *et al.* 1987).

Accurate bacterial and protistan biomass estimates are necessary for carbon and nutrient flow modelling and cell surface area could be an important parameter in models of nutrient and carbon fluxes. Fluorescence microscopy is one of

NATO ASI Series, Vol. G 25
Protozoa and Their Role in Marine Processes
Edited by P.C. Reid et al.

the best methods available for measuring pico- and nanoplankton biomass. This is a tedious task to perform visually when, at best, investigators measure only a few cells to calculate estimated mean volumes. Carbon or nitrogen biomass is then estimated from cell counts and conversion factors of average carbon or nitrogen per cell. Bacterial cell sizes, however, are known to vary over seasonal (Chrzanowski *et al.* 1988) and diel periods (Turley and Lochte 1986), and as a result of bottle incubation (Ferguson *et al.* 1984). Similar changes may occur in protozoa as well, and could reflect metabolic activity, growth rates or size-specific grazing pressure.

The need to examine microbial processes and population changes on finer spatial and temporal scales has contributed to the development of automated methods of microscopy. With advances in video and computer technology, affordable image analysis systems are now available, and are being applied to fluorescence microscopy in biomedicine (Arndt-Jovin *et al.* 1985, Bright and Taylor 1986) and marine microbial ecology (Sieracki *et al.* 1985, Bjørnsen 1986). They show much promise in enhancing the microbial ecologist's ability to acquire cell count, size and trophic classification data on planktonic microbial populations. This paper outlines the current state of image analysis instrumentation with illustrations of applications in plankton ecology and from our own colour system.

IMAGE ANALYSIS INSTRUMENTATION

The Colour Image Analysed Fluorescence Microscopy (CIAFM) system is schematically illustrated in Figure 1. The main components are typical of most small computer-based image analysis systems for microscopy (Bright and Taylor 1986). Our system, however, digitises full true-colour images so each image is stored in 3 separate colour planes (red, green and blue). A typical black-and-white system would use one gray-level memory plane for each image. The microscope (Universal, Carl Zeiss, Inc.) is equipped with a computer-controlled

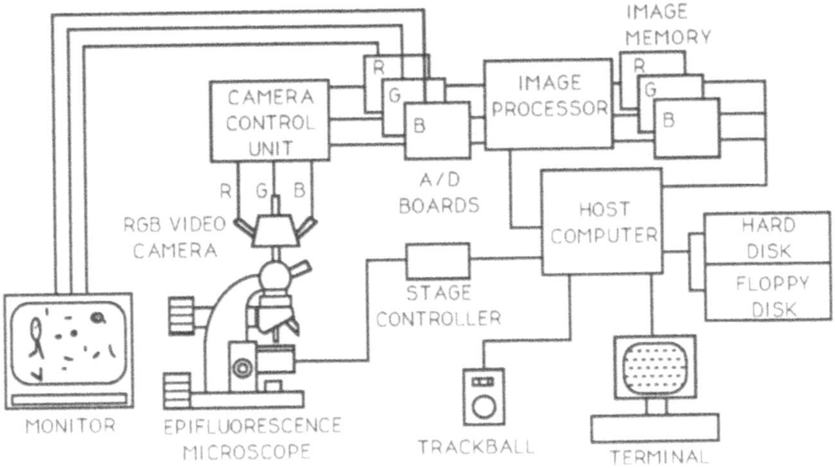

Fig. 1. Schematic diagram of the Color Image Analyzed Fluorescence Microscopy (CIAFM) system showing the major components and signal connections

motorised stage, epifluorescence illumination and Plan Neofluar^{TM} 63X and Neofluar^{TM} 100X objectives. Dual lamp housings allow either xenon or mercury lamp illumination for maximum blue or UV excitation, respectively. A colour video microscope camera (Hitachi DK-5053), equipped with three 2/3 inch chalnicon detector tubes, provides the signals and a camera control unit provides gain, black-level and colour balance adjustment. Separate red, green and blue video signals are digitised in 1/30 seconds and stored in three memory planes by a triple frame grabber. The resolution of the digitiser is 512 X 484 picture elements (pixels) with 8 bits of memory available for each of the three colour planes. This is equivalent to 256 possible 'gray' (brightness) levels for each pixel in each colour, and yields a high-resolution digitised colour image almost indistinguishable from a direct video image. Image analysis is done with a combination of a dedicated image processor (Intel 8086) and the host computer (Motorola 68000 cpu). Image and data storage is provided by a 20 Mb hard disk and a floppy disk drive. A trackball controller can operate the motorised microscope stage and simplifies user interaction with the host computer while the operator selects and focuses new fields for digitisation.

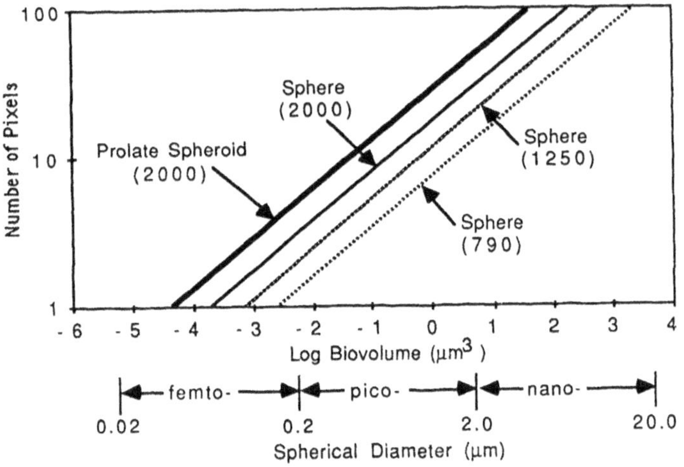

Fig. 2. Pixel coverage of fluorescing cells of different sizes and shapes using several magnifications (shown in parentheses) available with the CIAFM system. Sizes are given in terms of biovolume and 'equivalent spherical diameter'. The plankton size nomenclature of Sieburth *et al.* (1978) is also shown on the abscissa. For prolate spheroid calculation it was assumed that the length is 2 times the width

The CIAFM system has a maximum resolution of 14.3 pixels per μm (i.e. a pixel length of 0.07 μm) using the 100X objective and 2X magnifier on the microscope. The pixels on our system are rectangular with a length-to-width ratio of 1.25. This means that corrections must be made for all cell size and shape measurements. Square pixels are much easier to deal with computationally and most new systems have them. Figure 2 shows the number of pixels which will cover cells of various sizes, shapes and magnifications over the femto- (0.02 - 0.2 μm), pico- (0.2 - 2.0 μm) and nano-plankton (2.0 - 20.0 μm) size ranges (Sieburth *et al.* 1978) using our CIAFM system. For example, cells in the picoplankton size range would be 'covered' by 3 to 30 pixels.

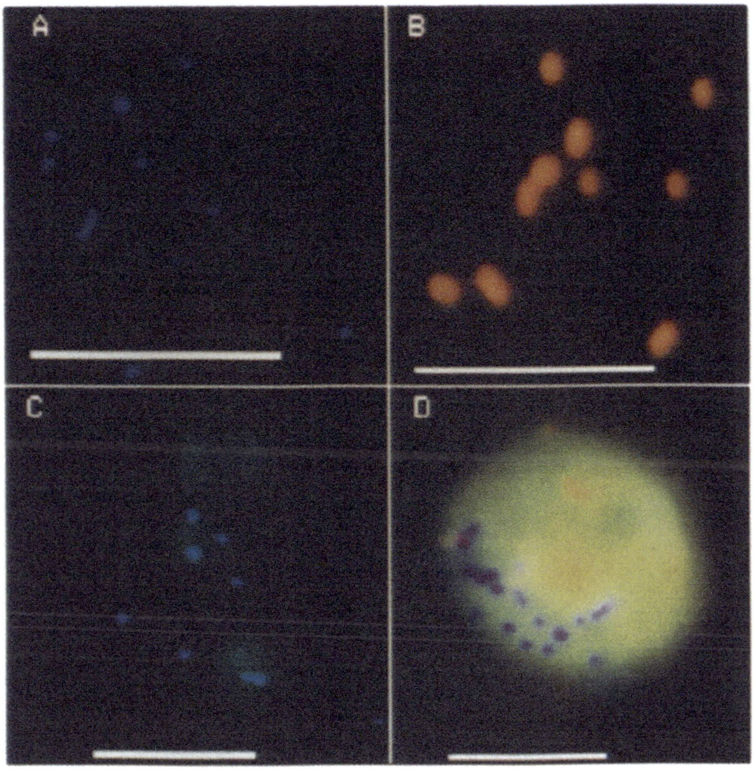

Fig. 3. Color digital images of (A) DAPI-stained bacteria, (B) autofluorescent *Synechococcus*-type cyanobacteria, and proflavine-stained (C) flagellates, and (D) a ciliate. Flagellates and ciliate contain ingested 0.5 µm blue-fluorescing latex beads. All scale bars are 10 µm. Images have been improved by linear contrast enhancement of each color plane

With this system we can accurately detect autofluorescing pigments in cyanobacteria and small flagellates, as well as the commonly used fluorochromes DAPI (4',6-diamidino-2-phenylindole, Porter and Feig 1980), acridine orange (Hobbie *et al.* 1977), and proflavine (Haas 1982) for bacteria and protists (Figure 3). The resolution of this system, and the usefulness of fluorescence microscopy in general, is demonstrated by the digitised images of a single *Myrionecta* (formerly *Mesodinium) rubrum* cell from a natural sample and stained with both DAPI and proflavine shown in Figure 4. These

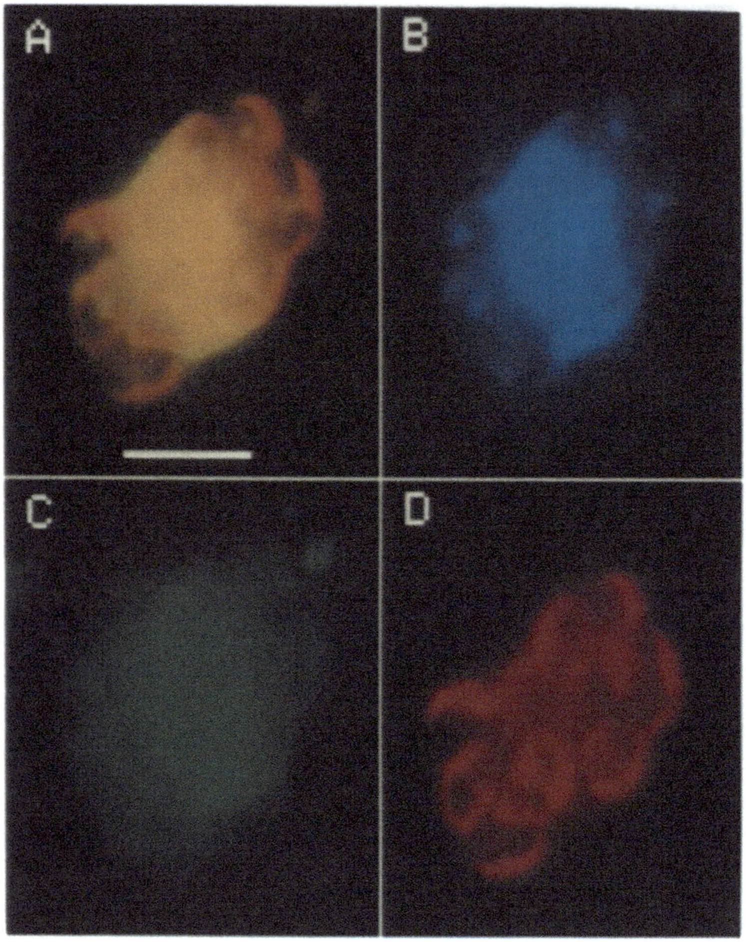

Fig. 4. Detection of a single *Myrionecta* (formerly *Mesodinium*) *rubrum* cell from Chesapeake Bay by Color Image Analyzed Fluorescence Microscopy under a variety of excitation - emission filter combinations. (A) Blue excitation (450 - 490 nm) shows proflavine-stained regions as green/yellow and pigment as orange; (B) UV excitation (365 nm) shows DAPI-stained nuclei and chloroplast DNA; (C) violet excitation (390 - 420 nm) and re-focused image shows cilia; and (D) green excitation (510 - 560 nm) shows red pigment autofluorescence of about 10 individual chloroplasts. Scale bar is ten µm and linear contrast enhancement has been used to improve the image

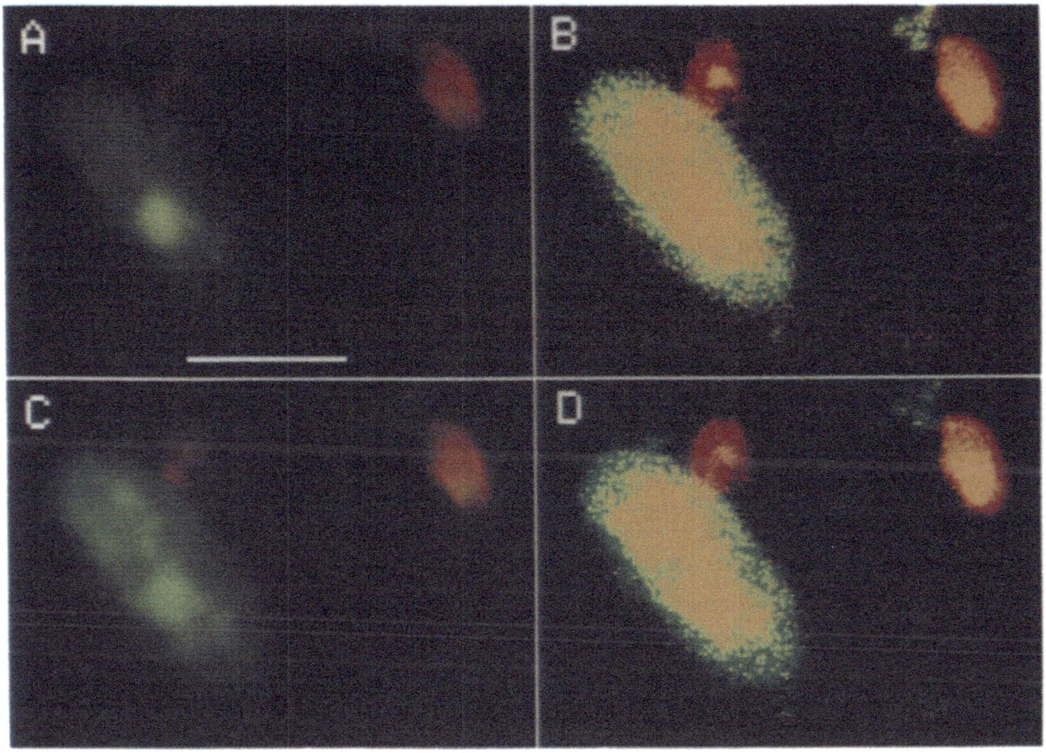

Fig. 5. Detection by the CIAFM system and effect of averaging an image containing a ciliate and two phototrophic flagellates. (A) A single frame has considerable background noise made more visible in (B) by brightening all pixels over a threshold brightness of 32. (C) The same image averaged 2 times shows significantly less noise (D) when brightened and thresholded as in (B). Sample was stained with proflavine and excited with blue light. Scale bar is 10 µm. Images (A) and (C) have been improved by linear contrast enhancement

images also show the advantage of using different excitation and emission filter sets to characterise a planktonic cell. Considerable intracellular detail can be seen, including blue-fluorescing nucleic acids associated with chloroplasts (Figure 4B), cilia (Figure 4C) and the 10 individual, endosymbiotic, cryptophyte chloroplasts (Figure 4D). Detection of a ciliate and phototrophic flagellates are shown in Figure 5.

The quality of the primary image is crucial to successful image analysis. While many image processing techniques are available to enhance poor images, analysis time and computer memory is best used for the measurement of cells in the best possible images. In this regard, the microscope and imaging camera are the most important components in an image analysis system.

Relative to other hardware components of an image analysis system, microscopes have had the longest period of development. Technically, they may be near their theoretical limit and advances in microscopy are slower than in computer or video technology. The optical characteristics of a microscope necessary for good fluorescent images need to be optimised for light transmission since fluorescent images tend to be of very low light intensity. In general transmission efficiency by transmitted light microscopy is less important, since illumination can easily be increased. Recently, microscopes have been re-designed for low-light applications by shortening light path lengths, using a minimum of glass and employing highly efficient lenses. These fluorescence microscopes (e.g. the Zeiss Axioplan^{TM}), produce significantly brighter and higher resolution images.

High quality video cameras are now available which perform well in low light environments. In addition several means have been developed to intensify low light images. These technologies have been reviewed by Inoue (1986) and include electrostatic-type, microchannel plate, and silicon-intensified target (SIT) tubes. New technology, solid-state chip cameras can, in many ways, out-perform conventional video cameras for quantitative image acquisition. For example, charge-coupled device (CCD) cameras designed for scientific imaging applications (e.g. Photometrics Ltd Tucson, Arizona, USA) can output digital images directly into a computer, eliminating the need for a video analog-to-digital converter (frame-grabber). These cameras, developed for astronomical imaging, can provide orders of magnitude greater sensitivity, and dynamic range than video cameras or CCDs adapted to standard broadcast video output (Kristian and Blouke 1982, Hiraoka *et al.* 1987). They have the advantage of ruggedness, stability, and compactness. High performance three-chip colour CCD cameras are now available which can acquire colour images comparable to colour video cameras at significantly lower cost.

Black-and-white (B/W) and colour systems each have advantages and disadvantages. Colour cameras are much more expensive, have lower resolution and less sensitivity. The higher resolution and better low-light sensitivity of B/W cameras allow more accurate detection for cell sizing since this depends upon detection of the relatively dim cell edge. Colour image information is required, however, in plankton samples prepared for fluorescence microscopy to discriminate live cells from detrital particles and autofluorescing, pigment-containing cells from obligately heterotrophic cells.

One means of retaining the high resolution of B/W cameras and gaining colour information is the use of sequential colour filters in the microscope. Since exact pixel-to-pixel registration is necessary for many image processing techniques; a major problem with this method is that vibration in the optical line due to changing filters will cause unpredictable pixel mis-registration of the different colour images. At the high magnifications necessary for imaging protists and bacteria, this problem has prevented engineers at Carl Zeiss, Inc from using this technique (Carl Zeiss Inc , personal communication). In applications where pixel-to-pixel alignment is not necessary, and the concomitant loss of resolution is tolerable, this technique could be useful. In cases where most cells in the image are similar and there is not much detrital colour material, is not needed to discriminate cells. Examples include oceanic samples, many DAPI-stained samples prepared for bacterial counts and many samples prepared for cyanobacteria counts. The majority of bright objects in these images are cells of interest and appropriate filter sets can then be used to highlight heterotrophic or phototrophic picoplankton cells. For these analyses B/W systems are probably adequate.

Sophisticated frame grabber boards are available for common computers and can include arithmetic logic units, co-processors, input and output look-up tables and video input and output multiplexers. Several frame grabbers support processing for rapid image averaging (see below). Digitised images require large amounts of memory for storage. For example, a typical image of 512 X 512 pixels and 8 bits of gray level per pixel requires 262 Kbytes of computer memory, while a high resolution, 1000 X 1000 pixel image would use 1 Mbyte of memory. For this reason we do not routinely archive all images that have been analysed. Equipment such as large-capacity (10 - 20 Mbytes) removable disks are useful for storing images and image data. The availability of writable optical disks will make routine archiving of digital images feasible.

ANALYSIS OF FLUORESCENCE MICROSCOPE IMAGES

Unlike some specific applications (e.g. counting blood cells), fully automated systems do not yet exist for image analysed microscopy of natural microbial samples. Typically, an interactive, semi-automatic procedure is used, requiring an experienced microscopist. Automated operation may be possible for analysis of specific populations such as bacteria or coccoid cyanobacteria where specific stains (e.g. DAPI) or autofluorescent signatures can be used and cells fall within a narrow size range. In coastal and estuarine environments, the large amount of detritus will confound fully automated image analysis procedures. The following is a brief description of techniques that have proved useful in analysing fluorescent microscope images of marine plankton organisms. A general procedure for the analysis of microscope images is shown in Figure 6. More detailed descriptions of image processing and analysis techniques are available elsewhere (Castleman 1979, Pavlidis 1982, Niblack 1985, Walter and Berns 1986).

Prior to analysis, images are usually processed to reduce noise, account for variation in the image background, or enhance certain features. A simple technique which can dramatically reduce noise is image averaging (Figure 5). Since much of the noise in amplified video images is random, averaging successive frames dramatically reduces noise. The human eye sees an integrated image equivalent to about six video frames (at the standard video rate of 30 frames per second), so averaging less than six frames will produce a noticeably degraded image (Bright and Taylor 1986). Image averaging is a computationally intensive operation since each pixel must be added and divided. Averaging more than 2 or 3 frames is only practical when processing is accelerated by an array processor, which can average images at video frame rates.

Another image processing technique is to manipulate the frequency distribution of pixel brightness values, e.g. histogram transformation. Information from the image histogram, or frequency distribution of pixel brightness values, is used to transform image gray level values. In Figures 3, 4 and 5 (A and C), for example, each colour plane (red, green and blue) has been individually contrast-stretched to cover all possible gray levels. The images in Figure 5B and 5D have been thresholded (i.e. backgrounds have been darkened by setting all pixels below a predetermined threshold gray level to zero. Histogram transformations can be used to correct for effects of the non-linear response of video cameras and to automatically scale images (Bjørnsen 1986). These operations can be computed quickly and are often implemented in image analysers through input or output look-up tables.

Another type of image enhancement is the use of image convolutions. These methods can blur or exaggerate detail in an image by low- or high-pass filtering, respectively. Often in low-light video images, the detail at the single pixel level is largely noise which needs to be removed. A high-pass filter can be used to correct for spatial variation in illumination across an image. Bjørnsen (1986) uses this method of shading correction. Image convolutions can also be used to enhance edges which aids in image segmentation, i. e. when cells are differentiated from background. Typically, procedures for

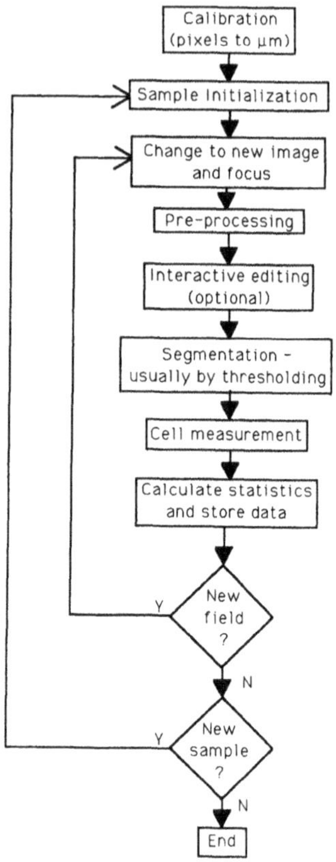

Fig. 6. Flow chart showing steps commonly used in analyzing microscope fields by image analysis

interactive image analysis include a step for manual image editing. In this step, objects in an image such as detrital particles can be removed, or overlapping cells can be separated. Most images do not require this type of editing.

Regions of interest in an image are differentiated from the background in the segmentation step (see review by Haralick and Shapiro 1985). In our application, this is the separation of brightly fluorescing cells from the dark background of the black-stained filter. The most common, and frequently easiest, way of segmenting a fluorescence microscope image is to apply a threshold. All pixels with brightness levels above the are considered part of a cell and all those below are background. Noise reduction by image averaging can improve the performance of thresholding (Figure 5). The large number of single background pixels above the threshold in Figure 5B is significantly reduced by averaging two frames (Figure 5D). Threshold values can be chosen manually (i.e. visually) or automatically, based on characteristics of the image. In the simple binary images analysed by Sieracki *et al.* (1985) a

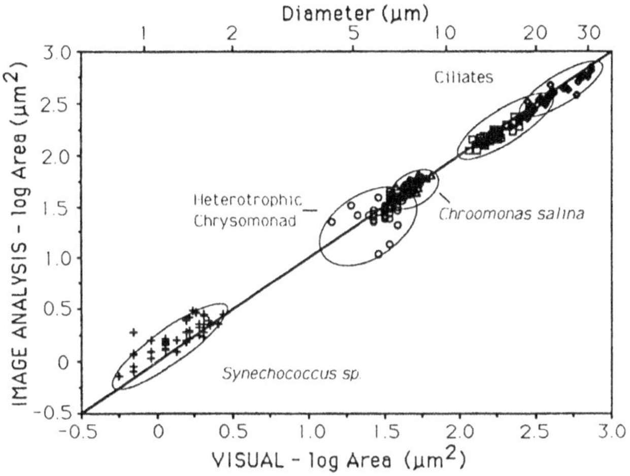

Fig. 7. Comparison of cell sizing by image analysis and visual microscopy. At least fifty cells from five different cultures were measured by eye at 2000X using an ocular micrometer and by our automated method of segmentation by image analysis at 400 or 630X magnification. Area of the cell image is a fundamental size measure by image analysis. Visual area was calculated for comparison from length and width using the formula of an ellipse. Scale across top shows diameter calculated from area assuming circular cell images (from Sieracki *et al.* 1989b)

visual threshold was interactively chosen to yield silhouettes that best matched the cells on the monitor. Bjørnsen (1986) chose to use a constant threshold of 100 (on a scale from 0 to 255) after several pre-processing steps. Neither of these methods takes into account the variation in fluorescence in a population of cells and the effect this may have on size measurements. We have developed an automated method of threshold selection that can accurately measure fluorescent microspheres of various brightnesses and sizes and a variety of common marine pico- and nanoplankton cells (Sieracki *et al.* submitted) (Figure 7).

CELL MEASUREMENTS

Once cells are segmented in an image a variety of measurements can be made. A list of measurements made by the CIAFM system is

Table 1. DATA AND MEASUREMENTS TAKEN ON EACH CELL BY CIAFM

Cell Identifiers		**Intensity Measurements**	
Image number		Integrated optical density (IOD)	
Cell number		Gray level:	minimum
Centroids (x,y)			maximum
Class			average
			variance
Size measurements		**Shape Factors**	
Area		Eccentricity	
Perimeter		Moments	
Radius:	minimum	Weighted moments	
	maximum	Compactness	
	average	Weighted compactness	
	variance	Orientation	
Diameter:	minimum		
	maximum		
Biovolume			

shown in Table 1. Cell identifiers include a cell number, an image identifier and the cell's location in the image. Each cell can be manually assigned to a class by the operator. Basic size measurements are area, calculated from the number of pixels covering the cell, and perimeter, or number of edge oints. For colour images individual area measurements are made for each colour plane as well as a total cell area. Radii are the distances from the cell centroid to each perimeter pixel. Diameter measurements are made by rotating the perimeter through 10 degree increments and measuring the horizontal and vertical extents. The biovolume measurement is unique to our system and is discussed in detail below. Intensity measurements are derived from the gray-level of all the pixels in each colour plane making up the cell image and relate to the amount of fluorescing material (i.e. stain or photopigment) in the cell. The integrated optical density (IOD, the sum of all pixel gray-levels) and gray-level statistics are calculated for each colour plane. Shape factors are calculated from the pixel (x,y) coordinates making up each cell and can also be weighted for pixel brightness.

Accurate measurement of cell biovolume is a primary interest in marine ecology. Cell carbon is estimated from biovolume using carbon conversion factors and these are summed to yield size- and trophic-specific biomass values for a given sample. The image of a fluorescing cell is a two-dimensional projection of a three-dimensional cell, so estimating biovolumes require certain assumptions about the third dimension. A common dimension. A common assumption made for single-celled organisms is that they have bi-lateral symmetry, so volume can be calculated from length and width. In a digitised image, biovolume can be calculated from area and perimeter assuming some shape factor (Bjørnsen 1986, Estep *et al.* 1986). Since many larger flagellates and ciliates are not bi-laterally symmetrical, we have developed an algorithm not dependent on this assumption (Figure 8) (Sieracki *et al.* 1989a). This method should yield more accurate biovolume estimates of these cells. Cell surface area may be a measurement of equal ecological importance to biovolume, especially in regard to bacterial influences on material flux and transformation in the sea (Williams 1984).

OTHER ECOLOGICAL APPLICATIONS

Visual discrimination between phototrophic and heterotrophic cells is made on the basis of the presence or absence of red- (for chlorophyll) or orange-autofluorescing (for phycoerythrin) pigments. This colour classification is relatively simple provided care is taken not to mask autofluorescence with fluorescence from added stains (Caron 1983). Colour image analysis uses similar colour criteria to classify cells by trophic type (Figure 9). Preliminary results indicate that phototrophic flagellates can be discriminated from

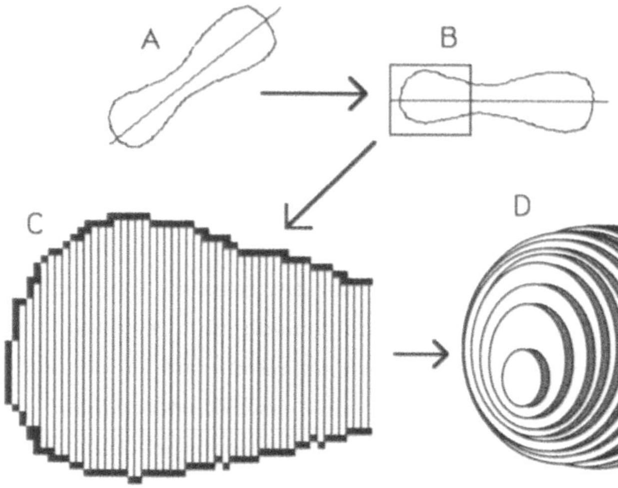

Fig. 8. Calculation of cell biovolume from a digitized cell image is made from (A) the digitized cell perimeter and orientation by: (B) rotating the perimeter to horizontal, (C) measuring the length of each one-pixel wide segment, (D) calculating the volume of each resulting segment, and then summing these volumes for the whole cell. This algorithm assumes symmetry in the third dimension (i.e., circular cross-sections) but can be easily corrected for compressed cells. (From Sieracki *et al.* 1989a)

heterotrophic flagellates by the ratio of green to red in proflavine-stained samples of either mixed cultures (Figure 9A) or natural samples (Figure 9B). The mixed culture (Figure 9A) includes a variety of heterotrophic and phototrophic flagellates. The phototrophic population in the Chesapeake Bay sample (Figure 9B) was dominated by cryptophytes and a small dinoflagellate, tentatively identified as a species of *Katodinium*. The condensed chromosomes of dinoflagellates cause these cells to have bright green-fluorescing nucleii, resulting in poorer separation from heterotrophic flagellates on the basis of the green to red ratio in this sample. The orange-fluorescing cryptophytes tended to fall between the green heterotrophs and red dinoflagellates causing further joining of the groups (Figure 9B). These preliminary results suggest that more detailed discriminant analysis or Bayesian classification techniques should lead to reliable discrimination between heterotrophs, and red- and orange-fluorescing phototrophs.

Fluorescent particles, added at tracer concentrations to a natural sample, are being used to study flagellate and ciliate grazing (McManus and Fuhrman 1986, Sieracki *et al.* 1987, Sherr *et al.* 1987). This technique uses short incubations (less than 1 h) with subsamples taken on the order of minutes. The number of ingested particles per grazer is counted over time to derive ingestion rates. Colour image analysis can be used to detect ingested fluorescing particles within cells. The flagellates in Figure 3C and the ciliate in Figure 3D show digital images of cells with ingested 0.5 µm diameter latex beads. With violet excitation (410 nm) the proflavine-stained cells appear green while the beads fluoresce blue. Measurements of blue area and IOD within flagellate cells were compared with visual counts of ingested beads in the same cells (Figure 10). On the basis of either area (Figure 10A) or IOD (Figure 10B) in the blue image plane, cells containing 0, 1, or 2 beads are readily discernible. Cells containing 3 beads are best discriminated from those containing 2 on the basis of blue area (Figure 10A). Cells containing more than 3 spheres per cell do not appear to be easily distinguishable on the basis of area or IOD. More than 3 or 4 ingested particles are difficult to discern visually as well. In these experiments, the most important samples are those taken soon after the addition of particles, when uptake is linear and egestion or digestion (in the case of fluorescently labelled bacteria) has not yet begun. In these early samples, relatively few cells have ingested particles and counting statistically significant numbers can be tedious. It appears that image analysis is a good method for counting ingested particles, especially when the number of ingested particles per cell is low.

Studies by Uhlmann *et al.* 1978 and Schlimpert *et al.* 1980 used Fourier transform analysis of digitised parameters of diatoms for automated taxonomic identification. In many ways this work was ahead of its time since computer and image analysis hardware and pattern recognition techniques available now should greatly improve this kind of analysis. One interesting recent

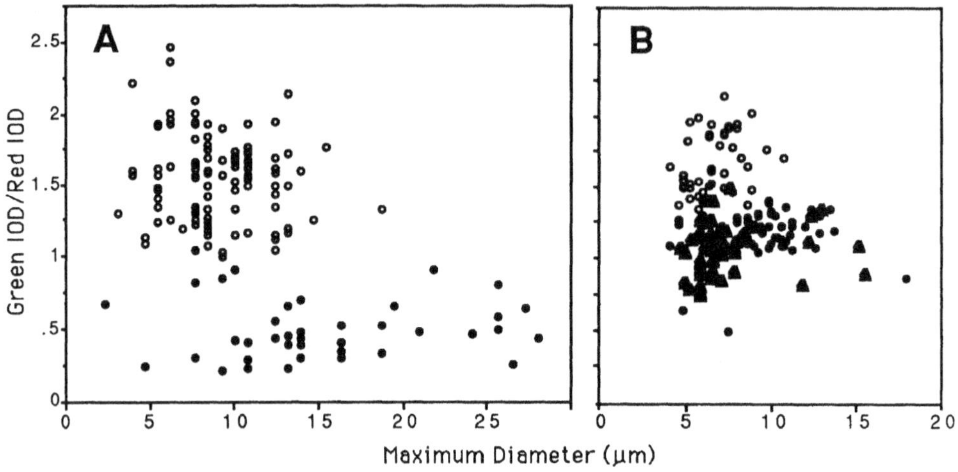

Fig. 9. Discrimination by color image analysis of mixed populations of heterotrophic (open circles) and phototrophic flagellates (closed symbols) stained with the fluorochrome proflavine. (A) An enrichment culture of York River water and (B) a Chesapeake Bay sample containing both orange-fluorescing cryptophytes (closed triangles) and red-fluorescing phototrophs (closed circles), which were predominantly dinoflagellates. The ratio of green to red integrated optical density (IOD) is one of the best discriminators of trophic type

application of image analysis to plankton ecology is the work of Chang and Carpenter (1988) and Carpenter and Chang (1988). They have used the IOD of DAPI-stained dinoflagellate nuclei as a measure of DNA content of individual cells. Since DNA must double for cell division the DNA content may relate to growth rate. The technique should allow estimations of *in situ* species-specific growth rates for phytoplankton, provided their cell division cycle is well-characterised.

LIMITATIONS OF IMAGE ANALYSIS

A major problem with fluorescent image analysis of microbial samples is the irreversible fading or bleaching of fluorescence over time, which is thought to be an oxidative reaction. The addition of the anti-oxidant n-propyl gallate, has been reported to reduce photobleaching (Giloh and Sedat 1982). Preliminary tests using this compound with natural marine samples stained with DAPI (P W Johnson and J McN Sieburth, personal communication) or proflavine (L Haas, personal communication) have not been successful. Further work with other anti-oxidants may be fruitful in this problem. Another limiting factor in analysing natural samples is the discrimination of detrital particles from living cells. Since non-living detrital particles overlap virtually the entire microbial cell size range, the problem can be severe, especially in near-shore and estuarine environments. With black-and-white systems discrimination must be done interactively by either removing detrital particles using image editing techniques (Sieracki et al 1985, Bjørnsen 1986) or identifying cells in the image for analysis. Automatic discrimination between cells and detritus may soon be possible with the use of specific staining methods and colour image analysis. Most commonly used fluorochromes have been chosen because they stain detritus a different colour than cells (Porter and Feig 1980, Haas 1982, Caron 1983). Additionally, nucleic acid specific stains (e.g. DAPI or Hoechst 33258) can clearly show blue-fluorescing cell nuclei under ultraviolet or violet excitation. These stain characteristics are used to differentiate detritus from living cells.

FUTURE DIRECTIONS

Continuing advances in computer technology is allowing powerful computers to become cheaper and widely available. Computers with faster processors and larger memories will allow larger, more sophisticated programmes to be developed for image analysis. Interest in graphic arts and, especially, biomedical applications should lead to continued development of image analysis hardware and software. Knowledge-based expert systems may allow easier implementation of more advanced pattern recognition techniques which could lead to fully automated identification and classification of plankton cells.

Solid-state CCD cameras designed for quantitative, scientific work are already available (Hiraoka *et al.* 1987) and will

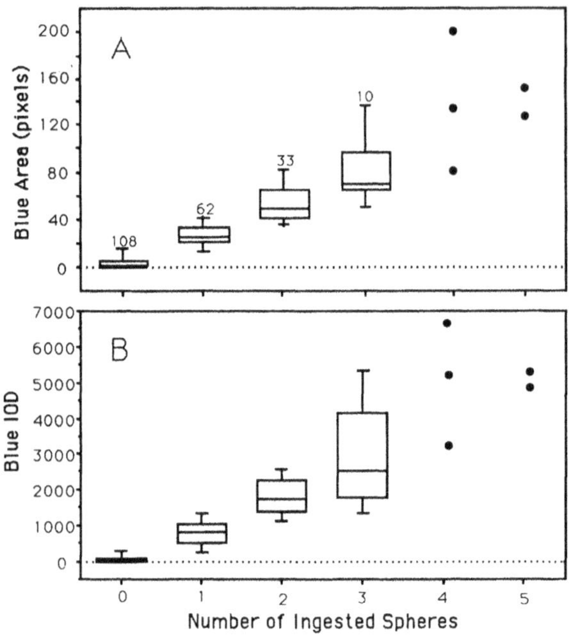

Fig. 10. Box plots showing the ability of two image analysis measurements, (A) area and (B) IOD, to discriminate various numbers of ingested blue-fluorescing beads (0.5 µm diameter) contained in proflavine-stained flagellate cells. The number of ingested spheres counted visually is shown on the X-axis. Each box encompasses the 25th to 75th percentiles, the line across the box is the median (50th percentile), and the vertical lines extend to the 10th and 90th percentiles. The numbers of cells measured are shown in (A). Only 3 and 2 cells containing 4 and 5 beads, respectively, were compared and are indicated by the single points

undoubtedly replace conventional video cameras for fluorescence microscopy work. High resolution CCD cameras can capture images with 1000 X 1000 pixels and will soon be available with 2000 X 2000 pixel resolution. Images produced by these cameras contain 4 and 16 times more information, respectively, than standard 512 X 512 images and will need comparably faster computers for analysis. These higher spatial resolutions will be useful for resolving bacterial shapes and sizes more accurately. New microscope technology, such as the scanning confocal microscope (Sheppard 1987), may revolutionise fluorescence microscopy. This system uses a scanning laser beam focused through a narrow aperture to excite the sample and the image is built up by scanning through a similar aperture. By narrowing the field of view, resolution significantly higher than conventional microscopy is achieved. In addition, a very narrow focal plane is produced allowing 3-dimensional optical sectioning. The image lends itself to digital analysis since it is built up from scan lines similar to a video image.

New methods of image segmentation need to be developed which are more computationally efficient and yield accurate cell data. One approach which may be useful in fluorescence microscopy is adaptive thresholding. The method of Vossepoel *et al.* (1979) optimises the threshold based on the shape of the resulting cell perimeter, or contour. Heuristic methods, based upon *a priori* knowledge of the image characteristics, could also prove useful (Walter and Berns 1986). For example, a relatively high threshold could be used to locate DAPI-stained cell nuclei in the blue image of a UV-excited field. Excitation would then be changed to blue and a 'region-growing' technique (Pavlidis 1982) would find the cell edge in the green (proflavine-stained cytoplasm) and red (autofluorescing chloroplasts) frames in the vicinity of the nucleus.

A variety of other applications of colour image analysis may prove useful. Quantitative fluorescence intensity measurements of phototrophic cells (e.g. red IOD) may yield a direct measurement of pigment content per cell. Additionally the arrangement of chloroplasts within the cell may indicate adaptation to its previous light history. Using image subtraction it is possible to quantify certain metabolic rates of individual cells on slide culture. For example, the reduction of tetrazolium salts to produce intra-cellular red formazan crystals by bacteria has been used as an indicator of electron transport system activity (Tabor and Neihof 1982). This process can be quantified in individual cells over time with image analysis (Sieracki and Catallo, unpublished data). Many fluorescence techniques being developed for flow cytometry can be applied to image analysed fluorescence microscopy. For example, bacterial characterisation by their GC/AT ratios described by Van Dilla *et al.* (1983) could be applied to natural bacterial populations using image analysed fluorescence microscopy. Fluorescent analogues of metabolically active compounds (such as amino acids, simple sugars or 1-carbon compounds) could be used to survey natural bacterial populations for uptake of these compounds. Also many fluorescent compounds are being developed for binding to specific intracellular macromolecules (Cowden 1985) which could be used to biochemically characterise plankton cells. The automation of fluorescent antibody techniques by image analysis could improve these methods for enumerating individual plankton species from natural waters.

ACKNOWLEDGMENTS

We would like to thank Stephen Reichenbach and Charles Viles for their considerable technical help, including software development and data analysis. Dr Larry Haas offered helpful suggestions on the manuscript, and Dr Frank Perkins is thanked for his administrative support which made this work possible. This work was partly funded by the US Office of Naval Research grant #00014-84-G-0180.

REFERENCES

Arndt-Jovin DJ, Robert-Nicoud M, Kaufman SJ, Jovin TM (1985) Fluorescence digital imaging microscopy in cell biology. Science 230:247-256

Azam F, Fenchel T, Field JG, Gray JS, Meyer-Reil LA, Thingstad F (1983) The ecological role of water-column microbes in the sea. Mar Ecol Prog Ser 10:257-263

Bird DF, Kalff J (1986) Bacterial grazing by planktonic lake algae. Science 231:493-495

Bjørnsen PK (1986) Automatic determination of bacterioplankton biomass by image analysis. Appl Environ Microbiol 51:1199-1204

Bright GR, Taylor DL (1986) Imaging at low light level in fluorescence microscopy. In: Taylor DL, Waggoner AS, Lanni F, Murphy RF, Birge RR (eds) Applications of Fluorescence in the Biomedical Sciences, Alan R Liss, Inc New York

Caron D (1983) Technique for enumeration of heterotrophic and phototrophic nanoplankton, using epifluorescence microscopy, and comparison with other methods. Appl Environ Microbiol 46:491-498

Caron DA, Pick FR, Lean DRS (1985) Chroococcoid cyanobacteria in Lake Ontario: vertical and seasonal distributions during 1982. J Phycol 21:171-175

Carpenter EJ, Chang J (1988) Species-specific phytoplankton growth rates via diel DNA synthesis cycles. I. Concept of

the method. Mar Ecol Progr Ser 43:105-111
Castleman KR (1979) Digital Image Processing. Prentice-Hall, Englewood Cliffs NJ
Chang J, Carpenter EJ (1988) Species-specific phytoplankton growth rates via diel DNA synthesis cycles. II. DNA quantification and model verification in the dinoflagellate *Heterocapsa triquetra*. Mar Ecol Prog Ser 44:287-296
Chrzanowski TH, Crotty RD, Hubbard GJ (1988) Seasonal variation in cell volume of epilimnetic bacteria. Microb Ecol 16:155-163
Cowden RR (1985) Fluorescent cytochemistry of living cells. Amer Biotech Lab Nov/Dec 1985, p. 8-14
Davis PG, Caron DA, Johnson PW, Sieburth J McN (1985) Phototrophic and apochlorotic components of picoplankton and nanoplankton in the North Atlantic: geographic, vertical, seasonal and diel distributions. Mar Ecol Prog Ser 21:15-26
Ducklow HW (1983) Production and fate of bacteria in the oceans. Bioscience 33:494-501
Estep KW, MacIntyre F, Hjorleifsson E, Sieburth JMcN (1986) MacImage: a user-friendly image-analysis system for the accurate mensuration of marine organisms. Mar Ecol Prog Ser 33:243-253
Ferguson RL, Rublee P (1976) Contribution of bacteria to standing crop of coastal plankton. Limnol Oceanogr 21:141-145
Ferguson RL, Buckley EN, Palumbo AV (1984) Response of marine bacterioplankton to differential filtration and confinement. Appl Environ Microbiol 47:49-55
Giloh H, Sedat JW (1982) Fluorescence microscopy reduced photobleaching of rhodamine and fluorescein protein conjugates by n-propyl gallate. Science 217:1252-1255
Goldman JC, Caron DA (1985) Experimental studies on an omnivorous microflagellate: implications for grazing and nutrient regeneration in the marine microbial food chain. Deep-Sea Res. 32:899-915.
Haas LW (1982) Improved epifluorescence microscopy for observing planktonic micro-organisms. Annal Inst Océanogr 58(S):261-266
Haralick RM, Shapiro LG (1985) Image segmentation techniques. Computer Vision, Graphics and Image Processing 29:100-132
Hiraoka Y, Sedat JW, Agard DA (1987) The use of a charge-coupled device for quantitative optical microscopy of biological structures. Science 238:36-41
Hobbie JE, Daley RJ, Jasper S (1977) Use of Nuclepore filters for counting bacteria by fluorescence microscopy. Appl Environ Microbiol 33:1225-1228
Inoue S (1986) Video Microscopy. Plenum Press, New York, USA
Johnson PW, Sieburth JMcN (1979) Chroococcoid cyanobacteria in the sea: a ubiquitous and diverse phototrophic biomass. Limnol Oceanogr 24:928-935
Kristian J, Blouke M (1982) Charge-coupled devices in astronomy Sci Amer 247(4):66-74
McManus GB, Fuhrman JA (1986) Bacterivory in seawater studied with the use of inert fluorescent particles. Limnol Oceanogr 31:420-426
Niblack W (1985) An Introduction to Digital Image Processing. Strandberg Publishing Co. Birkeroed, Denmark
Pavlidis T (1982) Algorithms for Graphics and Image Processing. Computer Science Press, Inc. Rockville, MD
Pomeroy L (1974) The ocean's food web, a changing paradigm. Bioscience 24:499-504
Porter KG, Feig YS (1980) The use of DAPI for identifying and counting aquatic microflora. Limnol Oceanogr 25:943-948
Schlimpert O, Uhlmann D, Schuller M, Hohne E (1980) Automated pattern recognition of phytoplankton - procedure and results. Int. Rev. ges. Hydrobiol. 65:427-437
Sheppard CJR (1987) Scanning optical microscopy. In: Barer R, Cosslett VE (eds) Advances in Optical and Electron Microscopy, Vol 10. Academic Press, New York
Sherr EB, Sherr BF, Fallon RD (1987) Use of monodisperse, fluorescently labelled bacteria to measure in situ protozoan bacterivory. Appl Environ Microbiol 53:958-965
Sherr EB, Sherr BF (1987) High rates of consumption of bacteria by pelagic ciliates. Nature 325:710-711
Sieburth JMcN (1977) International Helgoland symposium:

Convener's report on the informal session on biomass and productivity of microorganisms in planktonic ecosystems. Helgol wiss Meeresunters 30:697-794
Sieburth JMcN, Smetacek V, Lenz J (1978) Pelagic ecosystem structure: heterotrophic compartments of the plankton and their relationship to plankton size fractions. Limnol Oceanogr 23:1256-1263
Sieracki ME, Johnson PW, Sieburth JMcN (1985) Detection, enumeration and sizing of planktonic bacteria by image-analyzed epifluorescence microscopy. Appl Environ Microbiol 49:799-810
Sieracki ME, Haas LW, Caron DA, Lessard EJ (1987) The effect of fixation on particle retention by microflagellates: underestimation of grazing rates. Mar Ecol Prog Ser 38:251-258
Sieracki ME, Viles C, Webb KL (1989A) An algorithm to estimate cell biovolume using image analyzed microscopy. Cytometry 10:551-557
Sieracki ME, Reichenbach S, Webb KL (1989b) An evaluation of automated threshold selection methods for accurate sizing of microscopic fluorescent cells by image analysis. Appl Environ Microbiol. 55:2762-2772
Stoecker DK, Michaels AE, Davis LH (1987) Large proportion of marine plankton ciliates found to contain functional chloroplasts. Nature 326:790-792
Tabor PS, Neihof RA (1982) Improved method for determination of respiring individual microorganisms in natural waters. Appl Environ Microbiol 43:1249-1255
Turley C, Lochte K (1986) Diel changes in the specific growth rate and mean cell volume of natural bacterial communities in two different water masses in the Irish Sea. Microb Ecol 12:271-282
Uhlmann D, Schlimpert O, Uhlmann W (1978) Automated phytoplankton analysis by a pattern recognition method. Int. Rev. ges. Hydrobiol. 63:575-583.
Van Dilla MA, Langlois RG, Pinkel D, Yajko D, Hadley WK (1983) Bacterial characterization by flow cytometry. Science 220:620-621
Vossepoel AM, Smeulders AWM, Van den Broek K (1979) DIODA: delineation and feature extraction of microscopical objects. Computer Programs in Biomedicine 10:231-244
Walter RJ, Berns MW (1986) Digital image processing and analysis. In: Inoue' S. Video Microscopy. Plenum Press, New York, USA
Waterbury JB, Watson SW, Guillard RRL, Brand LE (1979) Widespread occurrence of a unicellular, marine, planktonic, cyanobacterium. Nature 277:293-294.
Watson SW, Novitsky TJ, Quinby HL, Valois FW (1977) Determination of bacterial number and biomass in the marine environment. Appl Environ Microbiol 33:940-946
Williams PJ LeB (1981) Incorporation of microheterotrophic processes into the classical paradigm of the planktonic food web. Kieler Meeresforsch S 5:1-28
Williams PJ LeB (1984) Bacterial production in the marine food chain: the emperor's new suit of clothes? In: Fasham MJR (ed) Flows of Energy and Materials in Marine Ecosystems. Plenum Press, New York, London

CULTURING MARINE PROTOZOA - SESSION SUMMARY

Andrew J Cowling
CCAP
Freshwater Biological Association
The Ferry House
Ambleside
Cumbria
LA22 OLP
UK

Andrew Cowling introduced the session and outlined general aspects of the routine maintenance of marine protozoa in long-term culture. There are three principle methods available for maintaining protist strains in a laboratory or an established collection: serial cultivation, continuous cultivation or cryopreservation. The latter method is arguably the best means to achieve genetic stability of strains maintained over time, but have yet to be successfully applied to most protozoa. The underlying principles of cultivation - enrichment and isolation from field samples, development of optimal culture media, and the establishment of 'permanent' clones or species' populations in culture - were illustrated with reference to methods employed for heterotrophic, free-living marine gymnamoebae and phagotrophic flagellate strains maintained at the Culture Collection of Algae and Protozoa (CCAP). both gnotobiotic and monoxenic cultures of protists have been established at CCAP using the associated techniques of antibiotic washes, and agar block or inoculation wire transfers. Examples of these simple and adaptable techniques were described. Recent work at CCAP (Rogerson, pers. comm.) has investigated the feasibility of a long-term maintenance regime for a range of marine gymnamoebae strains cultured on seawater-based agar media, employing a below-optimum temperature of 7°C. The results show little evidence of any inherent amenability in marine amoebae genera, with the possible exception of *Vannella*, to such culture conditions. These, and other such findings, suggest that new methods, in addition to continued improvements to serial cultivation techniques, will be required to facilitate more widely applicable or economic maintenance methods for large numbers of strains.

NATO ASI Series, Vol. G 25
Protozoa and Their Role in Marine Processes
Edited by P. C. Reid et al.

Dian Gifford spoke on the culture of marine planktonic oligotrich ciliates, summarising results from earlier investigations of the laboratory requirements of the genera *Strombidium* and *Strombidinopsis* collected from Halifax harbour (Nova Scotia). The fragile nature of these ciliate forms imposes severe limitations on experimental techniques for their collection, concentration and maintenance. The use of nets, and subsequent serial size-fractionation using a descending series of mesh sizes often results in substantial losses of aloricate ciliates from marine samples; even gentle reverse filtration techniques were shown to be unsuitable for concentrating purposes. A more reliable procedure was described involving transfer of individual ciliates by pipette from field samples into seawater media enriched with cultured phytoplankton isolates. Investigations utilising these techniques and employing multi-well plates to test a range of media conditions and food types showed that maximum ciliate reproductive rates could be obtained using natural seawater based media containing EDTA at concentrations between 10^{-8} and 10^{-6} M. The addition of trace metals to media further enhanced growth in all cases. Optimal food type differed, depending on the ciliate isolate tested, but two dinoflagellate strains - *Heterocapsa triquetra* and *Scrippsiella trochoidea* - were most effective in promoting reproduction. The spinose diatom *Thalassiosira* sp. was found to be a poor quality food, possibly because its morphology reduced prey-handling capacity. Successful maintenance of these oligotrichs depended on the regular, moderate provision of log phase phytoplankton as food and periodic (1-2 week) transfers of cultures to fresh media. Cyst production was never observed in cultures and although conjugation occurred, eventual loss of strains occurred through decreased vigour, although one strain of *Strombidinopsis acuminatum* was maintained for over a year.

Masachika Maeda outlined aspects of the cultivation of protozoa for use as food in mariculture and its associated problems.

The use of mass cultured live micro-organisms to enable the establishment of prawn and fish larvae in rearing ponds is more cost-effective than collecting wild plankton to use as food; it allows greater control over nutritional quality and is more convenient, as it is not limited by availability. Enclosure ponds were described in which larvae are reared in the absence of predators and where a bottom layer of soil in the ponds serves both to harbour prawn larvae during the day and to provide a source of microbial food organisms. Such conditions should result in high survival rates and good subsequent growth of commercial species. One difficulty, however, has been to identify microbial food of sufficient nutritional quality and of suitable dimensions which permits larvae to establish and survive beyond their early vulnerable development stages. Experiments have shown the benefits of encouraging certain microbial assemblages to establish in rearing ponds in order to provide, either directly or indirectly, the basic dietary components for the optimum growth of the commercial species. Oligotrich ciliates of the genera *Strombidium* and *Strobilidium* are commonly preyed on by prawn larvae in both natural and artificial marine environments. The results of feeding trials in ponds containing *Penaeus monodon* (tiger prawn) showed that the provision of *Strombidium sulcatum* as food (which in turn fed on the diatom *Navicula* sp. and associated indigenous bacteria) promoted a greater larval growth and survival than other oligotrich strains tested. Further experiments showed that only several bacterial strains (from a total of one hundred isolates) were able, under similar conditions, to produce high growth rates of larvae. Such findings have highlighted the need for careful consideration of the nutritional component provided by protozoa, algae and bacteria in promoting the growth of prawn and fish larvae in mariculture, and the potential for manipulating conditions in favour of specific microbial assemblages which enable maximum commercial yields.

A METHOD FOR THE CLONING AND AXENIC CULTIVATION OF MARINE PROTOZOA

Anthony T Soldo and Sylvia A Brickson
Research Laboratories of the VAMC and the Department of Biochemistry and Molecular Biology
University of Miami School of Medicine
Miami
Florida
USA

INTRODUCTION

Investigators working with marine protozoa often need to isolate, clone and maintain a particular protozoan species in laboratory culture. In the past the approach has been to physically isolate the desired specimen, usually with the aid of a micropipette, and transfer it through a series of washes to renive contaminating microorganisms and other particulate debris. The washed isolate is then placed in a volume of suitable culture medium where it may grow and divide. The method is tedious and requires a considerable degree of skill on the part of the operator. Protozoa too fragile or too small to be handled are not amenable to isolation by this method. Another major drawback is that excessive handling sometimes results in physical damage to the desired specimen thereby reducing the chances of a successful isolation.

We describe here a procedure that may be applied to the isolation and cloning of most protozoa, regardless of size, and which minimizes the handling of individuals. It takes advantage of certain intrinsic properties of the organisms themselves, i.e. positive or negative geotropism. The method utilizes the protozoan isolation apparatus shown in Figure 1, dubbed the 'hummingbird feeder', by Gene Small. It works by allowing the protozoa to swim free of contaminating microorganisms. Negatively geotropic organisms are introduced through the glass tube located at the bottom of the apparatus where they can migrate upwards and accumulate near the top. Positively geotropic protozoa are introduced at the top of the

NATO ASI Series, Vol. G 25
Protozoa and Their Role in Marine Processes
Edited by P. C. Reid et al.

apparatus and migrate downwards towards the bottom. In either case, glass beads serve as a barrier that prevents unwanted microorganisms and particulate debris as well as soluble components of the medium, carried by convection and/or gravity, from contaminating the desired specimen. Once the protozoa have completed their migration, they are removed and cloned in appropriate culture medium using the Silicone Oil Plating Procedure (SOPP Soldo and Brickson, 1980).

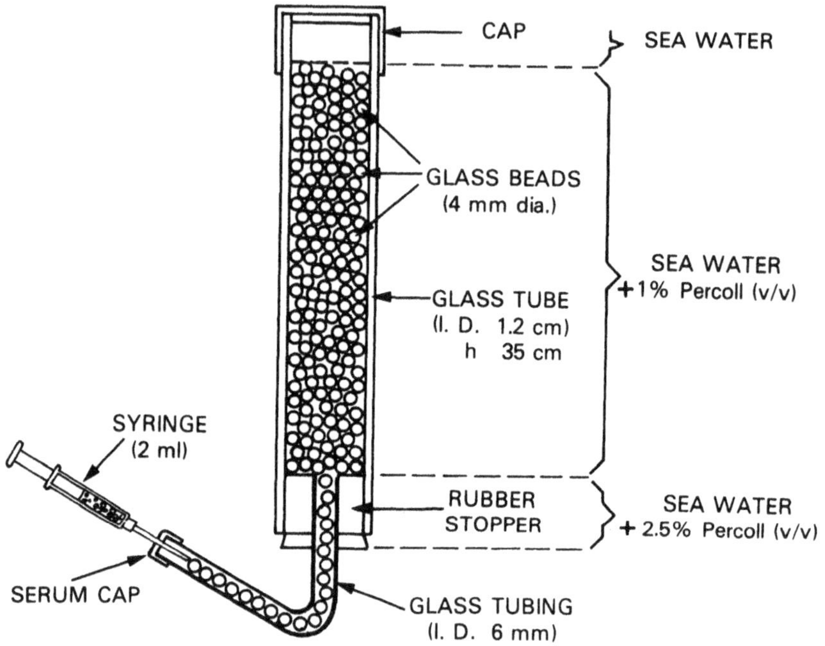

Fig. 1. Protozoan isolation apparatus

MATERIALS

Cerophyl may be obtained from Cerophyl Laboratories, Inc, Kansas City, Missouri, USA Proteose peptone, yeast extract, Bacto-agar and nutrient broth were from Difco, Inc. RNA, (Torula Yeast, grade VI) was purchased from Sigma, Inc, aquamarine salts (1 lb quantities), from Aquatrol, Inc, Anaheim, California, and the silicone oil (Viscosity standard, 9.6 centipoises) from Brookfield Engineering Laboratories,

Stoughton, Massachusetts, USA. Percoll is a product of Pharmacia LKB, Uppsala, Sweden.

COLLECTION AND PREPROCESSING OF SAMPLES

Small samples of sea water are taken by submerging 30 ml capacity screw-cap scintillation vials at the desired location. The samples may be enriched by adding small quantities of decaying material (fish, meat etc.) or nutrient agar plugs to perforated containers, i.e. holes may be drilled in plastic screw-caps and mounted on glass scintillation vials. These may be submerged at the original sampling sites for periods ranging from several hours to several days before retrieval. Large samples are first filtered through glass wool or nylon mesh to remove coarse debris and then concentrated to about 50 ml by millipore filtration (pore size - 0.45 µm diameter). If desired, both small and concentrated large samples may be enriched by adding small quantities (1-2 mg ml^{-1}) of dry cerophyl, hay, wheat germ, rice grains, etc. followed by incubation in the dark at 22-27°C for a few days. (It is necessary to heat the dry material at 180°C for one hour in order to kill any ciliates, cysts, or other protozoa that may be present). It is also advisable to plate out an unfiltered sample on nutrient or cerophyl agar prepared with source water to isolate a bacterium that may serve as a food source for the ciliates.

Prior to or after isolation of protozoa from natural or enriched samples, it is advisable to clone individuals and maintain them in non-axenic culture. This is most easily accomplished by the use of the Silicone Oil Plating Procedure (SOPP), as follows: the protozoan sample to be plated is mixed with an equal volume of Bacterized Cerophyl Medium (BCM, Table 1). Then, 0.2 ml portions of the mixture are added to 3 ml volumes of sterile silicone oil (Viscosity Standard, 9.6 cps) in 16 x 125 mm capped pyrex tubes. The tubes are mixed vigorously for about 15 to 20 seconds and the contents immediately poured into sterile plastic petri dishes (35 x 10 mm). The dishes are allowed to remain undisturbed for about 30 minutes before being placed in an incubator at 27°C. This allows time for the unstable emulsion, formed during mixing, to coalasce and separate out as discrete microdroplets, some of which contain individual protozoans entrapped and immobilized between the hydrophobic surfaces of the plastic petri plate and the oil. Individuals isolated by this procedure grow, divide and multiply to form clones. These are then picked up in micropipettes and transferred to 0.5 ml portions of fresh BCM medium in test tubes and are maintained by doubling the volume

Table 1. PREPARATION OF SOLUTIONS AND MEDIA FOR THE CULTIVATION OF MARINE CILIATES

Instant seawater (ISW)

Dissolve 1.0 lb (0.45 Kg) Aquamarine salts in 12 litres of distilled water at room temperature with constant stirring. After one hour determine the density of the water with a suitable hydrometer and adjust, by the addition of distilled water, to a final density of 1.015 g/cc. Filter through glass wool to remove undissolved materials and store at 4°C in a 20 litre capacity plastic container.

Bacterized Cerophyl Culture Medium (BCM)

To prepare 100 ml:

Add 0.5 g Cerophyl to 100 ml ISW and heat until boiling. Filter 3 times while hot through glass wool. Filtered source water may also be used. Sterilize at 121°C for 15 minutes, cool to room temperature and inoculate with a bacterium isolated from source water. Incubate at 37°C overnight before ready for use.

Proteose Peptone/Cerophyl medium (P/C)

To prepare 100 ml:

Add 0.5 g Cerophyl to 90 ml ISW and heat until boiling. Filter 3 times while hot through glass wool.

Dissolve in the filtered Cerephyl extract:

2.0 grams Proteose peptone
0.1 gram Yeast extract
0.1 gram RNA

Adjust the pH to 7.2 with 1 N NaOH and bring to volume with ISW. Dispense 5 ml per tube and sterilize at 121°C for 15 minutes.

---

periodically, usually weekly, until the volume reaches 4 to 8 ml. At this point, 0.5 ml of the culture is transferred to a fresh tube and the doubling process repeated.

USE OF THE PROTOZOA ISOLATION APPARATUS

The following protocol has proved effective for the sterilization of negatively geotropic ciliates:

Note: All solutions and other components must be **sterile**; aseptic techniques should be used throughout.

1. Prepare 50 to 100 ml of bacterized culture of the desired protozoan species as described above. Filter through nylon mesh or glass wool to remove clumps of bacteria and other debris.
2. Collect the protozoa by centrifugation, discard the supernatant and resuspend the pelleted organisms in an equal volume of sterile sea water. Allow the suspension to stand overnight in the incubator. This slows the growth of the food bacteria and permits the protozoa to complete digestion of ingested bacteria and expulsion of undigested microorganisms.
3. On the next day, pellet the protozoa again and resuspend in a volume (2 or 3 ml) of sea water. Add a volume of Percoll to bring the concentration to 2.5 % (v/v) and set aside. (Percoll adds viscosity without an accompanying osmotic effect, thereby facilitating layering of the ISW in subsequent steps).
4. Assemble the protozoan isolation apparatus, clamped to a ring stand, as shown in Figure 1, using individual components that have been sterilized in the autoclave and dried in a 180° C oven. Do not add the glass beads at this point.
5. Aseptically, add 5 to 10 ml of sea water containing 1 % Percoll (v/v) to the glass column and expel any air bubbles that may form. Then add the glass beads, a few at a time to a point a few cm from the top of the column.
6. Fill the column with sea water containing 1 % Percoll to a point just above the surface of the glass beads. Carefully overlayer 2-3 ml of sea water. Avoid disturbing the interface.
7. Slowly inject 1 or 2 ml of the concentrated protozoan

suspension through the serum port and allow the cells to swim upwards through the apparatus.

8. At various time intervals, remove samples from the overlayered sea water at the top of the apparatus to monitor the arrival of the ciliates. This is best accomplished by taking 5 µl aliquots of each sample and counting the number of protozoa present with the aid of a dissecting microscope.

9. When sufficient numbers have migrated, add 0.1 ml sterile P/C culture medium (Table 1) and 0.1 ml of the sample to 3 ml of the silicone oil in a test tube. Vortex and plate.

10. Incubate the plate at 22-27°C and observe periodically for growth. Isolates that produce clones containing sufficient numbers of ciliates are then picked up in micropipettes and transferred to 0.5 ml of full strength P/C medium in test tubes which are later used to inoculate 5 ml full-strength medium. The cultures are then maintained by transferring 0.1 ml portions of a seven day old culture to 5 ml of freshly prepared medium, weekly.

COMMENTS

The following experiment illustrates the effectiveness of the procedure in obtaining bacteria-free ciliate cultures from clones: *Parauronema acutum* strain 110-3 was cultured in bacterized medium by doubling the volume daily with fresh BCM until a final volume of 50 ml was obtained. The ciliates were then washed 3 x with 50 ml portions of sterile ISW by low speed centrifugation (500 x g for 3 min at 22°C). The washed, suspended cells were incubated overnight at 27°C; the next day the suspension was centrifuged at 500 x g for 3 min. at 22°C and the pelleted organisms resuspended in 1 ml of ISW. To the washed concentrated ciliates an equal volume of 5% Percoll in ISW was added. This suspension was injected into the serum port of the protozoa isolation apparatus and the protozoa allowed to migrate. At intervals, 0.1 ml samples were removed by propipette by positioning the end of the microtip a few millimeters below the surface of the ISW layer uppermost in the apparatus. Following the addition of 0.1 ml of fresh P/C medium, the samples were plated by the SOPP procedure. After 5 days at 27°C the plates were examined for the presence of ciliate clones and for the number of clones containing contaminating bacteria. Subsequent steps in the procedure were essentially those outlined above in 9 and 10. A total of 10 clonal isolates were made. The results are shown in Table 2. As a final check to ensure that the isolates were free from contaminating microorganisms, samples were plated on standard bacteriological media as well as the P/C medium. No growth was observed on any of the plates after 10 days at 37°C. The key to success in axenizing protozoa using the protozoan isolation apparatus rests in its use in conjunction with the silicone-oil plating procedure.

Table 2. ANALYSIS OF PLATING EXPERIMENT

Migration Time (min)	Microdroplets* Bacteria-free (%)	Number of Ciliates
0	100	0
15	99.3	13
30	99.5	20
60	98.3	50
120	97.1	78

*Approximately 1,000 per plate

Invariably, a few bacteria will co-migrate with the ciliates and will be deposited in some of the microdroplets following plating by the SOPP. Because the microdroplets are effectively compartmentalized by the SOPP, very few will contain both bacteria and protozoa. Contaminated microdroplets are easily detected after the plates have been incubated for a few days and appear as opaque white 'plaques'. Reject these contaminated microdroplets when transferring clones by micropipette to fresh medium. If microbial contamination becomes a problem, antibiotics may be added to the sea water and culture medium used in cloning and axenizing the protozoa. Neomycin and tetracycline at concentrations of 25 µg/ml have proved effective in this application. Other antibiotics and even anti-fungal agents may be tried but penicillin is to be avoided as it is unstable in sea water. P/C medium (Table 1) is a comparatively recent formulation for the axenic cultivation of marine ciliates which supports the growth of most of our laboratory strains. It is offered here because its components are relatively easy to obtain and it is simple to prepare. Other, more complex culture media for the axenic cultivation of marine ciliates have already been described (Soldo and Merlin 1972, 1977).

The procedure described above was designed primarily for the cloning and axenic cultivation of scuticociliates, most of which are negatively geotropic. Positively geotropic ciliates, including many hypotrichs may also be cloned by this procedure. In this case, starved, concentrated organisms are suspended in seawater and layered directly on top of the 1% Percoll-sea water layer. The organisms are removed by syringe after they migrate to the bottom of the apparatus. In this case do not add Percoll to the concentrated cells before layering.

REFERENCES

Soldo AT, Brickson AS (1980) A simple method for plating and cloning ciliates and other protozoa. J Protozool 27:328-331

Soldo At, and Merlin EJ (1972) The cultivation of symbiote-free marine ciliates in axenic medium. J Protozool 19:519-524

Soldo AT and Merlin EJ (1977) The nutrition of *Parauronema acutum*. J Protozool 24:556-562

POLLUTION - SESSION SUMMARY

Torbjørn Dale
Sogn og Fjordane College
Sogndal
Norway

The pollution session comprised only two contributions, a talk by **Torbjørn Dale** on "Effects from oil, dispersed oil and fertilized oil on planktonic ciliates" and a poster by **Juan Miguel Gonzáles** with J Iriberri, L Egea and I Barcina on the "Role of protozoa in the regulation of enteric bacteria populations in seawater". The oil effects study during spring and summer 1980 and summer 1981 at Lindåspollene, western Norway, used enclosures to which oil, oil/dispersant and oil/nutrients were added at the surface. The experiments with added oil alone showed that the upper 1-2 m layers were depopulated within 1-2 weeks when the oil concentrations exceeded 0.2-1 mg l^{-1}. The disappearance of the ciliates was probably due to downward migration, revealed as temporary density peaks of the ciliates. The effects were stronger in summer than in spring. Within the total assemblage of heterotrophic ciliates (including mixotrophic species), *Tontonia* spp. appeared to be among the most sensitive, whereas the tintinnids *Parafavella denticulata* and *Acanthostomella norvegica* appeared to be among the least sensitive. It was suggested that the lorica of the tintinnids may offer some protection against the oil, whereas the long tail of *Tontonia* spp. may increase their sensitivity due to a greater surface to volume ratio than non-tailed species of similar size. The autotrophic species *Mesodinium rubrum* appeared to be more sensitive to oil than heterotrophic ciliates. Negative effects generally increased when the dispersant Corexit 9527 was used, whereas additions of nutrients had an ameliorating effect.

In the study by Gonzáles *et al.*, the enteric bacteria *Escherichia coli* and *Enterococcus faecalis* were inoculated separately to seawater samples, with and without the natural biota. By quantifying the numbers of colony-forming units (CFU) of these bacteria, no changes were observed in the

NATO ASI Series, Vol. G 25
Protozoa and Their Role in Marine Processes
Edited by P. C. Reid et al.

samples devoid of microbiota, whereas marked decreases, concomitant with increased numbers of protozoa, were seen in the samples with the microbiota. Another experiment revealed that survival of the bacteria following predation by protozoa was significantly related to the incubation temperature through the Arrhenius model. Two mathematical models relating the numbers of CFU to the numbers of the protozoans were presented. It was concluded that predation by protozoa is the main cause of the elimination of enteric bacteria from marine ecosystems and that predation is directly related to temperature.

PROTISTS AND POLLUTION - WITH AN EMPHASIS ON PLANKTONIC CILIATES AND HEAVY METALS

Torbjørn Dale
Sogn og Fjordane College
Sogndal
Norway

INTRODUCTION

"Marine pollution is currently a problem of great concern". This was the first sentence of a review paper: "Pelagic protists and pollution. A review of the past decade", given by Curds (1982) at the first NATO-Workshop on marine protozoa in 1981. The sentence seems equally valid today. For example, he asked how long will it be before the organically polluted coastal regions extend into larger bodies of water such as the North Sea. The bloom of the toxic haptophycean alga *Chrysochromulina polylepis* in Scandinavian waters in late May 1988 might be an answer to this question as eutrophication could partly be responsible (Rosenberg *et al.* 1988, Sangfors 1988, Simen 1988). The explosion of the oil drilling platforms 'Piper Alpha' (July 1988), and 'Ocean Odyssey' (September 1988) in the North Sea also show the potential for oil pollution from accidents and normal operations of offshore oil wells. Fisher (1976) suggested that the large scale changes seen in North Sea phytoplankton communities could be due in part to the presence of persistent industrial pollutants and alteration in species composition through selective toxicity, but his arguments have been disputed by Eppley and Weiler (1979). Despite these concerns, Curds (1982) concluded that - "protozoologists have hardly begun to supply data (on the effects of pollution)". Parker (1983) and Dive and Persoone (1984) also noted the scarcity of data on effects from toxicants on marine protozoans. The few contributions to the pollution session of this workshop reflect that this topic still has not received much attention.

In contrast, pollution and especially organic pollution, has long received the attention of freshwater protozoologists such

NATO ASI Series, Vol. G 25
Protozoa and Their Role in Marine Processes
Edited by P. C. Reid et al.

as Kolkwitz and Marsson (1909) who introduced the 'Saprobien system', Lackey (1938), Cairns (1974), Legner (1975), Bringemann and Kühn (1980). Lackey (1961) noted that the knowledge of species and numbers in clean versus polluted ocean waters had lagged behind similar studies for inland waters. One likely explanation might be that effects of pollution in small lakes are more easily observed than in oceans. The poorer knowledge of cultivation techniques and taxonomy for the dominant marine forms have also been a problem. However, with the conceptual framework of the microbial loop of Azam *et al.* (1983) with later modifications (Sherr and Sherr 1988), improved microscopical (Sieburth 1984) and cultivation (Gifford 1985) techniques as well as better taxonomical knowledge (Montagnes and Lynn 1990) a better base for studying effects from pollution on both salt and freshwater protozoans now exists.

CATEGORIES OF POLLUTANTS

Curds (1982) discussed the effects from 5 categories of pollutants: sewage, oils and dispersants, organochlorine compounds metals, and thermal pollution. Radionuclides were deliberately omitted from the review. Although not documented in this presentation, the amounts from all these categories have probably increased in recent years. Also the number of pollutants has increased. For example in the wake of the rapidly growing aquaculture industry, inputs of antibiotics and medicines have increased markedly (Kinne 1986). The effects of these compounds on planktonic protozoans are virtually unknown; this overview concentrates on effects from metals, the area where most work has so far been concentrated. The Chernobyl accident in April 1986 also shows that radionuclides should not be forgotten. Within two months after the accident, an increased level of radiocaesium from Chernobyl was detected in the sediments at 1400 m depth at the Vøringen Plateau off the coast of northern Norway (Erlenkeuser and Balzer 1988). Radionuclides can be introduced to protists in the plankton through several routes. For example, surface reactive radionuclides may adsorb onto freeliving bacteria (Cho and Azam 1988), and thus pass into the microbial loop. Radioactive strontium might be introduced to the acantharians as they use strontium-sulphate (celestite) as the major inorganic constituent of their skeletons (Beers and Stewart 1970). In the Chernobyl fallout, the activity of Sr-90 amounted to about 1-2% of Ce-137 (T Berthelsen, National Institute of Radiation Hygiene, pers. comm.). Due to their different solubility, strontium remains in the water phase, while caesium is removed quickly (Saxén and Aaltonen 1987). However, so far the level of radioactivity in samples from fish, mussels and crustaceans from the Norwegian fjords are very low (Bøe 1987).

Some metals, such as Cu, Zn, Fe, and Co, are micronutrients, whereas metals such as Pb, Cd, Hg and As are not needed at all (Curds 1982). All these metals exist naturally in low concentrations in the sea. However, additional amounts are added through sources such as industrial spills, seepage from dumps, dumping of ash, precipitation, and from using metal-containing antifoulants such as tributyltin (TBT). Uptake of metals by protists may follow two main routes; ingestion of particles (bacteria, phytoplankton, zooplankton, detritus) exposed to metals or by direct uptake of dissolved metals or organometal complexes.

As pointed out by Cairns (1981) and Odum (1984), studies on effects of pollutants on organisms and their responses can be performed at three levels of ecological complexity: 1) simple laboratory microcosm studies, 2) the more complex mesocosm or enclosure studies and 3) at a further level of complexity, studies dealing with the open system of the real world.

MICROCOSMS

It was early suggested by Hutner (1964) that the use of protozoa as toxicological tools has a sunny future. Persoone and Dive (1978) gave a short review of the many methods available for toxicity testing using ciliates. These methods were broadly categorised as based on morphological, ultrastructural, ethological and metabolic criteria. The paper discussed the principles of each method and their advantages and disadvantages. Some new methods have appeared later. For example, dialysis bags suspended in test water in the laboratory or in the sea have successfully been used to test for effects from oil pollution on cultures of algae (Hegseth and Østgaard 1985). Similar methods might be adopted for protozoans, for example by using cages such as those used by Stoecker *et al.* (1983) or dialysis bags (Verity 1986). I shall not repeat the discussion on methods, but shall try to focus upon some of the results which may illustrate the complex responses of protozoans following exposure to various metals under different conditions.

Most papers dealing with studies of effects from metals on protists relate the responses of the organisms to the total concentrations of the metals. However, as pointed out by Stoecker *et al.* (1986), the availability of trace metals to organisms is controlled by the free metal ion activities rather than the total concentrations. In addition the toxicities of trace metals such as Cu and Zn are often affected by the activities of other metals due to competition among metals for binding to cellular ligands. Consequently metal ion ratios as well as individual ion activities are important in determining the overall effects (Stoecker *et al.* 1986). Dissolved organic matter released by the living organisms tested, may detoxify the medium by chelation, and dead cells which contain higher amounts of metals than living may cause partial detoxification (Rabsch and Elbrächter 1980). Also, film-forming bacteria may bind and precipitate metals (Corpe 1975). It has recently been estimated that greater than 99.7% of the total dissolved copper in surface water of the Central Northeast Pacific shallower than 200 m was associated with strong organic complexes (Coale and Bruland 1988). Due to the extremely low cupric ion activity (1.4×10^{-14} M) at the surface and higher activity at mid-depth (10^{-11} M), it was suggested that the ligand, presumably produced by phytoplankton, may hold free copper in a dynamic balance between copper limitation in the surface water and copper toxicity in the deeper water. Björnberg *et al.* (1988) have given a detailed discussion and proposed a theory on the mechanisms regulating the bioavailability of mercury in natural waters. Due to these factors, it is difficult to make direct comparisons between many of the experiments performed.

The responses of the freshwater ciliate *Tetrahymena pyriformis* towards Zn and Cd were studied by Chapman and Dunlop (1981) and Dunlop and Chapman (1981). According to these authors, it is well known that the toxicity of heavy metals is reduced in hard water. Chelation by organic material and particulate matter is also important for Cd, but apparently less for Zn. The authors found that the 8-hr LC_{50} of Zn and Cd were less than 1 ppm in the absence of soluble Ca and Mg, but that the LC_{50} were raised to 24 ppm of Zn and 19 ppm of Cd following the addition of 500 ppm (very hard water) Ca and Mg. Zinc was shown to exert a protective effect against Cd toxicity in terms of growth and ultrastructural appearance. Cells exposed to sublethal concentrations of Cd had damaged nuclei and the cells contained autophagic vacuoles.

Berk *et al.* (1985) used chemotaxis to study effects of Cu and Cd on both freshwater and marine ciliates. The method did not work well with Cu as sublethal concentrations of Cu apparently stimulated the chemotaxis through increased swimming rates, but their results indicate that Cu was more toxic than Cd for the marine scuticociliates (*Paranophrys, Miamiensis avidis*), but equally toxic to the freshwater *Tetrahymena* sp. The marine species were also more sensitive to Cu than the freshwater species. Concentrations of 5 ppb and 50 ppb Cu, respectively, significantly inhibited the chemotaxis of the two marine species.Stoecker *et al.* (1986) tested the effects of Cu and Zn on two marine planktonic ciliates, *Favella* sp. and *Balanion* sp. Short time experiments based on the swimming patterns of the ciliates, and long term experiment based on growth rates, showed that the responses to Cu and Zn were extremely complex with apparent antagonistic effects on the ciliates. Based on their results, and on calculations of the *in situ* cupric ion activity between 10^{-13} and 10^{-12} M in Perch Pond from which the ciliates were originally isolated, they suggested that the *in situ* concentration might have been sufficiently high to reduce the growth of *Favella* sp., but not of *Balanion* sp. In nature *Balanion* may have outcompeted *Favella*, as both normally coexist and have similar food requirements. Similar calculations and considerations of the concentrations of Zn in various environments indicated that the Zn concentrations in oceanic environments might be limiting to *Balanion* sp. and that the concentrations in some estuaries, especially those contaminated with trace metals, might be inhibitory to *Favella* sp. Comparison of their data and results obtained for dinoflagellates, did indicate that the tolerance of the ciliates to Cu was two orders of magnitude lower than that of the dinoflagellates. According to Stoecker *et al.* (1986) their data suggest that the free trace metal ion activities can thus be important niche parameters for marine planktonic ciliates.

In a study using Hg, Zn and Pb in a factorial design to examine growth rates as criteria of effects, Gray and Ventilla (1973) found significant synergistic effects on the marine, benthic ciliate *Cristigera* at all two factor and three factor combinations. When not tested in combinations, reduced growth rates were observed at 0.0025 ppm of $HgCl_2$, 0.15 ppm of $Pb(NO_3)_2$, and at 0.125 ppm of $ZnSO_4$. Parker (1979) also found significant synergistic effects on *Uronema marinum* exposed to chloride salts of the same metals as above. Acclimatisation did not increase the tolerance although this has been demonstrated for other species (see references in Parker 1979). Persoone and Uyttersprout (1975) tested the effects from 5 metals on the growth of the benthic ciliate *Euplotes vannus*. Hg was ranked as the most toxic, followed by Cu and then by Pb, Cd and Zn in the same rank.

Prevot and Soyer-Gobillard (1986) studied the combined effects of Cd and Se on the growth rates of the two marine dinoflagellates *Prorocentrum micans* (autotrophic) and

Crypthecodinium cohnii (heterotrophic). Their results showed that when the metals were present simultaneously in low concentrations, the toxicity of Cd and Se decreased (antagonistic), whereas if one of the elements was present in high concentrations, the toxicity increased (synergistic). At the ultrastructural level one effect seen was vacuolation of the cell.

Stebbing (pers. comm.) tested the effects of TBT on the marine ciliate *Euplotes* sp. and found increased growth rates at low concentrations of TBT and reduced at higher concentrations. The increased growth rate may be explained as a result of growth hormesis, or the tendency of subinhibitory levels of typical toxic agents to stimulate growth (see Stebbing *et al.* 1984, Stebbing 1987). It is possible that the apparent stimulation of chemotaxis noted by Berk *et al.* (1985) with Cu at high sublethal concentrations could be due to an effect similar to growth hormesis.

Cairns and co-workers have for many years used microbial communities grown in small units of polyurethane foam (PF) to evaluate the effects from various pollutants in the field and laboratory (Cairns *et al.* 1985). In one of these experiments, testing the effects of Cd on a freshwater community, they found that a Cd concentration of 4 ppm (4,000 µg/l) was required for a 50% reduction in species number relative to the control (Niederlehner *et al.* 1985). It is claimed that this technique is better to predict environmental effects from pollutants than single species tests (Yun-Fen *et al.* 1986). The technique, however, requires a significant taxonomical ability. The method which used the MacArthur-Wilson equilibrium model to explain the colonization of PF-pieces in presence of a toxicant (Lee 1986) has not been used much in the marine environment. As most of the species developing in the PF-units seem to be associated with surfaces, it is uncertain if the method can be used for strictly planktonic protozoans. It might prove useful for planktonic protozoans associated with detritus such as marine snow or dead phytoplankton.

From these microcosm studies it seems possible to draw several conclusions. Both synergistic and antagonistic effects may occur in protozoans exposed to different concentrations or combinations of metals. Hormesis has also been demonstrated. It is shown or known that compounds such as Ca or Mg, as well as organic chelators may considerably reduce the toxic effects from a given metal. Further, it is evident that the tolerance of the various protozoan species to metals may range over several orders. This has earlier been shown for freshwater flagellates and ciliates (Ruthven and Cairns 1973). In some waters, the free trace metal ion activities may be important niche parameters for marine protists. At the ultrastructural level, vacuolation is commonly observed. From the mentioned studies, a crude ranking of the relative toxicities of the various metals seems to emerge for freshwater and seawater ciliates with Hg as the most toxic and Zn as the least (Hg > Cu > Cd > Pb > Zn).

MESOCOSMS

The number of mesocosm experiments which have included studies on effects of metals on protozooplankton is small. The CEPEX-experiments included one with Cu (Beers *et al.* 1977b) and one with Hg (Beers *et al.* 1977a). Most other mesocosm experiments which have also included protozoans, have dealt with oil pollution (for example Lee *et al.* 1978, Dale 1988).

The Cu experiment (68 m^3) showed that the major ciliate groups dropped out at initial doses of 50 µg Cul^{-1} (50 ppb) and did not reappear over the period studied. Enclosures with 5 and 10 ppb added developed populations of oligotrichous ciliates different from that in the control container. It was assumed that this might reflect differences in the Cu tolerance of the species involved.

In the mercury-experiment (1300 m^3) the concentrations of 5 and 1 µg Hgl^{-1} used were around 500 and 100 times, respectively, above background level. In the container with the highest amount of mercury added, the oligotrich number dropped to approximately 10% of the initial concentration within two days, whereas only small changes occurred in the container with 1 ppb added or in the control. After the initial reduction of the densities in the 5 ppb container, a marked increase following subsidence of the level of Hg was seen. The dominant species *Lohmanniella* sp. (20-30 µm) was, however, different from those in the control (*Strombidium* sp., *Lohmanniella* sp. 40-50 µm).

In a set of bag experiments (1.3 m^3) in the Gulf of Riga testing both Hg and Cd at concentrations of 10 $\mu g l^{-1}$, Seisuma *et al.* (1985) also noted reduced densities of protozoans shortly after addition of the metals and increased densities after 9 days. Hg appeared to be more toxic than Cd. In another set where Pb (0.1, 1.0 mg l^{-1}) and Zn (0.1, 1.0 mg l^{-1}) were tested, Boikova (1986) found that addition of the metals markedly decreased the densities of *Strombidium* sp. The increases of *Euplotes* sp. in both control and test bags after 14 days were assumed to be effects, concealing the toxic responses.

From these enclosure experiments, it seems possible to conclude that addition of metals (Cu, Hg Cd, Pb, Zn) at the given levels may induce changed abundances of the dominant forms of the assemblages of planktonic ciliates. However, it was not certain whether the changes were due to direct toxic effects eliminating the most sensitive groups, or if they were due to other structural changes in the composition of the food sources of the ciliates or their predators (Beers *et al.* 1977b). For example, in the Cu treated container, the phytoplankton became progressively smaller and smaller.

According to Parker (1983), simple laboratory bioassays are unlikely to be of predictive value in assessing threats from pollution to the marine environment. However, the stronger negative effects on the ciliates in bags with added Hg than Cu or Cd, do fit the pattern of the relative toxicities of the metals that emerged from the microcosm experiments. Also the observation that ciliates are more sensitive to Cu than are dinoflagellates (Stoecker *et al.* 1986) may be of predictive value.

MACROCOSMS

It is clear that metal pollution might be a serious problem in some estuaries receiving large inputs of metal. The Minamata Bay tragedy in Japan showed the hazards from Hg to humans (Curds 1982). Due to industrial pollution from heavy metals, especially Cd, one of the largest fjords in Norway is now so contaminated that it can not be used for shellfish farming (Julshamn 1983). The use of the antifoulant TBT in bays such as Arcachon in France, where it caused failure of settlement of oyster larvae over several years (Alzieu *et al.* 1986), has now led to a ban or restrictions on use there and in several other countries. The transfer routes of the pollutants to fish and shellfish are not completely understood, but it is possible that a significant part could be chanelled through microzooplankton (Beers 1978). A better knowledge of the transfer routes will therefore also depend upon an improved understanding of the ecological roles of the planktonic protozoans.

According to Azam *et al.* (1984) the appearance of pollutants in animals may in part be due to pollution transfer along a microbial route. They calculated that the intracellular concentrations of Hg in bacteria in one CEPEX experiment would be 10 thousand fold higher than that found in the sea water. Some bacteria may accumulate so much Cd that it reaches 8% of the dry weight of the bacteria (Macaskie and Dean 1982). A transfer route of mercury from bacteria to copepods has already been demonstrated by Berk and Colwell (1981). By using mercury-labelled ($^{203}HgCl_2$) bacteria (*Vibrio, Pseudomonas*) as food for the marine ciliate *Uronema nigricans*, they showed a magnification of Hg in the ciliates. When mercury-labelled ciliates were fed to the copepod *Eurytemora affinis*, the mercury was transferred to the copepods. The concentrations in the copepods were higher than in the background but lower than in the ciliates.

Another route could be transfer of metals from sinking particles to associated protozoans. However, a recent study indicates that sinking particles may not be as efficient in carrying metals (Pb, Cu, Zn, Cd, Mn, Fe) from surface water to the deep ocean than was earlier believed (Anon. 1988).

Azam *et al.* (1984) put forward the hypothesis that pollution transfer via bacterioplankton is directly proportional to bacterial secondary production. It is probably not possible to put forward a similar hypothesis for the transfer along the phytoplankton route, as heavy blooms sometimes apparently sink out directly to the benthos without being grazed by zooplankton (Wassmann 1983).

PROTOZOANS AS INDICATORS OF POLLUTION

One result from better knowledge of the ecological roles of the pelagic protozoans and their responses to pollutants, could be that they might be used as indicators of pollution. According to Lee (1986) some protozoologists have long recognized the potential use of the organisms they study as ecological indicators, but however, there really is no agreement among protistan ecologists as to how best to apply the knowledge to assess ecosystem stress. A number of methods have been suggested. The first was probably the 'saprobiological indices' in freshwater introduced by Kolkwitz and Marsson (1909). This system has since been the subject of a considerable literature, for example Bick (1973). According to Fenchel (1987) the usefulness of protozoan communities for monitoring pollution by organic matter is overrated. By using the polyurethane foam technique, Buikema *et al.* (1983) have shown a correlation between an 'Autotrophic index' and protozoan colonization rates and suggest that the techniques are sensitive measures of pollution.

For seawater, Tacchi and Montanari (1985) have suggested that the ratio of diatoms/ciliates found after 48 h on coverslips submerged in tested water under laboratory conditions, could be used as an index of organic pollution. They found that dominance of ciliates was always associated with existing organic pollution. Seiglie (1968) found that some species of foraminifera were related to sewage outfall, and Friligos and Koussouris (1984) found dominance of dinoflagellates in polluted water, while diatoms dominated in clear water away from sewage outfalls. Revelante *et al.* (1985) showed that the abundances of ciliates in the North Adriatic were markedly increased at eutrophied sites and under stratified conditions. It has been suggested that assemblages of tintinnids and other ciliates (Coats and Heinbokel 1982) as well as evacuation of loricas by tintinnids (Beers *et al.* 1977b) may be used as indicators of pollution stress. In a review paper on the use of protozoa as test organisms in aquatic ecotoxicology, Dive and Persoone (1984) pointed out a number of arguments favouring the use of protozoa.

Benthic ecologists have long searched for species which can be used as pollution indicators. Pearson *et al.* (1983) rejected the possibility of finding a small range of benthic indicator species, even from a relatively similar range of environments. However, they suggested that the use of the log-normal distribution makes it possible to make an objective selection of sensitive species indicative of pollution-induced change in benthic communities. It has not yet been demonstrated if a similar method can be applied to marine planktonic protozoans.

CONCLUSIONS

Results from micro- and mesocosm experiments have increased our understanding of the effects of pollution on planktonic protozoans. But more work is needed to fully understand the effects of pollution on marine planktonic protists. At a symposium entitled "Protozoa as indicators of ecosystems" Lee

(1986) stated that the pollution problem is critical because of the pressure of increasing populations, industrial growth and intensification of agriculture. He raised a number of questions such as: "Are all changes harmful?...destructive? How do we measure it? Is there any urgency for us to find out and measure it? Can we reverse deterioration which has already taken place?" He further states: "These questions and many others are not a simple matter of commonsense, administrative judgement, or precise scientific endeavour...., they call for an amalgam of talents which is only now being developed in our society". In a recent paper Nilsson (1989) reviewed the cytological effects of the heavy metals Cu, Zn, Co, Ni, Cd, Hg and Pb on the freshwater *Tetrahymena*. Common 'pitfalls' in cytotoxicology are pointed out. A proposed scheme on intracellular handling by *Tetrahymena* of unwanted cations in relation to detoxification is also given.

ACKNOWLEDGEMENTS

A grant (D.51.41.013) from the Norwegian Research Council for Science and the Humanities, which is greatly appreciated, enabled me to participate in the workshop.

REFERENCES

Alzieu C, Sanjuan J, Deltreil JP, Borel M (1986) Tin contamination in Arcachon Bay: effects on oyster shell anomalies. Mar Poll Bull 17:494-498
Anon. (1988) MIT scientists studying trace metals in sinking particles at BBS. Bermuda Biological Station for Research, Inc. Currents. Summer 1988:3
Azam F, Ammerman JW, Fuhrman JA, Hagström Å (1984) Role of bacteria in polluted marine ecosystems. In: White HH (ed) Concepts in marine pollution measurements. A Maryland Sea Grant Publication, University of Maryland, College Park, p 431
Azam F, Fenchel T, Field JG, Gray JS, Mayer-Reil LA, Thingstad TF (1983) The ecological role of water-column microbes in the sea. Mar Ecol Prog Ser 10:257-263
Beers JR (1978) About microzooplankton. In: Sournia a (ed) Phytoplankton manual. UNESCO, Paris, p 288
Beers JR, Reeve MR, Grice GD (1977a) Controlled ecosystem pollution experiment: effect of mercury on enclosed water columns. IV. Zooplankton population dynamics and production. Mar Sci Comm 3:355-394
Beers JR, Stewart GL (1970) The preservation of Acantharians in fixed plankton samples. Limnol Oceanogr 15:825-827
Beers JR, Stewart GI, Hoskins HD (1977b) dynamics of micro-zooplankton populations treated with copper: controlled ecosystem experiment. Bull Mar Sci 27:66-79
Berk SG, Colwell RR (1981) Transfer of mercury through a marine microbial food web. J exp mar Biol Ecol 52:157-172
Berk SG, Gunderson JH, Derk LA (1985) Effects of cadmium and

copper on chemotaxis of marine and freshwater ciliates. Bull Environ Contam Toxicol 34:897-903
Bick H (1973) Population dynamics of protozoa associated with the decay of organic materials in fresh water. Amer Zool 13:149-160
Björnberg A, Håkanson L, Lundbergh K (1988) a theory on the mechanisms regulating the bioavailability of mercury in natural water. Environ Poll 49:53-61
Boikova E (1986) Protists - Biomonitors of marine environment. Symposia Biologica Hungarica 33:205-212
Bringemann G, Kühn R (1980) Bestimmung der biologischen Schadwirkung wassergefährender Stoffe gegen Protozoen II. Bakterienfressende Ciliaten. Z Wasser Abwasser Forsch 1:26-31
Buikema Jr AL, Cairns Jr J, Yongue Jr WH (1983) Correlation between the autotrophic index and protozoan colonization rates as indicators of pollution stress. In: Bishop WE, Cardwell RD, Heidolph BB (ed) Aquatic toxicology and hazard assessment: Sixth symposium American Society for testing and materials, Philadelphia, p 204
Bøe B (1987) Norsk fisk upåvirket av Tsjernobyl! Fiskets Gang 24:683
Cairns Jr J (1974) Protozoans (Protozoa). In: Hart Jr CW, Fuller SLH (ed) Pollution ecology of freshwater invertebrates. Academic Press, New York, p 1
Cairns Jr J (1981) Biological monitoring part VI - future needs. Wat Res 15:941-952
Cairns Jr J, Pratt JR, Niederlehner BR (1985) A provisional multispecies toxicity test using indigenous organisms. J Test Eval 13:316-319
Chapman G, Dunlop S (1981) Detoxication of zinc and cadmium by the freshwater protozoan *Tetrahymena pyriformis* I. The effect of water hardness. Environ Res 26:81-86
Cho BC, Azam F (1988) Major role of bacteria in biogeochemical fluxes in the ocean's interior. Nature 332:441-443
Coale KH, Bruland KW (1988) Copper complexation in the North-east Pacific. Limnol Oceanogr 33:1084-1101
Coats DW, Heinbokel JF (1982) A study of reproduction and other life cycle phenomena in planktonic protists using an acridine orange fluorescence technique. Mar Biol 67:71-79
Corpe WA (1975) Metal-binding properties of surface materials from marine bacteria. Developments in Industrial Microbiology 16:249-255
Curds CR (1982) Pelagic protists and pollution. A review of the past decade. Ann Inst océanogr, Paris 58(S):117-136
Dale T (1988) Oil pollution and plankton dynamics. VI. Controlled ecosystem experiments in Lindåspollene, Norway, June 1981: effects on planktonic ciliates following nutrient addition to natural and oil-polluted enclosed water columns. Sarsia 73:179-191
Dive D, Persoone G (1984) Protozoa as test organisms in marine ecotoxicology: luxury or necessity? In: Persoone G, Jaspers E, Claus C (ed) Ecotoxicological testing for the marine environment. State University of Ghent and Institute for Marine Scientific Research, Ghent, 1:281-305
Dunlop S, Chapman G (1981) Detoxication of zinc and cadmium by the freshwater protozoan *Tetrahymena pyriformis* II. Growth experiments and ultrastructural studies on sequestration of heavy metals. Environ Res 24:264-274
Eppley RW, Weiler CS (1979) The dominance of nanoplankton as an indicator of marine pollution: a critique. Oceanol Acta 2:241-245
Erlenkeuser H, Balzer W (1988) Rapid appearance of Chernobyl radiocesium in the deep Norwegian Sea sediments. Oceanol Acta 11:101-106
Fenchel T (1987) Ecology of protozoa. The biology of free-living phagotrophic protists. Science Tech Publisher, Madison. Springer, Berlin Heidelberg New York London Paris Tokyo
Fisher NS (1976) North Sea phytoplankton. Nature 259:160
Friligos N, Koussouris T (1984) Prelininary observations on sewage nutrient enrichment and phytoplankton ecology in the Thermaikos Gulf, Thessaloniki, Greece. Vie et Milieu 34:35-39
Gifford DJ (1985) Laboratory cultures of marine planktonic oligotrichs (Ciliophora, Oligotrichida). Mar Ecol Prog Ser

23:257-267
Gray JS, Ventilla RJ (1973) Growth rates of a sediment-living marine protozoan as a toxicity indicator for heavy metals. Ambio 2:118-121
Hegseth EN, Østgaard K (1985) Application of *in situ* dialysis cultures in studies of phytotoxicity of North Sea crude oils. Wat Res 19:383-391
Hutner SH (1964) Protozoa as toxicological tools. J Protozool 11:1-6
Julshamn K (1983) Analysis of major and minor elements in molluscs from western Norway. Thesis for the Dr. philos. degree. Institute of Nutrition. Directorate of Fisheries, Bergen, Norway
Kinne O (1986) Realism in aquaculture - the view of an ecologist. In: Bilio M, Rosenthal H, Sinderman CJ (ed) Realism in aquaculture: achievements, constraints, perspectives. European Aquaculture Society, Bredene, Belgium, p 11
Kolkwitz R, Marsson M (1909) Ökologie der tierischen Saprobien. Int Rev Ges Hydrobiol 2:1-126
Lackey JB (1938) Protozoan plankton as indicators of pollution in a flowing stream. Publ Health Rep 53:2037-2058
Lackey JB (1961) The status of plankton determination in marine pollution analysis. Engineering Progress at the University of Florida 15:1-11 (Leaflet no 140)
Lee JJ (1986) Protozoa as indicators of ecosystems. Insect Sci Applic 7:349-353
Lee RF, Takahashi M, Beers J (1978) Short term effects of oil in controlled ecosystems. In: Bates CC (ed) Proceedings of the conference on assessment of ecological impact of oil spills. American Institute of Biological Sciences, Arlington, p 635
Legner M (1975) Concentration of organic substances in water as a factor controlling the occurrence of some ciliate species. Int Rev ges Hydrobiol 60:639-65
Macaskie LE, Dean ACR (1982) Cadmium accumulation by microorganisms. Environ Technol Let 3:49-56
Montagnes DJS, Lynn DH (1990) Taxonomy of choreotrichs, the major marine planktonic ciliates. Mar Microb Foodwebs (submitted)
Niederlehner BR, Pratt JR Buikema Jr AL, Cairns Jr J (1985) Laboratory tests evaluating the effects of cadmium on freshwater protozoan communities. Environ Tox Chem 4:155-165
Odum EP (1984) The mesocosm. BioScience 34:558-562
Parker JG (1979) Toxic effects of heavy metals upon cultures of *Uronema marinum* (Ciliophora: Uronematidae). Mar Biol 54:17-24
Parker JG (1983) Ciliated protozoa in marine pollution studies. Ecotoxicol Environment Safety 7:172-178
Pearson TH, Gray JS, Johannessen PH (1983) Objective selection of sensitive species indicative of pollution-induced change in benthic communities. 2. Data analyses. Mar Ecol Prog Ser 12:237-255
Persoone G, Dive D (1978) Toxicity tests on ciliates - a short review. Ecotoxicol and Environ Safety 2:105-114
Persoone G, Uyttersprot G (1975) The influence of inorganic and organic pollutants on the rate of reproduction of a marine hypotrichous ciliate: *Euplotes vannus* Muller. Rev Intern Océanogr Méd 37-38:125-151
Prevot P, Soyer-Gobillard M-O (1986) Combined action of cadmium and selenium on two marine dinoflagellates in culture, *Prorocentrum micans* Ehrbg. and *Crypthecodinium cohnii* Biecheler. J Protozool 33:42-47
Rabsch U, Elbrächter M (1980) Cadmium and zinc uptake, growth and primary production in *Coscinodiscus granii* cultures containing low levels of cells and dissolved organic carbon. Helgolander Meeresunters 33:79-88
Revelante N, Gilmartin M, Smodlaka N (1985) The effects of Po River induced eutrophication on the distribution and community structure of ciliated protozoan and micrometazoan populations in the northern Adriatic Sea. J Plank Res 7:461-471
Rosenberg R, Lindahl O, Blanck H (1988) Silent spring in the sea. Ambio 17:289-290
Ruthven JA, Cairns Jr J (1973) Response of fresh-water

protozoan artificial communities to metals. J Protozool 20:127-135
Sangfors O (1988) Are synergistic effects of acidification and eutrophication causing excessive algal growth in Scandinavian coastal waters Ambio 17:296
Saxén R, Aaltonen H (1987) Radioactivity of surface water in Finland after the Chernobyl accident in 1986. STUK-A60. Finnish center for radiation and nuclear safety, Helsinki, Finland
Seiglie GA (1968) Foraminiferal assemblages as indicators of high organic carbon content in sediments and of polluted waters. Am Ass Petrol Geol Bull 52:2231-2241
Seisuma Z, Kulikova I, Boikova E, Marcinkevica S, Dzerve A (1985) A combined effect of mercury and cadmium on plankton *in situ*. Symposia Biologica Hungarica 29:207-219
Sherr EB, Sherr BF (1988) Roles of microbes in pelagic food webs: a revised concept. Limnol Oceanogr 33:1225-1227
Sieburth JMcN (1984) Protozoan bacterivory in pelagic marine waters. In: Hobbie JE, Williams PJLeB (ed) Heterotrophic activity in the sea. Plenum, p 405
Siemen RH (1988) Lessons from the North-Sea algae plague. German Research Service 27(8):6-7
Stebbing ARD (1987) Growth hormesis: a by-product of control. Health Physics 52:543-547
Stebbing ARD, Norton JP, Brinsley MD (1984) Dynamics of growth control in a marine yeast subjected to perturbation. J Gen Microbiol 130:1799-1808
Stoecker D, Davis LH, Provan A (1983) Growth of *Favella* sp. (Ciliata: Tintinnina) and other microzooplankters in cages incubated *in situ* and comparison to growth *in vitro*. Mar Biol 75:293-302
Stoecker DK, Sunda WG, Davis LH (1986) Effects of copper and zinc on two planktonic ciliates. Mar Biol 92:21-29
Tacchi B, Montanari M (1985) Research to determine the extent of organic pollution of the Genoese coastal waters-diatom/ciliated protozoa ratios. Oebalia 11(NS):875-878
Verity PG (1986) Growth rates of natural tintinnid populations in Narragansett Bay. Mar Ecol Prog Ser 29:117-126
Wassmann P (1983) Sedimentation of organic and inorganic particulate material in Lindåspollene, a stratified, land-locked fjord in western Norway. Mar Ecol Prog Ser 13:237-248
Yun-Fen S, Buikema Jr AL, Yongue Jr WH, Pratt JR, Cairns Jr J (1986) Use of protozoan communities to predict envionmental effects of pollutants. J Protozool 33:146-151

EFFECTS OF SALINE SEWAGE ON THE BIOLOGICAL COMMUNITY OF A PERCOLATING FILTER

Malcolm B Jones and Ian Johnson
Department of Biological Sciences
Polytechnic South West
Drake Circus
Plymouth PL4 8AA
UK

INTRODUCTION

The Sewage Treatment Works at the Cornish coastal resort of Looe (4°27'W, 50°21'N), opened in 1973, was designed to treat mainly domestic sewage and to serve winter and summer populations of 4500 and 11500 people respectively. The Works operates with biological filters, designed to remove the dissolved organic polluting load by providing conditions in which aerobic bacteria convert the unstable organic matter into stable inorganic substances (Bruce and Hawkes 1983). While conventional in design and process, the Looe plant is unusual in that the treatment operates on saline sewage. Sulphate reducing bacteria have colonised the Works due to the seawater intrusion and have proliferated under the anaerobic conditions present in the primary and secondary settlement tanks. At the pH levels generally found in seawater (i.e. 7.8 - 8.0) the reduction of sulphate to sulphide is enhanced, with resulting elevated rates of hydrogen sulphide production. This problem is ameliorated by the addition of lime to maintain the pH above 9.0 and thereby restrict the conversion process.

Although the efficiency of the biological filter has apparently not been affected by this saline operation, changes in species composition amongst the film community are apparent compared with freshwater treatment works. For example, the diverse macro-invertebrate grazing community normally associated with freshwater biological filters has been replaced by two species of estuarine amphipod crustaceans, *Gammarus duebeni* and *Orchestia gammarellus* (Jones and Wigham 1988). This study has been carried out to investigate the physico-chemical environment and the biology of the percolating filters at Looe with emphasis on the protozoan community.

NATO ASI Series, Vol. G 25
Protozoa and Their Role in Marine Processes
Edited by P. C. Reid et al.

MATERIALS AND METHODS

Quantitative monthly samples of both species of macroinvertebrates (*Gammarus duebeni* and *Orchestia gammarellus*) were taken over a year from various sites within the Works (i.e. intake, filter bed, humus tanks and outlet). The study investigated seasonal changes in abundance, population structure and reproduction of each species with the aim of identifying possible effects of this unique environment on life-history traits.

The physico-chemical fluctuations within the Works were assessed by measuring salinity, dissolved oxygen and heavy metal levels at hourly intervals during a tidal cycle at the input, filter beds and humus (secondary settlement) tanks. Salinities were determined using an Otago hand-held refractometer, and oxygen content measurements were made using a YSI Portable Dissolved Oxygen Meter and Probe. Total heavy metal concentrations were determined from unmodified water samples, whereas dissolved levels were obtained using water samples filtered through 0.45 µm cellulose nitrate membrane filters (Whatman Ltd). In both cases, the 100 ml water samples were acidified with 2 ml of Aristar Nitric acid to ensure the heavy metals present were in solution. The water samples were analysed for heavy metals using a Varian AA975 Atomic Absorption Spectrophotometer.

Protozoan samples were collected from the Works during an excursion to Looe organised as part of the field work associated with the NATO ASI (27th July 1988). Protozoan cultures were isolated and maintained (Soldo 1990), and identified (Small and Green 1990).

THE TREATMENT PROCESS

Domestic sewage is discharged to a pumping station (Bone Mill) on the quay in the harbour near the confluence of the West and

East Looe Rivers. Pumping is continuous except for a cut-off period of approximately 2 h at low water. The general flow rate into the works is about 50 l s^{-1}. Crude sewage is screened initially at the inlet to remove large extraneous matter and grit (Figure 1). The flow then passes to four primary settlement tanks, where the majority of solid wastes are removed as sewage sludge by sedimentation. In the settlement tanks anoxic conditions occur at between 3-4 m depth and, due to the relatively high concentrations of sulphate in the seawater, these conditions encourage the presence of sulphate reducing bacteria. To reduce sulphide production, lime is added to the sludge in the settlement tanks. The sludge passes to a storage tank prior to dumping, and the supernatent is pumped up to, and distributed through, four percolating biological filters (Figure 1).

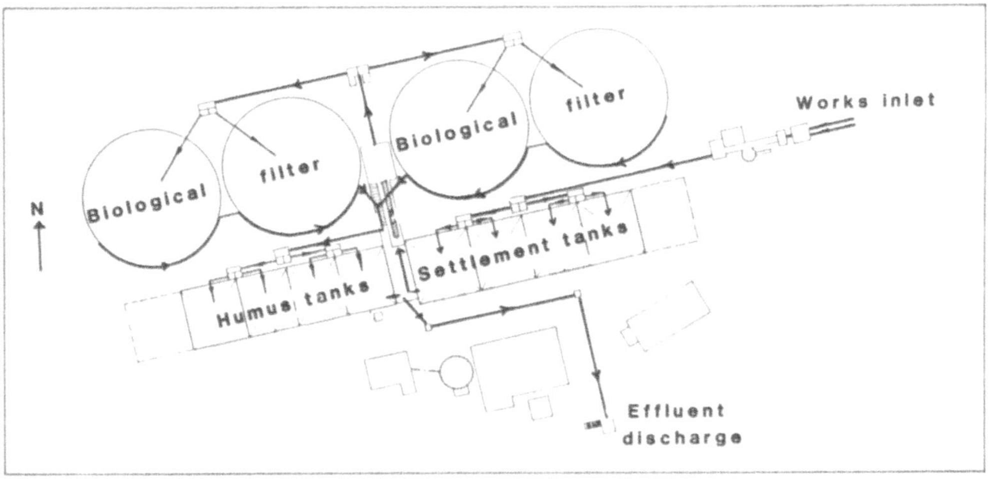

Fig. 1. Plan of the Looe Sewage Treatment Works

Sewage is distributed over the filter beds in the form of a jet by means of rotating metal arms with perforations along their lengths. The beds contain granite stones as a substratum on which the film of micro-organisms making up the 'biological filter' becomes established. The filter provides an extended ventilated surface over which sewage can flow uniformly and interact with the film community. The resulting purification by microbial activity is based on the process of aerobic biooxidation.

Following biological filtration, the flow leaves from the base of the filter beds and passes to a series of secondary settlement or humus tanks in which further sedimentation occurs resulting in an effluent suitable for discharge. Lime is again added in the humus tanks to alleviate hydrogen sulphide production. The final, treated effluent is discharged into a culverted section of a stream feeding into the West Looe River.

The design of the system of flow to the percolating filters allows for the diversion of sewage with high salt concentrations directly into the humus tanks. This feature was incorporated during planning to ensure the survival of the presumed stenohaline bacteria in the percolating filters as it was anticipated that seawater would enter the Works, though only at high spring tides.

WATER CHEMISTRY

Measured changes in salinity reflect the tidally based nature of the sewage input. In each tidal cycle there is an influx of seawater, most marked around high tide which causes a high salinity sewage input at the inlet. In contrast, from approximately 3 h before low tide, the sewage input becomes considerably less saline due to the reduced seawater influx. The consequences of these inputs are the cyclical salinity fluctuations within the Works shown in Figure 2 for a representative spring tidal cycle. Salinity extremes are most marked at the inlet (13-35°/₀₀) and are reduced in magnitude as the sewage passes through the Works due to mixing. However, within the filter beds the film community is exposed to considerable salinity fluctuations (15-34°/₀₀), although the efficiency of the Treatment Works is evidence of the adaptation of both the primary and secondary feeders to the cyclical salinity regime.

Dissolved oxygen levels recorded within the Works at 1 and 5 h after high tide (i.e. as given in Figure 2) are shown in Figure 3. Sewage input around high tide has considerably higher dissolved oxygen content than that of the brackish sewage entering before low tide. Lowest dissolved oxygen levels at a given site were found at the lowest salinity of a tidal cycle.

Although fluctuations in sewage dissolved oxygen levels were evident in the percolating filters these were less marked (i.e. 80 - 90% of O_2 saturation for a given temperature and salinity) than at other sites within the Works. Maintenance of an

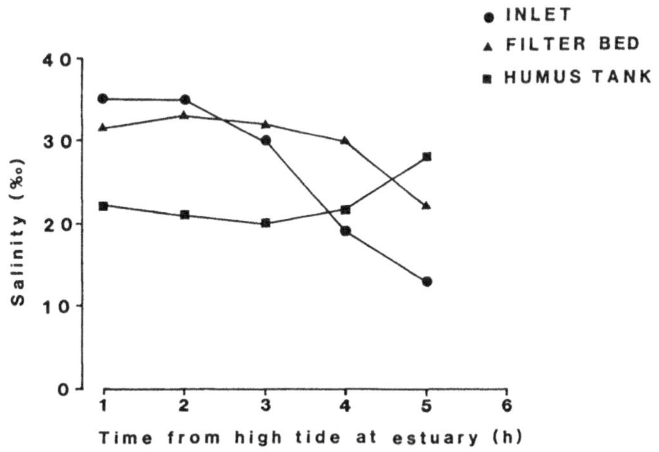

Fig. 2. Salinity changes within the Sewage Treatment Works during a spring tidal cycle

oxygenated substratum within the beds is necessary for the functioning of the biological filter (Bruce and Hawkes 1983) and presumably reflects the additional aeration of the sewage during application to the beds. Indeed, oxygen saturation has been identified as one of the most important ecological factors determining protozoan distribution, since the majority are obligate aerobic species requiring free oxygen (Curds 1975).

In the settlement tanks anoxic conditions occur at depths of 3-4 m even though the surface water may contain significant levels of dissolved oxygen. However, within the summer months, these tanks, and the Works in general, may become increasingly anoxic due to the elevated operating temperature (Agnew and Jones 1986).

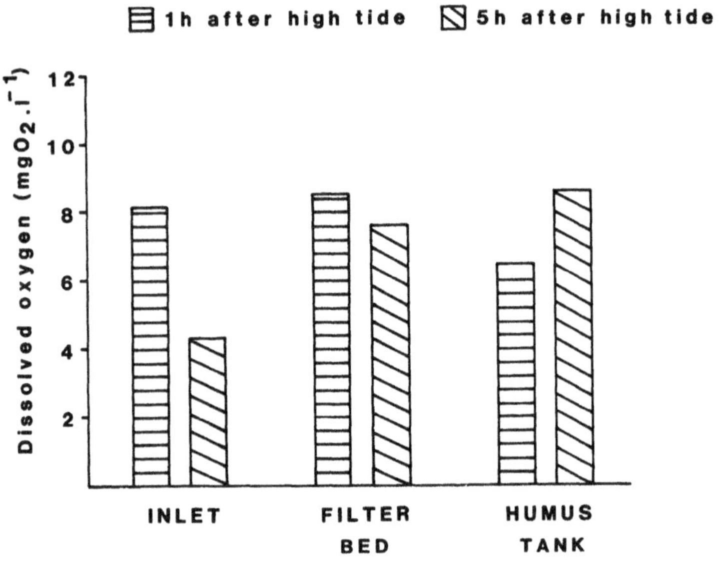

Fig. 3. Dissolved oxygen levels within the Sewage Treatment Works at 1 h and 5 h after a spring tide

A variety of heavy metals (e.g. Cd, Cr, Cu, Ni, Pb and Zn) are associated with the sewage input although the concentrations of particular metals in a given loading may vary considerably. Research at this site, and at other sewage treatment works (Lester *et al.* 1979, Stoveland *et al.* 1979), has shown that Zn is generally the most elevated heavy metal present in comparison with background non-polluted levels. The nature of heavy metal concentrations in the humus tanks is shown in Table 1, where the ranges of Cu and Zn are given for the salinity range of 15-34°/∞, which is the recorded range for this site.

Table 1. DISSOLVED HEAVY METAL CONCENTRATION RANGES IN SEAWATER FROM A TREATMENT WORKS HUMUS TANK AND THE ESTUARY AT LOOE

Heavy metal	Dissolved concentration range ($\mu g\ l^{-1}$) in seawater	
	Humus tank	Estuary
Cu	20.0 - 67.5	5.0 - 22.5
Zn	30.0 - 144.5	14.0 - 28.5

Within the Works, heavy metals in the sewage show salinity dependent fluctuations, such that the highest total water loadings at a given site correspond with the lowest salinity. Figure 4 shows the total Zn levels in the sewage at the filter bed and the humus tanks throughout a tidal cycle. The cyclical salinity dependent fluctuations in total Zn loading are a consequence of the dilution of the raw sewage and associated heavy metals by the seawater influx. During passage of sewage through the Works total water levels of Zn, and other heavy metals, decrease as a result of adsorption/complexation of the metal with particulate matter, and hence removal via sedimentation (Forstner and Wittman 1978).

At the Looe Works, there are marked heavy metal concentrations in the sewage entering the percolating filter. The effect on the film community and the protozoan population in particular, however, will be dependent on the balance between the dissolved and particulate fractions, since heavy metals are only available to organisms in their ionic form. In the percolating filters the dissolved Zn fraction generally accounted for 60% of the total Zn loading. The efficiency of the film community appears to be maintained despite exposure to elevated heavy metal levels. Indeed, Curds

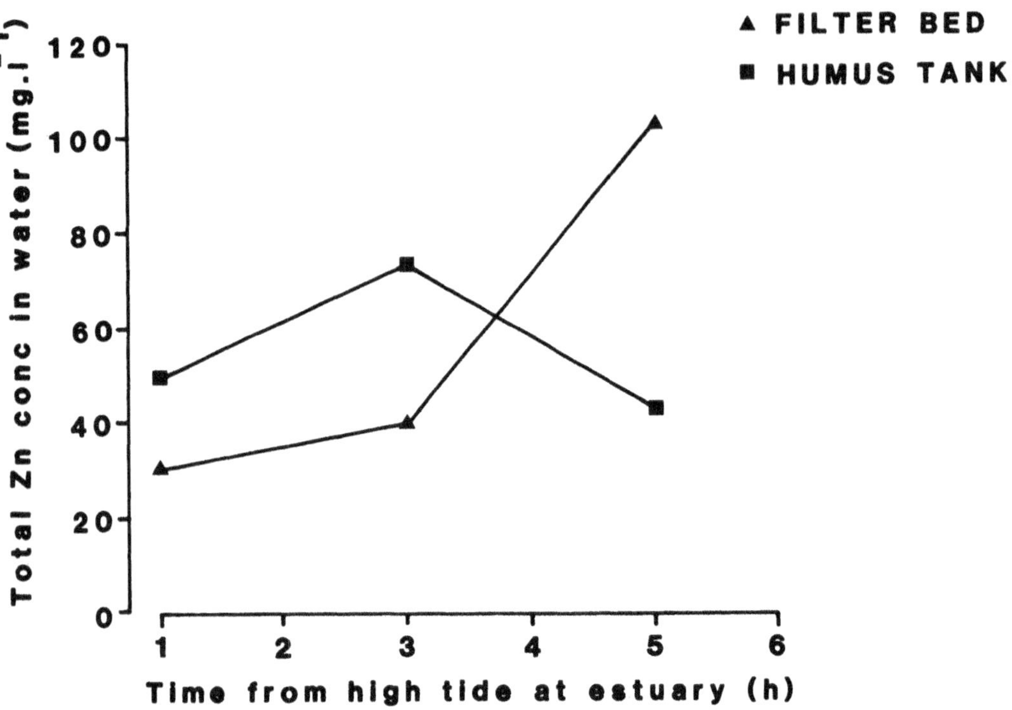

Fig. 4. Total zinc concentrations in seawater within the Sewage Treatment Works during a spring tidal cycle

and Cockburn (1970) found no evidence that sewage containing heavy metal laden industrial wastes had any detrimental effects on the protozoan population of the perculating filters.

THE BIOLOGY OF THE PERCOLATING FILTER

The secondary invertebrate grazers of freshwater percolating filter communities are well known and include the micro-fauna (protozoa), the meio-fauna (mainly nematode worms and rotifers) and the macro-fauna (typically tubificid and lumbricid worms and insect larvae). At Looe, the seawater intrusion has created a unique environment which has resulted in the replacement of the freshwater species by euryhaline forms.

Only two species of macro-invertebrates occur and these are the amphipods *Gammarus duebeni* and *Orchestia gammarellus*. Each is a permanent resident and breeds successfully under the special conditions prevailing within the Works. Immigration appears to have ceased, suggesting that these amphipods have been isolated within the Works for some twelve years. Physiological comparisons of 'sewage' *Gammarus duebeni* with individuals of the presumed donor population in the East Looe River Estuary have identified differences in tolerance to hypoxia (Agnew and Jones 1986) and external zinc concentrations (Johnson and Jones 1989). These studies, together with ongoing work comparing osmoregulation of 'sewage' and 'estuarine' amphipods when exposed to elevated zinc concentrations, suggest that 'sewage' amphipods are physiologically distinct from 'estuarine' individuals, presumably as a response to prolonged exposure to environmental pollution at the Works.

The freshwater protozoa of a percolating film community are usually dominated by ciliates such as *Carchesium polypinium*, *Chilodenella unicinata*, *Cinetochilum margaritaceum*, *Opercularia* sp. and *Vorticella* sp. (Curds 1975). There is little information on the composition of the protozoa of a saline sewage plant. In this study the protozoan community was found to be deporporate, paralleling the reduction in numbers of the grazing macro-invertebrate species of the percolating filter (Jones and Wigham 1988). A list of species inhabiting the Looe Works is given in Table 2; none has been recorded previously from a percolating filter. The majority of species present are euryhaline forms, which is consistent with the physico-chemical environment presented here, and are typically found in a wide range of habitats including fresh, brackish and marine waters. For example, the flagellate *Pavlova lutheri* (Table 2) occurs in rock pools in Finland and can tolerate from 0.1-33°/₀₀ with a preference for the higher extremes of salinity (Droop 1953). Although no details are available on the meio-faunal component of the Looe filter, initial findings indicate that the nematodes are not the typical species found in freshwater percolating filters and are probably marine forms (J Laybourn-Parry pers. obs.). In addition the densities of nematodes collected were

Table 2. PROTOZOA IDENTIFIED FROM THE LOOE SEWAGE WORKS, 27TH JULY 1988 BY G SMALL AND J GREEN

Species	Filter beds	Humus tanks
Pseudocohnilembus hargassi		*
Pseudocohnilembus sp	*	
Pseudocohnilembus cysts		*
Uronema (nigricans)	*	
Uronema sp		*
Euplotes vannus		*
Euplotes sp	*	
Oxyrrhis sp		*
Carteria sp		*
Cohnilembus sp		*
Phacus pyrum		*
Entripiella sp		*
Pavlova lutheri		*
Bodinid flagellates		*

estimated to be only 10 - 20% of those found in freshwater filters. These findings are consistent with available data on the operation of saline works, which have shown that although a continuous salinity input has no long-term effects on functioning, fluctuating salinity sewage inputs can result in marked changes in both species composition and abundance, and the efficiency of the works (Lawton and Eggert 1957, Mills and Wheatland 1962). The effects of fluctuating saline sewage on the functioning and succession of the film community of biological filters is an area in which further research is required.

ACKNOWLEDGEMENTS

We thank the delegates to the NATO ASI for their enthusiasm and dedication shown when visiting and sampling at the Looe Sewage Works, and express our appreciation of the patience and time given by Tony Tulk and his team at the Looe Works over several years.

REFERENCES

Agnew DJ, Jones MB (1986) Metabolic adaptations of *Gammarus duebeni* Liljeborg (Crustacea, Amphipoda) to hypoxia in a sewage treatment plant. Comp Biochem Physiol 84A:475-478

Bruce AM, Hawkes HA (1983) Biological filters. In: Curds CR, Hawkes HA (eds) Ecological aspects of used-water treatment, Vol III. Academic Press London, p 1

Curds CR (1975) Protozoa. In: Curds CR, Hawkes HA (eds) Ecological aspects of used-water treatment, Vol I. Academic Press London, p 203

Curds CR, Cockburn A (1970) Protozoa in biological sewage-treatment processes. I, A survey of the protozoan fauna of British percolating filters and activated sludge plants. Wat Res 4:225-236

Droop MR (1953) On the ecology of flagellates from some brackish and freshwater rockpools of Finland. Acta Bot Fenn 51:3-52

Forstner U, Wittman GTW (1978) Metal pollution in the aquatic environment, 2nd edn. Springer, Berlin Heidelberg

Johnson I, Jones MB (1989) Effects of zinc/salinity combinations on zinc regulation in *Gammarus duebeni* from the estuary and the sewage treatment works at Looe, Cornwall. J Mar Biol Ass UK 69:249-260

Jones MB, Wigham GD (1988) Colonization by estuarine amphipods of a sewage treatment works. EBSA Bull 50:29-33

Lawton GW, Eggert CV (1957) Effect of high sodium chloride concentration on trickling filter slimes. Sewage Industr Wastes 29:1228-1236

Lester JN, Harrison RM, Perry R (1979) The balance of heavy metals through a sewage treatment works. I, Lead, Cadmium and Copper. Sci Total Environ 12:13-23

Mills EV, Wheatland AB (1962) Effect of a saline sewage on the performance of a percolating filter. Water and Waste Treatment 9:170-172

Soldo AT, Brickson SA (1990) A method for the cloning and axenic cultivation of marine protozoa. In: Reid PC, Turley CM, Burkill PH (eds) Protozoa and their role in marine processes. Springer, Berlin Heidelberg New York

Small EB, Green JC (1990) Taxonomy (46 - or more - protistan phyla). In: Reid PC, Turley CM, Burkill PH (eds) Protozoa and their role in marine processes. Springer, Berlin Heidelberg New York

Stoveland S, Astruc M, Lester JN, Perry R (1979) The balance of heavy metals through a sewage treatment works. II, Chromium, Nickel and Zinc. Sci Total Environ 12:25-34

ENDOSYMBIOSIS IN THE PROTOZOA - SESSION SUMMARY

Michèle Laval-Peuto
Laboratoire de Protistologie Marine,
Université de Nice-Sophia, Antipolis,
Campus Valrose, F-06034 NICE Cedex,
France

INTRODUCTION

Symbiosis is a general and widespread biological phenomenon, in which two partners are associated with various levels of exchanges ranging from mutualism to parasitism. This wide definition, which is preferred by most biologists (Henry 1966, Lewis 1973, 1985, Smith and Douglas 1987, Starr 1975) corresponds to the initial definition by de Bary (1879). It is restricting, if not erroneous, to consider symbiosis simply as providing a reciprocal benefit to each partner or no harm, as applied respectively to mutualistic or commensalistic associations. But even when a benefit seems probable, different steps in the relationships can be demonstrated between some species and their various partners. Parasitic associations must then be accepted as one of the possible situations in symbiosis. It is likely that a large diversity in the quality of symbiotic exchanges will be shown, when more examples of symbiosis are studied.

During the preceeding NATO workshop on 'Marine Pelagic Protozoa and Microzooplankton Ecology' (Villefranche-sur-Mer, France, 1981) Taylor (1982) showed the tremendous variety of symbioses in marine microplankton, especially from tropical areas. He also pointed out that few examples were described in detail while alive. In general,the degree of interactions between both partners was not clearly established, and morphological and physiological adaptations mostly remain to be studied. After six years, this second workshop offers an opportunity to demonstrate the increasing development of our knowledge in this field.

This session on Endosymbiosis in the protists particularly addressed recent advances in the topic due to the application

NATO ASI Series, Vol. G 25
Protozoa and Their Role in Marine Processes
Edited by P. C. Reid et al.

of new techniques of investigations: ultrastructural observations, biological data and/or physiological experiments from cultivated strains. Our aim was to consider all kinds of symbiotic associations between endosymbionts (bacteria, cyanobacteria, algae or isolated plastids) and various protistean hosts, ranging from mutualistic to parasitic associations. The endobionts, or cytobionts, are intracellular in their hosts within membrane-bound vacuoles or directly in the cytoplasm.

The session was divided into three parts: 'Endosymbiosis of protozoa with algae or cyanophyceae', 'Bacterial endobionts' and 'Plastid symbiosis in the protozoa'. The keynote lecture by **Diane Stoecker** concerned plastid-retention in the oligotrich ciliates.

ENDOSYMBIOSIS OF PROTOZOA WITH ALGAE OR CYANOPHYCEAE

John Lee examined the role of endosymbionts in the nutrition of larger foraminifera in the Red Sea. Two species, *Amphisorus hemprichii* and *Amphistegina lobifera*, reared in the laboratory (Lee *et al.* submitted), play host respectively to endosymbiotic dinoflagellates and to diatoms. Neither species is capable of growth in the dark, even when the foraminifera, cultured under the best conditions, are given the food organisms they are known to ingest and assimilate. All specimens of *A. hemprichii* died before eight weeks in the dark. *A. lobifera* bleached in the dark, but remained alive after eight weeks, if they were fed and if the medium was changed weekly. These results suggest that the host-symbiont association is obligate in the former species and not in the latter. Both species require pulses of NO_3^- and PO_4^{3-} for sustained growth in the light. *A. hemprichii* grows much better in the light when it is fed a mixture of selected species of algae. For *A. lobifera*, feeding is less important if NO_3^- and PO_4^{3-} are replenished weekly. When a low concentration of GeO_2 (0.5 mg/l) was used to selectively inhibit the endosymbiotic diatoms within *A. lobifera*, fed hosts incubated in the light died within three weeks, even under the best possible culture conditions. In a carbon/phosphorus double labelling pulse-chase experiment (Kuile *et al.* 1987), *A. lobifera* retained after a week 33% of the phosphorus but only 8% of the carbon. In *A. lobifera* there was no meaningful difference between the rates of retention of labelled C or P. These observations indicate that feeding in *A. hemprichii* may be a major source of organic carbon and energy, while *A. lobifera* uses food mainly as a source of inorganic nitrate and phosphate.

Benno ter Kuile, as an adendum to John Lee's lecture, compared two models representing the differences in the intracellular functioning of both symbiont-bearing foraminifera, *Amphisorus hemprichii* and *Amphistegina lobifera*.

Werner Reisser reviewed the role Ciliophora play as microhabitats for different green algae species. The lecture was based on freshwater ciliates, which are frequently observed to form endocytobiotic units with chlorophycean coccoid algae (Reisser 1986, 1987). These so-called green ciliates have been extensively studied, while little is known about similar cases among marine ciliates. The associations differ significantly in stability, from the hereditary symbiosis of permanently green ciliate species, to temporary associations where algal partners can be lost under stress conditions with subsequent repeated reinfections. These constitute marvellous examples for the study of possible steps during the evolution of an increasing symbiotic integration of partners. In both stable and temporary associations, hosts are able to distinguish between algae suitable and unsuitable for symbiosis formation. Algae taken up via phagocytosis are either digested in food vacuoles or stored in perialgal vacuoles, which protect them against host lytic enzymes at least for some time. Research has centred on the analysis of underlying cell-to-cell recognition processes. Reisser *et al.* (1982) suggested a strong host-partner specificity, because one given host strain usually selects only one species among different offered algae for symbiosis, but recent studies (Reisser et al. in press) show that this rule may no longer be true. In *Paramecium bursaria* a typical symbiotic algal strain *per se* does not exist. Different strains of green *P. bursaria*, collected from various localities in Europe and the U.S.A., harbour different chlorophycean algal strains, assignable to the *Chlorella vulgaris* and *Chlorella fusca* groups, and to *Graesiella sp.* Each algal strain or species forms, with the corresponding host, a stable hereditary association. Symbiotic algae isolated from them were classified by light and electron microscopy, as well as cytochemical, physiological and molecular (DNA) characteristics. These included cell surface structures (Figure 1), shape of plastids, tolerance to acid pH, and different NaCl-concentration, activity of hydrogenase, growth with different C- and N-sources, amount of excreted carbohydrates, and mol % G+C of DNA. Interestingly, symbiotic algae differ from free-living counterparts only by a pH-dependent excretion of glucose, maltose or fructose. Other features of algae do not show any specific adaptation to the symbiotic milieu.

A new hypothesis for symbiosis formation was suggested. The constitution of a stable ciliate-algae endocytobiotic unit is a multi-step process with different levels of host-partner interactions. Suitable algae are recognized by the host from their cell surface structures 'fingerprints of algae' (Reisser *et al.* 1982) which trigger the individual enclosure of algae in special 'perialgal' vacuoles, protecting them against host lytic enzymes permanently or temporarily. Algae once enclosed can divide and populate the host according to the ecophysiological conditions offered by their special habitat (amount of available organic and inorganic nutrients, light and CO_2). Less adapted algae can be replaced by better adapted ones, which is a type of ecological regulation (Reisser 1986, 1987). From ecological studies (Reisser and Herbig 1987) symbiosis formation appears to be a primary advantage mainly for the algal partner, the ciliate representing a sort of evolutionary resting-place for old or transitional forms as well as for new lines, at least within the genus *Chlorella*.

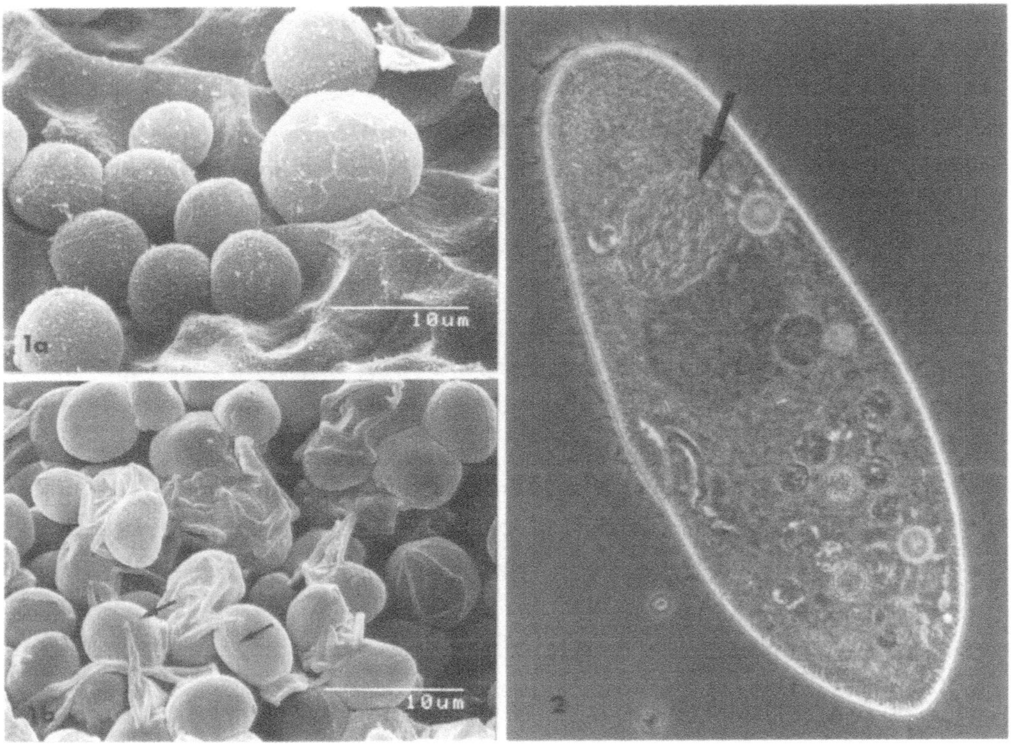

Fig. 1 (from Reisser). Two examples of cell surface structures of algae: (a) network of ribs on free-living *Chlorella fusca* var. *vacuolata*; (b) meridional rib (arrow) on symbiotic *Chlorella* isolated from *Paramecium bursaria*

Fig. 2 (from Gortz) *Paramecium caudatum* with *Holospora undulata* in the nucleus. The numerous sigmoid bacteria are the cause of the wavy pattern of the micronucleus, hypertrophied (arrow) by the infection

Malte Elbrächter presented a poster showing phototrophic symbionts in dinophysoid dinoflagellates. Dinoflagellates, when photosynthetic, are remarkable in having a large variety of types of plastids. In most cases plastids contain three-thylakoid lamellae and they are surrounded by three membranes, which is unusual among the other algae except the euglenoids. Some dinokaryotic species harbour eukaryotic photosynthetic symbionts derived from chrysophycean algae or diatoms, chlorophycean algae and cryptomonads. These observations have led to stimulating, sometimes controversial, evolutionary interpretations. Photosynthetic dinoflagellates might have evolved from phagotrophic ancestors which were invaded by an algal symbiont which became more or less integrated and subsequently reduced. Two phototrophic species of *Dinophysis*, *D. acuta* and *D. acuminata* were studied through electron microscopy. Their chloroplasts are cryptophycean-like, having pairs of thylakoids filled with electron dense material. They are bound by only a double membrane, in contrast to the blue-green plastids of *Amphidinium wigrense* enveloped by three membranes, and their primary fluorescence is orange indicating the presence of phycobilins (Schnepf and Elbrächter 1988). The possible evolution of the *Dinophysis* plastids, compared to the cryptophyte plastids of *Gymnodinium acidotum* and the blue-green plastids of *Amphidinium wigrense* were discussed with various interpretations. The acquisition of plastids in *Dinophysis* might come from the phagocytosis of a eukaryotic cell, or by the phagocytosis of a prokaryotic cell, or alternatively by myzocytosis (the extraction of the cytoplasm of the host cell by means of a peduncle without ingesting the host plasma membrane).

BACTERIAL ENDOBIONTS

Anthony Soldo outlined the nature of infectious xenosomes in the marine ciliate *Parauronema acutum*. The term Xenosome was adopted to describe a group of obligate infectious, Gram negative, bacteria-like endosymbionts (Soldo 1983, 1988, Corliss 1985). They were found in several strains of the marine ciliate *Parauronema acutum*. The symbiont-host association is a highly stable one. Several xenosome-bearing stocks of *P. acutum* have been maintained in axenic cultures for several years without loss of the symbionts. Xenosomes appear to be dependent upon the host for their survival, since their cultivation extracellularly in artificial media has been unsuccessful. Symbionts of each strain are capable of infecting homologous and heterologous protozoan stocks. Xenosomes from only certain strains can kill other marine protozoa, notably ciliates of the genus *Uronema*. Because this system is effectively a host-parasite analog, it is extensively used for model studies of infection (the killer effect and the interaction of the xenosome with the host) as well as to try to establish the phylogenetic origins of the symbionts. A variety of chemical procedures and many of the most recently developed techniques in molecular biology have been applied to xenosomes.

Both non-killer and killer xenosomal chromosomal DNA is organized in the form of nine duplexes, each circularly permuted, and of 515 kilobase (KBP) pairs in length. Non-killer and killer xenosomes possess extrachromosomal DNA configured as covalently-closed circular molecules. Two plasmids are present in the non-killer, each 63 KBP in length with a single BAM HI site and multiple BGLI, Bst IIE, PstI and Sal I sites. Killer extrachromosomal DNA is comprised of four plasmids each 63 KBP in length. Two possess single BAM HI sites and two contain two BAM HI sites each. All contain a region of homology of about 17 KBP in length.

Infection of non-killer xenosomes into a symbiont-free host ciliate, which previously harboured killer xenosomes, tranforms the non-killer into a killer. This change is accompanied by an alteration in the restriction pattern of the extrachromosomal DNA from that of the non-killer to that of the killer. Thus the composition of the extrachromosomal DNA appears to be the genetic determinant responsible for the non-killer to killer transformation. In a subsequent experiment, a clone bank consisting of restriction fragments generated from killer extrachromosomal DNA, using pBR322 and gt 11 as vectors, was prepared in the expectation of achieving genetic expression in *Escherichia coli*. Cloning a substantial portion of the DNA was obtained, but no transformation of *E. coli* cells into killers could be observed. Indirect evidence suggests that the killer effect is due to the presence of a protein in the killer itself. Other data show that xenosomal genomic DNA is a chimera with respect to *dam* methylation, since all the adenine residues present in GATC sequences derived from extrachromosoal DNA are methylated, whereas none of the adenines in these sequences is methylated in the chromosome itself. Implicit here is that a unique mechanism for methylation of these adenines may be operative in the symbiont. Because *dam* methylation is considered to have occured in Gram negative bacteria about 200,000,000 years ago, Soldo postulates that the acquisition of the plasmid by xenosomes may be a relatively recent evolutionary event.

Hans-Dieter Görtz described the mode of infection of endonucleobionts in ciliates. Various microorganisms live in the nuclei of marine, brackish water and freshwater ciliates. This form of symbiosis is a parasitism. And certainly, the extent and the nature of such endonucleobioses are of importance for the population dynamics of the host organisms. Endonucleobionts can easily be detected, either unfixed phase contrast microscopy, or fixed. (e.g. osmic vapor then a 3:1 mixture of ethanol and acetic acid) and stained (e.g. lactoaceto orcein or fast-green) are necessary.

Examples of endonucleobionts were given, such as the bacteria of the genus *Holospora* which often infects species of *Paramecium*. The life-cycle of *Holospora* comprises an infectious form and a reproductive form, which is not infectious. The route of infection is: phagocytosis of the infectious form, escape from the phagosome, transport through the cytoplasm, entrance into the nucleus, and multiplication by the reproductive form (see review in Ossipov 1981, Görtz 1986, Schmidt *et al.* 1987). Some species infect only the micronucleus (Figure 2), others only the macronucleus of the host ciliates.

Infection, which is the beginning of this endocytobiosis, requires a series of recognition events between the host cell membrane and the invading microorganism. Ultrastuctural observations suggest several steps in the communication (Ossipov and Podlipaev 1977, Podlipaev and Ossipov 1979, Görtz and Wiemann 1989). One of the aims in the study of endonucleobionts is to find the signals and receptors of this communication. Görtz *et al.* (1988) demonstrated that the infectious form of *Holospora* differs from the non-infectious reproductive form in morphology and in protein patterns. It may, therefore, be possible to discover the signal-substances used in the communication, by investigating the proteins found only in the infectious form. A demonstration, using phase-contrast microscopy, of the infection of various strains of paramecia by *Holospora* species, was also given.

PLASTID SYMBIOSIS IN THE PROTOZOA

Dian Stoecker presented the keynote lecture (Stoecker 1990) of this session and gave a demonstration (also related to the session Energetics and Production) on techniques for measuring ^{14}C uptake by ciliates.

Michèle Laval-Peuto offered another approach to the same topic: Mixotrophy in marine planktonic ciliates emphasising ultrastructural and cytochemical observations on plastidic Strombidiidae. The presence of isolated plastids in the cytoplasm of planktonic ciliates has been demonstrated ultrastructurally (Blackbourn *et al.* 1976, Laval-Peuto and Febvre 1986, Laval-Peuto *et al.* 1986, unpublished results, Stoecker and Silver 1987, Stoecker *et al.* in press,). It has been postulated that these plastids could remain functional, owing to their excellent state of preservation; thus they are considered to be 'symbiotic'. With epifluorescence microscopy Laval-Peuto and Rassoulzadegan (1988) demonstrated that this situation occurs in about 40% of the species of Oligotrichina in mediterranean coastal areas, all belonging to the family Strombidiidae, and proposed to name them 'plastidic ciliates'.

The plastids observed do not belong to complete symbiotic algae. Isolated, or in small groups at the periphery of the ciliate, they appear to come mostly from unicellular chromophyte algae (Figure 3). They are always surrounded by three membranes: two plastidial membranes and a periplastidial membrane derived from the vesicular endoplasmic reticulum (VER), which is well developed in these ciliates. Symbiotic plastids are rarely seen degenerating, and do not divide. They do not appear to be genetically integrated in the ciliates, like in euglenids and dinoflagellates which also have plastids bound by three membranes. Their obligatory maintenance has been demonstrated for one species (Stoecker *et al.* 1988), but the way they are selected, tolerated, maintained in activity and renewed remains to be studied. Heterotrophy is maintained. The digestive vacuoles in plastidic ciliates contain various algal and bacterial prey in the course of digestion. They are centrally located and surrounded by a continuous network of dense endoplasmic reticulum (DER). Thus the digestive system and the plastids are juxtaposed but in totally separated cell compartments. Acid phosphatase activity can be demonstrated inside digestive vacuoles. But neighbouring plastids are never marked by the technique, which demonstrates that they are not being digested.

Chlorophyll pigments in plastids within both intact algae and plastidic ciliates are autofluorescent, when they are excited by short wavelength blue light in epifluorescence microscopy. This capability was used to demonstrate the presence of photosynthetic pigments in an atlantic plastidic oligotrich (McManus and Fuhrman 1986), to estimate the biomass of plastidic ciliates in the Atlantic (Stoecker *et al.* 1987), and to record the Mediterranean plastidic species (Laval-Peuto and Rassoulzadegan 1988). The intensity of the fluorescence suggests that the plastids remain photosynthetically active. This is confirmed by recent publications on the photosynthetic assimilation of inorganic carbon by plastidic ciliates (Jonsson 1987, Stoecker *et al.* 1987). During the workshop, a separate demonstration of chlorophyll autofluorescence in plastidic planktonic ciliates was given using epifluorescence microscopy.

Polysaccharide plates, unusual among the Ciliophora, lie at the base of the ciliate pellicle (Figure 3). Their nature has been analysed by various cytochemical techniques and control tests

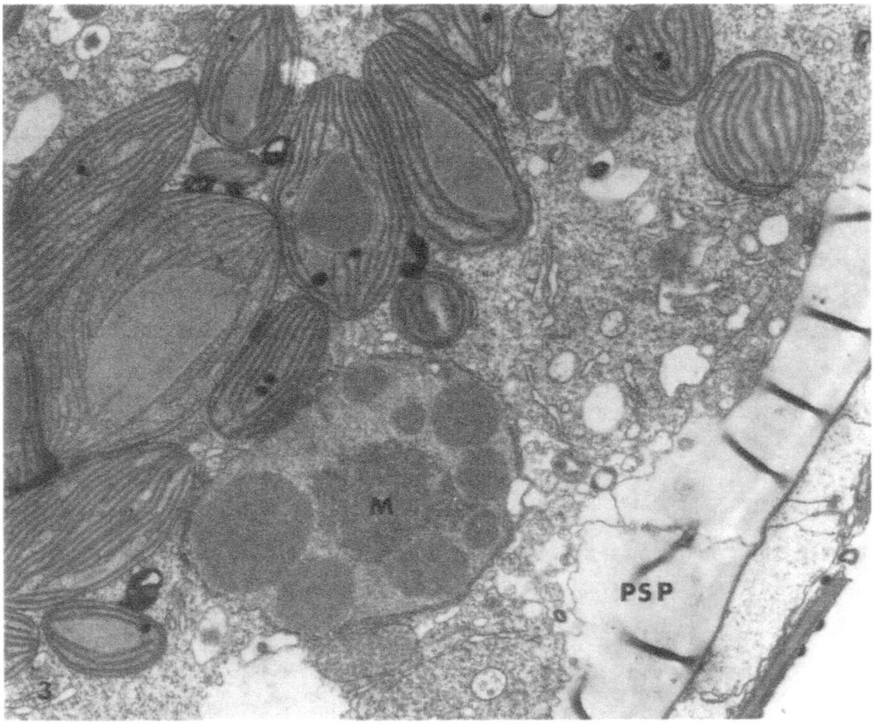

Fig. 3 (from Laval-Peuto). Plastid symbiosis in the oligotrich *Tontonia appendiculariformis*. Various plastids lie in the cytoplasm, close to the ciliate macronucleus (M), under cortical polysaccharide plates (PSP)

(Laval-Peuto *et al.* submitted). It is likely that these polysaccharides are formed by the plastids, just as they were formed originally in the chromophyte algae. They might represent a metabolic reserve.

In conclusion, the plastidic Strombidiidae ciliates are mixotrophic protists owing to their double heterotrophic and autotrophic activity. On account of the abundance of their biomass in the microplankton, their role in marine food webs should henceforth be considered. Both cytochemical and biochemical data are needed in the future.

Malte Elbrächter presented a video demonstration of the endosymbionts of oligotrich ciliates, a superb illustration which complemented both previous lectures. *Laboea strobila* is a species of oligotrich ciliates, widely distributed in coastal and shelf/slope waters. Live *L. strobila* from the North Sea, were shown in succession under transmitted and epifluorescence microscopy. First it appeared yellow-green, then, when excited by blue light it fluoresced orange, a bright colour which faded progressively after one minute. Results from Stoecker *et al* (1988) and Laval-Peuto (unpublished) demonstrate the presence of symbiotic plastids in this species. **Malte Elbrächter** suggested that the natural orange fluorescence may come from cryptophycean symbionts.

John Lee outlined recent results on chloroplast-retention in Elphiid Foraminifera. Freshly collected specimens of the foraminifer *Elphidium crispum* from Eilat (Israel) and Mombasa (Kenya) contain in their cytoplasm plastids which are probably derived from partially digested algae. Some vacuoles containing plastids also had additional algal remnants, e.g. pyrenoids, mitochondria and occasionally nuclei. Measurements of carbon uptake with ^{14}C tracer methods suggest that these plastids are still capable of carbon fixation, at the same rate as some larger species with intact algal endosymbionts. Fine structural studies of the cytoplasm of many specimens indicate that, even though the plastids may function as 'symbionts', they are slowly but actively being digested by the host (Lee *et al.* 1988). Preliminary experiments with *Elphidium incertum*, from salt marshes in the USA, indicate some selectivity in plastid retention. *E. incertum* fed on a diet of two chlorophytes retained less than 23 chloroplasts per average foraminifera. Those fed diatoms retained between 32-65 plastids. A high mortality rate was experienced among those foraminifera fed only chlorophytes during a five-week experiment.

Elphiids are morphologically complex foraminifera with a well-developed canal system. The apertures are reduced to the size of fossettes. Very small tooth-like projections (tuberculate borders of septal pits) seem to bar the entry of intact rigid food, such as diatom frustules, into the cell; but they could let cell fragments pass through. The authors speculate that Elphiids might be morphologically highly adapted for plastid husbandry even though they cannot yet interpret all aspects of the specialization.

DISCUSSION AND CONCLUSION

A knowledge of symbiosis, and particularly of endosymbioses, is essential to understand the role of protists in marine processes. Planktonic protists appear to develop a larger variety of symbioses than benthic protists. It is important to consider symbiosis among all the protists and not only among the protozoa. Due to the increasing efforts of Corliss and Patterson, it is now clear that the protozoa and protophytes have no real frontier between them and that they represent a single evolving group, the protists. Many symbiotic associations established between protozoa, unicellular algae, cyanobacteria and/or bacteria, bear witness to this concept. The recent theories of evolution, explaining the creation of eucaryotic cells, are based on the occurence of successive endosymbiotic associations which became permanent (Margulis 1981, Cavalier-Smith 1987, Schwemmler and Schenk 1980). So endosymbiosis appears to be a general biological phenomenon (Lee *et al.* 1985, Reisser *et al.* 1985, Schenk and Schwemmler 1983) and it represents attempts during evolution, with a diversity of adaptations between various partners, which allowed the survival of species in the environment and/or the creation of new, sometimes more efficient, physiological units.

Endosymbioses between algae and protozoa are better known from freshwater habitats than from the marine environment. Although the algal symbionts of marine foraminifera and radiolaria have received considerable study, much less is known about marine ciliates, heliozoa and acantharia. Their possible symbioses with algae or cyanobacteria need to be identified and studied biologically and physiologically. Recent results in the field show that the associations are variable in stability, from temporary to permanent hereditary states. A given protozoan species may harbour different species of algae with a large diversity of host-partner interactions. The balance between the partners, from mutualism to parasitism, is an important point but not easy to define.

Bacteria in the marine food web are most often considered to represent food or parasites, but they also play other roles through symbiotic associations. Electron microscopic studies demonstrate their frequency within the cytoplasm or the nuclei of numerous marine protists. Are these bacterial endobionts mutualistic in their relationships or mostly parasites? More physiological studies are needed.

Plastid symbiosis in the protozoa offers an excellent subject for consideration. Discovered in marine molluscs, where its function was studied in detail (Trench 1975, 1980, Muscatine *et al.* 1983), this peculiar association also named chloroplast maintenance was first described among marine protists in some ciliates (Blackbourn *et al.* 1973). Taylor (1982) stressed that it could be more common and correspond to some of the pigmented bodies described by Kahl (1935). Today we know that this phenomenon, hidden up to now, is not rare; it is represented within all the classes of free-living protozoa, in marine and freshwater habitats. Some scientists do not accept the expression 'plastid symbiosis', because plastids are only organelles. But if one considers that plastids are partially independant organelles, which might have evolved from cyanobacterial endobionts, that their biochemical components are peculiar and that their capabilities of survival outside the cell are long *in vitro* as well as in the sea (Gieskes and Elbrächter 1986), this phenomenon has wider implications than the simple maintenance of a foreign organelle. One needs also to remember that symbioses are not always mutualistic. Plastid symbiosis comprises selection, recognition and maintenance as well as many symbioses. In many cases described above the association seems obligatory and essential to the temporary survival of both 'partners'.

The notion of mixotrophy as defined by Pfeffer (1881) has been recently applied to phagotrophic phytoflagellates (Gaines and Elbrächter 1987, Sanders and Porter 1988). It is also valid for the protozoa in symbiosis with plastids. Plastidic ciliates, heliozoa and foraminifera are at the same time heterotrophic by their digestive vacuoles and autotrophic by their symbiotic plastids. The physiological and ecological importance of both simultaneous activities within the marine food web is worth studying.

Dinoflagellates are unique in comprising true protophytes and true protozoa. Some are able to feed on larger prey than themselves by a pseudopodal mechanism (Jacobson and Anderson 1986), others are parasites. In the phototrophic species plastids are genetically integrated, but they are unusual in having three membranes. Since they also present large differences in their morphologies, they are supposed to have evolved from various algal symbionts, more or less reduced to the plastid itself. This diversified group of protists is important for the study of both symbiosis and evolution.

The above comments lead one to consider symbiosis in a more general context. Though remarkable, endosymbiosis may be understood simply as a specialized, sophisticated case of trophic behaviour. Parasitism, with its fatal end for one of the partners, is typically a mode of nutrition for predators smaller than their prey, promoting their growth in the host (see bacteria, but also dinoflagellates). Plastid symbiosis may also be interpreted as either a parasitic or a trophic relationship: plastids are exploited or 'enslaved' by a large predator and their production used up to their death by rejection or perhaps digestion by the host cell. But can we conclude that mutualism, with its harmonious exchanges between both partners, might be a superior form of a 'struggle for life'?

In conclusion, a need to inventory all types of symbioses and to develop their study was identified. Within this large field, it may be more urgent to focus on the diversity of bacterial endobionts, on the extent and impact of plastid symbiosis, and finally on the different aspects of parasitism in the marine environment.

REFERENCES

Blackbourn DJ, Taylor FJR, Blackbourn J (1973) Foreign organelle retention by ciliates. J Protozool 20:286-288

Cavalier-Smith T (1987) The simultaneous origin of mitochondria, chloroplasts and microbodies. In: Lee JJ, Fredericks J (eds) Endocytobiology III. Ann NY Acad Sci 503:55-71

Corliss JO (1985) Concept definition prevalence and host-interactions of xenosomes (cytoplasmic and nuclear endosymbionts). J Protozool 32:373-376

de Bary A (1879) Die Erscheinung der Symbiose. Naturforschung Versammlung Cassel, LI p 121

Gaines G, Elbrächter M (1987) Heterotrophic nutrition. In Taylor FJR The Biology of dinoflagellates. Blackwell Sci Publ:224-269

Gieskes WWC, Elbrächter M (1986) Abundance of nanoplankton-size chlorophyll containing particles caused by diatom disruption in surface waters of the Southern Ocean (Antarctic Penninsula region).Netherl J Sea Res 20:291-303

Görtz H-D (1986). Int rev Cytol 102:167-213

Görtz H-D, Wiemann M (1989) Route of infection of the bacteria *Holospora elegens* and *Holospora obtusa* into the nuclei of *Paramecium caudatum*. Eur J Protistol 24:101-109

Görtz H-D, Freiburg M, Wiemann M (1988) (in press). Endocyt Cell Res

Henry SM (1966) Symbiosis, 1. Academic Press NY p 478

Jonsson PR (1987) Photosynthetic assimilation of inorganic carbon in marine oligotrich ciliates (Ciliophora Oligotrichina). Mar microb Food Webs 2:55-67

Kuile BH ter, Erez J, Lee JJ (1987) The role of feeding in the metabolism of larger symbiont bearing foraminifera. Symbiosis 4:335-350

Laval-Peuto M, Febvre M (1986) On plastid symbiosis in *Tontonia appendiculariformis* (Ciliophora Oligotrichina). BioSystems 19:137-158

Laval-Peuto M, Rassoulzadegan F (1988) Autofluorescence of marine planktonic Oligotrichina and other ciliates. Hydrobiologia 159: 99-110

Laval-Peuto M, Salvano P, Gayol P, Greuet C (1986) Mixotrophy in planktonic ciliates: ultrastructural study of *Tontonia appendiculariformis*. Mar microb Food Webs 1:81-104

Laval-Peuto M, Salvano P, Gayol P, Greuet C (submitted) Cytochemistry of ciliates in symbiosis with plastids

Lee JJ, Erez J, Kuile BH ter, Lagziel A, Burgos S (submitted) Feeding rates of two species of larger foraminifera *Amphistegina lobifera* and *Amphisorus hemprichii* from the Gulf of Elat (Red Sea)

Lee JJ, Hallock P (1987) Algal symbiosis as the driving force in the evolution of larger foraminifera. In: Lee JJ, Fredericks J (eds) Endocytobiology III. Ann NY Acad Sci 503:330-347

Lee JJ, Lanners E, Kuile B ter (1988) The retention of chloroplasts by the foraminifer *Elphidium crispum*. Symbiosis 5:45-68

Lee JJ, Lee MJ, Weis DS (1985) Possible adaptative value of endosymbionts to their protozoan hosts. J Protozool 32:380-382

Lee JJ, Sang K, Strauss E, Lee PJ (submitted) Culture of two species of symbiont-bearing larger foraminifera *Amphistegina lobifera* and *Amphisorus hemprichii*, in the laboratory

Lee JJ, Soldo AT, Reisser W, Lee MJ, Jeon KW, Görtz H-D (1985) The extent of algal and bacterial endosymbioses in protozoa. J Protozool 32:391-403

Lewis D H (1973) Concepts in Fungal Nutrition and the Origin of Biotropy. Biol Rev 48 : 261-278
Lewis D H (1985) Symbiosis and mutualism : crisp concepts and soggy semantics in Boucher D H The biology of mutualism : Ecology and Evolution, Croom-Helm, London : 29-39
Lopez E (1979) Algal chloroplasts in the protoplasm of three species of benthic foraminifera: Taxonomic affinity viability and persistence. Mar Biol 53:201-211
Margulis L (1981) Symbiosis in cell evolution. Freeman, San Francisco
McManus GB, Fuhrman JA (1986) Photosynthetic pigments in the ciliate *Laboea strobila* from Long Island Sound USA. J Plankton Res 8:317-327
Muscatine L, Falkovski PG, Dubinsky Z (1983) Carbon budgets in symbiotic associations In: Schenk HEA, Schwemmler W (eds) Endocytobiology II. de Gruyter, Berlin, :649-658
Ossipov DJ (1981) Problems of nuclear heteromorphism of the unicellular organisms. NAUKA, Leningrad
Ossipov DV, Podlipaev SA (1977) Acta Protozool 16:289-308
Patterson DJ, Dürrschmidt M (1987) Selective retention of chloroplasts by algivorous heliozoa : Fortuitous chloroplast symbiosis? Euro J Protist 23:51-55
Pfeffer W (1881) Handbuch der Pflanzenphysiologie. Wilhelm Engelmann, Leipzig
Podlipaev SA, Ossipov DV (1979) Acta Protozool 18:465-480
Reisser W (1986) Endosymbiotic associations of freshwater protozoa and algae. Progr Protistol 1:195-214
Reisser W (1987) Studies on the ecophysiology of endocytobiotic associations of ciliates and algae. II. Potential features of adaptation of symbiotic and free-living *Chlorella* sp to the endocytobiotic habitat formed by *Paramecium bursaria*. Endocyt Cell Res 4:317-329
Reisser W, Herbig E (1987) Studies on the ecophysiology of endocytobiotic associations of ciliates and algae. I; Carbohydrate budgets of green and alga-free *Paramecium bursaria* under laboratory and natural growth conditions. Endocyt Cell Res 4:305-316
Reisser W, Meier R, Görtz H-D, Jeon KW (1985) Establishment maintenance and integration mechanisms of endosymbionts in Protozoa. J Protozool 32:383-390
Reisser W, Radunz A, Wiessner W (1982) The participation of algal surface structures in the cell recognition process during infection of aposymbiotic *Paramecium bursaria* with symbiotic chlorellae. Cytobios 33:39-50
Reisser W, Vietze S, Widowski M (in press) Taxonomic studies on endosymbiotic chlorophycean algae isolated from different american and european strains of *Paramecium bursaria*. Symbiosis
Sanders RW, Porter KG (1988) Phagotrophic phytoflagellates. Adv Microb Ecol 10:167-192
Schenk HEA, Schwemmler W (1983) Endocytobiology II. Intracellular space as oligogenetic ecosystem. W de Gruyter, Berlin, p 1070
Schmidt HJ, Feriburg M, Görtz H-D (1987) Comparison of the infectious forms of two bacterial endonucleobionts *Holospora elegans* and *H obtusa* from the ciliate *Paramecium caudatum*. Microbios 49:189-197
Schnepf E, Elbrächter M (1988) Cryptophycean-like double membrane bound chloroplast in the dinoflagellate *Dinophysis ehrenb*: evolutionary phylogenetic and toxicological implications. Botanica Acta 101:196-203
Schwemmler W, Schenk HEA (1980) Endocytobiology I. Endosymbiosis and cell biology. A synthesis of recent research. W de Gruyter, Berlin, p 1060
Smith DC, Douglas AE (1987) The biology of Symbiosis (Contemporary Biology), Arnold, London :1-302
Soldo AT (1983) The biology of the xenosome an intracellular symbiont. Int rev Cytol S 14:79-109
Soldo AT (1988) The ciliated protozoan and its guests. BioEssays 4:86-90
Starr M B (1975) A Generalized Scheme for Classifying Organismic Associations. Symposia of the Society for Experimental Biology 29 : 1-20
Stoecker DK, Michaels AE, Davis LH (1987) A large fraction of marine planktonic ciliates can contain functional chloroplasts. Nature 326:790-792

Stoecker DK, Silver MW (1987) Chloroplast retention by marine planktonic ciliates. Ann New York Acad Sci 503:562-565
Stoecker DK, Silver MW, Michaels AE, Davis LH (1988) Obligate mixotrophy in *Laboea strobila* a ciliate which retains chloroplasts. Mar Biol 99:415-423
Stoecker DK, Silver MW, Michaels AE, Davis LH (in press) Enslavement of algal chloroplasts by four *Strombidium* sp (Ciliophora Oligotrichida). Mar Micr Food Webs
Taylor FJR (1982) Symbioses in marine microplankton. Ann Inst océanogr Paris 58(S):61-90
Trench RK (1975) Of "leaves that crawl": Functional chloroplasts in animal cells. Soc Exp Biol 29:229-265
Trench RK (1980) Uptake retention and function of chloroplasts in animal cells In: Schwemmler W, Schenk HEA, Endocyto biology. W de Gruyter, Berlin :703-728

MIXOTROPHY IN MARINE PLANKTONIC CILIATES: PHYSIOLOGICAL AND ECOLOGICAL ASPECTS OF PLASTID-RETENTION BY OLIGOTRICHS

Diane K Stoecker
Biology Department
Woods Hole Oceanographic Institution
Woods Hole
Massachusetts
USA

INTRODUCTION

Although oligotrichous ciliates (subclass Choreotrichia, order Choreotrichida) are often regarded as strict phagotrophs, deriving their nutrition from the ingestion of other cells, many species contain pigmented bodies which have long been thought to be residues of digestion of algal cells and only recently have been shown to be algal chloroplasts (Kahl 1932, Burkholder *et al.* 1967, Taylor 1982, McManus and Fuhrman 1986). Transmission electron microscopy has demonstrated that many marine oligotrichs retain isolated chloroplasts derived from phytoplankton (Blackbourn *et al.* 1973, Laval-Peuto and Febvre 1986, Jonsson 1987, Stoecker and Silver 1987). A substantial proportion of the planktonic oligotrich fauna contains chloroplasts and is probably mixotrophic, deriving nutrition from both phagocytosis and photosynthesis (Laval-Peuto *et al.* 1986, Stoecker *et al.* 1987, Laval-Peuto and Rassoulzadegan 1988).

These data suggest that the role of planktonic oligotrichs in marine processes is more varied and complex than we once thought. Physiological and ecological data are needed to understand the role of plastidic ciliates in trophodynamics, particle transformation, and nutrient cycling in the oceans. In addition, chloroplast-retention in oligotrichs, although this phenomenon is not unique to this taxon (Trench 1975, Lopez 1979, Patterson and Dürrschmidt 1987), is a model system in which organelle-cell interactions can be experimentally investigated from ecological, cellular and molecular viewpoints. Presently, our interest in, and knowledge of, plastidic ciliates is increasing rapidly. In the

NATO ASI Series, Vol. G 25
Protozoa and Their Role in Marine Processes
Edited by P. C. Reid et al.

following sections, I will describe the current status of our knowledge of plastidic oligotrichous ciliates, discuss their possible roles in marine processes, and when possible, suggest potentially fruitful areas for future research.

FREQUENCY AND ABUNDANCE

Various criteria have been used to identify plastidic ciliates. Epifluorescence microscopy is a convenient and reliable technique for categorizing ciliates as potentially photosynthetic (due to isolated chloroplasts or algal endosymbionts) or strictly heterotrophic (McManus and Fuhrman 1986, Stoecker *et al.* 1987, Laval-Peuto and Rassoulzadegan 1988). Based on light microscopy alone, it is difficult to distinguish between small algal endosymbionts and isolated algal plastids. With electron microscopy, isolated algal chloroplasts have been observed in *Laboea strobila* (Blackbourn *et al.* 1973, Jonsson 1987, Stoecker *et al.* 1988), several *Strombidium* species (Blackbourn *et al* 1973, Stoecker and Silver 1987, Stoecker *et al.* submitted), and in *Tontonia appendiculariformis* (Laval-Peuto and Febvre 1986, Laval-Peuto *et al.* 1986). Based on light microscopy, some marine and freshwater oligotrich species have been reported to contain whole algal cells (Burkholder *et al.* 1967, Hecky and Kling 1981, Jonsson 1987). The freshwater oligotrich, *Strombidium viride*, had been thought to contain zoochlorellae (Hecky and Kling 1981) but recently transmission electron micrographs have revealed that isolated chloroplasts are present rather than algal cells (Rogerson *et al.* in press). Algal endosymbiosis and plastid-retention are physiologically and ecologically quite different and transmission electron microscopy should be used, whenever possible, to differentiate between them.

The term 'plastidic' was proposed by Laval-Peuto and Rassoulzadegan (1988) for all ciliates that contain intact plastids, either isolated or in algal endosymbionts. I use this definition because most studies employ epifluorescence microscopy and thus cannot always distinguish between small algal cells and isolated plastids. Laval-Peuto and Rassoulzadegan (1988) report that 41% of the oligotrichous ciliate species in nearshore Mediterranean waters are plastidic and that the presence of plastids appears to be limited to the family Strombidiidae which includes the genera *Laboea, Strombidium* and *Tontonia* (Fauré-Fremiet 1924, 1969, Small and Lynn 1985). Members of this family usually dominate the marine planktonic ciliate fauna (Sorokin 1981, Montagnes 1986). Although many members of the Strombidiidae are strictly heterotrophic, some species are consistently plastidic (Table 1). The evidence for the presence of plastids is contradictory for *S. vestitum* and *S. conicum* (Table 1), but this may reflect a taxonomic problem {Oligotrichs can be difficult to identify to species (Laval-Peuto and Rassoulzadegan 1988, Montagnes *et al.* 1988)}, rather than the presence of plastidic and non-plastidic forms in the same species. Questions of this type should be resolved by combining taxonomic studies, using modern

Table 1 MARINE PLANKTONIC OLIGOTRICH SPECIES REPORTED TO CONTAIN PLASTIDS OR ALGAL ENDOSYMBIONTS

Species	**References**
Laboea strobila Lohmann	2-3, 6, 8, 10
Several Unidentified *Strombidium* spp	1, 2, 6, 8, 10
Strombidium capitatum (Leegard) Busch	10, 11, 12
Strombidium reticulatum (Leegard) Kahl	7, 8, 10
Strombidium elegans (Florentin)	10
Strombidium vestitum (Leegard) Kahl	10 (but see 2, 8)
Strombidium delicatissimum (Leegard)	10
Strombidium crassulum (Leegard) Kahl	9
Strombidium reticulatum (Leegard)	7, 8, 10
Strombidium conicum (Lohmann)	8, 11 (but see 2+10)
Strombidium acutum (Leegard) Kahl	11, 12
Strombidium chlorophilum n. sp.	11, 12
Unidentified *Tontonia* sp.	6
Tontonia appendiculariformis Fauré-Fremiet	4, 5, 10
Tontonia ovalis (Leegard) Laval-Peuto and Rassoulzadegan	10
Tontonia gracillima Fauré-Fremiet	10

1. Burkholder *et al.* 1967, 2. Blackbourn *et al.* 1973, 3. McManus and Fuhrman 1986, 4. Laval-Peuto and Febvre 1986, 5. Laval-Peuto *et al.* 1986, 6. Stoecker *et al.* 1987, 7. Dale and Dahl 1987, 8. Jonsson 1987, 9. Reid 1987, 10. Laval-Peuto and Rassoulzadegan 1988, 11. Montagnes *et al.* 1988, 12. Stoecker *et al.* submitted.

techniques, with studies of chloroplast-retention in ciliates collected from different geographic areas/seasons.

Oligotrichs with isolated algal chloroplasts are documented to occur in Pacific (Blackbourn *et al.* 1973), Atlantic (Jonsson 1987, Stoecker *et al.* 1987) and Mediterranean waters (Laval-Peuto and Rassoulzadegan 1988). They probably occur in other areas as well, but data are lacking. In open waters, the tintinnids and oligotrichs usually comprise over 90% of the microplanktonic ciliate fauna in the upper water column and thus are the dominant ciliates (Rassoulzadegan 1977). Although the microplankton in neritic and shelf/slope waters differ from that in subtropical oceanic waters, plastidic oligotrichs contribute an important proportion of the ciliate

Table 2. PERCENT CONTRIBUTION OF PLASTIDIC OLIGOTRICHS TO THE CILIATE FAUNA OF SURFACE WATERS

Neritic Waters**		
Woods Hole	Spring and Summer	47-51% (1)*
MA USA	Fall and Winter	22% (1)*
Bay of Villefranche -sur-Mer, France	Oct - Feb	41% of Oligotrichina sp. (2)
Shelf/Slope Waters		
Georges Bank NW Atlantic	Early Summer	39% (3)*
Shelf, NW Atlantic	Late Spring	48% (4)*
Oceanic Waters		
Sargasso Sea	Late Spring	37% (4)*
Gulf Stream	Late Spring	25% (4)*

1. Stoecker *et al.* 1987, 2. Laval-Peuto and Rassoulzadegan 1988, 3. Stoecker *et al.* in press, 4. Unpub. data, *Tintinnids and Oligotrichs only, **Blooms of pigmented oligotrichs can occur in coastal waters (Blackbourn *et al.* 1973, Reid 1987, Dale and Dahl 1987), at these times the % contribution of plastidic ciliates may be much greater.

fauna in both assemblages (Table 2). During the spring and summer, plastidic ciliates can comprise 25-51% of the tintinnid plus oligotrich fauna (Table 2). In neritic and shelf/slope waters of the Northwest Atlantic, microplanktonic plastidic oligotrichs occur at average densities of around 10^{-3} L^{-1} in the upper water column (Table 3). *Laboea strobila*, because of its relatively large size and whorls (Montagnes *et al.* 1988), is the most conspicuous plastidic oligotrich in temperate waters. It is also one of the most important, often contributing 40% or more of the total biomass of plastidic oligotrichs (Stoecker *et al.* 1987, Stoecker *et al.* in press). This species usually occurs at densities of ~10^2-10^3 cells L^{-1} in temperate coastal waters during the spring and summer (Table 3). Two plastidic *Strombidium* species, *S. reticulatum* and *S. crassulum*, occasionally bloom in coastal waters, with peak densities of over 10^3 L^{-1} (Table 3). At times, plastidic

Table 3 ESTIMATED ABUNDANCE OF PLASTIDIC OLIGOTRICHS

Species-specific estimates	**Season**	**Cells l^{-1}**
Laboea strobila	Mid-June (top 10m)	<10^{-2} -10^3(1)
" "	April (surface waters)	$1.1x10^3$(2)
" "	Early Summer	~$1.3x10^2$(3)
Strombidium reticulatum	May (peak density surface waters)	$1.2x10^6$(4)
" "	May (top 10 m)	$0.4-13.7x10^3$(4)
Strombidium crassulum	April & May (top 10 m)	$7.5x10^2$-$9x10^3$(5)
Strombidium reticulatum & s cf conicum combined	April (surface waters)	$8.1x10^3$(2)
Community estimates	**Season**	**Cells l^{-1}**
ciliates >20 µm	Yearly average (surface waters)	$1.2x10^3$ (6)
"	June (peak density surface waters)	$3.6x10^3$ (6)
ciliates >20 µm	Early Summer (surface to depth of 1% light level)	~$1.0x10^3$ (3)

1. McManus & Fuhrman 1986, 2. Jonsson 1987, 3. Stoecker *et al.* in press, 4. Dale and Dahl 1987, 5. Reid 1987, 6. Stoecker *et al.* 1987.

oligotrichs can reach high enough densities to visibly discolor surface waters (Blackbourn *et al.* 1973, Dale and Dahl 1987).

Although we now know that plastidic oligotrichs can be an important component of the planktonic ciliate fauna, we still know little about the factors that control their temporal and spatial distributions. In temperate waters these ciliates seem to be more abundant in summer than in winter (Stoecker *et al.* 1987). In general, planktonic protozoa with algal endosymbionts (with the exception of *Myrionecta rubra* = *Mesodinium rubrum*) are most common in warm oligotrophic oceans (Taylor 1982, Lee *et al.* 1985). Plastidic oligotrichs are common in nearshore eutrophic or mesotrophic waters as well as in more oligotrophic waters (Table 2). Plastidic ciliates are usually only abundant in the euphotic zone and often have diel vertical distribution patterns that would tend to maximize photosynthesis and minimize photoinhibition (McManus and Fuhrman 1986, Stoecker *et al.* 1987, Dale 1987, Stoecker *et al.* in press). In culture, plastidic oligotrichs are not phototaxic and the mechanisms by which they regulate their vertical distribution are not understood. Water column structure and frontal boundaries are known to have important influences on the species composition and size distribution of phytoplankton. It would be interesting to know if these physical features have similar influences on plastidic ciliate assemblages.

CHLOROPLAST SPECIFICITY AND TURNOVER

An important physiological and ecological question is whether plastidic ciliates specifically retain only certain types of chloroplasts. Marine phytoplankton contain a great diversity of chloroplasts that differ in their structure and molecular biology as well as in their pigment composition. Chromophytic chloroplasts (those with chlorophylls *a* and *c*, including the chloroplasts of diatoms, chrysophytes, prymnesiophytes, cryptophytes, and of most dinoflagellates) are surrounded by 3-4 membranes whereas chlorophytic chloroplasts (those with chlorophylls *a* and *b*, which includes chloroplasts of prasinophytes and chlorophytes) are surrounded by 2 membranes. In the chlorophytes, photosynthate is stored as starch within the chloroplast, but in chromophytes, photosynthate is usually transported into the cytoplasm for storage except in the cryptophytes where starch is stored in the periplastidial compartment (reviewed in Dodge 1973). Recent data indicate that chromophyte chloroplasts may be somewhat more 'independent' of the plant cell nucleus than chlorophyte chloroplasts; both subunits of ribulose biphosphate carboxylase (the main dark reaction enzyme) are coded for by chloroplast DNA in a chromophyte but in chlorophytes, one subunit is coded for by the plant cell nucleus (Reith and Cattolico 1985 a, b). These differences could be expected to influence recognition of chloroplasts by ciliates, survival and continuing function of these organelles in the hosts' cytoplasm, and availability of photosynthate to ciliates.

Most plastidic oligotrichs retain chloroplasts derived from a wide range of algal taxa (Blackbourn *et al.* 1973, Laval-Peuto *et al.* 1986, Jonsson 1987, Stoecker *et al.* 1988, Stoecker *et al.* submitted) but at least one *Strombidium* sp. either selectively ingests cryptophytes or selectively retains cryptophyte chloroplasts (Stoecker and Silver 1987). Some *Strombidium* spp. isolated from estuarine waters have a preference for chlorophytic chloroplasts (Stoecker *et al.* submitted; also see Jonsson 1987).

The species composition of the phytoplankton may have an important influence on the population dynamics of ciliates that retain chloroplasts and on their complement of chloroplasts.

In order to understand these phenomena, we need to understand the factors that control chloroplast specificity. Selective feeding, selective uptake of chloroplasts from ingested algae, digestion of certain chloroplast types, and differences in the survival of chloroplasts in the ciliate cytoplasm may all contribute to the specificity which is observed. Collaboration between plant cell biologists and protozoologists will be necessary to answer these basic questions about chloroplast uptake and retention.

There are several reasons for expecting that an isolated chloroplast would not survive for long nor divide in a ciliate, these are: 1. *In vitro*, higher plant chloroplasts rapidly lose their ability to fix inorganic carbon and to synthesize proteins (reviewed in Leech 1980); 2. Nuclear genes within algae code for many chloroplast components and are involved in the regulation of chloroplast function (reviewed in Steinback *et al*. 1985). At least one chloroplast protein, the Q_B protein, is rapidly degraded in the light during photosynthetic electron transport (Mattoo *et al*. 1986); and 3. The polymerases necessary for the replication of chloroplast DNA are coded for by nuclear genes (reviewed in Ellis 1985). When suitable algae are available as a source of chloroplasts, chloroplast turnover occurs within less than a day in *Strombidium capitatum* (Stoecker and Silver unpub. data). However, isolated intact chloroplasts can survive at physiological temperatures for at least 14 days in plastidic *Strombidium* sp. (Stoecker and Silver 1987) and for 6 days or more in *Laboea strobila* (Stoecker *et al*. 1988). Prolonged survival of algal chloroplasts also occurs in sarcoglossan molluscs (reviewed in Trench 1980) and in benthic foraminifera (Lopez 1979). We need to better understand the mechanisms which allow chloroplasts to remain functional when isolated for extended periods of time from the algal nucleus. Furthermore, we need to know how environmental factors such as photon flux, plastid availability and nutrition affect the turnover rates of plastids in ciliates. It is possible that chemical reactor models which have been used to describe food processing in animal guts (Penry and Jumars 1987) can be adapted for prediction of optimal chloroplast turnover times in ciliates. Models of this type would be particularly useful for evaluating the role of ciliates in pigment budgets and their effects on bio-optical properties in the oceans.

PHOTOSYNTHESIS

Measurement of rates of photosynthesis in plastidic ciliates with ^{14}C presents several problems and standard procedures for measuring productivity (Parsons *et al.* 1984) cannot be used without modification. Because most ciliates are phagotrophic, they must be separated from algae before addition of label (Stoecker *et al.* 1987, Jonsson 1987). This is a tedious process and involves picking and washing single cells. However, with high specific activity ^{14}C, measurements can be made using a few cells (Rivkin and Seliger 1981). Second, because most oligotrichs are very fragile (Gifford 1985), filtration and rinsing at the end of incubations (Goldman and Dennett 1985) probably results in the loss of fixed carbon. Fixation makes ciliates less fragile, but also can result in the loss of radioisotope (Silver and Davoll 1978). Filtration and rinsing can be eliminated by acidifying whole samples in scintillation vials after incubation and evaporating them to dryness (for procedure, refer to Li and Goldman 1981). Because

Table 4 AVERAGE CHLOROPHYLL *a* CONTENT (pg chl *a* cell^{-1}) AND RATE OF PHOTOSYNTHESIS (pg C fixed cell^{-1}h^{-1}) IN PLASTIDIC OLIGOTRICHINA

Species	pg chl *a*. cell^{-1}	pg C fixed cell^{-1}h^{-1}
Laboea strobila(1-3)	-	41
	187	586
	248	925 P_{max}
Strombidium reticulatum(1)	-	31
Strombidium conicum(1,4)	-	12
	15	60
Strombidium capitatum(4)	49	78
Strombidium acutum(4)	50	35
Strombidium chlorophilum(4)	108	98
Strombidium vestitum(1)	-	0.1 (n.s)

(1) Jonsson 1987 (Incubation: 24h; 190 µE m^{-2}s^{-1})
(2) Stoecker *et al.* 1987 (Incubation: 6h, 150-200 µE m^{-2}s^{-1})
(3) Stoecker *et al.* 1988 (Incubation: 2h; 850 µE m^{-2}s^{-1})
(4) Stoecker *et al.* submitted (Incubation: 2h; 400-500 µE m^{-2}s^{-1})

plastids or endosymbionts may preferentially take up metabolic carbon dioxide produced by the host, estimates of photosynthesis in these systems using ^{14}C methodology must be regarded as minimum estimates.

Photosynthesis has been documented in *Laboea strobila* and six plastidic *Strombidium* spp. (Table 4). One species which is reported to have plastids, *S. vestitum*, may not be photosynthetic (Table 4). Comparison of rates of photosynthesis among investigations (Table 4) is difficult because variation in incubation conditions strongly influences the measurement. For the plastidic oligotrichs, ^{14}C uptake rates usually remain linear for 4-6 h when ciliates have been separated from their algal foods (Stoecker *et al.* 1987), longer incubations can result in a reduction in uptake rate (Stoecker unpub. data). After being deprived of algal prey for several hours, plastidic ciliates appear to respire recently fixed carbon rapidly (Taniguchi and Stoecker unpub. data), which may result in a reduction in the net rate of uptake during longer incubations. It is also possible that rates of photosynthesis decline in ciliates deprived of a source of new plastids. However the initial linearity of ^{14}C uptake suggests that rates from short-term incubations closely mimic the rates which occur in natural populations and in culture with food present.

Plastidic ciliates have photosynthesis vs. irradiance curves similar in shape to those found in phytoplankton (Stoecker *et al.* 1988). Light-saturated chlorophyll specific rates of photosynthesis (P_{max}) in plastidic ciliates have been observed to range from about 0.7 to 4.0 a similar range to phytoplankton (Table 5). P_{max} decreases in nutrient-limited phytoplankton (stationary phase cultures in Table 5) and is often higher in high-light adapted phytoplankton than in low-light adapted phytoplankton. In ciliates, the nutritional state of the ciliate cell as well as the previous history of the retained chloroplasts may affect photosynthetic rates. Experimental investigation of the physical, chemical and biological factors which determine rates of photosynthesis in plastidic oligotrichs are needed.

Volume specific chlorophyll contents have been estimated for *Laboea strobila* and for four plastidic *Strombidium* species (Stoecker *et al.* 1988, Stoecker *et al.* submitted). These

Table 5 COMPARISON OF LIGHT-SATURATED RATES OF PHOTOSYNTHESIS (P^{chl}_{max}) IN CULTURED PLASTIDIC OLIGOTRICHS TO RATES IN CULTURED PHYTOPLANKTON

Oligotrichous Ciliates	P^{chl}_{max} = pg C pg chl.a^{-1}h^{-1}	**Ref**
Laboea strobila	3.7; 1.0-4.0	(1) (2)
Strombidium capitatum	~1.6	(3)
S. conicum	~4.0	(3)
S. acutum	~0.7	(3)
S. chlorophilum	~0.9	(3)
Phytoplankton		
Diatoms	(a) (b)	
Thalassiosira pseudonana (3H)	4.2/0.7	(4)
Phaeodactylum tricornutum	3.3/0.7	(4)
Coscinodiscus spp.	1.6/0.5	(4)
Green Flagellate		
Dunaliella tertiolecta	2.2/0.9	(4)
Coccolithophore		
Coccolithus pelagicus	20.8/3.8	(4)
Cryptophyte		
Chroomonas salina	5.6/0.3	(4)
Dinoflagellate		
Gonyaulax tamarensis	4.3/0.6	(4)

a. exponential - phase culture
b. stationary - phase culture

1. Stoecker *et al.* 1988, 2. Putt unpub. data, 3. Stoecker *et al.* submitted, 4. Glover 1980

estimates range from 0.7 to 7.2 fg chl. *a* μm^{-3}. Volume specific chlorophyll contents in phytoplankton are very variable, but 1.0 to 10.0 fg chl. *a* μm^{-3} is about mid-range for microphytoplankton (see review by Malone 1980). Plastidic ciliates appear to contain about as much chlorophyll per unit volume and to be about as photosynthetic as microphytoplankton.

What contribution do plastidic ciliates make to community rates of photosynthesis? McManus and Fuhrman (1986) estimated that *Laboea strobila*, when most abundant, contributed 2% or less of the total primary productivity in Long Island Sound. On Georges Bank in early summer, this species is estimated to contribute ~1-7% of the primary productivity in surface waters, with the average contribution being close to 2% (Stoecker *et al.* in press). If we assume that the total biomass of other plastidic oligotrichs was about equal to that of *L. strobila* (Stoecker *et al.* 1987, Stoecker *et al.* in press), and that, per unit cell volume, these other plastidic ciliates were about as photosynthetic as *L. strobila*, then the total contribution of plastidic oligotrichs to photosynthesis was typically less than 10% of the total. However, in the microplankton (≥20 µm) size class, plastidic oligotrichs are estimated to have contributed as much as 20% or more of the primary productivity (Stoecker *et al.* in press). This may have important trophodynamic implications (see Role of Plastidic Oligotrichs in - Marine Processes).

NUTRITION AND GROWTH

The plastidic oligotrichs both ingest particulate food and have intact chloroplasts and thus are regarded as mixotrophic (Laval-Peuto *et al.* 1986, Stoecker *et al.* 1987, Laval-Peuto and Rassoulzadegan 1988). In culture, some plastidic oligotrichs only grow well when provided with both suitable algal food and proper illumination (Stoecker *et al.* 1988, Stoecker *et al.* submitted). However, the plastidic *S. reticulatum* can grow in the dark when supplied with suitable particulate food (Jonsson 1986). *Laboea strobila* and some, but not all plastidic *Strombidium* species are probably obligate mixotrophs, whereas other plastidic species are probably facultative mixotrophs.

There are several potential reasons why plastidic oligotrichs require algal foods. Firstly, they need a source of chloroplasts. Sequestered chloroplasts probably do not divide (Blackbourn *et al.* 1973, Laval-Peuto and Febvre 1986) and would be diluted out as the ciliates divide unless they were added between cell divisions. Secondly, most animal cells cannot take up nitrogen as nitrate (Eppley *et al.* 1979). The plastidic oligotrichs probably depend on ingestion to satisfy most of their requirements for nitrogen and perhaps phosphorus, but this needs to be investigated. Thirdly, many protozoa require trace nutritional factors derived from their prey (Provasoli 1977, Lee 1980). And lastly, photosynthesis alone may not provide sufficient carbon for respiration and growth.

The reasons why some plastidic oligotrichs require light, even in the presence of sufficient food, are less obvious. Is their basic carbon metabolism different from that of strict heterotrophs? Photosynthate may be necessary for the production of cortical polysaccharide plates which typically occur in these species (Laval-Peuto and Febvre 1986, Laval-Peuto *et al.* 1986). However, some non-plastidic oligotrichs, for example *Strombidium sulcatum*, also have cortical plates, although the chemical composition of the plates has not been investigated (see review of species description in Maeda and Carey 1985). The carbon metabolism of mixotrophic oligotrichs needs to be compared to that of closely related, strictly heterotrophic species. With the serial solvent extraction and liquid scintillation counting techniques which have been developed for use with phytoplankton cells (Rivkin 1985), it is possible to investigate partitioning of carbon among metabolic pools in ciliates (Putt unpub. data).

In symbiotic relationships between whole algal cells and animals, photosynthate may both support the growth of the symbiont and be translocated to the host. Photosynthate not translocated to the host can be used by the host if and when the endosymbionts are phagocytosed (Muscatine *et al.* 1983). In plastid retention involving chromophytic chloroplasts, most of the net photosynthate is probably transported to the cytoplasm of the ciliate because photosynthate is generally not stored within these types of plastids. However, chlorophytic chloroplasts store starch internally and in ciliates with these types of chloroplasts, photosynthate is sometimes stored within enlarged chloroplasts (Stoecker *et al.* submitted). It is not known whether carbon stored in chlorophytic chloroplasts in ciliates is eventually translocated or whether the ciliates digest whole, enlarged chlorophytic chloroplasts.

Some plastidic ciliates appear to be more dependent on photosynthesis than others (Table 5). *Laboea strobila* and *Strombidium chlorophilum* have relatively high chlorophyll contents and photosynthetic rates per unit volume, whereas some species have lower volume specific chlorophyll contents and rates of photosynthesis (Stoecker *et al.* 1987, Jonsson 1987, Stoecker *et al.* 1988, Stoecker *et al.* submitted). From a trophodynamic viewpoint, it is interesting to know what proportion of the carbon budget of plastidic ciliates is contributed by photosynthesis. Rough estimates of the contribution of photosynthate range from <5% to 37% or more (Jonsson 1987, Stoecker *et al.* 1988).

Ecologically relevant estimates of the contribution of photosynthesis to the carbon budget of ciliates are difficult to make. The contribution must depend not only on the ciliate species, but on food and light availability. Even under defined conditions, it is difficult to make accurate measurements of the contribution of photosynthate to the carbon budget of plastidic ciliates because metabolic carbon may be preferentially fixed by the plastids, some fixed carbon may be stored or used in the plastids, and accurate estimates of the carbon content of ciliates are generally not available (Choi ms, Putt pers. comm.). In addition, it is possible that photosynthate is preferentially respired, particularly when the ciliates are deprived of food (as in ^{14}C uptake incubations). Preferential respiration of photosynthate may "spare" ingested carbon and increase trophic efficiency.

An aspect of the nutrition of mixotrophic oligotrichs which has not received attention is their nitrogen and phosphorous metabolism. They may obtain all their N and P from ingested food, but this needs to be investigated. Nitrate reductase occurs in the autotrophic ciliate, *Myrionecta rubra* (= *Mesodinum rubrum*) (reviewed in Lindholm 1985), so it is conceivable that some oligotrichs also have this enzyme. It is also possible that plastidic oligotrichs take up dissolved nitrogen as ammonia or as dissolved amino acids as found in *M. rubra* (Smith and Barber 1979), but this has not been investigated.

In heterotrophic protozoa, nutrient remineralization is directly related to respiration and inversely related to gross growth efficiency during balanced growth (Caron and Goldman 1988). Mixotrophic oligotrichs that gain carbon from both photosynthesis and respiration should have higher ratios of production to ingestion than strictly heterotrophic forms; as a consequence, their nutrient regeneration rates should be lower than in strict heterotrophs. Mixotrophy may have important consequences for nutrient cycling in the euphotic zone.

ROLE OF PLASTIDIC OLIGOTRICHS IN MARINE PROCESSES

The role of planktonic ciliates in trophodynamics (Conover 1982, Sherr *et al.* 1986), nutrient regeneration and particle transformation in the upper ocean is more varied and complex than we once thought. Data indicate that a large proportion of the microplanktonic ciliates contain functional plastids and are primary producers as well as consumers of fixed carbon. Plastidic oligotrichs usually only contribute a few percent to total primary production but may, at times, contribute as much as 20% of the primary production in the microplankton size class. Twenty-five percent or more of total primary production is usually in the less than 3-5 µm size class and is not directly utilized by most macrozooplankton (reviewed in Malone 1980 and Sherr *et al.* 1986). Cells in the microplankton size category, into which most of the plastidic oligotrichs fall, are readily consumed by many macrozooplankters and some fish larvae (Conover 1982). Carbon fixed by ciliates should be more directly available to larger consumers than that fixed by the more abundant small phytoplankton. Recent evidence suggests that many macrozooplankters have significantly higher clearance rates for ciliates than for most phytoplankton (Stoecker and Egloff 1987, Gifford and Dagg ms.). Thus primary production by ciliates, even if it is only a few percent of total primary production, may make a significant contribution to higher trophic levels.

Mixotrophic ciliates should have higher production/ingestion ratios than strictly heterotrophic ciliates and thus they may be extremely efficient producers of animal biomass in the oceans (Blackbourn *et al.* 1973, McManus and Fuhrman 1986). As a corollary, they should have relatively low nutrient regeneration rates. Since plastidic oligotrichs often make up 25-50% of the microplanktonic ciliate fauna, their metabolic activities may have important consequences for trophodynamics and nutrient cycling within the euphotic zone.

The grazing activities of ciliates, and other microzooplankton, are thought to be important in the degradation of plant pigments in the oceans (SooHoo and Kiefer 1982, Klein *et al.* 1986, Burkill *et al.* 1987) as well as being a source of fine, non-living particulate matter (Stoecker 1984). Grazing by plastidic ciliates can result in the repackaging of plant pigments into larger size cells as well as degradation of these pigments. The magnitude of this 'repackaging' is not known, it is possible that it may influence the distribution of bio-optical properties in the oceans (for a discussion of bio-optics, refer to Morel and Bricaud 1986).

Most models of ecological processes in the oceans are based on the premise that producers and consumers are different organisms. Mixotrophy is not limited to plastidic ciliates. Some photosynthetic flagellates consume bacteria (Estep *et al.* 1986) and many oceanic sarcodines have algal endosymbionts (Anderson 1983, Lee *et al.* 1985, Michaels 1988). Mixotrophy appears to be frequent among marine protists and should be taken into account when their role in marine processes is considered.

ACKNOWLEDGEMENTS

Contribution No. 6853 from the Woods Hole Oceanographic Institution. The writing of this review was supported by grants NSF OCE-8709961 and OCE-8800684. Discussions with D A Caron, C A Price, M E Putt, M W Silver and A Taniguchi were important in the formulation of many of the ideas presented herein. Suggestions from M Laval-Peuto, M E Putt and an anonymous reviewer contributed to important improvements in the manuscript.

REFERENCES

Anderson OR (1983) Radiolaria. Springer-Verlag New York
Blackbourn DJ, Taylor FJR, Blackbourn J (1973) Foreign organelle retention by ciliates. J Protozool 20:286-288
Burkill PH, Mantoura RFC, Llewellyn CA, Owens NJP (1987) Microzooplankton grazing and selectivity of phytoplankton in coastal waters. Mar Biol 93:581-590
Burkholder PR, Burkholder LM, Almodovar LR (1967) Carbon assimilation of marine flagellate blooms in neritic waters of southern Puerto Rico. Bulletin of Marine Science 17:1-15
Caron DA, Goldman JC (1988) Dynamics of protistan carbon and nutrient cycling. J Protozool 32:247-249
Conover RJ Jr (1982) Interrelations between microzooplankton and other plankton organisms. Ann Inst Oceanogr Paris 58 (S):31-46
Dale T (1987) Diel vertical distribution of planktonic ciliates in Lindaspollene, western Norway. Mar Microbial Food Webs 2:15-28
Dale T, Dahl E (1987) Mass occurrence of planktonic oligotrichous ciliates in a bay in southern Norway. J Plankton Res 9:871-897
Dodge JD (1973) The fine structure of algal cells. Academic Press New York
Fauré-Fremiet E (1924) Contribution à la connaissance des infusoires planktoniques. Bull Biol Fr Belg 19 Suppl 6:1-171
Fauré-Fremiet E (1969) Remarques sur la systematique des cilies Oligotrichida. Protistol 5:345-352
Eppley RW, Coastsworth JL, Solorzano L (1979) Studies of nitrate reductase in marine phytoplankton. Limnol Oceanogr 14:194-205
Estep KW, Davis PD, Keller MD, Sieburth JMcN (1986) How important are oceanic algal nanoflagellates in bacterivory? Limnol Oceanogr 31:646-650
Gifford DJ (1985) Laboratory culture of marine planktonic oligotrichs (Ciliophora, Oligotrichida). Mar Ecol Prog Ser 23:257-267
Gifford DJ, Dagg MJ (ms) The microzooplankton-mesozooplankton link: Consumption of protozoa by calanoid copepods. MS
Glover HE (1980) Assimilation numbers in cultures of marine phytoplankton. J. Plankton Res 20:69-79
Goldman JC, Dennett MR (1985) Susceptibility of some marine phytoplankton species to cell breakage during filtration and post-filtration rinsing. J Exp Mar Biol Ecol 86:47-58
Hecky RE, Kling HJ (1981) The phytoplankton and protozooplankton of the euphotic zone of Lake Tanganyika: Species composition, biomass, chlorophyll content, and spatio-temporal distribution. Limnol Oceanogr 26:548-564
Jonsson PR (1986) Particle size selection, feeding rates and growth dynamics of marine planktonic oligotrichous ciliates (Ciliophora: Oligotrichina). Mar Ecol Prog Ser 33:265-277
Jonsson PR (1987) Photosynthetic assimilation of inorganic

carbon in marine oligotrich ciliates (Ciliophora, Oligotrichina). Mar Microbial Food Webs 2:55-68
Kahl A (1932) Urtiere oder Protozoa I. Wimpertiere oder Ciliata (Infusion) s. Spirotricha. In: Dahl F (ed) Die Tierwelt Deutschlands und der angrenenden Meeresteile. Gustav Fisher Jena 18:1-180
Klein B, Geiskes WWC, Kraay G (1986) Digestion of chlorophylls and carotenoids by the marine protozoan *Oxyrrhis marina* studied by HPLC analysis of algal pigments. J Plank Res 8:827-836
Laval-Peuto M, Febvre M (1986) On plastid symbiosis in *Tontonia appendiculariformis* (Ciliophora Oligotrichina). BioSystems 19:137-158
Laval-Peuto M, Rassoulzadegan F (1988) Autofluorescence of marine planktonic Oligotrichina and other ciliates. Hydrobiologia 159:99-110
Laval-Peuto M, Salvano P, Gayol P, Greuet C (1986) Mixotrophy in marine planktonic ciliates: Ultrastructural study of *Tontonia appendiculariformis* (Ciliophora, Oligotrichina). Mar Microbial Food Webs 1:81-104
Lee JJ (1980) Informational energy flow as an aspect of protozoan nutrition. J Protozool 27:5-9
Lee JJ, Lee MJ, Weis DS (1985) Possible adaptive value of endosymbionts to their protozoan hosts. J Protozool 32:391-382
Lee J, Soldo AT, Reisser W, Lee MJ, Jean KW, Görtz HD (1985) The extent of algal and bacterial endosymbioses in protozoa. J Protozool 32:380-382
Leech RM (1980) The survival, division and differentiation of higher plant plastids outside the leaf cell, In: Reinert J (ed) Chloroplasts Spinger-Verlag, Berlin, p 255
Li WK, Goldman JC (1981) Problems in estimating growth rates of marine phytoplankton from short-term ^{14}C assays. Microbial Ecol 7:113-121
Lindhölm T (1985) *Mesodinium rubrum* - a unique photosynthetic ciliate. Adv In Aquatic Microbiology 3:1-48
Lopez E (1979) Algal chloroplasts in the protoplasm of three species of benthic foraminifera: Taxonomic affinity, viability and persistence. Mar biol 53:201-211
Malone T (1980) Algal Size, In: Morris I (ed) The Physiological Ecology of Phytoplankton. Univ of California Press Berkeley California, p 433
Mattoo AK, Hoffaman-Falk H, Marder JB, Edelman M (1984) Regulation of protein metabolism: Coupling of photosynthetic electron transport to *in vivo* degradation of the rapidly metabolized 32-kilodalton protein of the chloroplast membrane. PNAS, USA 81:1380-1384
McManus GB, Fuhrman JA (1986) Photosynthetic pigments in the ciliate *Laboea strobila* from Long Island Sound, USA. J Plankton Res 8:317-327
Michaels AF (1988) Vertical distribution and abundance of Acantharia and their symbionts. Mar Biol 97:559-569
Montagnes DJS (1986) The annual cycle of planktonic ciliates in the waters surrounding the Isles of Shoals, Gulf of Maine: Estimates of biomass and production. Ms Thesis, University of Guelph Guelph Ontario Canada, pp 114
Montagnes DJS, Lynn D H, Stoecker DK, Small EB (1988) Taxonomic descriptions of one new species and redescription of four species in the family Strombidiidae (Ciliophora, Oligotrichida). J Protozool 35:189-197
Morel A, Bricaud A (1986) Inherent optical properties of algal cells including picoplankton: Theoretical and experimental results, pp 521-559 In: Platt T, Li WK (eds) Photosynthetic Picoplankton. Can Bull Fish Aqu Sci 214
Muscatine L, Falkowski PG, Dubinsky Z (1983) Carbon budgets in symbiotic associations, pp649-658 In: Shenk HEA, Schwemmler W (eds) Endocytobiology III. Walter de Gruyter Berlin
Parsons TR, Maita Y, Lalli CM (1984) A manual of chemical and biological methods for seawater analysis, Pergamon Press New York
Patterson DJ, Dürrschmidt M (1987) Selective retention of chloroplasts by algivorous helioza (fortuitous chloroplast symbiosis?). European Journal of Protistology 23:51-55
Penry DL, Jumars PA (1987) Modeling animal guts as chemical reactors. The American Naturalist 129:69-96
Provasoli L (1977) Ch. 5 Cultivation of animals. 5.1 Research

cultivation. 5.11 Axenic cultivation, pp 1295-1320 In: Kinne O (ed) Marine Ecology, Vol 3 Wiley, New York
Rassoulzadegan F (1977) Evolution annuelle des cilies pelagiques en Mediterranee nordoccidentale. I. Cilies oligotriches (non-tintinnides) (Oligotrichina). Ann Inst Oceanogr Paris 53:125-134
Reid PC (1987) Mass encystment of a planktonic oligotrich ciliate. Mar Biol 95:221-230
Reith ME, Cattolico RA (1985a) *In vitro* chloroplast protein synthesis by the chromophytic alga *Olisthodiscus luteus* Biochemistry 24:2550-2556
Reith ME, Cattolico RA (1985b) In vivo chloroplast protein synthesis by the chromophytic alga *Olisthodiscus luteus* Biochemistry 24:2556-2561
Rivkin RB, Seliger HH (1981) Liquid scintillation counting for ^{14}C uptake of single algal cells isolated from natural samples. Limnol Oceanogr 26:780-784
Rivkin RB (1985) Carbon-14 labelling patterns of individual marine phytoplankton from natural populations. Mar Biol 89:135-142
Rogerson A, Finlay BJ, Berninger U-G (in press) Sequestered chloroplasts in the freshwater ciliate *Strombidium viride* (Ciliophora: Oligotrichida). Trans Am Micrsc Soc
Sherr EB, Sherr BF, Paffenhöfer GA (1986) Phagotrophic protozoa as food for metazoans: a "missing" link in marine pelagic food webs. Mar Microbial Food Webs 1:61-80
Silver MW, Davoll PJ (1978) Loss of ^{14}C activity after chemical fixation of phytoplankton: error source for autoradiography and other productivity measurements. Limnol Oceanogr 23:362-368
Small EB, Lynn DH (1985) Phylum Ciliophora, pp 393-575 In: Lee JJ, Hutner SH, Bovee EC (eds) Illustrated Guide to the Protozoa. Soc Protozool Special Pub Allen Press Lawrence Kansas
Smith WO Jr, Barber RT (1979) A carbon budget for the autotrophic ciliate *Mesodinium rubrum*. J Phycol 15:27-33
Soo Hoo JB, Kiefer DA (1982) Vertical distribution of phaeopigments - 1. Simple grazing and photooxidative scheme for small particles. Deep-Sea Research 29:1539-1551
Sorokin YI (1981) Microheterotrophic organisms in marine ecosystems, pp 293-342. In:Longhurst AR (ed) Analysis of Marine Ecosystems. New York Academic Press
Steinback KE, Arntzen CJ, Bogorad L (1985) The physical organization and genetic determinants of the photosynthetic apparatus of chloroplasts, pp 1019 In: Steinback KE *et al* (eds) Molecular Biology of the Photosynthetic Apparatus. Cold Spring Harbour Laboratory
Stoecker D (1984) Particle production by planktonic ciliates. Limnol Oceanogr 29:930-940
Stoecker DK, Egloff DA (1987) Predation by *Acartia tonsa*. Dana on planktonic ciliates and rotifers. J Exp Mar Biol Ecol 110:53-68
Stoecker DK, Michaels AE, Davis LH (1987) Large proportion of marine planktonic ciliates found to contain functional chloroplasts. Nature 326: 79-792
Stoecker DK, Silver MW (1987) Chloroplast retention by marine planktonic ciliates. Endocytobiology III. Annals of the New York Academy of Sciences 503:562-565
Stoecker DK, Silver MW, Michaels AE, Davis LH(1988) Obligate mixotrophy in *Laboea strobila*, a ciliate which retains chloroplasts. Mar Biol 99:415-423
Stoecker DK, Silver MW, Michaels AE, Davis LH (submitted) Enslavement of algal chloroplasts by four *Strombidium* spp (Ciliophora, Oligotrichids). Mar Microbial Food Webs
Stoecker DK, Taniguchi A, Michaels AE (in press) Abundance of autotrophic, mixotrophic and heterotrophic planktonic ciliates in shelf and slope waters. Mar Ecol Prog Ser
Taylor FJR (1982) Symbioses in marine microplankton. Ann Inst oceanogr Paris 58 (S):61-90
Trench RK (1975) Of 'leaves that crawl,: Functional chloroplasts in animal cells. Soc Exp Biol 29:229-265
Trench RK (1980) Uptake, retention and function of chloroplasts in animal cells, In: Schwemmler W, Schenk HEA (eds) Endocytobiology. Walter de Gruyter, Berlin, p 703

BRIEF PERSPECTIVE ON THE AUTECOLOGY OF MARINE PROTOZOA

John J Lee
Dept of Biology
City College of CUNY
Convent Avenue
at 138 St
New York
NY 10031
USA

INTRODUCTION

At this vantage point in the late 20th century there are good arguments for believing that the term 'protozoa' is an outmoded taxonomic concept. I believe it is, but nevertheless it still conveys meaning as a collective term for those phyla which protozoologists once culled from the spectrum of eukaryotic protists (Lee *et al.* 1985). When we use the term 'protozoa' we are reminded at once of the breadth of the spectrum of these organisms which seem to be able to utilise every type of resource and are found in all habitats available in the sea.

Autecology is an intellectual concept which focuses on individuals or individual species and emphasises how their life histories and behaviour are adaptive to their environment. In contrast, synecology focuses on the study of groups of associated organisms which are perceived as a unit. Planktonic organisms and associations seem never to be in equilibrium in time or space. The nature of the medium, our sampling methods, and frequencies of sampling, suggest that populations of planktonic protozoa are transient without the significant biotic interactions generally recognised as characterising communities. This leads to the general perception that autecological facets of 'protozoalplankton' are much more important than the synecological. As ecologists move their focus from the purely descriptive towards the analytical, the need for recognising patterns, functional groups, and or the unique or special features in the ecology of particular groups has become a more important issue. On the basis of the

NATO ASI Series, Vol. G 25
Protozoa and Their Role in Marine Processes
Edited by P. C. Reid et al.

information available today, is it reasonable for us to generalise that niches of marine protozoa are framed by autecological factors? More narrowly focused questions follow. What are the forms of protozoan population responses to changes in habitat criteria such as temperature, salinity, O_2, light and depth? How are they adapted to unfavourable periods? What are the relative niche widths of different protozoan species along habitat criteria in the sea? How much niche overlap is there among marine protozoa? Although none of these questions can be easily or simply answered, some of them will be probed in this 'perspective'.

HABITAT DIMENSIONS AND NICHE CRITERIA

With respect to their broad distribution in the various habitats and water masses in the world's oceans, foraminifera are perhaps the most extensively studied group of protozoa. Because their tests are generally preserved in marine sediments, are well represented in the fossil record, and because foraminifera are abundant in most habitats and diverse taxonomically (~4.0 x 10^4 described species), the study of their distribution became important in petroleum exploration. For example: A kilogram of well-core might yield >2x10^5 foraminifera to study (Boltovskoy and Wright 1976) and a 10ml core sample of outer-shelf muds might yield several thousand specimens represented by some 50 species (Haynes 1981). Generalisations have been made about the distributions of both planktonic and benthic foraminiferal species in modern oceans (critically reviewed in books Boltovskoy and Wright 1976, Murray 1973, and Bé 1977, earlier literature reviewed by Phleger 1960).

Although the ability to culture planktonic foraminifera in the laboratory is still in its formative stages, careful studies using plankton tows, surface sediments, and sediment traps have refined our knowledge of their distributions in relationship to important water mass properties (e.g. temperature, salinity, nutrients) (e.g. Bé and Tolderlund 1971, Tolderlund and Bé 1971,

Bé and Hutson 1977, Fairbanks *et al.* 1980, 1982, Thunell and Reynolds 1984). These studies of habitat criteria and relative niche widths have laid the groundwork for using planktonic foraminifera to interpret major global climatic changes in the Neogene (e.g. Thunell and Belyea 1982 and Kennett *et al.* 1985). For example, the work of Bé and Hutson (1977) which defined present day faunal provinces based on habitat criteria has been used by Wright and Thunell (1988) to reconstruct surface circulation patterns for five Neogene time-slices from cores in the Indian Ocean.

Distinctive assemblages of benthic populations are distributed in different zones of salt marshes, mangrove swamps, estuaries and lagoons, deltas, shelf seas, deep seas, coral reefs, tropical carbonate environments, and all the major oceanic water masses are characterised by distinctive assemblages of planktonic species. Species are found within each of these enormous world-wide zones of distribution bounded by observed habitat criteria. However, we have scant information to go the next step of refinement which would be to define niche parameters as few foraminiferal species have been studied in enough detail for us to be able to understand their niches. The best known exception is the study of Muller (1975) who considered the niches of four species of salt marsh foraminifera (*Allogromia laticollaris, Rosalina leei, Spiroloculina hyalina* and *Ammonia beccarii*). While the above species differed greatly in their abilities to tolerate, survive, and reproduce within the ranges of temperature, salinity, and pH tested, they were all widely adapted to the littoral and sublittoral habitat in which they naturally occur. Food quality was an important aspect which separated their niches. Two of the algal species eaten in large numbers by *Allogromia laticollaris* were not eaten by the other species in great quantity. The same was true for *Rosalina leei*. *Spiroloculina hyalina* ate very few of the algal species tested but did consume quantities of one species of *Amphora* and several of the bacterial species tested. Interspecific and intraspecific (crowding) competition were two other important biological aspects of their niches. *A. laticollaris* was a poor competitor. It withdrew its pseudopods in the presence of other organisms which, in turn, reduced its feeding rate. This correlated with observations on its natural distribution. It is a rare species which becomes locally abundant when dominant species are missing. *Spiroloculina hyalina* blooms in nature in communities with large populations of bacteria. *Rosalina leei* was a good competitor for food and has a relatively broad tolerance for the physical attributes of its niche (e.g. temperature, salinity, pH. It was common in most natural populations.

Broad generalisations about marine protozoa usually have their exceptions. A recent discovery at the 'Oasis' deep sea vent (20°50'N; 109°06'W. 2,603 m depth) makes the point. Dr Karl Berg had placed a number of panels at the vent for the study of recruitment. When they were retrieved by DSRV Alvin, some of them were densely populated by foraminifera. One expects to find agglutinating foraminifera at depths below the lysocline but these trochaminid foraminifera had built their unique shells out of volcanic ash (probably obsidian judging from the fracture surfaces on the individual particles) (Figures 1-3). Since their whole test is built from ash particles there is no doubt they spent their entire lives in the vent community. since these unusual protozoa have not been described before and their characteristic shells are built from a very restricted resource, it is possible that the new species of foraminifera may be unique to only one vent in all the seas; a very narrow habitat indeed!

SEASONAL NICHE OCCURRENCE AND MICROHABITATS

The habitats of many marine protozoa are seasonal. As conditions in their environments become more favourable and within limits for growth, niches open up. Although it is tempting to delineate realised niches along abiotic habitat axes, in many cases we know that this is overly simplistic. There are many protozoa whose propagules are quite

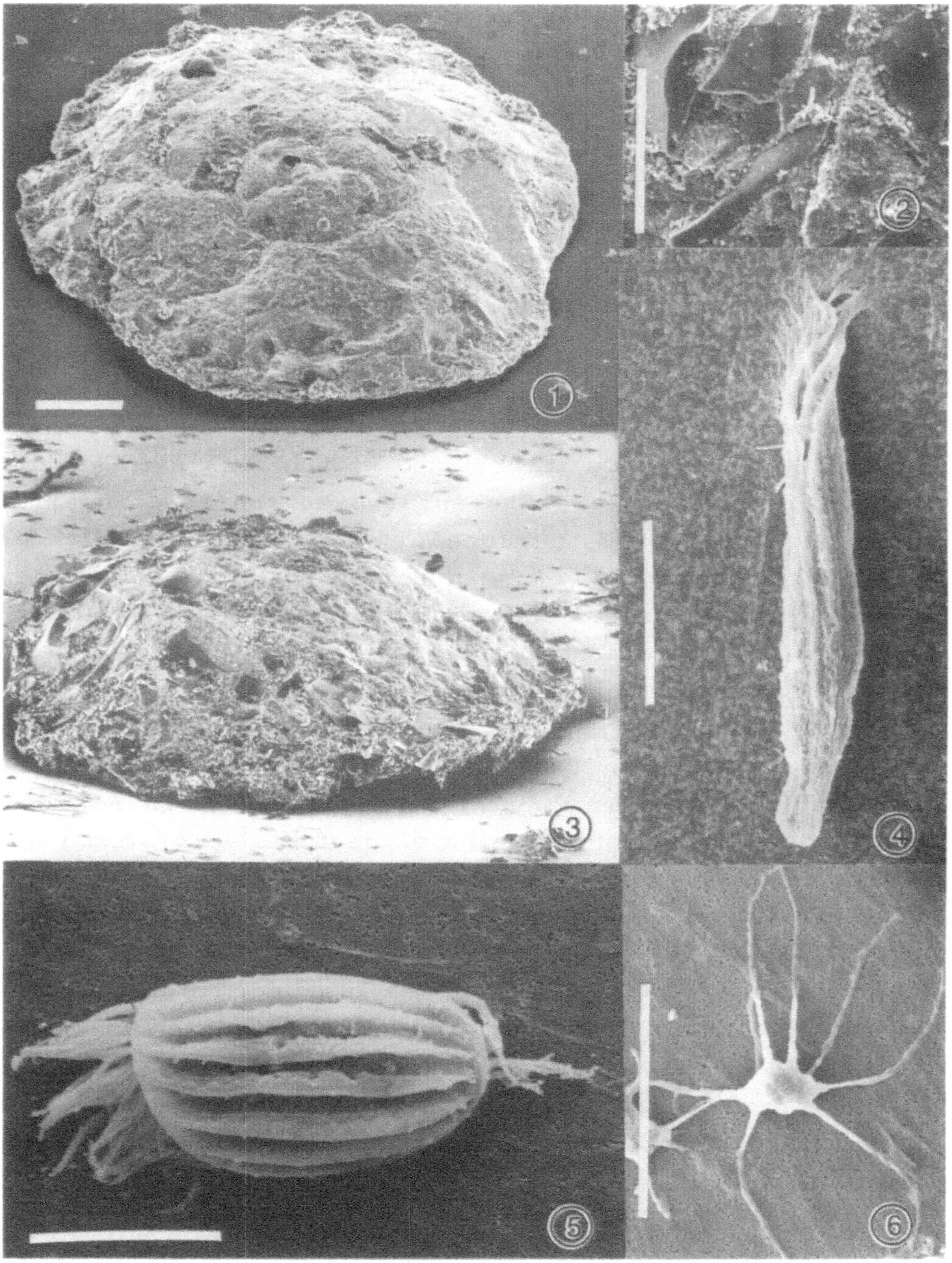
1
2
3
4
5
6

Plate 1. All figures are Scanning Electron Micrographs.

Figures 1-3 New trochaminid from the 'Oasis' hydrotherman vent. Figure 1. Dorsal view. Figure 2. enlargement showing obsidian crystals. Figure 3. Lateral view illuminated obliquely to emphasize the obsidian crystals on one side of the protozoan.

Figures 4-6 Representative protozoa from a coastal embayment which bloomed in baited-bloom procedure. Figure 4. *Vexillifera* sp. Figure 5. *Euplotes* sp. Figure 6. *Condylostoma* sp.

Scale bars Figures 1-3 100 µm, Figures 4 and 6 40 µm, Figure 5 20 µm.

opportunistic (r strategists) and whose abundance, while seasonally driven, are interlinked with others with similar seasonal trajectories. The interwoven themes of spatial interconnection and seasonality as elements of the autecology/synecology of marine protozoa are not easily apparent to workers outside our protistological field. The diversity of populations, particularly in the plankton, and their seemingly ephemeral appearances in time and space, challenge the uninformed to find organisational elements and a sense of community. Yet they are there. Much of our appreciation for these connective links has grown from the fundamental nutritional studies of gnotobiotic protozoa and micrometazoan cultures by Provasoli, Hutner, Droop, Dougherty, and their students and coworkers (e.g. Conklin and Provasoli 1977, Dougherty 1959, Droop 1966, Lee 1980, Lee *et al*. 1976, Muller 1975 and Provasoli 1977). The technical complexities of the isolation and cultivation procedures, the intense efforts which some of the studies have consumed, and the technical lexicon (jargon) associated with these types of studies has undoubtedly discouraged some ecologists from becoming interested in the results of this type of research.

One of the problems we face as protozoan ecologists is how to translate the detailed findings on dietary requirements to broader concepts of food quality and energy flow which are easily grasped by a wider group of our colleagues. Since protozoa are generally single cells, it is easy to construct a conceptual framework which relates food quality and energy flow to their nutritional needs. We can imagine the diet of the protozoan to be conceptually divisible into categories of nutritional needs: 1) bulk; 2) stimulatory; and 3) essential. The bulk of 'macronutrient' sources satisfy the energetic needs of the protozoan and provide most of the general metabolites needed for synthesis, maintenance, and growth. These are nutrients obtained by digesting food and funnelling the products through anabolic and catabolic pathways as they are needed. But that is not the whole story. To a certain extent the bulk nutrients must be matched to the protozoans digestive and synthetic abilities. For example, an excess of one particular amino acid might lead to the shutting down or slowing of a multiple-branched biosynthetic pathway, of which it is a product, and thereby slow or stunt growth and reproduction. Stimulatory metabolites are those which are not required by the protozoan, in the sense that it can synthesise them when they are needed, but are used if present, saving the energy needed for their synthesis and thereby speeding up growth and reproduction. Essential nutrients are those which the protozoan is totally unable to synthesise itself and which must be supplied in the diet. Essential nutrients could be vitamins, particular unsaturated fatty acids or amino acids.

Food quality is a property of both the food organisms themselves and the ability of the protozoan to recognise and funnel essential and stimulatory molecules from its catabolic to its anabolic pathways at the least energetic costs. While one has to be careful in making analogies, let us look at the building blocks of books and of cells. On one hand we have individual letters assembled into words. Then words are organised into meaningful sentences, which have context in a paragraph, of a chapter in a book. Cells, on the other hand, are built of carbon, hydrogen, nitrogen, oxygen, sulphur, etc. atoms which must be assembled into molecules such as amino acids. These, in turn, are organised into peptides, proteins, membranes, etc. Each level costs energy to build. Some of that energy is represented as negentropy or information, in the form of cellular organisation (Morowitz 1968, 1970, Kondepudi 1988, Harrison 1988). We can continue the rough linguistic analogy to the synthetic phase. Suppose our input is in magazines and our output is books written in a certain style. As the magazines enter our machine, their covers and binding are stripped off and then they are fed page by page into parser. The covers and binding are waste and are lost to the system. The sentences begin to be chopped up by the parser and are briefly passed through a recogniser/scanner. The recogniser/scanner can either identify the parsed fragments as something meaningful to its eventual output or it passes the fragment back for processing. In time, the input and its processing results in the production of a new book or books written in a particular style.

The experimental evidence gained from a comparative nutritional study of two inshore marine ciliates, *Euplotes vanus* and *Uronema marinum*, supports the concept that informationally encoded energy in the form of particular molecules can be scavenged by efficient consumers and can play an important role in ecosystem trophic dynamics (Rubin and Lee 1976). The growth and reproductive rates of each species of ciliate were measured on diets (monoxenic) of 13 different species of unicellular algae. In the experiments each ciliate was fed non-limiting quantities of various food species of equal calorific value. Because both ciliates grew at different rates on the various diets, this was taken as direct evidence that informational energy in the form of constituent molecules of the algal prey could be scavenged by efficient consumers. *Uronema marinum* was the ecologically more efficient of the two species tested. Its ecological growth efficiency ranged from 0-21% on different diets. It did not eat or grow on some of the algal species tested. Presumably, the latter algae were lacking in molecules which also could not be synthesised by the ciliate. The ecological growth efficiencies for *Euplotes vannus* ranged from 0-12%. It too could not grow on diets of some species of algae. An interesting aspect of the energy/information transfer process was revealed when we rank ordered the algae with respect to the efficiency of the growth of each ciliate species. Although two isolates of *Chlamydomonas* spp (S-93 and S-94) were the top ranked diets for both ciliate species the rest of the rank ordering was quite random. For example, the third ranked food species for *E. vannus* was last ranked for *U. marinum*. Another way of showing qualitative matching of consumer and consumed was shown by experiments in which the ciliates were grown in synchronous culture and then switched from one species of food organism to another. Controls involved no switching of food organisms. Induction time lags were found when switching both animals from one diet to another. This was reasonably interpreted as time needed for metabolic 'retooling' (e.g. adaptive enzyme formation) for the metabolic processing of the new food organisms.

The data clearly shows that nutritional quality factors (Molecular information) are part of the energetics and autecology of these two ciliates in marine food webs since their reproduction was neither a strict function of feeding rates nor strictly coupled with the calorific values of the food organisms themselves. We (Rubin and Lee 1976) developed a theoretical model to make quantitative estimates of the informational gain in food/consumer relationships. The difference between the energetic cost of growth for the consumer on the diet most informationally impoverished (lowest non-zero ecological efficiency) and other foods was taken as the base line to which all other food species could be compared. Any energy saved, or acceleration in generation time, was translated through fundamental respiration relationships into a new unit of energy flow.

We have recently speculated that this aspect of nutritional quality could have broader implications in the autecology of marine protozoa (Rubin and Lee 1989, in press). A model was constructed in which we placed one or both ciliate species in environments with randomly distributed food species, environments with randomly distributed 'high I' food and patches of 'low I' food or the converse, randomly distributed 'low I' food and patches of 'high I' food. In these simulations we observed rapid increases when the ciliates encountered 'high I' patches and when they encountered 'low I' patches they grew slowly. 'High I' patches were grazed down quickly and then the ciliates turned to 'low I' patches upon which they barely maintained their populations. When both 'high and low I' foods were randomly distributed neither ciliate species grew well. The ciliates 'processing machinery' was not 'tooled up' for highest efficiency. In simulations involving patches of 'high I' food organisms for both species, initial collisions between the ciliates and food organisms were important determinants of the outcome. When the food was randomly distributed the competition was fierce and the simulated populations reached only one-half the biomass of a similar patchy system. It also follows that some of these types of protozoa have niches which are ephemeral. That is, they largely maintain themselves as propagules on 'low I' food from the time they exhaust patches of 'high I' food to the next time they encounter some (Muller and Lee 1977).

It seems reasonable to ask ourselves whether food quality in given seasonal contexts is an important aspect of the niches of many marine protozoa or a relatively minor one. Recently we had an opportunity to probe this question. We were searching for ciliates which could be raised in large numbers and had other characteristics which made them candidates to test as potential food for the first feeding larvae of marine fish which had potential for mariculture. We used a 'baited-bloom' approach (Lee *et al.* 1987). Axenic cultures of *Dunaliella salina, Monochrysis lutheri,* and *Isochrysis galbana* were grown to late log phase and then distributed as 45 ml aliquotes into sterile 125 ml microfernbach flasks. In early June we aseptically collected samples of surface water in shallow embayments (Lackey's Bay, Naushon Island, Massachusetts; Greater Sippwissett March, Falmouth, Massachusetts, North Sea Harbour, Peconic Bay, Long Island, NY) and inoculated 5 ml into each of the flasks. In 2 days explosive blooms of ciliates or radiate amoebae (Figures 4-6) consumed almost all of the algal inocula in 18% (N=250) of the flasks. This result confirms that food quality can be a rather common niche parameter for herbivorous protozoa in these coastal habitats.

As mentioned earlier, food quality seems also to be an important niche ordering factor for the foraminifera living on and in the benthos of the same coastal habitats. One of these, *Allogromia laticollaris*, has such interesting life-cycle options (Lee *et al.* 1975) that it is likely to challenge even the most imaginative modellers. This foraminifer can reproduce by binary fission, budding, cytotomy, shizogony, and sexually (McEnery and Lee 1976). Life cycle options of this protozoan seems to be diet dependent. Binary fission is most common when the organism is kept largely on a bacterial diet (Lee *et al.* 1969). Diets of particular algae and mixtures of algae favour other asexual life cycle options (Lee *et al.* 1969) and rates of fecundity (Lee *et al.* 1969, Muller, 1975). Although we do not know of any other protozoan which has such complex community contextual relationships, this organism gives us perspective to be cautious in our tendencies to simplify our views of nature by using the characteristics of single organisms to model assemblages or guilds of organisms.

As a final comment on seasonality, the stability of populations of dinoflagellate hypnocysts in sediments needs to be compared to the transient blooms from which they are derived. Because of the importance of these organisms in Red tides they have received a great deal of attention (e.g. Conferences edited by Anderson *et al.* 1985, Taylor and Seliger 1979). The length of dormancy of hypnozygotes is highly variable (reviewed by Pfiester and Anderson 1987). Excystment may occur in some species within hours of hypnozygote formation or there may be a mandatory dormancy period which lasts for months. At cold temperatures, hypnocysts may remain quiescent for years. While temperature seems to be the most obvious parameter controlling dormancy and excystment, other factors contribute to the timing and numbers which germinate (Pfiester and Anderson 1987).

ENDOSYMBIOSIS-SURVIVAL IN NUTRIENT DESERTS

Another protozoan autecological phenomenon which is receiving increased attention in recent years is endosymbiosis (see Lee *et al.* 1985, 1988 and 1990). The phenomenon is widespread in the larger sarcodines and some ciliates of oligotrophic waters. These adaptations promote survival and growth under conditions which might not favour other protozoa. Some larger foraminifera (mm to cm size range) have very long life spans for protozoa. Reproduction may be seasonally driven and occurs in different species semiannually, annually, or even biannually. The relative constancy of their well illuminated oligotrophic tropical and semitropical habit (e.g. Hottinger and Reiss 1984) seems to have favoured the evolution and diversification of these algal endosymbiont-bearing protozoa (Lee and Hallock 1987). Different families of larger foraminifera are the hosts for endosymbiotic chlorophytes, rhodophytes, dinoflagellates, or small pennate diatoms. Data from the field suggest that the type of algal symbiont in a particular species is an important factor determining optimal depth and thereby niche separation (Letenegger 1984). Other foraminifera (Elphidids) and ciliates scavenge and sequester algal chloroplasts.

The message of this brief perspective is clear. Protozoa are ecologically too diverse for us to simplify our models by emulating the behaviour of a small number of species. There are exceptions to almost every generalisation we try and make. Some of the 'black box' modellers may wish to throw their hands up in disgust, others may wish to slam the 'box' shut and never again think about the complexities and puzzles writhing around inside. Fortunately there are others who view the complexities as a challenge to give them greater insight into the conceptual fabric of marine ecosystems.

REFERENCES

Anderson DM, White AW, Baden DG (eds) (1985) Toxic Dinoflagellates. Proc 3rd Int Conf on Toxic Dinoflagellates. Elsevier, NY
Bé AWH, Tolderlund DS (1971) Distribution and ecology of living

planktonic foraminifera in surface waters of the Atlantic and Indian Oceans. In: Funnell BM, Riedel WR (eds) The Micropaleontology of the Oceans. Cambridge University Press, London p 105-149
Bé AWH, Hutson WH (1977) Ecology of planktonic foraminifera and biogeographic patterns of life and fossil assemblages in the indian Ocean. Micropaleontology 23:369-414
Conklin DE, Provasoli L (1977) Nutritional requirements of the water flea *Moina macrocopa. Biol Bull* 152:337-350
Dougherty EC (1959) (ed) Symposium on Acenic Culture of Invertebrate Metazoa: A Goal. Ann NY Acad Sci 77:25-406
Droop M (1966) The role of algae in the nutrition of *Heteramoeba clara* Droop with notes on *Oxyrrhis marina* Dujardin and *Philodina roseola* Ehrenberg. In: Barnes, H (ed) Some Contemporary Studies in Marine Science. Geo Allen and Unwin, London
Fairbanks, RG, Wiebe PH, Bé AWH (1980) Vertical distribution and isotropic composition of living planktonic foraminifera in the estern North Atlantic. Sci 209:61-63
Fairbanks RG, Sverdlou M, Free R, Wiebe PH, Bé AWH (1982) Vertical distribution and isotope fracionation of living planktonic foraminifera from the panama Basin. Nature 298:841-846
Harrison, LG (1988) Kinetic Theory of Living Pattern and Form and Its Possible Relationship to Evolution. In: Webber BH, Depew DJ, Smith JD (eds) Entropy Information and Evolution. M I T Press, Cambridge, Massachusetts
Hottinger L, Reiss Z (1984) The Gulf of Aaba-Ecological Micropaleontology. Springer-Verlag, Berlin Heidelberg
Kennett JP, Keller G, Srinivasan MS (1985) Miocene planktonic foramineral biogeography and paleoceanographic development of the Indo-Pacific region. In: Kennett JP (ed) The Miocene Ocean: Paleoceanography and Biogeography Geol Soc Amer Mem 163:197-236
Kondepudi D (1988) Parity violation and the origin of Biomolecular Chirality. In: Weber BH, Depew DJ, Smith JD (eds) Entropy Information and Evolution. M I T Press Cambridge, Massachusetts
Lee JJ (1980) Informational energy flow as an aspect of protozoan nutrition. J Protozool 27:5-9
Lee JJ (1990) Foraminifera. In: Algal symbioses. Reisser W (ed) Biopress Ltd, Bristol UK
Lee JJ, Douglas AC, Kahan D (1987) a baited-bloom approach to the isolation of protozoa and meiofauna for potential use in mariculture. Annual Meeting Society of Protozoologists Abstract 42
Lee JJ, Hallock PM (1987) Algal symbiosis as the driving force in the evolution of larger foraminifera. In: Lee JJ, Fredrick J Endocytobiology III. Ann NY Acad Sci 503:330-347
Lee JJ, Hutner SH, Bovee EC (1985) An Illustrated Guide to the Protozoa. Society of Protozoologists, Lawrence, Kansas
Lee JJ, Lanners E, ter Kuile B (1988) The retention of chloroplasts by the foraminifer *Elphidium crispum.* Symbiosis 5:45-60
Lee JJ, McEnery ME, Rubin H (1969) Quantitative studies on the growth of *Allogromia laticollaris* (foraminifera). J Protozool 16:377-395
Lee JJ, Tietjen JH, Garrison JR (1976) Seasonal switching in the nutritional requirements of *Nitocra typica* an harpaticoid copepod from salt marsh aufwuchs communities. Trans Amer Micros Soc 95:628-637
Lee JJ, Tietjen JH, Saks NM, Ross GG, Rubin H, Muller WA (1975) Educing and modeling the functional relationships within sublittoral salt-march sufwuchs communities - inside one of the black boxes. Estuarine Res 1:710-734
Lutenegger S (1984) Symbiosis in benthic foraminifera: specificity and host adaptations. J Foram Res 14:16-35
McEnery ME, Lee JJ (1976) Allogromia laticollaris: A foraminiferan with an unusual apogamic metagenic life cycle. J Protozzol 23:94-108
Morowitz HJ (1968) Energy flow in Biology. Academic Press, NY
Morowitz HJ (1970) Entropy for Biologists. Academic Press, NY
Muller WA (1975) Competition for food and other niche-relted studies of three species of salt-marsh foraminifera. Mar Biol 31:339-351

Muller WA, Lee JJ (1977) Biological interactions and the realized niche of *Euplotes vannus* from the Salt Marsh Aufwuchs. J Protozzol 24:523-527
Murray JW (1973) Distribution and Ecology of Living Benthic Foraminiferids. Heinemann Educational Books, London
Pfiester LA, Anderson DM (1987) Dinoflagellate reproduction. In: Taylor, FJR (ed) The Biology of Dinoflagellates. Blackwell Sci Publ, Oxford
Phleger FB (1960) Ecology and distribution of Recent Foraminifera. Johns Hopkins Press, Baltimore
Provasoli L (1977) Cultivation of animals: Axenic cultivation. In: Kinne O (ed) Marine Biology. John Wiley and sons, NY 3:1295-1230
Rubin HA, Lee JJ (1976) Informational energy flow as an aspect of the ecological efficiency of marine ciliates. J Theor Biol 62:69-91
Rubin HA, Lee JJ (1989) Food quality and microzooplankton patchiness - an automation model. In: Capriulo 6 (ed) Ecology of Marine Protozoa. Hutchinson Ross Doud New York (in press)
Taylor DL, Seliger HH (eds) (1979) Toxic Dinoflagellate Blooms. Proc. 2nd Int conf on Toxic dinoflagellates. Elsevier, NY
Thunell R, Belyea P (1982) Neogene planktonic foraminiferal biogeography of the Atlantic Ocean. Micropaleon 28:281-398
Thunel R, Reynolds LA (1984) Sedimentation of planktonic foraminifera: seasonal changes in species flux in the Panama Basin. Micropaleon 30:243-262
Tolderlund DS, Bé AWH (1971) Seasonal distribution of planktonic foraminifera in the western North Atlantic. Micropaleon 17:297-329
Wright JD, Thunell RC (1988) Neogene planktonic foraminiferal biogeography and paleoceanography of the Indian Ocean. micropaleon 34:193-216

TROPHIC BEHAVIOUR - SESSION SUMMARY

Victor Smetacek
Alfred-Wegener-Institute for Polar and Marine Research
2850 Bremerhaven
FRG

INTRODUCTION

While preparing this summary from the recording of the session, I was surprised by the many interesting details I had since forgotten. What struck me most, however, was the enthusiasm generated by the lecturers and participants in this session. The degree of empathy for protozoan "pets" that was shown was remarkable considering that protists are supposed to be comparatively "simple" organisms. The quality of this empathy was most clearly reflected in the choice of words and gestures employed by the speakers during their presentations. Thus, vivid visual impressions conveyed by these means were greeted by audible appreciation from the audience. Liberal use of adjectives such as "enormous", "huge" and "giant" (with accompanying hand-waving) applied to the protozoa clearly demonstrate that the contributors had transported themselves - and the audience - to the size scales of their "animals", an obvious pre-requisite for empathy. As expressions such as these tend not to appear in scientific journals and are quickly forgotten by the audience, I have taken the liberty of sprinkling some of them in the text (in inverted commas). Such terms convey a tangible sense of behaviour that is difficult to reproduce in a standard scientific article. In this summary I shall try and concentrate more on the meat of the lectures and discussions rather than the cut and dried bones that will subsequently appear in the various written versions. Meat is digested quickly, but it provides the energy to continuously sort and resort the longer lived bones of "hard facts". I have also added some of my own fat (accumulated on the rich Plymouth diet) in the form of afterthoughts at the ends of various paragraphs.

Food for thought was provided by the intricate picture of

NATO ASI Series, Vol. G 25
Protozoa and Their Role in Marine Processes
Edited by P. C. Reid et al.

protozoan feeding ecology that has emerged in the last few years. Until recently, we have implicitly assumed that small is simple, and that dealing with the protozoans according to phylogeny would provide us with the necessary knowledge of their feeding behaviour. Now it appears that protozoa are more versatile than e.g. birds (who are stubbornly heterotrophic) when foraging within "their" environment. To the modellers' dismay, the taxonomists' tools are steadily being improved and the ecological boxes allocated to protozoans are dividing at an alarming rate.

The study of protozoan trophic behaviour is a rapidly growing field still anticipating fundamental new discoveries; in fact, the connotations of the term are not yet clear. Can protozoa, obviously without a nervous system, exhibit behaviour? Is an individual capable of learning, e.g. differentiating between different types of food? What are the limits of selective feeding? What do the species that posses a highly developed ocellus experience? Do they use it for capturing food? These provocative questions must be kept in mind when designing experiments to assess protozoan feeding behaviour. We might well learn a lesson from the copepod feeding experiments of the past, based on assumptions that copepods filter mechanically and that scrutiny of feeding appendages would suffice to characterise them. We now know that copepod feeding response is a complex matter. Did we have to take so long to find out, and how do we deal with such complexities?

SUMMARIES

Gerry Capriulo, in his keynote lecture (see Capriulo *et al.* 1990) portrayed the protozoans as a vast complex of feeding types and behaviour. We have indeed come a long way since phagotrophy, osmotrophy and saprotrophy were the only categories available for classifying feeding behaviour. Now, further differentiation is called for. Obviously fascinated by "weird" species, he described some flagellates which "writhe, twist and grind" while attacking their food; a ciliate whose mouth had not yet been located which was reputed to "unzip its entire ventral surface", a capability enabling it to "engulf huge particles". The sarcodines harbour the most "powerful carnivores" in their ranks. Radiolarians will eat anything up to and including larval fish. Prey organisms are captured with their formidable array of spine-supported pseudopodia. After capture, they "rip the food apart, tearing it into shreds", prey tissue is "pried off" from the exoskeleton and the "chunks engulfed". Foraminifera exhibit similar behaviour. This "gluttony" contrasts strongly with the delicate table manners of the Suctoria that feed only with the tips of their tentacles, by invagination. Undoubtedly, the bulk of the protozoa are less remarkable in their feeding habits. These are the organisms that create feeding currents by means of flagella or cilia or those that wander along substrates, ingesting whatever food particles they come across.

Dean Jacobson provided a graphic account of the feeding behaviour of *Protoperidinium* and of other dinoflagellates. Typically, these organisms swim along straight paths but "hungry" individuals exhibit a characteristic "looping swimming pattern" and "twitchy, jerky" movement. Details of the feeding behaviour and the mechanisms involved are already published (Jacobson and Anderson 1986). A feeding bout lasts from 0.5 to 1 hour and an individual consumes about 20 diatom cells - in 15 meals - per day. The entire plasma is retrieved after feeding and the assimilatory efficiency of the predator is quite high, about 40 - 50%. Many types of particles, and not just diatoms, are taken as food. Thus, individuals have been seen feeding on flocculent debris emanating from a declining diatom bloom. The average division rate of these dinoflagellates is approximately one per day which means that they can grow as fast as their food. What, therefore, controls their annual cycles and why are they not more abundant? At the end of his talk Dean reminded us that small can also be beautiful. Does it not follow that more is not always better?

Gottram Ühlig described how the phagotrophic dinoflagellate

Noctiluca uses its tentacle as a raptorial appendage and appears to ingest whatever comes its way; it seems to have "no taste". Even raw petroleum added to the top of a *Noctiluca* culture attracted many individuals which were seen to have ingested, but not digested, oil droplets. The mucoid net filtration mode resembles "social activity". At the surface of the culture vessel, each organism produces a mucoid thread which proceeds to entangle with that of neighbours resulting in a thick net of slime in which the organisms are suspended. This unsightly mass then commences to sink to the bottom of the vessel where the organisms subsequenlty lie for some time, presumably "digesting their meal". Efficient though it might seem, this mucoid net does not remove particles quantitatively from the water column. Organisms as large as *Scrippsiella* were not effectively collected. Whether the dinoflagellate is capable of avoiding the net is a tantalizing question certainly worthy of further pursuit. These laboratory observations confirm those made in the field by Omori and Hamner (1982).

Mary-Jo Sibbald outlined the response of the nutritionally versatile flagellate *Ochromonas danica* to changing environmental factors. This phagotrophic autotroph can utilize either light, food particles or dissolved organic matter (only at artificially high concentrations) as an exclusive form of nutrition. When growing in an autotrophic mode, bacteria can be taken up to supplement various nutrient deficiencies, thus short-circuiting the microbial pathway. This versatility of resource acquisition should put *Ochromonas* at a competitive advantage over the less flexible autotrophs and bacterivores. Nevertheless, organisms with these capabilities tend to play a secondary role, at least in terms of biomass, in the marine environment. The only record of a sizeable bloom to my knowledge, is that of the much dramatized *Chrysochromulina polylepis* bloom that occurred in the transition zone between the North and Baltic Seas in May/June 1988. This outstanding exception proves the rule, so should we conclude that resource acquisition strategies and the resulting competitive edges conferred by greater ability are not all that important in shaping pelagic annual cycles after all? Or are these organisms handicapped in other ways by their diverse talents - "jacks of all trades"? Here again, the *Chrysochromulina* bloom proves that versatile autotrophs are indeed capable of outstripping competitors. Is a certain, unknown combination of factors necessary to launch these species into bloom dimensions?

Gerry Capriulo, in a second talk, presented data on vacuole passage time in a ciliate. Regardless of food concentration, the passage time of fluorescent plastic beads offered as food was 120 min. at the experimental temperature. Digestion time estimated from decolouration of chlorophyll was about half as long. This does not appear to be short when compared to the food processing rate of small metazoa. A conclusion to be drawn from these observations is that protozoan grazing does not necessarily increase community turnover time when compared to that of metazoa. However, the use of plastic beads is a "worst case scenario" as they are not digested. Possibly, more digestible food is processed more rapidly. Further, the species examined - *Fabrea salina* - might turn out to be 'lethargic' in comparison to other unexamined species. The wide fluctuation in species-specific division rates of ciliates documented by Zaika (1973) suggests similarly fluctuating metabolic rates, a feature which needs to be confirmed.

Dian Gifford presented the results of shipboard experiments to assess consumption of planktonic protozoa by calanoid copepods from two widely differing regions. Grazing by the small neritic copepod *Acartia tonsa* from a eutrophic site in the Gulf of Mexico did not impact protozoan populations, particularly during summer. However, the large species *Neocalanus plumchrus* from the oligotrophic Subarctic Pacific consumed 30 - 80% of ambient protozoan carbon per day. Phytoplankton biomass in terms of chlorophyll differed by two orders of magnitude between the regions but this was not the only reason explaining the differences in copepod diets. Thus, clearance rates of *Neocalanus* on protozoa were much higher than on chlorophyll indicating distinct preference for this prey item.

DEBATE

The lively general discussion at the end of the session ranged over many topics. **Dian Gifford**'s talk had left a strong impression, so metazoan grazing of protozoans was first discussed at length. Copepods and euphausiids are known to feed heavily on protozoan populations and hence on the microbial assemblage. Thus, it was suggested that the question of whether the 'microbial loop' acted as a link or sink for the metazoan food chain could be resolved on a seasonal basis. However, the argument of **Evelyn** and **Barry Sherr** and several others that the link/sink controversy was a non-issue was not contested by the audience. The role of protozoans as food for fish larvae was next discussed; most participants would have welcomed evidence that protozoa are indeed an excellent and coveted food source for fish. However, **T Dale**'s observation that cod and turbot larvae did not grow on a diet of strombidiid ciliates and actually "spat out" captured animals was received with some disconcertedness. The discussion was closed by **E Small** who appealed to the audience to avoid over-generalization. He maintained that the ciliates are as complex as any class or even phylum of metazoans and hence should not be packed into a single box of convenience.

The discussion next shifted to whether there are recurrent seasonal patterns of occurrence in the protozoa comparable to the better known annual species succession of phytoplankton. To date, detailed annual cycles of the protozoa have been recorded from only a few temperate, coastal localities; the cycle in Chesapeake Bay was reported by **J Dolan** to be similar to that described in detail for the Kiel Bight (Smetacek 1981). Food supply was considered by **M Elbrächter** to at least partly influence protozoan succession. Thus, **D Montagnes** stated that large, "vicious-looking" ciliates are present in association with the spring bloom but small, "teeny" ones tend to dominate in summer.

The driving forces shaping seasonal patterns of the protozoa were discussed next. **G Uhlig**, drawing on his 20 year study of *Noctiluca* in the German Bight, suggested that light was the major controlling factor governing the sequence of plankton succession including that of the protozoa. The role of temperature in initiating germination of resting spores was raised by **D Jacobson. G Capriulo** invoked competitive exclusion in conjunction with contemporaneous disequilibrium to explain seasonal cycles. Thereafter, the discussion grew increasingly complex as additional factors and considerations were added to the accumulating pile. Calls for differentiation: regional/seasonal, proximate/ultimate, annual/interannual, stochastic/deterministic helped structure the discussion and prevent it from veering into chaos. Predation of protozoans had captured the imagination of the session and the discussion swung back with a comment by **D Stoecker** that some copepods can "clean-up" the ciliate population specifically. When predation by "jellies" on copepods increases, the ciliates return. The Kiel Bight example, where metazoan grazing was suggested to regulate summer protozoan populations, was again invoked in this connection.

Continuing the discussion, **T Dale** pointed out the role of intrinsic life history cycles in which resting stages figure prominently. The suggestion was made that species succession is a continuous phenomenon (presumably driven by organism intrinsic factors) skewed by a numerical response to stochastic perturbations of the physical environment. The implication is that only the quantitative patterns are variable and that the degree of numerical success in a given year is of little bearing for the next. Hypotheses such as these can best be tested by field observations. Thus, we have little or no information on annual cycles from the tropics or the open oceans. **M Elbrächter** summed up by stressing the obvious need for more long-term monitoring studies in different ecosystems to elucidate plankton cycles. The discussion eventually had to be broken up, albeit reluctantly, for our own meal.

The field of protozoan trophic behaviour is obviously in an exponential growth phase as evidenced by the wealth of new information and the many intriguing questions now being raised. The impression we gained from the session was that protozoan feeding behaviour is more complex and, in many cases, can be much more selective than previously thought; the following questions arise:

1 Can we adequately systematize the available and future information?
2 Is it possible to develop methods to assess feeding types and measure rates when dealing with a suite of trophic behavioural patterns?
3 What impact, if any, do the different feeding types have on pelagic food web functioning and magnitude of vertical flux?
4 How do we incorporate this wealth of information into meaningful ecological models?

Gerry Capriulo stressed the theme of "central tendency" in his talk and even urged the modellers to guide their models in accordance with the reigning "central tendency". Applying this dictum to our own times, it seems evident to me that all sciences are undergoing a radical rethinking in their attitude towards complexity. The "central tendency" now appears to be shifting from the search for simple laws and relationships to a struggle with the increasing complexities of the real world as revealed to us by new generations of instruments and techniques. Finding simplicity in complexity has its reward as demonstrated by artistic endeavour through the centuries. Unfortunately, I sometimes feel that we biologists, and particularly ecologists, have not yet whole-heartedly embraced the new "central tendency". Rather than search for simple relationships at the brink of reality we should plunge into the complexities as they are revealed to us and rely on our own ingenuity to perceive and record patterns where we might yet be sensing chaos.

ACKNOWLEDGEMENTS

I thank the organizers of the NATO workshop for creating the amiable atmosphere that enabled the free flow of ideas. These were recorded in admirable quality by the staff of the Polytechnic. Ulf Riebesell and Renate Scharek provided useful comments to the manuscript for which I am most grateful.

REFERENCES

Capriulo GM, Sherr EB, Sherr BF (1990) Trophic behaviour and related community feeding activities of heterotrophic marine protists. In: Reid PC, Turley CM, Burkill PH (eds) Protozoa and their role in marine processes. Springer, Berlin Heidelberg, New York

Jacobson DM, Anderson DM (1986) Thecate heterotrophic dinoflagellates: Feeding behavior and mechanisms. J Phycol 22:249-258

Omori M, Hamner WM (1982) Patchy distribution of zooplankton. Behavior, population assessment and sampling problems. Mar Biol 72:193-200

Smetacek V (1981) The annual cycle of protozooplankton in the Kiel Bight. Mar Biol 63:1-11

Zaika VE (1973) Specific production of aquatic invertebrates. John Wiley and Sons New York

COMMUNITY GRAZING IN HETEROTROPHIC MARINE PROTISTA - SESSION SUMMARY

Gerard M Capriulo
Division of Natural Sciences
State University of New York
Purchase, New York 10577
USA

INTRODUCTION

Various estimates suggest that the earth is populated by perhaps 10^7 species, with in the order of 10^{40} simple (i.e. first order) interactions between them (Wilson 1988, May 1988). Faced with such complexity, ecologists (numbering perhaps in the 10^4 range, worldwide) can have little hope of identifying and describing more than a small percentage of those interactions. Such research is fuelled by at least two forces including: 1. artistic fascination and the intrinsic excitement associated with discovery and 2. the desire and necessity, to understand how ecosystems function. The latter seeks to understand ecosystems to the point where we can predict, through mathematical models, how a system will behave in the future. Models, of necessity, are based on simplifications of reality, with emphasis on unifying themes rather than individual differences. Simplification of the interactions within the microbial component of marine food webs, for the purposes of modelling, requires either that we treat various components of the protista as groups or communities based on things other than taxonomy (e.g. size, Dussart 1965, Sieburth *et al.* 1978, or trophic level) or that we consider individual species activities and estimate how the whole functions based on data concerned with its seemingly most important parts. Both of these strategies have been successfully employed in studies of community grazing (Capriulo 1990, Capriulo *et al.* 1990).

Estimation of protistan community grazing rates represents a second tier of inquiry (the first being the study of individual

NATO ASI Series, Vol. G 25
Protozoa and Their Role in Marine Processes
Edited by P. C. Reid et al.

trophic behaviours and interactions) leading to the construction of predictive models. Community feeding data, to date, have demonstrated that protists are important consumers of bacterial, primary, secondary and tertiary production (Sherr *et al.* 1984, 1986a-c, Capriulo 1990, Capriulo *et al.* 1990, Sieburth 1984, Porter *et al.* 1985), and represent an integral component of the ecology of the seas.

In this session, both oral and poster presentations examined feeding in all the major groups of heterotrophic protists and the microzooplankton as an operational size group. **Barry** and **Evelyn Sherr** in their keynote address pointed out the stimulus that Pomeroy's 1974 paper had given to research on the microbial component of aquatic food webs, and questioned whether or not our current convictions about that microbial component are based on solid research findings or merely shaky ground. In partial answer to their own question, they pointed out the ubiquitous distribution of heterotrophic protistan biomass, the fact that the heterotrophic protista ingest a wide range of food types, controlling, at times, algal and bacterial production, have low energetic costs for their motility, are capable of chemosensory activity, grow as fast or faster than their prey, and, with their high weight specific metabolic and excretion rates, and high growth efficiencies, have a major impact on nutrient cycling. The factor most retarding progress in the estimation of feeding rates is the lack of generally accepted methods for their measurement, (Capriulo *et al.* 1990).

A need for standardization of methods to measure community grazing was underscored by many of speakers who presented data collected using widely different methodologies including: various radiotracer techniques (Lessard, Lee, Becquevort and Servais, Bjornsen), direct microscope observation (Joslin and Laybourne-Parry, Caron and Swanberg, Ayukai), various combinations of stain-induced or auto-epifluorescence techniques including use of inert particles or fluorescently labelled bacteria (FLB) (Sanders, Sherr and Sherr, Capriulo, Galvao and Fritz), size fractionation (Verity), serial dilution

(Strom, Paranjape, Gifford) or combinations of techniques (Kemp, Bjornsen).

FLAGELLATE AND CILIATE BACTERIVORY

Evelyn and **Barry Sherr**, in a second contribution, explained that they found no selectivity differences in uptake of Fluorescent Labelled Bacteria (FLB) versus natural bacteria among the protists they studied (Sherr's unpublished data), and that FLB's are both rapidly and randomly digested. Some saltmarsh and pelagic ciliates eat and survive on bacteria alone (e.g. in some tidal creeks of Georgia, USA). Nonetheless, they believe that flagellates are generally the dominant bacterivores in the marine environment. Bacterial production is difficult to measure in estuaries. While the thymidine incorporation method (Fuhrman and Azam 1982) is an attractive way to measure production, it must be shown to be relatable to actual growth in terms of cell numbers. They estimated the coefficient of variation to be about 50% for bacterial production, and 64% for protist feeding rates, with the magnitude of these parameters averaging 1.4-2.8 x 10^6 cells ml^{-1} d^{-1} and 0.7-1.4 x 10^6 cells ml^{-1} d^{-1}, respectively.

Peter Bjornsen described three methods of measuring community grazing on bacteria which were used to follow grazing losses over a two week period in experimental enclosures of estuarine water, including: 1. Carbon-budget method: unmanipulated samples of bacteria and flagellates are incubated (8-24 hr) and regularly subsampled to measure bacterial cell production and grazing. The method assumes that grazing is the only process by which bacteria are lost. 2. Antibiotics method: (modified after Fuhrman and McManus 1984) unfractionated samples of bacteria and flagellates are incubated (8-24 hr) in the presence of penicillin and streptomycin. Bacterial numbers are followed over time. 3. ^3H Thymidine label method: (modified after Servais *et al.* 1985) - bacterioplankton minus flagellates (from culture or from 1 µm filtration of natural samples) are labelled and mixed with flagellates (from culture or water samples). Grazing is estimated from the activity levels of Thymidine against time. He found that on average grazing losses were balanced by cell production, with flagellates ingesting 5-9 bacteria flagellate^{-1} hr^{-1}. He also described a method to determine bacterial cell volume by automated image analysis.

Robert Sanders considered the seasonal feeding impact of individual members of the bacterivorous plankton community in a eutrophic lake by measuring uptake of bacteria-sized (0.57 µm) microspheres as well as FLB. Flagellates were the major bacterivores accounting for 55-99% of the feeding impact with obligate heterotrophs accounting for 49-81% of the grazing. Mixotrophs were the major feeders during winter and spring blooms, when they were responsible for up to 45% of community feeding. In late spring to early summer rotifers and ciliates were responsible for up to 25-30% of the bacterivory (average = 3-11%, annually). Overall, 11-162% (average = 79%) of the bacterial production was lost to feeding protists. He also indicated that feeding rate measurement inaccuracies, produced by egestion of beads upon fixation (Sieracki *et al.* 1987) could be eliminated, at least in certain species, by use of alternative fixation methods (e.g. addition of sample to an equal volume of ice cold 4% gluteraldehyde).

Sylvie Becquevort followed the disappearance of a radioactive, natural bacterial DNA tracer (^{3}H thymidine) under a combination of conditions involving eucaryote inhibitors (colchicine and cycloheximide) with subsequent filtration through a 2 µm filter, to estimate grazing rates. Experiments were carried out on both natural samples and cultures of heterotrophic nanoflagellates (HNAN). Feeding rates of 35-53 bacteria flagellate^{-1} hr^{-1} were noted for protists in the Scheldt estuary, and 65-108 bacteria flagellate^{-1} hr^{-1} for heterotrophs from Belgian coastal waters. Results from experiments carried out on cultured organisms varied from 20-36 bacteria flagellate^{-1} hr^{-1}.

Helena Galvao described bacteria/protist trophodynamics in a

Chesapeake Bay (USA) salt marsh. Heterotrophic nanoplankton (HNAN) mean growth rates of 0.02 - 0.07 hr^{-1} were observed (maximum growth rates = 0.19 hr^{-1}). Grazing rates varied from 40-150 bacteria consumed $flagellate^{-1}$ hr^{-1} (mean = 80). Of noteworthy importance was the fact that microprotistans, including cryptomonads and tintinnids, significantly predated on and digested flagellated bivalve gametes (e.g. American oyster *Crassostrea virginica*). As many as 30-100% of the HNAN were found to contain ingested gametes during active mollusk spawning seasons, and their growth rates were enhanced by the gametic food. Feeding rates of 40-120 gametes $flagellate^{-1}$ hr^{-1} were recorded.

Paul Kemp considered the fate of bacterial production and the microbial loop in marine sediments. Potential consumers of bacteria in sediments include macrofauna, meiofauna and protistan grazers (microfauna). Macrofauna directly consume only a small fraction of bacterial production (< 10%). Meiofaunal estimates (only infrequently measured) range from 100% to almost no bacterial production removed by this group. For both of these groups bacterivory was most intense when restricted to surface sediments. Among the protists, the microflagellates may represent important bacterivores in organic rich sediment, where their numbers are extremely high, with ciliates likely to have a strong impact in sandy aerobic bottom sediments where they achieve high numbers relative to bacterial density. He concluded that most bacterial production in sediments is ultimately respired and remineralized and thus represents a sink for energy and a source of nutrients. This interpretation is likely to stimulate intense future debate. Certainly, more detailed quantitative data on distribution, abundance, feeding rates and microbial production are needed before the feeding dynamics of benthic protists can be integrated into the larger food web picture.

DINOFLAGELLATE FEEDING

Evelyn Lessard pointed out that heterotrophic dinoflagellates, at times, dominate protistan biomass in coastal and oceanic waters (Lessard and Swift 1985). They feed by a variety of mechanisms (see Capriulo *et al.* 1990, Capriulo 1990) on both large (e.g. diatom chains) or small (e.g. bacteria) food, and may represent important consumers of diatom blooms. In particular, thecate forms have been found associated with diatom blooms in large and small estuarine embayments, as well as along the Gulf Stream front (Lessard unpublished data). Evelyn and her colleagues have used radioisotopically labelled prey to demonstrate that heterotrophic dinoflagellates ingest bacteria and bacteria feeding protists as well as phytoplankton. Her results suggest that dinoflagellates are relatively non-selective feeders particularly with respect to a preference for auto versus heterotrophic food. She reported that *Protoperidinium antarcticum* (from McMurdo Sound) feeds exclusively on bacteria when phytoplankton concentrations are low. Dinoflagellate numbers varied in surface waters from several to 15 ml^{-1} and increased in number in deeper waters to 20 ml^{-1}, and even to 45 ml^{-1} just above the bottom. These numbers represent a biomass range of 1-50 µg l^{-1} which is a significant contribution to protistan biomass. Clearance rates for the dinoflagellates studied varied from about 0.5-28 µl $dinoflagellate^{-1}$ hr^{-1}, with oceanic species exhibiting somewhat lower average rates (0.5-8). These rates are equivalent to those reported for ciliates (Capriulo 1982, Jonsson 1986).

SARCODINE FEEDING

Johanna Laybourn-Parry reported on feeding rates, growth rates and digestion times of the salt marsh amoeba *Thecamoeba pulchra*, feeding on cyanobacteria. This amoeba selects the cyanobacterial species it can ingest, based on prey size and filament pliability. Different food types resulted in different growth rates (Laybourn-Parry *et al.* 1987). For *T. pulchra* feeding on *Nostoc* there was no functional relationship between ingestion rate and temperature between 8-24°C. Above 24°C ingestion rate increased dramatically. Population growth rate varied from 0.001-0.040 hr^{-1} over a temperature range of 8-32°C, and ceased above 36°C. Such a wide range of temperature tolerance with respect to physiological activity makes ecological sense for an organism living in a salt marsh.

Dave Caron presented work, carried out jointly with Neil Swanberg, on feeding in planktonic sarcodines including acantharians, foraminiferans, and radiolarians. In all cases the protists were collected by SCUBA divers in jars and were fixed *in situ* for identification and quantification of the sarcodines and their prey. Tintinnid ciliates represented numerically important prey for all sarcodines studied (about 20 species to date). Other ciliates, diatoms, copepods and mollusk larvae were also important. On a numerical basis foraminifera consumed more diatoms and copepods, radiolarians more tintinnids and mollusk larvae, and acantharians mostly tintinnids. In terms of biomass, however, copepods and mollusk larvae represented two thirds to three quarters of sarcodine food. All species examined appear to be generalists in feeding. This makes evolutionary sense for organisms living in nutritionally dilute environments. Some partitioning of prey did occur; this partitioning followed apparent morphological rather than taxonomic features, suggesting that prey selectivity may be mechanically controlled. An assessment of the quantitative impact of this group is a future hope rather than a present reality given the lack of detailed study of community feeding rates.

John Lee discussed feeding in two benthic foraminiferan species, *Amphisorous hemprichii* and *Amphistegina lobifera* based on radiotracer work carried out in the Gulf of Elat (Israel). Concentrations of locally isolated species of food organisms in the 1×10^6 ml^{-1} range were presented to the foraminifera. Selectivity in choice of food organisms was observed for both species. *A. hemprichii* had higher ingestion rates (about 1.7% of its body weight per hour) than did *A. lobifera* (about 0.83% per hour) with no apparent light effects on feeding despite the fact that both species contain algal endosymbionts. Estimation of community ingestion rates for these and other benthic foraminiferans are complicated by the fact that feeding (as well as egestion) is episodic. Food is gathered in large amounts over a short time interval after which digestion occurs. Additional complicating factors include feeding behavioural differences and prey preferences among species (see Capriulo 1990 for a review), that all species have not been studied, and that captivity effects the feeding rate of many forms including those considered in this study. All of the above considerations make estimation of community feeding rates difficult. For this reason the role of the benthic foraminiferans in the marine food web remains an open question.

CILIATE FEEDING

Gerard Capriulo described the use of quantitative microscope epifluorescence to measure the ingestion rates of ciliates feeding on algae. This method provides an instantaneous measure of ciliate feeding based on previous feeding history, without experimental incubations, if one has previously determined digestion times for the species in question. If not previously determined, digestion time measurements require incubations of about two hours. This method is well suited for work with fragile ciliates that cannot be studied using standard methods. Ingested chlorophyll *a* as pigment derived fluorescence in single ciliate cells is measured by quantitative, microscope based, monochromatic spectrum analyses. Initially 522 individual single cell measurements were made on four major taxonomic groups of algae to determine the relative fluorescence (RF) at 682 nm (the *in vivo* chl *a* peak) to unit chlorophyll *a* ratio. This ratio varied by a maximum of 3.4 x among algal divisions (grand mean for all species = 3.31×10^7 RF 682 nm per µg chl *a* ± S.E. = 3%). Next, ciliate vacuole passage time (VPT) and digestion time (DT) must be measured or obtained from a previous data set. Measurements of VPT or DT, respectively, are made by pulse feeding ciliates, under constant food conditions, either fluorescent beads or the red colored flagellate *Rhodomonas lens*, as visual tracers. At room temperature (~ 24°C) VPT and DT were constants and for *Fabrea salina* (a marine heterotrich coastal ciliate) non-functionally related to food concentration (120 and 71 min, respectively). Measurements of total 'gut fullness' with respect to ingested pigment based on previously determined fluorescence to chl *a* ratios and adjusted for vacuole acidification are made next. Assuming that the ciliate approximates a steady state system with respect to chlorophyll *a* then division of gut fluorescence by digestion time provides an estimate of ingestion rate in units of relative fluorescence or chl *a* ciliate^{-1} per time. Experiments on *F. salina* feeding on *Dunaliella tertiolecta* (at food concentrations in the 10^5 cells ml^{-1} range) indicated ingestion rates of 0.5-3.6 cells ciliate^{-1} min^{-1} with corresponding clearance rates of 0.3 to 1.2 µl ciliate^{-1} hr^{-1}.

MICROZOOPLANKTON COMMUNITY FEEDING RATES

Suzanne Strom, Mahdu Paranjape and **Dian Gifford** each discussed microzooplankton community feeding rates from different ecosystems, measured (with occasional minor variation) by means of the Landry and Hassett (1982) serial dilution technique.

Suzanne Strom described the field relationship between light intensity and microzooplankton grazing at Station P in the subarctic North Pacific Ocean. Parallel dilution series were incubated at light intensities ranging from 0-50% of surface irradiance (I_0), in deck mounted incubators. Grazing rates (units of I/time, based on Frost's 1972 equations) were highest in the dark (0.26 d^{-1}) and lowest (about 0) at 10% I_0. The percentage of the grazing organisms containing algal endosymbionts and the effect this might have had on the feeding rates measured under varying light conditions, was not known. The ciliate *Strombidium* sp. isolated from Station P (September, 1987) was studied in the laboratory by Suzanne, to determine its growth and feeding rates under different light regimes (100% to 0.1% of average Station P I_0 levels).

Filtration rates of 1-20 µl $ciliate^{-1}$ hr^{-1} were found with rates proportional to light intensity.

Madhu Paranjape estimated the feeding impact of microzooplankton over the Grand Banks Canadian Atlantic. Samples were chlorophyll fractionated into less than and greater than 20 µm subsamples. Experiments were run for 48 hr in the presence of excess nutrients, and subsamples were taken at 24 hr. The phytoplankton community was dominated by diatoms in April and coccolithophores and flagellates in mid-summer and late fall. For all the seasons studied oligotrichs and tintinnid ciliates were the dominant predators. Grazing rates for the microzooplankton feeding on < 20 µm food varied from g = 0.15-0.25 d^{-1}. This represents about 14-22% of the phytoplankton chlorophyll a standing crop removed per day or 33-80% of the daily production, with coccolithophores, small dinoflagellates, microflagellates as well as small diatoms all accepted as food.

Dian Gifford measured the feeding rates of natural micro-zooplankton communities in Halifax Harbor, Nova Scotia, Canada. During her studies, the microzooplankton were dominated by oligotrich ciliates (2,600 to 11,000 ciliates l^{-1}). The main phytoplankton species varied with season as did the microzooplankton grazing impact. About 74% of the standing crop of chlorophyll a per day (50% of the potential production) was removed by the microzooplankton in June at which time the waters were dominated by less than 12 µm flagellates. The feeding impact reached 100% of the potential daily production in March, over a period in which grazing and growth of phytoplankton appeared to be in balance. In November during a bloom of large diatoms grazing was negligible. The two main assumptions of the serial dilution method (i.e. that phytoplankton nutrients are not limiting and that threshold feeding does not occur) were also tested and it was determined that, at times, threshold feeding effects and some nutrient limitation might be occurring, but that the experimental results were largely unaffected by this.

Peter Verity reported on the distribution of phytoplankton, bacteria, cyanobacteria, heterotrophic nanoplankton and ciliates (weekly sampling), as well as on the growth and grazing rates of the size fractionated (< 2, 2-8 and 8-200 µm) components of the plankton of the Southern Atlantic Bight. Peter also estimated grazing rates by the serial dilution method and examined ciliate preferences in association with various algal food types. The abundances of all groups were temperature related with average bacterial densities of 1-5 x 10^6 cells ml^{-1}, nanoplankton densities of 1-15 x 10^3 ml^{-1} and ciliates numbering in the range of 1-25 ml^{-1}. Bacterial production averaged 1-3 x 10^6 cells ml^{-1} d^{-1}, of which 50% to 70% were grazed by the 2-8 µm size fraction which itself exhibited 2-4 doublings per day. 50-75% of the 2-8 µm heterotrophic nanoplankton production and 50% of the autotrophic production was grazed by the 8-200 µm size fraction (mostly ciliates and dinoflagellates). Ciliates grew at rates of 1-3 doublings per day. Feeding rates determined by means of the fractionation method were not significantly different (due to high variabilities) from those measured by the dilution technique. Peter pointed out that due to problems associated with the techniques, data collected by fractionation experiments should be viewed cautiously and the method employed only when field conditions are such that minimum error can be expected. Methodological problems include: 1. extreme fragility of certain protists with unmonitored lysis of cells occurring with even gentle mechanical agitation (e.g. Gifford 1985) and associated measurement inaccuracies and 2. predator/prey size overlap confounding attempts to decouple trophic interactions by sieving (e.g. < 2 µm bacterivores, Fuhrman and McManus 1984).

NON-LORICATE CILIATE/COPEPOD DISTRIBUTIONAL CORRELATIONS

Tenshi Ayukai motivated by published reports of the importance of ciliates to the diet of copepods. (Heinle *et al.* 1977, Robertson 1983, Stoecker and Sanders 1985) examined the temporal (seasonal and in some instances daily) and spatial (3 stations) relationship between non-loricate ciliates and copepods in Onagawa Bay Japan and found interesting positive correlations between the two groups, particularly with respect to harpacticoid copepods (and including nauplii and copepodite stages). He recommended that these be followed up by detailed feeding studies, to ascertain the functional relationship between the two groups.

CONCLUSIONS

Quantitative feeding rate data is most complete for planktonic ciliate and heterotrophic flagellate communities. Many techniques have been employed and results derived from widely differing methodologies are often in good agreement. This has resulted in a degree of consensus among researchers, concerning the trophodynamic importance of these protists. No such consensus exists concerning the under-investigated sarcodine and benthic protist groups. Quantification of dinoflagellate community feeding, although only begun recently, points to the potential importance of this group of protists as grazers.

At present, it is justifiable to say that flagellate, ciliate and likely dinoflagellate groups represent major components of marine food webs (as important, at least at times, as the microcrustaceans), and must be included in food web models. In the future, continued research on the relatively well understood groups must be carried out in a wider range of environments. Additionally, intensified efforts concentrated on benthic protistan and planktonic sarcodine activities are needed in order that their trophodynamic importance to marine food webs might be determined.

REFERENCES

Burkill PH, Mantoura RFC, Llewellyn CA, Owens NJP (1987) Microzooplankton grazing and selectivity of phytoplankton in coastal waters. Mar Biol 93:581-590

Capriulo GM (1982) Feeding of field-collected tintinnid microzooplankton on natural food. Mar Biol 71:73-86.

Capriulo GM (1990). Feeding Related Ecology of Marine Protozoa. In: Capriulo GM (ed) The Ecology of Marine

Protozoa, Oxford University Press, New York, Oxford
Capriulo GM, Sherr EB, Sherr BF (1990) Trophic behaviour and related community feeding activities of heterotrophic marine protists. In: Reid PC, Turley CM, Burkill PH (eds) Protozoa and their role in marine processes. Springer, Berlin Heidelberg New York
Ducklow H, Hill SM (1985) The growth of heterotrophic bacteria in the surface waters of warm core rings. Limnol Oceanogr 30:239-259.
Dussart BM (1965) Les differentes categories de plancton. Hydrobiologia 26:72-74
Frost BW (1972) Effects of size and concentration of food particles on the feeding behavior of the marine planktonic copepod *Calanus pacificus*. Limnol Oceanogr 17:805-815
Fuhrman JA, McManus GB (1984) Do bacteria-sized marine eukaryotes consume significant bacterial production? Science 224:1257-1260
Fuhrman JA, Azam F (1982) Thymidine incorporation as a measure of heterotrophic bacterioplankton production in marine surface waters: evaluation and field results. Mar Biol 66:109-120
Gifford DJ (1985) Laboratory culture of marine planktonic oligotrichs (Ciliophora), Oligotrichida). Mar Ecol Prog Ser 23:257-267
Jonsson PR (1986) Particle size selection, feeding rates and growth dynamics of marine planktonic oligotrichous ciliates (Ciliophora: Oligotrichina). Mar Ecol Prog Ser 33:265-277
Landry MR, Hassett RP (1982) Estimating the grazing impact of marine micro-zooplankton. Mar Biol 67:283-288
Laybourn-Parry J, Jones K, Holdich JP (1987) Grazing by *Mayorella* sp. (Protozoa; Sarcodina) on cyanobacteria. Functional Ecol 1:99-104
Lessard EJ, Swift E (1985) Species-specific grazing rates of heterotrophic dinoflagellates in oceanic waters, measured with a dual-label radioisotope technique. Mar Biol 87:289-296
Linley EAS, Newell RC, Lucus MI (1983) Quantitative relationships between phytoplankton, bacteria and heterotrophic microflagellates in shelf waters. Mar Ecol Prog Ser 12:77-89
May RM (1988) How many species are there on earth? Science 241:1441-1449
McManus GB, Furhman JA (1988) Control of marine bacterioplankton populations: Measurement and significance of grazing. Hydrobiologia 159:51-62
Pace ML (1988) Bacterial mortality and the fate of bacterial production. Hydrobiologia 159:41-49
Paranjape M (1987) Grazing by microzooplankton in the eastern Canadian arctic in summer, 1983. Mar Ecol Prog Ser 40:239-246
Pomeroy LR (1974) The ocean's food web, a changing paradigm. Bioscience 24:499-504
Porter KG, Sherr EB, Sherr BF, Pace M, Sanders RW (1985) Protozoa in planktonic food webs. J Protozool 32:409-415
Rassoulzadegan F, Laval-Peuto M, Sheldon RW (1988) Partitioning

of the food ration of marine ciliates between pico and nanoplankton. Hydrobiologia 159:75-88
Servais P, Billen G, Rego JV (1985) Rate of bacterial mortality in aquatic environments. Appl Environ Microbiol 49:1448-1454
Sieburth J McN, Semtacek V, Lenz J (1978) Pelagic ecosystem structure: heterotrophic compartments of the plankton and their relationship to plankton size fractions. Limnol Oceanogr 23:1256-1263
Sieburth J McN (1984) Protozoan bacterivory in pelagic marine waters. In: Hobbie JE, Williams PJ leB (eds) Heterotrophic activity in the sea. Plenum Press, New York, p 405
Sieracki ME, Haas LW, Caron DA, Lessard EJ (1987) The effects of fixation on particle retention by microflagellates: underestimation of grazing rates. Mar Ecol Prog Ser 38:251-258
Sherr BF, Sherr EB, Andrew TA, Fallon RD, Newell SY (1986a) Trophic interactions between heterotrophic protozoa and bacterioplankton in estuarine water analysed with selective metabolic inhibitors. Mar Ecol Prog Ser 32:169-180
Sherr EB, Sherr BF, Fallon RD, Newell SY (1986b) Small aloricate ciliates as a major component of the marine heterotrophic nanoplankton. Limnol Oceanogr 31:177-183
Sherr EB, Sherr BF, Paffenhofer G-A (1986c) Phagotrophic protozoa as food for metazoans: a "missing" trophic link in marine pelagic food webs? Mar Microb Food Webs 1:61-80
Sherr BF, Sherr EB (1984) Role of heterotrophic protozoa in carbon and energy flow in aquatic ecosystems. In: Klug M, Reddy CA (eds) Current Perspectives in Microbiol Ecology. Amer Soc Microbiol (Washington DC)
Stoecker DK, Sanders NK (1985) Differential grazing by *Acartia tonsa* on a dinoflagellate and a tintinnid. J Plankton Res 7:85-100
Taylor GT, Pace ML (1987) Validity of eukaryotic inhibitors for assessing production and grazing mortality of marine bacterioplankton. Appl Environ Microbiol 53:119-128
Wikner J, Anderson A, Normark S, Hagstrom A (1986) Use of genetically marked minicells as a probe in measurement of predation on bacteria in aquatic environments. Appl Environ Microbiol 52:4-8
Wilson EO (1988) The Current state of biological diversity. In: Wilson EO (ed) Biodiversity. p 521
Wright RT, Coffin RB (1984) Measuring microzooplankton grazing on planktonic marine bacteria by its impact on bacterial production. Microb Ecol 10:137-149

Trophic Behaviour and Related Community Feeding Activities of Heterotrophic Marine Protists

Gerard M Capriulo*, Evelyn B Sherr▪ and Barry F Sherr▪

*Division of Natural Sciences
State University of New York
Purchase, New York 10577

▪University of Georgia Marine Institute
Sapelo Island, Georgia 31327

INTRODUCTION

As a consequence of their evolutionary diversity in form and function the protista of aquatic ecosystems exhibit a great variety of complex trophodynamic interactions, which confounds attempts to produce gross estimates of their *in situ* feeding rates. It is our view that studies of trophic behaviour at species and taxon level and quantification of community ingestion rates, are needed to develop an understanding of how food webs operate. New behavioral observations (e.g. veil/pallium feeding in dinoflagellates, Gaines and Taylor 1984, Jacobson and Anderson 1986, or small, aloricate planktonic ciliates feeding on bacteria, Sherr and Sherr 1987) suggest possible directions for future quantitative studies.

This contribution reviews (see Capriulo 1990) trophic behaviours of marine heterotrophic protists (defined as any protist that obtains all or part of its nutrition by ingesting previously synthesised organic matter). From this starting point, we progress to a quantitative assessment of the importance of different protistan groups to the ecology of the seas. Finally, we consider trophic interactions and feeding impact in relation to evolutionary forces. In so doing we point out the likely existence of a unifying theme or framework for the microbial food web. It is this underlying theme that represents the modeller's 'best route' to produce quantitative budgets at the ecosystem level.

NATO ASI Series, Vol. G 25
Protozoa and Their Role in Marine Processes
Edited by P. C. Reid et al.

TROPHIC BEHAVIOUR AND INTERACTIONS

Protists that are at least partially dependent on heterotrophic activity for their nutritional requirements have developed numerous detailed, and at times even species-specific, mechanisms for obtaining required organics which can be categorised as:

1. uptake of dissolved organic compounds of small, and perhaps large, molecular weight through the cell membrane or part of it (this sometimes follows previous capture of particulate food with subsequent external digestion)
2. feeding directly on both living and detrital particles by means of endocytosis (reviewed in Nisbet 1984)
3. metabolic exchange with endosymbionts
4. combinations of all or some of the above

Flagellates

Heterotrophic flagellates include both phyto- and zooflagellate groups (Lee *et al.* 1985, Anderson 1988, Sleigh 1989) which exhibit the following general characteristics:

1. a cytostome (mouth) which is not found in all species
2. food and nutrient uptake occurs either by diffusion, active transport or endocytosis
3. certain species can switch from autotrophy to heterotrophy in the absence of light (perhaps also in the absence of inorganic nutrients)
4. individuals possess one or more flagella which are used for locomotion, attachment to a substrate the creation of feeding currents or for wafting food into the cytostome.
5. individuals of certain species associate into colonies

In general, food capture is accomplished by filtration, as is found in the monads, bicoecids (= choanoflagellates), actinomonads, pseudobodonids and paraphysomonads, or by direct encounter as in many bodonids (Fenchel 1987, 1986a, b, Sleigh 1989). It is likely that chemoreception is an important mechanism for locating food, at least in certain species, (Sibbald *et al.* 1987, 1988). In most instances water currents move particles to a point on the cell surface where phagocytosis takes place. The actual site of ingestion varies with group and may occur by endocytosis at the flagella base (e.g. *Monas*) by pseudopodial trap (e.g. *Actinomonas*), by pseudopodial engulfment at the base of a collar of tentacles, as in the choanoflagellates (e.g. *Codosiga*) or by endocytosis at the base of an organized cytostome (e.g. *Pleuromonas jaculans*) (Sleigh 1964, Fenchel 1982 a-d, 1986a, b, 1987, Leadbeater and Morton 1974, Anderson 1988).

The phytoflagellates exhibit many nutritional modes ranging from complete autotrophy to mixotrophy and obligate heterotrophy (Leedale and Hibberd 1985). Sanders and Porter (1988) provide a good review of the phagotrophic phytoflagellates. Phagocytosis has never been observed in the volvocids, silicoflagellates and prasinophytes. There is scant published evidence and more anecdotal evidence indicating phagocytosis in colorless cryptomonads (e.g. Mignot 1966, Sherr and Sherr unpublished data). Particle feeding has been observed in numerous chrysophytes as well as in euglenids, dinoflagellates, haptophytes and chloromonads. Some euglenids are colorless and possess a permanent mouth (Leedale 1967). Members of the genus *Heteronema* feed on bacteria and coccoid algae (Loefer 1931). Some forms such as *Peranema trichophorum* and *Entosiphon sulcatum* in addition to a mouth possess a specialised feeding organelle, the rodorgan in *Peranema* (Chen 1950, Leedale 1967, Mignot 1966, Nisbet 1974, 1984) and a siphon tube based feeding structure in *Entosiphon sulcatum* (Triemer and Fritz 1987). Feeding in *Peranema* is accompanied by violent writhing and twisting of the organism. *Ochromonas* sp. are primarily mixotrophic, capable of being phagotrophic (Aaronson 1980) and at times cannibalistic (Fenchel 1982b). In certain chrysochromulinids as well as some haptophytes, particle feeding is achieved by means of rhizopodial engulfment with the rhizopods originating from between body scales. Some dinoflagellates capture and ingest particulate food by engulfing it at special regions of the sulcal groove (Barker 1935, Droop 1953, Kimor 1981, Morey-Gaines and Elbrachter 1987, Frey and Stoermer 1980, Dodge and Crawford 1970). Particles drawn in by flagella-generated currents are caught in some species by fine cytoplasmic nets or by means of a feeding veil (Gaines and Taylor 1984) or a tow filament and pseudopodial pallium (Jacobson and Anderson 1986). This pallium/veil feeding mechanism is used to feed on diatoms including long chains (up to 58 cells long observed to date) and species with external spiny projections (Jacobson and Anderson 1986, Gaines and Taylor 1984). *Noctiluca* is an obligate heterotrophic dinoflagellate that possesses a permanent mouth at the base of a tentacle, which it uses to capture food (Droop 1954, 1959). It is a voracious and indiscriminate feeder which has been shown to have a quantitatively significant effect on copepod production through predation on eggs in Ise Bay, Japan (Sekiguchi and Kato 1976). *Gyrodinium* possesses exploding trichocysts which are likely used in prey capture. *Gyrodinium pavillardi* ingests bacteria, algae and larger protists including the ciliate *Strombidium* (Biecheler 1952, Sieburth 1979) which itself feeds on smaller dinoflagellates. *Polykrikos* produces nematocysts similar to those of coelenterates (Greuet 1972, Greuet and Hovasse 1977). *Oxyrrhis marina* ingests bacteria, diatoms and flagellates (Barker 1935, Dodge and Crawford 1974, Sieburth 1979). Certain dinoflagellates, such as *Dissodinium*, parasitise the eggs of copepods (Chatton 1952, Elbrachter and Drebes 1978, Drebes 1969) or are ectoparasitic on marine invertebrates (Taylor and Seliger 1979), while others (e.g. *Dubscquella*) parasitise tintinnids. In general, particle feeding appears to be widespread among the dinoflagellates, which have been shown to be quantitatively significant grazers in the planktonic environment with filtration rates similar to those reported for ciliates (Lessard and Swift 1985).

Zooflagellates are all obligate heterotrophs which lack plastids and include saprozoic, holozoic and parasitic forms (Lee and Capriulo 1990, Capriulo 1990). The principal free living marine groups include the choanoflagellates (colored forms may also exist, E B Small pers. comm.), the ebriids and the kinetoplastides (which also include many parasitic species). The single flagellum of choanoflagellates creates feeding currents which bring food particles in contact with the collar where they are trapped and subsequently phagocytized by pseudopodial engulfment (Anderson 1988, Fenchel 1986a, b, 1987). The kinetoplastids (e.g. bodonids) (Brooker 1971, Vickerman 1976, and Fenchel 1982a-d, 1986, 1987) generally have a well formed cytostome and associated cytopharynx. Food items are wafted into the mouth by the action of mastigoneme-covered flagella. *Rhynchomonas* is unique in that it possesses a cytostome at the end of a mobile, scavenging proboscis, which is attached to the anterior flagellum (Nisbet 1974, Burzell 1973, 1975). Fenchel (1982a) described feeding in *Pseudobodo tremulans* which is similar to that of *Ochromonas* and *Paraphysomonas*. *Pleuromonas jaculans* is also a typical representative of the free living kinetoplastids which use their anterior flagellum to drive water and food to the cytostome.

Feeding rates in phagocytotic flagellates are, in general, controlled by:

1. time to phagocytize (handling time)
2. food concentration
3. prey size, shape and level of motility
4. velocity of the feeding current
5. geometry of the feeding current
6. chemosensory capabilities and associated feeding selectivity
7. vesicle recycling capability, as well as synthesis of enzymes and membrane material involved in food vacuole formation
8. concentration and type of suspended particles (most flagellates must attach to particles before feeding to improve the efficiency of food collection during generation of feeding currents)

Additionally, phagotrophic flagellates are important to the decomposition cycle, in as much as they stimulate bacterial growth through grazing (Sherr *et al.* 1982, 1983, Fenchel and Harrison 1976). This has been demonstrated in the field for the Sea of Japan (Sorokin 1977), the English Channel (Newell and Linley 1984), and in Lake Kinneret, Israel (Sherr *et al.* 1983).

Sarcodines

For the most part naked amoebae appear to be primarily bacterivores and herbivores (Lee *et al.* 1985), and are typically found associated with surfaces including those of various seaweeds (Lee *et al.* 1985), and the air-sea interface (Davis *et al.* 1978), where bacteria and detritus accumulate. Sieburth (1979) gives a good review of the trophic ecology of this group. Little has been added to the ecological literature beyond that summary. Growth, reproduction and locomotion of at least certain species have been shown to be dependent on food concentration and their feeding has been shown to stimulate bacterial growth and activity (Pussard and Rouelle 1986).

Testaceans are primarily freshwater dwellers, but are also found in the marine pelagial, associated with beach sand (Wailes 1927, 1937), and with bacterial-rich surfaces as well as the textured surfaces of various invertebrates such as bryozoans. Haberey (1973a and b) described the feeding habits of some forms.

Filose amoebae (both monolocular testate and naked forms exist) possess slender pseudopods and are found among the holdfasts of kelp and under rocks (Lee *et al.* 1985). The Acarpomyxea are much branched and are found associated with coral reefs or rich bottom sediments of deep fjords, where they are thought to act primarily as scavengers. The food of naked amoebae includes bacteria, cyanobacteria, diatoms, flagellates, other protista (including intraspecific cannibalism) and detritus (Anderson 1988, Lee *et al.* 1985). Pseudopods can be used as 'pincerlike pseudopods' or a feeding cup can be formed to capture prey

(Anderson 1988). Large prey are captured at the anterior end and smaller prey (e.g. bacteria) on the ventral side, including the uroid region (Anderson 1988). Scale bearing forms associated with *Trichodesmium* in the open ocean ingest bacteria (Anderson 1977, 1988).

Foraminifera obtain their nutrition by the uptake of DOM, herbivory, carnivory, omnivory, cannibalism (including their own gametes), parasitism, scavenging and mixotrophy, with a wide variety of food sources (Lipps 1982, 1983, Christiansen 1971, Capriulo 1990). Feeding in all species is passive (i.e. there are no self-generated feeding currents), and is achieved by suspension or deposit feeding (Lipps 1982, 1983, Hemleben *et al.* 1985, Bè 1982, Nisbet 1984, Capriulo 1990). Prey are captured by random encounter, which is, at least in some species, enhanced by chemically attracting prey and/or potential symbionts by what has been called the 'Circean effect' (Lee *et al.* 1961, 1963).

Anderson and Bè (1976a, b) described prey capture by the planktonic species *Hastigerina pelagica* feeding on *Artemia* nauplii. Capture of larger prey is initiated by prey contact with the spines of the foraminiferan, followed by rhizopodial flow to ensnare the prey. This is followed by death due to exhaustion. Prey death signals the onset of penetration of the prey cuticle and subsequent engulfment of tissue by the advancing rhizopodia, and formation of food vacuoles. Smaller prey, in similar but simpler fashion, are trapped by rhizopodial flow after which food vacuoles are produced. Spinose planktonic species are generally carnivorous (or omnivorous) and include calanoid copepods as a main part of their diet, while non-spinose forms rely more on herbivory (Caron and Bè 1984, Bè *et al.* 1977, Anderson *et al.* 1979). It appears that spines support the rhizopodial network, thus allowing for the capture of larger prey (Bè 1982).

Delaca (1982) reported the use of dissolved amino acids by the benthic foraminiferan *Notodendrodes antarctikos*. Benthic species appear to be more selective in their feeding habits than planktonic ones. Benthic suspension feeders at times extend their pseudopods above the substratum into the overlying water column where they may act as a seine, or capture food by direct interception, inertial impaction, mobile particle deposition or gravitational deposition. These types of general suspension feeding are described by Rubenstein and Koehl (1977); other benthic forms are deposit feeders.

Benthic foraminifera are found associated with mud bottoms and to a lesser extent sandy and rocky bottoms. Deeper dwelling species rely on dead organic particles for food. *Pilulina* shapes its test into a pit which it then uses to trap food by gravitational deposition. *Elphidium crispum* (as well as certain other species) weaves itself into a feeding cyst formed from food trapped in a pseudopodial network. *Strebulus* sp. sends up pseudopods from deeper in the sediment to the mud surface. *Nemogulmia longivariabilis* contains numerous nematocysts obtained from various hydrozoans (Nyholm 1956). Some species stand up off the bottom and take in suspended organic debris (e.g. *Dendrophrya erecta*).

Sandy and rock bottoms have less DOM due to the decreased surface area associated with coarser sediment. Fewer foraminifera are found in these habitats. *Astrorhiza limicola* extends pseudopods into the sand interstitia for food (Lipps 1983). *Elphidium crispum* is also capable of suspending itself between stipes of coralline algae on sandy bottoms, and suspension feeding. On rocky bottoms, foraminifera are found among the holdfasts of algae and associated with algal mats and/or algal slimes, where they feed on bacteria, diatoms and the like. They are also found in abyssal benthic environments (Christiansen 1971, Lipps 1983). Some species (e.g. *Placopsilina cerromana* and *Bdelloidina vincentoronensis*) can be found associated with the pore opening of burrowing sponges or with brachiopods, where they make use of their hosts' movements and feeding activity to enhance their own passive food collection (Bromley and Nordmann 1971, Zumwalt and Delaca 1980, Lipps 1983).

Actinopods

This group of protists includes the heliozoans, acantharians, and radiolarians. Little is known of the feeding ecology of the acantharians, although axopodial movement has been studied (Febvre-Chevalier and Febvre 1986).

Trophodynamics in the few marine heliozoans has been little studied; they are generally passive feeders unable to create feeding currents, and often contain algal symbionts. Some data exists on preferred prey and feeding in the Heliozoa (Greissman 1914, Looper 1928, Bovee and Cordele 1971). Known prey include gastrotrichs, ciliates, yeasts, flagellates and metazoans, with some heliozoans feeding cooperatively, temporarily fusing together to feed and later separating again (Sieburth 1979). Heliozoan feeding (Bovee and Cordele 1971, Patterson 1979, Patterson and Hausmann 1981, Hausmann and Patterson 1982) involves special pseudopods for different prey (i.e. small, straight pseudopods which form cups upon contact with small prey, and large, wide pseudopodial outgrowths that advance over prey from all directions, for larger prey). Prey contact the adhesive axopodial tips and are then drawn in towards the predator by axopodial retraction. Extrusomes are often employed during feeding.

Radiolarians have a catholic diet (Anderson 1980, 1983ab, 1978a, Anderson *et al.* 1983, 1984, Anderson and Botfield 1983, Swanberg *et al.* 1986, Swanberg 1983, Swanberg and Harbison 1980, Swanberg and Anderson 1985, Caron and Swanberg 1988). Most are omnivores with some preference for zooplankton over phytoplankton. Some form colonies cm to m in size (Swanberg and Harbison 1980, Anderson *et al.* 1985) which feed on a variety of larger prey while also playing host to algal endosymbionts (Anderson 1976, 1978b). Small radiolarian species eat, among other small prey, bacteria (e.g. the Nasselarians and Spumellarians).

Food capture takes place by random, passive encounters with prey which are trapped by rhizopodial adherence. Subsequent pseudopodial flow and engulfment occurs, and food vacuoles are formed. Prey die by exhaustion, or dismemberment by rhizopodial shearing action. After capture, prey are carried toward the central capsule where digestion occurs. In larger skeletal bearing species, a contractile response of the axopods which bring prey to the pericapsular cytoplasm (similar to what is found in the heliozoans) has been observed. Large prey (e.g. copepods) represent variations on the capture theme. More rhizopodial engulfment occurs, which causes the prey to become progressively more tangled. The rhizopods flow along the exoskeletal surface and engulf the appendages, while the prey is continuously drawn in deeper, and increased pressure is exerted on the exoskeleton, which ultimately ruptures. This rupture is followed by rhizopodial penetration to the underlying tissue, which is slowly pried from the exoskeleton and enclosed in a food vacuole for later digestion (Anderson 1980, 1983).

Much differentiation in trophic patterns exists among radiolarians. Those with long spines use them as a reinforcement for the advancing pseudopods. This represents a mechanical advantage which enables the pseudopods to be deployed to greater distances, while simultaneously providing more adhesive surface area for use in prey capture. Other species possess a large gelatinous spherical envelope or an alveolar-like frothy mass of extracapsular cytoplasm to support rhizopodial networks.

Ciliates

Most free living ciliates have mouths and feed on particulate food (Fenchel 1968, Small and Lynn 1985), particularly since the concentration of dissolved organics needed to support their growth rarely, if ever, exists in nature. Dissolved organics adsorbed onto particles could, at times, support the growth of some ciliates, as well as other protists, but likely represents an incidental uptake of nutrients. The principal food of ciliates are particles, including bacteria, chroococcoid cyanobacteria, microalgae, diatoms, dinoflagellates, heterotrophic microflagellates and other ciliates. Food collection is achieved in a number of ways including:

1. suspension feeding - oral membranelles may act as filters removing particles based on their size and shape (Fenchel 1980a-d, 1986a, b, 1987) and potentially their charge. These membranelles may or may not contain extrusomes (Gold *et al.* 1979). It is likely that the extrusomes are involved in prey capture, as has been discussed for tintinnids (Capriulo *et al.* 1986). Ciliates may utilize food items, such as certain chain-forming diatoms, normally unavailable to them due to handling problems associated with inappropriate food size, if the food size is first modified in nature by the feeding activities of other organisms (e.g. microcrustaceans) or the ciliates, themselves, or by environmental turbulence-related fragmentation (Capriulo *et al.* 1988)
2. Active predatory hunting - (Salt 1967) as is found in ciliates such as *Didinium* (Wessenberg and Antipa 1970, Luckinbill 1974). Such hunting may be by random encounter or may be stimulated by chemoreception (Verity 1988, Levandowsky *et al.*1984)
3. Deposit feeding - such as is found in certain benthic ciliates e.g. *Strombidium* sp. ingests sediment to capture associated microbial biomass, *Tracheloraphis* lives in the interstitial spaces of sediment, and 'unzips' a large section of its ventral surface to engulf food (Lenk *et al.* 1984)
4. Mixotrophy - either due to the presence of algal endosymbionts or by retention of functional chloroplasts (McManus and Fuhrman 1986b, Laval-Peuto and Febvre 1986, Stoecker *et al.* 1987, Lindholm 1985, Jonsson 1986). *Mesodinium rubrum* is a completely autotrophic ciliate (Lindholm 1985)
5. Parasitism - nutrition derived from a host on or in which the ciliate lives (e.g. apostome ciliate *Collinia*

beringensis living in the krill (*Thysanoessa inermis*) haemocoel (Capriulo and Small 1986)

Ciliates produce feeding currents whose patterns and velocities are species specific and dependent on body shape and size, as well as determined by arrangement and spacing of oral and somatic ciliature (Fenchel 1980a-d, 1986a, b, 1987). Mouths in ciliates vary greatly from simple types which can be shallow or deep (Fenchel and Small 1980, Sieburth 1979, Small and Lynn 1985, Small 1984) with or without a well developed cytopharynx and elaborate associated oral ciliature. Interciliary spacing, at least in benthic or surface loving ciliates, largely determines the size of food that can be retained. Such rules do not apply to voracious, raptorial feeding carnivorous ciliates (e.g. litostome ciliates, Small and Lynn 1985). Bacterivorous holotrichs can retain food down to 0.1 µm size, while many planktonic forms are limited to food >1 µm in size (Fenchel 1986a, b, 1987, 1980a-d).

Data of Porter *et al.* (1979), Berk *et al.* (1976) and Fenchel (1980a, b) demonstrated that most bacterivorous ciliates require bacterial food concentration in the 10^7-10^9 ml^{-1} range to achieve positive growth. These concentrations are typical of benthic communities and are found associated with suspended, planktonic aggregates (i.e. snow and fluff) where bacterivorous ciliates routinely occur, but rarely are reached in open planktonic habitats. For this reason it has been believed that bacterivorous ciliates should be generally absent from the plankton where larger ciliates (particularly choreotrichs, Montagnes and Lynn 1988, Montagnes *et al.* 1988) dominate. The food of these planktonic ciliates primarily consists of nanoplankton-sized food (e.g. autotrophic and heterotrophic microflagellates, diatoms, dinoflagellates and other protists (Rassoulzadegan and Etienne 1981, Capriulo 1982, Capriulo and Carpenter 1983, Verity 1985, 1987, Jonsson 1986). However, Rivier *et al.* (1985) and Albright, *et al.* (1987) found that certain choreotrichs and free swimming peritrichs can grow on bacterioplankton at concentrations in the range of 10^5 to 10^7

ml^{-1}, levels often observed in various planktonic environments (Ducklow 1983). Additionally, Sherr *et al.* (1986a) and Sherr and Sherr (1987) pointed out the importance of nanoplankton-sized aloricate ciliates in the plankton. They routinely observed bacteria and coccoid cyanobacteria in the food vacuoles of these ciliates, which exhibited extremely high specific clearance rates on bacterial food.

Some ciliates such as *Blepharisma* (Repak 1968, Pierce *et al.* 1978) and *Fabrea salina* (Repak and Anderson in prep.) produce giants during starvation, probably by increased vacuolization (Repak pers. comm.). Larger size in *Blepharisma* is accompanied by a switch to cannibalistic feeding. Such an adaptation represents a niche widening that enables the ciliate to survive unfavorable food periods, and has also been observed in marine planktonic oligotrichs (Gifford 1985).

Other ciliate adaptations have also allowed for the utilization of alternate food resources. For example cyrtophorine ciliates possess a complex cytopharyngeal basket (in some with teethlike capitula) of microtubular rods and filaments. This basket is alternately dilated and constricted by twisting, to break off or fold pieces of algal or cyanobacterial chains (Nisbet 1984).

Suctorian ciliates possess no permanent mouth but instead use their tentacles as temporary mouths, during feeding (Rudzinska 1973, Bardele 1974, Hackney and Butler 1981, Nisbet 1984). Haptocysts (extrusomes found at the tip of each tentacle) are used to capture and anchor prey which appear to randomly contact a tentacle(s). After capture of prey a tentacle is transformed into a feeding tentacle, with microtubules pushing up into and rupturing the prey's cell membrane which has been 'fused' with the tentacle membrane. The tentacle knob then invaginates, carrying prey cytoplasm and its contents down into the suctorian body where food vacuoles are formed. Numerous other specialized feeding arrangements can be found in ciliates (e.g. certain hymenostomes feed specifically on bacteria associated with decaying tissue while the colpodid ciliate *Pseudoplatyophrya nana* possesses a tubular structure which breaks open the cell wall of prey and then sucks out the cytoplasm).

Ciliates, as is the case with the flagellates, are important to the decomposition cycle particularly in their role as stimulators of bacterial growth by grazing (Butterfield *et al.* 1931, Porter *et al.* 1979, Finlay 1978, Johannes 1964, 1965), and prolongers of growth through the release of stimulatory substances (Straskrabova-Prokesova and Legner 1966, Curds and Cockburn 1968).

The question of feeding selectivity in ciliates still remains unanswered, due to the variety and complexity of forms and functions. In some cases (as evidenced above) feeding behavior is taxonomically or even species specific. The role of chemoreception remains underinvestigated. Also, the possibility that as a general rule, feeding selectivity is greater in eutrophic versus oligotrophic waters, must be considered in a more detailed fashion, as must the question of how widespread is planktonic ciliate bacterivory.

REVIEW AND COMPARISON OF METHODS TO MEASURE COMMUNITY GRAZING

Because most primary and secondary production in the sea is microbial (Pomeroy 1974, Sorokin 1981, Azam *et al.* 1983) and because protists are capable of growing even more rapidly than bacteria and algae *in situ* (Goldman 1984, Verity 1986a, b), protozoan feeding should, in theory, play a significant role in marine food webs. During the past decade, empirical evidence that protists are in fact important consumers of primary and secondary production in both marine and freshwater systems has rapidly accumulated (Sherr and Sherr 1984, Sherr *et al.* 1986c, Sieburth 1984, Porter *et al.* 1985). The following is a brief review of some of this work pertaining mostly to marine pelagic food webs.

Understanding the roles of protists in marine systems has been hampered by the lack of widely accepted methods to estimate protozoan standing stocks and feeding rates *in situ*. Sampling, preservation, and enumeration techniques to quantify the numbers and standing stocks of flagellates, ciliates, and sarcodines are discussed in several other papers in this volume. Here we describe methods currently used to estimate community feeding by protists on bacterioplankton (see also McManus and Fuhrman 1988b) and phytoplankton.

One approach to the determination of feeding impact by planktonic protists is to measure cell-specific rates of grazing, and then to apply those rates to protozoan assemblages in the field. The variations of this approach lie in the techniques used to measure the per-cell grazing rates. The traditional method is to monitor the rate of disappearance of prey organisms in the presence of a known quantity of protists. Such experiments have yielded estimates of grazing rates of flagellates on bacteria (Fenchel 1982a, Sherr *et al*. 1983, Davis and Sieburth 1984, Andersson *et al*. 1986, Caron 1987, Mitchell *et al*. 1988), of ciliates on algae and natural food (Heinbokel 1978a, b, Rassoulzadegan 1978, Rassoulzadegan and Etienne 1981, Capriulo 1982, Stoecker *et al*. 1981, Jonsson 1986, Verity 1985), and of ciliates on bacteria (Berk *et al*. 1976, Rivier *et al*. 1985) (see Table 1). Because accurate determination of the rate of decrease of prey generally requires sequential sampling over long time periods (6 to >24 hr, either in batch or chemostat culture) potential growth of both the predator and the prey populations must be accounted for in the analysis of the results. Most determinations of cell-specific grazing rates via prey disappearance experiments have used mono-specific laboratory cultures of phagotrophic protists.

A second approach to estimating mortality of specific prey populations due to protozoan grazing is measurement of the rates of decrease of prey cells in the presence of protists, and/or rates of increase of prey cells in the absence of protistan grazing. These methods depend upon manipulation of water samples to either reduce or eliminate protozoan feeding, or to inhibit growth of prey populations, after which sampling for prey abundance is carried out during long-term (of the order of 24 hr) incubations. Manipulations which have been used include: 1) selective filtration to physically separate protists from their prey (Wright and Coffin 1984, Servais *et al.* 1985), 2) a range of dilutions to gradually reduce protozoan grazing (Landry and Hassett 1982, Landry *et al.* 1984, Ducklow and Hill 1985), and 3) selective metabolic inhibition of either prokaryotic prey or eukaryotic predators (Fuhrman and McManus 1984, Sherr *et al.* 1986a, Sanders and Porter, 1986, Taylor and Pace 1987, Tremaine and Mills 1987a). Each of these manipulations has problems with required assumptions. Selective filtration generally does not yield a complete separation of predator and prey (Cynar *et al.* 1985, Servais *et al.* 1985). Dilution experiments can be compromised by the presence of residual prey in the water used for dilution (Li and Dickie 1985), or by changes in the grazing rate of predators due to changes in prey abundance (Gallegos submitted). Selective inhibitors often adversely affect non-targeted populations (Sherr *et al.* 1986a, Tremaine and Mills 1987b, Bloem *et al.* 1988). In addition, as noted above, significant growth of both predator and prey can occur during long incubations.

A more recent approach to cell-specific grazing rate has been to assay the rate of appearance of food (specific prey or prey analogues) in protozoan cells. A major advantage of such methods is that grazing rates can be determined during protozoan 'real-time', i.e. during periods (10-60 min) of much shorter duration than the average cell generation time of both predator and prey. In addition, prey appearance experiments are easily carried out using natural assemblages of protists. Investigators have measured uptake by protists of microspheres (approximately the same size as bacteria 0.5 µm to 1.0 µm diameter) or of small phytoplankton (3.0 to 5.0 µm diameter) (Fenchel 1980a-d, Borsheim 1984, Bird and Kalff 1986, McManus

Table 1. Range of observed ingestion, clearance and growth rates, as well as gross growth efficiencies of various marine protistan groups

Ingestion Rate	Clearance Rate ul/protist/hr	Growth Rate doublings day	Gross Growth Efficiency (%)	References
Microflagellates				McManus & Furhman 1988a and b
240 - 7,200 bact/protist/day	0.0002 - 0.08	0.4 - 9	24 - 49	Sherr et al. 1986a McManus & Furhman 1986a Fenchel 1982 a-d Davis & Sieburth 1984
				Sherr et al. 1983 Sherr & Sherr 1983 Daggett & Nerad 1982 Kopylov et al. 1980 Lucas et al. 1987 Nygaard et al. 1988
Phagotrophic algae				
----	0.002 - 0.026	----	----	Bird & Kalff 1986 Sanders & Porter 1988
Dinoflagellates				
----	0.4-8 on bacteria to 28 on algae	----	----	Lessard & Swift 1985
Tintinnid ciliates 0.4 - 68 ng carbon per ciliate/day on flagellates & algae	0.5 - 65	0.14 - 3.4	17 - 76	Stoecker et al. 1983 Taniguchi & Kawakami 1983 Lessard and Swift 1985 Rassoulzadegan 1978 Rassoulzadegan & Etienne 1981 Capriulo 1982 Spittler 1973 Verity 1985, 1986b Heinbokel 1978 a & b
----	0.04 on bacteria	----	----	Hollibaugh et al 1980
Planktonic aloricate ciliates 5 - 21 ng carbon per ciliate/day 0.06×10^6 um^3 per ciliate/day	0.11 - 213	0.4-2.7	25 - 70	Rassoulzadegan 1982 Lessard & Swift 1985 Jonsson 1986 Stoecker et al 1983 Gifford 1985 Sherr et al. 1986b on algae
----	0-7.6 on bact. typically in the 0 - 0.3 range	up to 1	----	Fenchel 1980 a & b Lessard & Swift 1985 Hollibaugh et al 1980 Borsheim 1984 Sherr & Sherr 1987
Benthic and surface associated ciliates 10% - 120% body volume per hour 16,000 bacteria per ciliate/hr 0.024-0.19 $\times 10^6$ um^3 per ciliate/day	0.003 - 0.7	1.8 - 4.8	10 - 27	Fenchel 1980 a & b Taylor 1986 Fenchel 1986 a & b Turley et al. 1986
Spinose planktonic foraminifera 0.3 - 1 copepod per foram/day	----	Survival time days to a month or more before production of swarmers		Caron & Be 1984 Spindler et al. 1984 Caron et al. 1987 a & b

and Fuhrman 1986a, b, 1988a, b). It has been demonstrated, however, that many species of bacterivorous flagellates and ciliates have lower grazing rates on artificial particles (the microspheres) than they do on bacterial cells (Sherr *et al.* 1987, Pace and Bailiff 1987, Nygaard *et al.* 1988, Bloem *et al.* 1988, Sieracki *et al.* 1987).

Alternative techniques involve measuring rates of uptake of prey cells which have been tagged either by staining with a fluorescent dye (Sherr *et al.* 1987, Nygaard *et al.* 1988, Bloem *et al.* 1988, Rublee and Gallegos 1989), or with radioisotopes (Hollibaugh *et al.* 1980, Lessard and Swift 1985, Lessard *et al.* submitted). The initial studies using tagged prey have obtained per-cell grazing rates comparable to those estimated from prey disappearance experiments (Table 1). Prey appearance experiments have also yielded valuable, often surprising, insights into specific trophic links in aquatic food webs. For instance, although it was originally thought that heterotrophic flagellates < 5 µm in size were the only important bacterivores in the plankton, the potential for significant bacterivory by pigmented flagellates (Bird and Kalff 1986, Sanders and Porter 1988), by heterotrophic dinoflagellates (Lessard and Swift 1985), and by pelagic ciliates (Sherr and Sherr 1987) has been demonstrated via prey appearance experiments.

A technique which utilizes a bacterial analogue in a prey disappearance experiment has been proposed by Wilkner *et al.* (1986); rates of decrease of radioactively labeled *E. coli* minicells are measured in various size fractions of seawater. The advantage of their method is that the minicells are not growing, and the incubations required (4-6 hr) are shorter than in alternative techniques. The technology involved in producing isotopically labeled minicells has limited the use of the method by other investigators.

Although a variety of methods has been developed for measuring grazing rates of planktonic protists, no one method has yet been identified as optimal for *in situ* estimates of the impact of protozoan grazing. Each approach has shortcomings. Determination of cell-specific grazing rates assumes that laboratory cultured organisms will have grazing responses similar to those of protists in the field. Measuring uptake of tagged prey by field populations assumes that the protists do not discriminate for or against the added prey analogues. Long-term manipulative experiments in which changes in prey abundance are monitored are prone to serious effects of the manipulations and to changes in cell growth or grazing rates during the experiments.

However, in spite of the problems inherent in each of the methods, rates of protozoan clearance of bacteria and algal prey estimated from the various techniques are generally within one order of magnitude: 0.001-0.010 µl $cell^{-1}$ hr^{-1} for bacterivorous flagellates, 0.04-0.4 µl $cell^{-1}$ hr^{-1} for bacterivorous ciliates, and 2-26 µl $cell^{-1}$ hr^{-1} for algivorous ciliates (Table 1). Rates of bacterivory for field populations tend to be of the order of 0.002-0.004 µl $cell^{-1}$ hr^{-1} for nonpigmented flagellates (Sherr *et al.* 1986a-c, Bloem *et al.* 1988, Daggett and Nerad 1982), and 0.1-0.2 µl $cell^{-1}$ hr^{-1} for small aloricate ciliates (Sherr and Sherr 1987). The rates for field populations are lower than rates measured in laboratory culture, which may be explained by slower growth rates of field protozoa, by alternate food resources, or because some species in the flagellate or ciliate assemblage are not bacterivorous. Between-method variations in grazing rates may be due in part to variability in the functional response of specific protozoan assemblages exposed to different conditions of prey abundance and environmental parameters.

Grazing on bacteria

There is now a general consensus that the major consumers of suspended bacteria in most marine systems are small (< 5 µm) nonpigmented flagellates (Haas and Webb 1979, Fenchel 1982a-d, 1987, Sieburth 1984, Wikner and Hagstrom 1988, Anderson and Fenchel 1985, Anderson and Sorensen 1986). Flagellates less than 5 µm numerically dominate the assemblages of heterotrophic protists which have been examined (Fenchel 1982a-d, Wikner and Hagstrom 1988, Sherr and Sherr submitted), and since phagotrophic protozoa generally feed on cells smaller than themselves (Goldman and Caron 1985), the appropriate-sized food for nonpigmented flagellates would be picoplankton, including heterotrophic bacteria and < 2 µm autotrophs. Studies of rates of bacterivory in size-fractionated water have demonstrated that the largest share of bacterial mortality is due to organisms smaller than 5 µm (Wright and Coffin 1984, Fuhrman and McManus 1984, Servais *et al.* 1985, Wikner and Hagstrom 1988). Fuhrman and McManus (1984) proposed that flagellates even smaller than 2 µm may be significant bacterivores.

However, other pelagic protists also graze bacteria (Turley *et al.* 1986). In eutrophic waters such as those of salt marsh tidal creeks, in which bacterial cell concentrations exceed about 5×10^6 ml^{-1}, aloricate ciliates < 20 µm in size are often the dominant grazers of suspended bacteria (Sherr *et al.* 1986b, Sherr and Sherr 1987). Pelagic choreotrichous ciliates are capable of clearing bacteria at rates of up to 0.3 µl $cell^{-1}$ hr^{-1} in meso- to oligotrophic waters of the coastal Mediterranean, although their low abundance (<10-20 cells ml^{-1}) precludes their having a major impact on bacterial standing stocks (Sherr *et al.* submitted). Lessard and Swift (1985) found that several species of heterotrophic dinoflagellates and ciliates ingested radiolabeled bacterioplankton. Phagotrophic phytoflagellates can also ingest bacteria at rates comparable to those measured for nonpigmented flagellates, and may be important bacterivores in some systems, particularly in freshwater (Porter *et al.* 1985, Estep *et al.* 1986, Borass *et al.* 1988, Sanders and Porter 1988).

Estimates of *in situ* bacterivory by protozoan assemblages range from 20% to 100% of daily bacterial production (Table 2). On average, community grazing rates, assayed over long duration experiments, indicate that bacterial mortality, presumed to be due to protozoan feeding, is approximately equal to average daily bacterial production. However, studies in which bacterivory by specific groups of protists is assayed often yield data suggesting that protozoan grazing cannot by itself balance bacterial production (Table 2, see Pace 1988). There may be other sources of bacterial mortality in the sea, for example cell lysis due to attack by viruses, which are not yet understood (Fuhrman and McManus 1984, Pace 1988, Proctor *et al.* 1988).

Grazing on phytoplankton

Perhaps the largest body of work concerning grazing by a single group of planktonic protists is that on phytoplankton consumption by tintinnids, (Tables 1 and 3). Tintinnids prey primarily on nanoplanktonic (2-20 µm) single-celled diatoms and phytoflagellates (Spittler 1973, Johansen 1976, Heinbokel 1978a, b, Capriulo and Ninivaggi 1982, Capriulo 1982, Capriulo and Carpenter 1983, Stoecker *et al.* 1981, Verity and Villareal 1986, Taniguchi and Kawakami 1983). They do not appear to be able to support their growth on bacteria, cyanobacteria, diatoms with threads or setae, or zooflagellates (Blackbourn 1974, Verity and Villareal 1986). At least one species of phytoflagellate, *Olisthodiscus luteus*, is toxic to tintinnids (Verity and Stoecker 1982). Estimates of the impact of tintinnid community grazing in coastal waters range from 4 to 60% of phytoplankton production (Heinbokel and Beers 1979, Rassoulzadegan and Etienne 1981, Burkill 1982, Capriulo and Carpenter 1983, Verity 1985) (Table 3). Tintinnids can be the dominant grazers during episodic blooms, such as those of *Phaeocystis pouchetii* in northern temperate waters (Admiraal and Venekamp 1986). Tintinnids have been shown to remove an average of about 25% of the yearly primary production in Long Island Sound (Capriulo and Carpenter 1983), Narragansett Bay (Verity 1985, 1987) and the northern Baltic Sea (Leppanen and Brunn 1986). For Long Island Sound, tintinnid yearly ingestion rates paralleled those of the copepod community (Capriulo and Carpenter 1983).

Other protists in the same size range as tintinnids, i.e. microplanktonic (20-200 µm) organisms, (e.g. aloricate ciliates

Table 2. Grazing impact of phagotrophic protozoa on bacteria in various marine pelagic waters.

Site	Protists	% of production grazed daily	Reference
Limfjord sound, Denmark	Flagellates	100%	Fenchel 1982d
English Channel, Benguela upwelling	Flagellates	66%	Linley et al 1983
Massachusetts coastal waters	Protozoan community	100%	Wright & Coffin, 1984
Kaneohe Bay Hawaii	Protozoan community	47% - 78%	Landry et al, 1984
Long Island Sound	Protozoan community	100%	Fuhrman & McManus 1984
Gulf Stream warm core rings	Protozoan community	100%	Ducklow and Hill 1985
Georgia estuary	Nanoplanktonic protozoa	40% - 45%	Sherr et al, 1986a
Salt marsh tidal creek	Flagellates and ciliates	80%	Sherr et al, submitted
Open estuary	Flagellates and ciliates	50%	Sherr et al, submitted
Chesapeake Bay plume, winter	Flagellates	100%	McManus and Fuhrman 1988a
Chesapeake Bay plume, summer	Flagellates	23%	McManus and Fuhrman 1988a
Bothnian and Mediterranean Seas	Protozoan community	100%	Wikner and Hagstrom 1988

and heterotrophic dinoflagellates), may have an even greater impact than tintinnids as consumers of phytoplankton, due to greater cell plasticity (e.g. aloricate ciliates) and extracellular feeding mechanisms (e.g. some colorless dinoflagellates). It also appears that aloricate ciliates and heterotrophic dinoflagellates predominantly consume nanophytoplankton (Fenchel 1982a-d, Rassoulzadegan 1982, Burkill 1982, Stoecker *et al.* 1984, Lessard and Swift 1985, Jonsson 1986, Stoecker *et al.* 1986, Verity 1986a, Burkill *et al.* 1987, Barlow *et al.* 1988). Estimates of total grazing

Table 3. Grazing impact of phagotrophic protists/microzooplankton on algal primary production and standing crops in various marine waters.

Location	Predator	Method	% of Primary Production or standing crop removed	Reference
Eastern Tropical Pacific	microzooplankton	theoretical calculations	70% of primary production per day	Beers & Stewart 1971
Long Island Sound	microzooplankton & bacteria	estimates from changes in O_2 & CO_2	43% annual primary production	Riley 1956
Long Island Sound	35-202 um microzooplankton	disappearance of chlorophyll *a*	up to 41% standing crop per day	Capriulo & Carpenter 1980
Coastal Waters of Washington, USA	microzooplankton	serial dilution	17%-52% of daily production	Landry and Hassett 1982
Northeastern Atlantic	microzooplankton	serial dilution	13%-65% of standing crop	Burkill et al. 1987
Canadian Arctic	microzooplankton	serial dilution	8%-15% of daily production	Paranjape 1987
Southern California Bight	tintinnids	disappearance of food/corn starch	4% to 21% of daily production	Heinbokel and Beers 1979
Long Island Sound	tintinnids	disappearance of food	27% of yearly primary production	Capriulo and Carpenter 1983
Narragansett Bay	tintinnids	disappearance of food	16%-26% of yearly production	Verity 1985, 1987
North Baltic Sea	ciliates	estimated from model	25% of primary production	Leppanen and Bruun 1986
Villefranche Bay	ciliates	estimated	59% of primary production	Rassoulzadegan & Etienne 1981
Southampton Estuary	tintinnids	estimated	60% of primary production	Burkill 1982

impact of microzooplankton, mainly phagotrophic ciliates and dinoflagellates but also including copepod nauplii and rotifers, range from about 20 to 60% of phytoplankton production (Table 3).

Phagotrophic flagellates other than dinoflagellates are also able to feed on phytoplankton. Colorless flagellates are known to ingest and grow on large diatoms (Suttle *et al.* 1986), several species of nano-phytoplankton (Goldman and Caron 1985), a 2 µm phytoflagellate (Parslow *et al.* 1986) and a 1 µm coccoid cyanobacteria (Johnson *et al.* 1982). Information on the importance of flagellate herbivory *in situ* is scarce. Campbell and Carpenter (1986) reported that from 37% to 100% of the daily production of the coccoid cyanobacterium *Synechococcus* was grazed in northeastern U.S. coastal waters,

Table 4 Predators of Protozoan Biomass
See text for references

Prey	Predators
Heterotrophic microflagellates	naked amoebae; ciliates; dinoflagellates; microzooplankton including copepods; cannibalism.
Dinoflagellates	ciliates; larger dinoflagellates; copepods.
Radiolarians	
Solitary forms	planktonic foraminiferans
Colonial forms	amphipods (Hyperietta and Oxycephalus clausi; juvenile Brachyscelus sp.; juvenile Lycaea sp.; harpacticoid copepod Miracia efferata.
Foraminiferans	Selective consumers: other forams; nematodes; polychaetes; gastropods; scaphopods; crustaceans. Incidental consumers; flatworms. polychaetes; chitons; gastropods; nudibranchs; ammonities; bivalves; crustaceans; holothuroids; asteroids; ophiuroids; echinoids; crinoids; tunicates; fish.
Ciliates	Cannibalism; other ciliates, crustaceans including cladocerans, copepods; snails, midges, rotifers, turbellarians, oligochaete worms; fish such as larval menhaden; ctenophores; jellyfish; dinoflagellates.

presumably by < 20 µm flagellates and ciliates. Kuosa and Kai (submitted) suggested that heterotrophic flagellates in the northern Baltic Sea might satisfy at least half of their carbon requirements by grazing on small algae. Phagotrophic flagellates are likely to be important consumers of phytoplankton in oligotrophic seawater (Smith *et al.* 1984), where the bulk of primary production is accomplished by < 3 µm cells (Glover *et al.* 1986, Iturriaga and Marra 1988, Iturriaga and Mitchell 1986, Li *et al.* 1983).

Grazing by protists on surfaces and in benthic habitats

Most work to date has dealt with protists feeding in pelagic systems. Research on the importance of protists in other habitats has lagged, in part because of methodological problems with enumeration and estimation of grazing rates. Protists are abundant both on suspended particles (e.g. Caron *et al.* 1982, Silver *et al.* 1984, Taylor *et al.* 1986) and in the benthos (e.g. Fenchel 1967, Lighthart 1969, Alongi 1987, Turley *et al.* 1988, Patterson *et al.* 1988). The species composition of particle-associated and benthic protists differs from that of planktonic assemblages, in that species of flagellates (e.g. *Bodo*) and ciliates (e.g. *Euplotes*) adapted to surface-feeding, as well as sarcodines such as naked amoebae and Foraminifera, are common (Caron *et al.* 1982, Silver *et al.* 1984, Gooday 1988, Turley *et al.* 1988). The little information available concerning the grazing impact of marine protists on surfaces and in the benthos suggests that they can be an important component of the food webs of these habitats (Fenchel 1967, Taylor *et al.* 1986, Gooday 1988, Turley *et al.* 1988, Patterson *et al.* 1988).

Effect of feeding specificity on community grazing rates

As information on the feeding behavior of individual species of protists accumulates, it is becoming clear that phagotrophic protists cannot be treated as a single unit in grazing rate assays. Some of the characteristics of protistan feeding behavior which must be considered in future work are: 1) chemosensory responses (Antipa *et al.* 1983, Levandowsky *et al.* 1984, 1988, Capriulo *et al.* 1986, Hellung-Larsen *et al.* 1986, Sibbald *et al.* 1987, 1988, Van Houten 1988, Verity 1988) 2) differential grazing and growth rates of protists depending on size, motility, or biochemical composition of prey cells (Stoecker *et al.* 1981, 1986, Sherr *et al.* 1983, Andersson *et al.* 1986, Turley *et al.* 1986, Mitchell *et al.* 1988), 3) selective feeding on attached or free bacteria (Caron 1987, Albright *et al.* 1987, Sibbald and Albright 1988, Mitchell *et al.* 1988), 4) mixotrophy (Estep *et al.* 1986, Stoecker *et al.*

1987, Borass *et al.* 1988, Sanders and Porter 1988, Laval-Peuto and Febvre 1986, Laval-Peuto and Rassoulzadegan 1988), and 5) simultaneous ingestion of, or switching between, alternate sources of food, including bacteria, phytoplankton, other phagotrophic protists, and even non-living particulate and dissolved organic matter (Goldman and Caron 1985, Gast 1985, Marchant 1985, Townsend and Cammen 1985, Glaser 1988, Rassoulzadegan *et al.* 1988, Sherr 1988).

In addition, investigators must consider the effects of abiotic parameters on grazing by protists *in situ* (Lee 1982). These factors include temperature (Caron *et al.* 1986, Sherr *et al.* 1988), water turbulence (Stoecker *et al.* 1984, Levandowsky *et al.* 1988), and light levels (Lessard *et al.* 1987, Strom submitted).

Implications for grazing rate assays

The diversity of behavioral and abiotic factors which can affect protistan grazing in natural environments presents a challenge to obtaining accurate estimates of the impact of grazing by protists on specific types of prey. It is also likely that there will be significant variability in grazing impact both seasonally and diurnally. As well, there may be large variations in grazing activity on short time (minutes to days) and spatial (mm to m) scales. More ecological studies are needed on all groups of protists, to include measurements of grazing rates both seasonally and on short time and fine spatial scales, to adequately understand the impact of protistan grazing within marine food webs.

It will also be necessary to consider the natural history of protistan feeding behavior *in situ*. A detailed investigation of every protistan species found in a habitat is not possible (e.g. Fenchel and Jonsson 1988). For research conducted in the near future, protists may be more broadly divided into feeding guilds, i.e. diverse types of protists which feed on the same group of prey (Pomeroy and Wiebe 1988), or into size ranges, since it appears that the generalization of Sheldon *et al.* (1972), that predators tend to feed on the next smaller size class of prey, can be used as a reasonable, first approximation, assumption with respect to phagotrophic flagellates and ciliates (Sheldon *et al.* 1986, Rassoulzadegan and Sheldon 1986, Rassoulzadegan *et al.* 1988, Sherr and Sherr submitted, Wikner and Hagstrom 1988). However, as pointed out below such assumptions are not valid in all instances, and can lead to inaccurate assessment of secondary production.

Fate of Heterotrophic Protistan Biomass

An analysis of the fate of protistan biomass is made in Capriulo (1990) and in Sherr *et al.* (1986). Many invertebrates and some vertebrates are dependent on heterotrophic protista for at least part of their food. However, transformation of the existing data into an assessment of the quantitative importance of protistan biomass to higher trophic levels of marine food webs is at best difficult, and is confounded by at least four factors:

1. Intraprotistan trophic links appear to be widespread. For example, heterotrophic flagellates which feed primarily on bacteria and bacterial sized eukaryotes are preyed upon by microzooplanktonic protozoa (Azam *et al.* 1983, Sheldon *et al.* 1986). Dinoflagellates are an important part of the diet of certain ciliates but also prey on other ciliates (Biecheler 1952, Jacobson unpublished data) which in turn eat smaller dinoflagellates. Various ciliates are food for other ciliates (Corliss 1979, Small 1973, Fenchel 1980a-d, Faure-Fremiet 1924, Kahl 1930-1935, Salt 1967, Fenchel 1986a, b, 1987, Gifford 1985). Many foraminiera feed on other foraminifera (Christiansen 1971, 1964). Additionally, in many protista, within species cannibalism occurs (e.g. microflagellates, Goldman and Caron 1985, ciliates, Pierce *et al.* 1978, Gifford 1985), as does interspecies parasitism (Cachon 1964, Cachon and Cachon 1987, Coats 1988, Coats and Heisler submitted). Such energy exhanges may cycle many times within the protista before providing biomass to higher trophic levels

2. Intracellular autophagy during starvation also occurs in various protists (e.g. flagellates, Fenchel 1982e)
3. Food size is not always a good index to who eats whom, since many protists eat food larger than themselves (e.g. Kahl 1930-1935, Faure-Fremiet 1924, Smetacek 1981, Jacobson and Anderson 1986, Gaines and Taylor 1984, Suttle *et al.* 1986)
4. Gross growth efficiencies vary widely among species and taxonomic groups, although most are routinely high (30-40% range on average, Table 1)

Evolution of trophic behavior in marine microbial food webs

The first 'primordial soup' organisms on earth were most probably anaerobic, heterotrophic prokaryotes (Kamshilov 1976 and Oparin 1957). These first cells evolved in dissolved organic matter (DOM) rich, sub-surface water, away from the harmful, high ultra violet (UV) light intensities (which bombarded the earth unchecked by the ozone free, anoxic, primitive atmosphere) of the surface ocean, under conditions that might be thought of as analagous to a global batch culture. Growth initially was likely to be exponential, and free of intra/inter species competition, with state variables as the predominant growth limiting factors. At some point DOM should have become limiting, thus creating competition for dwindling resources among organisms, as well as stationary growth conditions. Cells with lower half-saturation constants (K_S values) for required nutrients, that could also survive under the higher light intensities of the organic rich surface waters (including the surface air-sea interface microlayer), would have fared better as competition intensified. This competition, coupled with high, UV light-mediated, rates of mutation may have resulted in rapid evolution and the appearance of pigmented cells (e.g. cyanobacteria) as well as primitive phagocytizing procaryotic cells (Spoon 1986, reported the discovery of present day phagocytizing colonial-procaryotes). The ability to initially use pigment to shield cells from high light-intensity-associated sub-cellular damage, and later to derive biochemical reducing power and energy for carbon fixation through photosynthesis, would have conferred a selective advantage on pigmented cells. Additionally, the ability to phagocytize would have at least partially freed cells from nutrient limitations. The combination of both photosynthesis and particle feeding in a single cell must truly have been of great value with regards to competition. Cells not possessing these characteristics would have continued to compete among themselves, while the newly evolved cells would have achieved a level of competitive release, and likely renewed global exponential growth.

With the above players in place, the stage was set for serial symbiosis (Margulis 1970, 1981). Cells, both pigmented and non-pigmented types, may have ingested and retained without digestion other, more primitive, heterotrophic and mixotrophic forms, resulting in the evolution of both colorless and pigmented phagocytizing eukaryotes. Some pigmented forms might have secondarily lost the ability to phagocytize. These pioneering organisms would have achieved yet a new level of competitive release and a new round of global exponential growth, at a time when photosynthesis was changing the atmosphere from a reducing to an oxidizing one, with concomitant production of ozone and associated reduction in UV radiation, and therefore mutation rates as well as adaptive radiations. As global eukaryote densities increased, competition at this organization level also intensified. This provided for continued natural selection and evolutionary adaptations, which continuously modified form and function in the eukaryotic cells, with appropriate extinctions for failed adaptations. Such continuing pressures are likely to have resulted in the evolution of colonial organisms and eventually to true multicellularity.

During each step of this process cells used what was available to them until competitive pressures became too great and adaptive radiation or extinction occurred. Since primitive prokaryotes were the only forms of life present on earth initially, phagotrophic cells must have used these as a food resource. The successful nature of such nutrition is witnessed today in the feeding activities of colorless and pigmented flagellates, small ciliates, as well as in many sarcodines. That pigmentation and phagocytosis can co-exist is clear (e.g. found in dinoflagellates, chrysochromulinids etc.).

As inter-eukaryote competition intensified, larger cells evolved which could ingest both prokaryote as well as eukaryote food items. Each major evolutionary shift to a new trophodynamic mode was rewarded by competitive release (or alternatively by extinction). In this way we see the evolutionary basis for what we identify as trophic levels, in ecology. These evolutionarily derived trophic levels represent the skeletal framework or 'central tendency' of the microbiol food web. It should be kept in mind, however, that competitive releases of this nature are unidirectional, to the side of increasing complexity, and are followed by renewed competition, after a new round of global exponential growth has run its course, each time at a different but higher level of complexity. The competing organisms have different competitive abilities with some behaviors being more beneficial than others. Those organisms (or groups of organisms) competing poorly are the ones most likely to change or become extinct in the future. Changes can be subtle or drastic variations on a theme (e.g. flagellates eating diatom chains), and represent the 'noise' around the skeletal framework or central tendency of the food web. If the changes are successful for the organism, they might represent a new competitive release, and subsequently become a genetically fixed, new component of the existing framework.

When ecologists have made enough isolated observations on feeding behavior in protists, basic patterns appear (e.g. bacteria take up DOC, flagellates photosynthesize and/or feed on prokaryotes or bacterial sized eukaryotes, ciliates feed on nano-sized flagellated and non-flagellated eukaryotes, some prokaryotes and each other, sarcodines feed on prokaryotes, various protista as well as copepods, etc.) The transfers most often reported outline the major pathways of food transfer and the skeleton. When they appear to be important enough, these interactions are quantified.

Mathematical modelers, whose job it is to best describe in predictive and quantitative fashion how the food web works, can only be concerned with the skeletal framework of trophic interactions. We arrive at an understanding of this skeleton by consensus built on numerous observations and on quantitative experimentation. As new anecdotal and observational data stand the test of time, multi-investigator scrutiny, and are then shown to be quantitatively significant, only then should they be included in our models. In this way models improve with time, become increasingly more predictive and move closer to the underlying truth of how the marine ecosystem works.

ACKNOWLEDGEMENTS

We are grateful to OR Anderson, EB Small, JJ Lee and an anonymous reviewer for their critical examination of this work.

REFERENCES

Aaronson J (1980) Descriptive biochemistry and physiology of the Chrysophycea. In: Levandowsky M, Hutner SH (eds.) Biochemistry and Physiology of Protozoa. Vol 2, Academic Press, New York

Alongi DM (1987) The distribution and composition of deep-sea microbenthos in a bathyal region of the western Coral Sea. Deep-Sea Res 34: 1245-1254

Admiraal W, Venekamp LAH (1986) Significance of tintinnid grazing during blooms of *Phaeocystis pouchetii* (Haptophyceae) in Dutch coastal waters. Neth J Sea Res 20:61-66

Albright LJ, Sherr EB, Sherr BF, Fallon RD (1987) Grazing of ciliated protozoa on free and particle-attached bacteria. Mar Ecol Prog Ser 38:125-129

Andersen P, Fenchel T (1985) Bacterivory by microheterotrophic flagellates in seawater samples. Limnol Oceanogr 30:198-202

Andersen P, Sorensen HM (1986) Population dynamics and trophic coupling in pelagic microorganisms in eutrophic coastal waters. Mar Ecol Prog Ser 33:99-109

Anderson OR (1976) Ultrastructure of a colonial radiolarian *Collozoum inerve* and a cytochemical determination of the role of its zooxanthellae. Tissue and Cell 8:195-208

Anderson OR (1977) Fine structure of a marine amoeba

associated with a blue-green alga in the Sargasso Sea. J Protozool 24:370-376
Anderson OR (1978a) Light and electron microscopic observations of feeding behavior, nutrition, and reproduction in laboratory cultures of *Thalassicolla nucleata*. Tissue Cell 10:401-412
Anderson OR (1978b) Fine structure of a symbiont-bearing colonial radiolarian, *Collosphaera globularis*, and ^{14}C isotopic evidence for assimilation of organic substances from its zooxanthellae. J Ultrastruct Res 62:181-189
Anderson OR (1980) Radiolaria. In: Levandowsky M, Hutner S (eds.), Biochemistry and Physiology of Protozoa. 2nd ed, Vol 3, Academic Press, New York, p 1
Anderson OR (1983a) Radiolaria. Springer-Verlag, New York
Anderson OR (1983b) The Radiolarian Symbiosis. In: Goff LJ (ed.), Algal Symbiosis. Cambridge Univ Press, p 69
Anderson OR (1988) Comparative Protozoology: Ecology, Physiology, Life History. Springer-Verlag, New York
Anderson OR, Be AWH (1976a) A cytochemical fine structure study of phagotrophy in a planktonic foraminifera, *Hastigerina pelagica* (d'Orbigny). Biol Bull 151:437-449
Anderson OR, Be AWH (1976b) The ultrastructure of a planktonic foraminifera *Globigerinoides sacculifer* (Brady) and its symbiotic dinoflagellates. J Foraminiferal Res 6:1-21
Anderson OR, Botfield M (1983) Biochemical and fine structure evidence for cellular specialization in a large spumellarian radiolarian *Thalassicolla nucleata*. Mar Biol 72:235-241
Anderson OR, Spindler M, Be AWH, Hemleben C (1979) Trophic activity of planktonic foraminifera. J Mar Biol Ass UK 59:791-799
Anderson OR, Swanberg NR, Bennett P (1983) Assimilation of symbiont-derived photosynthates in some solitary and colonial radiolaria. Mar Biol 77:265-269
Anderson OR, Swanberg NR, Bennett P (1984) An estimate of predation rate and relative preference for algal versus crustacean prey by a spongiose skeletal radiolarian. Mar Biol 78:205-207
Anderson OR, Swanberg NR, Bennett P (1985) Laboratory studies of the ecological significance of host-algal nutritional associations in solitary and colonial radiolaria. J Mar Biol Ass UK 65:263-272
Andersson A, Larsson U, Hagstrom A (1986) Size-selective grazing by a microflagellate on pelagic bacteria. Mar Ecol Prog Ser 33:51-57
Antipa GA, Martin K, Rintz MT (1983) A note on the possible ecological significance of chemotaxis in certain ciliated protozoa. J Protozool 30:55-57
Azam F, Fenchel T, Field JG, Grey JS, Meyer-Reil LA, Thingstad F (1983) The ecological role of water-column microbes in the sea. Mar Ecol Prog Ser 10:257-263
Bardele C (1974) Transport of materials in the Suctorian tentacle. Symp Soc Exp Biol #28. Transport at the cellular level. Cambridge Univ Press, p 191
Barker HA (1935) The culture and physiology of the marine dino-flagellates. Arch Microbiol 6:157-181

Barlow RG, Burkill PH, Mantoura RFC (1988) Grazing and degradation of algal pigments by the marine protozoan *Oxyrrhis marina*. J exp Mar Biol Ecol 119:119-129
Bé AWH (1982) Biology of planktonic foraminifera. In: Broadhead TW (ed.) Foraminifera: notes for a short course. Univ Tennessee Dept Geol Sci Studies in Geology 6
Bé AWH, Hemleben C, Anderson OR, Spindler M, Hacunda J, Tuntivate-Choy S (1977) Laboratory and field observations of living planktonic foraminifera. Micropaleontology 23:155-179
Beers JR, Stewart GL (1971) Microzooplankters in the plankton communities of the upper waters of the eastern tropical Pacific. Deep-Sea Res 18:861-883
Berk S, Colwell RR, Small EB (1976) A study of feeding responses to bacterial prey by estuarine ciliates. Trans Am Microsc Soc 95:514-520
Biecheler B, (1952) Recherches sur les Peridiniens. Bull biol Fr Belg 36:1-149
Bird DF, Kalff J (1986) Bacterial grazing by planktonic lake algae. Science 231:493-494
Blackbourn DJ (1974) The feeding biology of tintinnid Protozoa and some other inshore microzooplankton. PhD Thesis, Univ British Columbia, Vancouver, p 244
Bloem J, Starink M, Bar-Gillissen M-JB, Cappenberg TE (1988) Protozoan grazing, bacterial activity, and mineralization in two-stage continuous cultures. Appl Environ Microbiol 54:3113-3121
Borass ME, Estep KW, Johnson PW, Sieburth JMcN (1988) Phagotrophic phototrophs: the ecological significance of mixotrophy. J Protozool 35:249-252
Borsheim KY (1984) Clearance rates of bacteria-sized particles by freshwater ciliates, measured with mono-disperse fluorescent latex beads. Oecologia 63:286-288
Bovee EC, Cordele DC (1971) Feeding on gastrotrichs by the heliozoan *Actinophyrys sol*. Trans Amer Micros Soc 10:365-369
Bromley RG, Nordmann E (1971) Maastrichtian adherent foraminifera encircling clionid pores. Bull Geol Soc Den 20:362-368
Brooker BE, (1971) Fine structure of *Bodo saltans* and *Bodo candatus* (Zoomastigophora:Protozoa) and their affinities with the Trypanosomatidae. Bull British Museum Nat Hist 22:81-102
Burkill PH (1982) Ciliates and other microplankton components of a near-shore food-web: standing stocks and production processes. Annls Inst Oceanogr (Paris) 58:335-350
Burkill PH, Mantoura RFC, Llewellyn CA, Owens NJP (1987) Microzooplankton grazing and selectivity of phytoplankton in coastal waters. Mar Biol 93:581-590
Burzell LA (1973) Observations on the proboscis-cytopharynx and flagella in *Rhynchomonas metabolita* Pshenen 1964 (Zoomastigophora Bodonidae). J Protozool 20:385-393
Burzell LA (1975) Fine structure of *Bodo curvifilus* Griessmann. J Protozool 22:35-39
Butterfield CT, Purdy WC, Theriault EJ (1931) Studies on natural purification in polluted waters. IV The influence of plankton on the biochemical oxidation of organic matter.

Public Health Report No 46:393-426
Cachon J (1964) Contribution a l'etude des peridiniens parasites. Cytologie, cycles evolutifs. Ann Sci Nat Zool Biol Animal 6:1-158
Cachon J, Cachon M (1987) Parasitic dinoflagellates. In: Taylor, FJR (ed), The biology of dinoflagellates, Blackwell Sci Pub, Oxford, p 571
Campbell L, Carpenter EJ (1986) Estimating the grazing pressure of heterotrophic nanoplankton on *Synechococcus* sp using the sea water dilution and selective inhibitor techniques. Mar Ecol Prog Ser 33:121-129
Capriulo GM (1982) Feeding of field-collected tintinnid microzooplankton on natural food. Mar Biol 71:73-86
Capriulo GM (1990) Feeding related ecology of marine protozoa. In: GM Capriulo (ed), Ecology of Marine Protozoa, Oxford Univ Press. In press
Capriulo GM, Carpenter EJ (1980) Grazing by 35 to 202 µm microzooplankton in Long Island Sound. Mar Biol 56:319-326
Capriulo GM, Ninivaggi DV (1982) A comparison of the feeding activities of field collected tintinnids and copepods fed identical natural particle assemblages. Annls Inst Oceanogr (Paris) 58:325-334
Capriulo GM, Carpenter EJ (1983) Abundance, species composition, and feeding impact of tintinnid microzooplankton in central Long Island Sound. Mar Ecol Prog Ser 10:277-288
Capriulo GM, Small EB (1986) Discovery of an apostome ciliate (*Collinia beringensis*, n sp) endoparasitic in the Bering Sea euphausiid *Thysanoessa inermis*. Dis Aquat Organ 1:141-146
Capriulo GM, Taveras J, Gold K (1986) Ciliate feeding: effect of food presence or absence on occurrence of striae in tintinnids. Mar Ecol Prog Ser 30:145-158
Capriulo GM, Schreiner RA, Dexter BL (1988) Differential growth of *Euplotes vannus* fed fragmented versus unfragmented chains of *Skeletonema costatum*. Mar Ecol Prog Ser 47:205-209
Caron DA (1987) Grazing of attached bacteria by heterotrophic microflagellates. Microb Ecol 13:203-218
Caron DA, Be AWH (1984) Predicted and observed feeding rates of the spinose planktonic foraminifera *Globigerinoides sacculifer*. Bull Mar Sci 35:1-10
Caron DA, Davis PG, Madin LP, Sieburth JMcN (1982) Heterotrophic bacteria and bacterivorous protozoa in oceanic macroaggregates. Science 218:795-797
Caron DA, Goldman JC, Dennett MR (1986) Effect of temperature on growth, respiration, and nutrient regeneration by an omnivorous microflagellate. Appl environ Microbiol 52:1340-1347
Caron DA, Swanberg NR (1988) Prey and feeding selectivity of planktonic sarcodines. Abstract NATO ASI, Protozoa and their role in marine processes
Chen YT (1950) The biology of *Peranema trichophorum*. Quart J Microscop Sci 91:279-308
Christiansen B (1964) *Spiculosiphon radiata*, a new foraminifera from northern Norway. Astarte 25:1-8
Christiansen B (1971) Notes on the biology of foraminifera.

Vie et milieu Troisieme Sym European de Biologie Marine S 22:465-478
Coats DW (1988) *Duboscquella cachoni* n sp a parasitic dinoflagellate lethal to its tintinnine host *Eutintinnus pectinis*. J Protozool 35: 607-617
Coats DW, Heisler JJ. Spatial and temporal occurrence of the parasitic dinoflagellate *Duboscquella cachoni* and its tintinnine host *Eutintinnus pectinis* in Chesapeake Bay. submitted
Corliss JO (1979) The Ciliated Protozoa: Characterization, Classification, and Guide to the Literature. 2nd ed Pergamon Press, New York, p 455
Curds CR, Cockburn A (1968) Studies on the growth and feeding of *Tetrahymena pyriformis* in axenic and nonaxenic culture. J Gen Microbiol 54:343-358
Cynar FJ, Estep KW, Sieburth JMcN (1985) The detection and characterization of bacteria-sized protists in 'protist-free' filtrates and their potential impact on experimental marine ecology. Microb Ecol 11:281-288
Daggett P, Nerad TA (1982) Axenic cultivation of *Bodo edax* and *Bodo ancinatus* and some observations on feeding rate in nonoxenic culture. J Protozool 29: 290-291
Davis PG, Caron DA, Sieburth JMcN (1978) Oceanic amoebae from the North Atlantic: culture, distribution, and taxonomy. Trans Amer Micros Soc 97:73-88
Davis PG, Sieburth JMcN (1984) Estuarine and oceanic microflagellate predation of actively growing bacteria: Estimation by frequency of dividing-divided bacteria. Mar Ecol Prog Ser 19:237-246
Delaca TE (1982) Use of dissolved amino acids by the foraminifera *Notodendrodes antarctikos*. Am Zool 22:683-690
Dodge JD, Crawford RM (1970) The morphology and fine structure of *Ceratium hirundinella* (Dinophyceae). J Phycol 6:137-149
Dodge JD, Crawford RM (1974) Fine structure of the dinoflagellate *Oxyrrhis marina* III Phagotrophy. Protistologica 10:239-244
Drebes G (1969) *Dissodinium pseudocalani* sp nov lin parasitischer Dinoflagellat auf Copepodeneiem. Helgol Wiss Meeresu 19:58-67
Droop MR (1953) Phagotrophy in *Oxyrrhis marina*. Nature 172-250-252
Droop MR (1954) A note on the isolation of small marine algae and flagellates in pure culture. J Mar Biol Ass UK 33:511-514
Droop MR (1959) Water soluble factors in nutrition of *Oxyrrhis marina*. J Mar Biol Ass UK 38:605-620
Ducklow HW (1983) Production and fate of bacteria in the oceans. Bioscience 33:494-501
Ducklow H, Hill SM (1985) The growth of heterotrophic bacteria in the surface waters of warm core rings. Limnol Oceanogr 30:239-259
Elbrachter M, Drebes G (1978) Life cycles, phylogeny and taxonomy of *Dissodinium* and *Pyrocystis* (Dinophyta). Helgol Wiss Meeresu 31:347-366
Estep KW, Davis PG, Keller MD, Sieburth JMcN (1986) How important are oceanic algal nanoflagellates in bacterivory?

Limnol Oceanogr 31:646-650
Faure-Fremiet E (1924) Contribution a la connaissance des infusoires planktoniques. Bull Biol France Belgique 6:1-171
Febvre-Chevalier C, Febvre J (1986) Motility mechanisms in the actinopods (Protozoa). A review with particular attention to axopodial contraction/extension, and movement of nonactino-filament systems. Cell Motility ad the Cytoskeleton 6:198-200
Fenchel T (1967) The ecology of marine microbenthos, I. The quantitative importance of ciliates as compared with metazoans in various types of sediments. Ophelia 4:121-137
Fenchel T (1968) The ecology of marine microbenthos II. The food of marine benthic ciliates. Ophelia 5:73-121
Fenchel T (1980a) Suspension feeding in ciliated protozoa: structure and function of feeding organelles. Arch Protistenk 123:239-260
Fenchel T (1980b) Suspension feeding in ciliated protozoa: functional response and particle size selection. Microb Ecol 6:1-11
Fenchel T (1980c) Suspension feeding in ciliated protozoa: feeding rates and their ecological significance. Microb Ecol 6:13-25
Fenchel T (1980d) Relation between particle size selection and clearance in suspension-feeding ciliates. Limnol Oceanogr 25:733-738
Fenchel T (1982a) Ecology of heterotrophic microflagellates I. Some important forms and their functional morphology. Mar Ecol Prog Ser 8:211-223
Fenchel T (1982b) Ecology of heterotrophic microflagellates II. Bio-energetics and growth. Mar Ecol Prog Ser 8:225-232
Fenchel T (1982c) Ecology of heterotrophic microflagellates III. Adaptations to heterogeneous environments. Mar Ecol Prog Ser 9:25-33
Fenchel T (1982d) Ecology of heterotrophic microflagellates IV. Quantitative occurrence and importance as bacterial consumers. Mar Ecol Prog Ser 9:35-42
Fenchel T (1982e) The bioenergetics of a heterotrophic microflagellate. Annls Inst Oceanogr 58:55-60
Fenchel T (1986a) Protozon filter feeding In: Corliss JO, Patterson DJ (eds), Progress in Protistology Vol 1. Biopress Ltd, Bristol, England, p 65
Fenchel T (1986b) The ecology of heterotrophic microflagellates. In: Marshall KC (ed), Advances in Microbial Ecology. Plenum Pub Corp 9, p 57
Fenchel T (1987) Ecology of Protozoa. Sci Tech Pub and Springer-Verlag, Berlin
Fenchel T, Harrison P (1976) The significance of bacterial grazing and mineral cycling for the decomposition of particulate detritus. In: Anderson JM, Macfadyen A (eds), The role of terrestrial and aquatic organisms in decomposition processes. Blackwell Sci Pub Ltd, Oxford, p 285
Fenchel T, Jonsson PR (1988) The functional biology of *Strombidium sulcatum* a marine oligotrich ciliate (Ciliophora, Oligotrichina). Mar Ecol-Prog Ser 48:1-15
Fenchel T, Small EB (1980) Structure and function of the oral cavity and its organelles in the hymenostome ciliate

Glaucoma. Trans Am Microscop Soc 99:52-60
Finlay BJ (1978) Community production and respiration by ciliated protozoa in the benthos of a small eutrophic loch. Freshwater Biol 8:327-341
Frey LC, Stoermer EF (1980) Dinoflagellate phagotrophy in the upper Great Lakes. Trans Am Microscop Soc 99:439-444
Fuhrman JA, McManus GB (1984) Do bacteria-sized marine eukaryotes consume significant bacterial production? Science 224:1257-1260
Gaines G, Taylor FJR (1984) Extracellular digestion in marine dinoflagellates. J Plank Res 6:1057-1061
Gallegos (Submitted) Microzooplankton grazing on phytoplankton in the Rhode River, Maryland: Nonlinear feeding kineticis. Mar Ecol Prog Ser
Gifford DJ (1985) Laboratory culture of marine planktonic oligotrichs (Ciliophora, Oligotrichida). Mar Ecol Prog Ser 23:257-267
Glaser D (1988) Simultaneous consumption of bacteria and dissolved organic matter by *Tetrahymena pyriformis*. Microb Ecol 15:189-201
Glover HE, Campbell L, Prezelin BB (1986) Contribution of *Synechococcus* sp to size fractionated primary productivity in three water masses in the Northwest Atlantic. Ocean Mar Biol 91:193-203
Gold K, Storm E, Laval-Peuto M (1979) Scanning electron microscopy of *Tintinnopsis parva*: studies on particle accumulation and the striae. J Protozool 26:415-419
Goldman JC (1984) Conceptual role for microaggregates in pelagic waters. Bull Mar Sci 35:462-476
Goldman JC, Caron DA (1985) Experimental studies on an omnivorous microflagellate: Implications for grazing and nutrient regeneration in the marine microbial food chain. Deep-Sea Res 32:899-915
Gooday AJ (1988) A benthic foraminiferal response to the deposition of phytodetritus in the deep-sea. Nature 332:70-73
Gast V (1985) Bacteria as a food source for microzooplankton in the Schlei Fjord and Baltic Sea with special reference to ciliates. Mar Ecol Prog Ser 22:107-120
Greissman K (1914) Uber marine Flagellaten. Arch Protistenk 32:1-78
Greuet C (1972) La nature trichocystaire due cnidoplaste dans le complexe cnidoplaste nematocyste du *Polykrikos schwartzi* Butschli. CR Hebd Seances Acad Sci Paris, Ser D 275:1239-1242
Greuet C, Hovasse R (1977) A propose de la genese des nematocystes de *Polykrikos schwartzi* Butschli. Protistologica 13:145-149
Haas LW, Webb KL (1979) Nutritional mode of several non-pigmented microflagellates from the York River Estuary, Virginia. J exp mar Biol Ecol 39:125-134
Haberey M (1973a) Die Phagocytose von Oscillatorien durch *Thecamoeba sphaeronucleolus* I. Lichtoptische Untersuchung Arch Protistenk 115:99-110
Haberey M (1973b) Die Phagocytose von Oscillatorien durch *Thecamoeba sphaeronucleolus* II. Electronmikroscopische Untersuchung Arch Protistenk 115:111-124

Hackney CM, Butler RD (1981) Tentacle contraction in glycerinated *Discophyra collini* and the localization of HNN-Binding filaments. J Cell Sci 47:65-75
Hausmann K, Patterson DJ (1982) Feeding in *Actinophrys* II. Pseudopod formation and membrane production during prey capture by a heliozoan. Cell motility 2:9-24
Heinbokel JF (1978a) Studies on the functional role of tintinnids in the southern California Bight I. Grazing and growth rates in laboratory cultures. Mar Biol 47:177-189
Heinbokel JF (1978b) Studies on the functional role of tintinnids in the southern California Bight II. Grazing rates of field populations. Mar Biol 47: 191-197
Heinbokel JF, Beers JR (1979) Studies on the functional role of tintinnids in the southern California Bight III. Grazing impact of natural assemblages. Mar Biol 52:23-32
Hellung-Larsen P, Leick V, Tommerup N (1986) Chemoattraction in *Tetrahymena*: on the role of chemokinesis. Biol Bull 170:357-367
Hemleben C, Spindler M, Breitinger I, Deusen WG (1985) Field and laboratory studies on the ontogeny and ecology of some floboratalid species from the Sargasso Sea off Bermuda. J Foram Res 15:254-272
Hollibaugh JT, Fuhrman JA, Azam F (1980) Radioactive labeling of natural assemblages of bacterioplankton for use in trophic studies. Limnol Oceanogr 25:172-181
Iturriaga R, Mitchell BG (1986) Chroococcoid cyanobacteria: a significant component in the food web dynamics of the open sea. Mar Ecol Prog Ser 28:291-297
Iturriaga R, Marra J (1988) Temporal and spatial variability of chroococcoid cyanobacteria *Synechococcus* sp specific growth rates and their contribution to primary production in the Sargasso Sea. Mar Ecol Prog Ser 44:175-181
Jacobsen DJ, Anderson DM (1986) Thecate heteterotrophic dinoflagellates:feeding behavior and mechanisms. J Phycol 22:249-258
Johannes RE (1964) Phosphorus excretion and body size in marine animals: microzooplankton and nutrient regeneration. Science 146:923-924
Johannes RE (1965) Influence of marine protozoa on nutrient regeneration. Limnol Oceanogr 10:434-442
Johansen PL (1976) A study of tintinnids and other Protozoa in eastern Canadian waters, with special reference to tintinnid feeding, nitrogen excretion, and reproductive rates. PhD Thesis, Dalhousie Univ, Halifax, Nova Scotia, p 156
Johnson PW, Xu H-S, Sieburth JMcN (1982) The utilization of chroococcoid cyanobacteria by marine protozooplankters but not by calanoid copepods. Annls Inst Oceanogr (Paris) 5:297-308
Jonsson PR (1986) Particle size selection, feeding rates and growth dynamics of marine planktonic oligotrichous ciliates (Ciliophora: Oligotrichina). Mar Ecol Prog Ser 33:265-277
Jonsson PR (1987) Photosynthetic assimilation of inorganic carbon in marine oligotrich ciliates (Ciliophora, Oligotrichina). Mar Microb Food Webs 2:55-68
Kahl A (1930-1935) Urtiere oder Protozoa I. Wimpertiere

oder Ciliata (Infusoria) In: Dahl F (ed), Die Tierwelt Deutschlands, Teil 18, 21, 25 and 30 G Fisher, Jena

Kamshilov MM (1976) Evolution of the Biosphere. MIR Publishers, Moscow p 269

Kimor B (1981) The role of phagotrophic dinoflagellates in marine ecosystems. Kieler Meeresforsch 5:164-173

Kopylov AI, Mamayevia RI, Botsanin SF (1980) Energy balance of the colorless flagellate *Parabodo attenuatus*. Oceanology 20:705-708

Kuosa H, Kivi K (submitted) Bacteria and heterotrophic flagellates in the pelagic carbon cycle in the northern Baltic Sea

Landry MR, Hassett RP (1982) Estimating the grazing impact of marine micro-zooplankton. Mar Biol 67-283-288

Landry MR, Haas LW, Fagerness VL (1984) Dynamics of microbial plankton communities: experiments in Kaneohe Bay, Hawaii. Mar Ecol Prog Ser 16:127-133

Laval-Peuto M, Febvre M (1986) On plastid symbiosis in *Tontonia appendiculariformis* (Ciliophora, Oligotrichia). Biosystems 19:137-158

Laval-Peuto M, Rassoulzadegan F (1988) Autofluorescence of marine planktonic oligotrichina and other ciliates. Hydrobiologia 159:99-110

Leadbeater BSC, Morton C (1974) A microscopical study of a marine species of *Codosiga* (James-Clark) (Choanflagellata) with special reference to the ingestion of bacteria. Biol J Limn Soc 6:337-347

Lee JJ (1982) Physical, chemical and biological quality related food-web interactions as factors in the realized niches of microzooplankton. Annls Inst Oceanogr 58:19-29

Lee JJ, Capriulo GM (1990) The Ecology of Marine Protozoa: An Overview In: Capriulo GM (ed), The Ecology of Marine Protozoa Oxford Univ Press, (in press)

Lee JJ, Hunter SH, Bovee EC (eds) (1985) Illustrated Guide to the Protozoa Society of Protozoologists and Allen Press, Lawrence, Kansas

Lee JJ, Freudenthal, HD, Muller WA, Kossoy V, Pierce S, Grossman R (1963) Growth and physiology of foraminifera in the laboratory III. Initial studies of *Rosalina floridana* (Cushman). Micropaleontology 9:449-466

Lee JJ, Pierce S, M Tentchoff, McLaughlin JA (1961) Growth and physiology of foraminifera in the laboratory I. Collection and maintenance. Micropaleontology 7:461-466

Leedale GF (1967) Euglenid flagellates. Prentice Hall Inc New Jersey

Leedale GF, Hibberd DJ (1985) Class 1 Phytomastigophorea Calkins, 1909. In: Lee JJ, Hunter SH, Bovee EC (eds) An Illustrated Guide to the Protozoa, Soc of Protozoologists and Allen Press, Lawrence Kansas

Lenk SE, Small EB, Gunderson J (1984) Preliminary observations of feeding in the psalmobiotic ciliate *Tracheloraphis* sp. Origins of Life 13:229-234

Leppanen J-M, Bruun JE (1986) The role of pelagic ciliates including the autotrophic *Mesodinium rubrum* during the spring bloom of 1982 in the open northern Baltic proper. Ophelia 4:147-157

Lessard EJ, Swift E (1985) Species-specific grazing rates of

heterotrophic dinoflagellates in oceanic waters, measured with a dual-label radioisotope technique. Mar Biol 87:289-296
Lessard EJ, Caron DA, Ho K, Voytek MA (1987) Grazing impact and food selection of nano-, micro-, and macrozooplankton in natural estuarine communities. EOS 68:1782
Lessard EJ, Caron DA, Voytek M, Ho K (Submitted) Grazing impact and food selection of nano-, micro- and macrozooplankton in natural estuarine communities. J Plank Res
Levandowsky M, Cheng T, Kehr A, Kim J, Gardner A, Silvern L, Tsang L, Lai L, Chung C, Prakash E (1984) Chemosensory responses to amino acids and certain amines by the ciliate *Tetrahymena*: a flat capillary assay. Biol Bull 167:322-330
Levandowsky M, Klafter J, White BS (1988) Feeding and swimming behavior in grazing microzooplankton. J Protozool 35:243-246
Li WKW, Dickie PM (1985) Growth of bacteria in seawater filtered through 0.2 µm Nuclepore membranes: implications for dilution experiments. Mar Ecol Prog Ser 26:245-252
Li WK, Subba Rao DV, Harrison WG, Smith JC, Cullen JJ, Irwin B, Platt T (1983) Autotrophic picoplankton in the tropical ocean. Science 219:292-295
Lighthart B (1969) Planktonic and benthic bacterivorous Protozoa at eleven stations in Puget Sound and adjacent Pacific Ocean. J Fish Res Bd Can 26:299-304
Lindholm T (1985) *Mesodinium rubrum* - a unique photosynthetic ciliate. In: Janhasch HW, Williams PJLeB (eds), Advances in Aquatic Microbiology Academic Press, London, p 1
Lipps JH (1982) Biology/Paleobiology of Foraminifera. In: Broadhead TW (ed), Foraminifera: notes for a short course. Univ of Tennessee Dept of Geol Sci Studies in Geology 6, p 1
Lipps JH (1983) Biotic interactions in benthic foraminifera. In: Tevesy MJS, McCall PL (eds), Biotic interactions in recent and fossil benthic communities. Plenum, p 331
Loefer JB (1931) Morphology and binary fission in *Heteronema acus* (Ehrb). Stein Arch Prostistenk 74:449-470
Looper JB (1928) Observations on the food reactions of *Actinophrys sol*. Biol Bull 54:485-502
Lucas MI, Probyn TA, Painting SJ (1987) An experimental study of microflagellate bacterivory: further evidence for the importance and complexity of microplanktonic interactions. S Afr J Mar Sci 5:791-808
Luckinbill LS (1974) The effects of space and enrichment on a predator-prey system. Ecology 55:1142-1147
Marchant HJ (1985) Choanoflagellates in the antarctic marine food chain In: Siefreid WR, Cody PR, Laws RM (eds), Antartic Nutrient Cycles and Food Webs. Springer-Verlag, Berlin, p 272
Margulis L (1970) Origin of eukaryotic cells Yale Univ Press, New Haven
Margulis L (1981) Symbiosis in Cell Evolution Freeman, San Francisco
McManus GB, Fuhrman JA (1986a) Bacterivory in seawater studies with the use of inert fluorescent particles. Limnol Oceanogr 31:420-426

McManus GB, Fuhrman JA (1986b) Photosynthetic pigments in the ciliate *Laboea strobila* from Long Island Sound, USA. J Plankton Res 8:317
McManus GB, Fuhrman JA (1988a) Clearance of bacteria-sized particles by natural populations of nano-plankton in the Chesapeake Bay outflow plume. Mar Ecol Prog Ser 42:199-206
McManus GB, Furhman JA (1988b) Control of marine bacterioplankton populations: Measurement and significance of grazing. Hydrobiologia 159:51-62
Mignot JP (1966) Structur et ultrastructure de guelgues Englenomonadines. Protistologica 2:51-117
Mitchell GC, Baker JH, Sleigh MA (1988) Feeding of a freshwater flagellate, *Bodo saltans*, on diverse bacteria. J Protozool 35:219-222
Montagnes DJS, Lynn DH (1988) Taxonomy of choreotrichs, the major marine planktonic ciliates. Abstract NATO ASI Protozoa and their role in marine processes
Montagnes DJS, Lynn DH, Roff JC, Taylor WD (1988) The annual cycle of heterotrophic planktonic ciliates in the waters surrounding the Isles of Shoals, Gulf of Maine: An assessement of their trophic role. Mar Biol 99:21-30
Morey-Gaines G, Elbrachter M (1987) Heterotrophic nutrition. In: Taylor FJR (ed), The Biology of Dino-flagellates. Blackwell Sci Ltd, Oxford
Newell RC, Linley EAS (1984) Significance of microheterotrophs in the decomposition of phytoplankton: estimates of carbon and nitrogen flow based on the biomass of plankton communities. Mar Ecol Prog Ser 16:105-119
Nisbet B (1974) An ultrastructural study of the feeding apparatus of *Peranema trichophorum*. J Protozool 21:39-48
Nisbit B (1984) Nutrition and feeding strategies in protozoa Croom Helm, London
Nyholm K-G (1956) On the life cycle of the foraminiferan *Nemogullmia longivariabilis*. Zool Biol Upps 31:483-496
Nygaard K, Borsheim KY, Thingstad TF (1988) Grazing rates on bacteria by marine heterotrophic microflagellates compared to uptake rates of bacterial-sized monodisperse fluorescent latex beads. Mar Ecol Prog Ser 44:159-165
Oparin AI (1957) The Origin of Life on Earth. Pub House Acad Sci USSR, Moscow
Pace ML (1988) Bacterial mortality and the fate of bacterial production. Hydrobiologia 159:41-49
Pace ML, Bailiff MD (1987) An evaluation of a fluorescent microsphere technique for measuring grazing rates of phagotrophic microorganisms. Mar Ecol Prog Ser 40:185-193
Paranjape M (1987) Grazing by microzooplankton in the eastern Canadian arctic in summer, 1983. Mar Ecol Prog Ser 40:239-246
Parslow JS, Doucette GJ, Taylor FJR, Harrison PJ (1986) Feeding by the zooflagellate *Pseudobodo* sp on the picoplanktonic prasinomonad *Micromonas pusilla*. Mar Ecol Prog Ser 29:237-246
Patterson DJ (1979) On the organization and classification of the protozoan *Actinophrys sol* Ehrenberg, 1830. Microbios 26:165
Patterson DJ, Hausmann K (1981) Feeding by *Actinophrys sol* (Protista, Heliozoa): I. Light microscopy. Microbios

31:39-55
Patterson DJ, Larsen J, Corliss JO (1989) The ecology of heterotrophic flagellates and ciliates living in marine sediments. Progress in Protistology 3:185-277
Pierce E, Isquith IR, Repak AJ (1978) Quantitative study of cannibal-giantism in *Blepharisma*. Acta Protozoologica 17:493-501
Pomeroy LR (1974) Significance of microorganisms in carbon and energy flow in marine ecosystems. In: Klug MJ, Reddy CA (eds), Current Perspectives in Microbial Ecology Am Soc Microbiol Washington DC, p 405
Pomeroy LR, Wiebe WJ (1988) Energetics of microbial food webs. Hydrobiologia 159:7-18
Porter KG, Pace ML, Battey JF (1979) Ciliate protozoans as links in freshwater planktonic food chains. Nature 277:563-565
Porter KG, Sherr EB, Sherr BF, Pace M, Sanders RW (1985) Protozoa in planktonic food webs. J Protozool 32: 409-415
Proctor LM, Fuhrman JA, Ledbetter MC (1988) Marine bacteriophages and bacterial mortality. EOS 69:1111
Pussard M, Rouelle J (1986) Predation de la microflore effet des protozoaines sur la dynamique de population bacterienne. Protistologica 22:105-110
Rassoulzadegan F (1978) Dimensions et taux d'ingestion des particules consommees par un tintinnide: *Favella ehrenbergii* (Clap et Lachm). Jorg Cilie pelagique. Annl Inst Oceanogr Paris 54:17-24
Rassoulzadegan F (1982) Dependence of grazing rate, gross growth efficiency and food size range on temperature in a pelagic oligotrichous ciliate *Lohmanniella spiralis* Leeg, fed on naturally occurring particulate matter. Annls Inst Oceanogr, Paris 58:177-184
Rassoulzadegan F, Etienne M (1981) Grazing rate of the tintinnid *Stenosemella ventricosa* (Clap & Lachm). Jorg on the spectrum of the naturally occurring particulate matter from a Mediterranean neritic area. Limnol Oceanogr 26:258-270
Rassoulzadegan F, Sheldon RW (1986) Predator-prey interactions of nanozooplankton and bacteria in an oligotrophic marine environment. Limnol Oceanogr 31:1010-1021
Rassoulzadegan F, Laval-Peuto M, Sheldon RW (1988) Partition of the food ration of marine ciliates between pico- and nanoplankton. Hydrobiologia 159:75-88
Repak AJ (1968) Encystment and excystment of the heterotrichous ciliate *Blepharisma stoltei* Isquith. J Protozool 15:407-412
Riley GA (1956) Oceanography of Long Island Sound, 1952-1954 IX. Production and utilization of organic matter. Bull Bingham Oceanographic Coll 15:324-341
Rivier A, Brownlee DC, Sheldon RW, Rassoulzadegan F (1985) Growth of microzooplankton: a comparative study of bactivorous zooflagellates and ciliates. Mar Micro Food Webs 1:36-51
Rublee PA, Gallegos CL (1989) Use of fluorescently labelled algae (FLA) to estimate microzooplankton grazing rate. Mar Ecol Prog Ser (in press)
Rubenstein DI, Koehl MAR (1977) The mechanisms of filter

feeding: some theoretical considerations. Am Nat 111:981-994
Rudzinska MA (1973) Do Suctorians really feed by suction? Bioscience 23:87-94
Salt GW (1967) Predation in an experimental protozoan population (*Woodruffia-Paramecium*). Ecol Monogr 37:113-144
Sanders RW, Porter KG (1986) Use of metabolic inhibitors to estimate protozooplankton grazing and bacterial production in a monomictic eutrophic lake with an anaerobic hypolimnion. Appl Environ Microbiol 52:101-107
Sanders RW, Porter KG (1988) Phagotrophic phytoflagellates. In: Marshall KC (ed), Advances in Microbial Ecology, Vol 10. Plenum Publishing Corp
Sekiguchi H, Kato T (1976) Influence of Noctiluca's predation on the *Acartia* population in Ise Bay, Central Japan. J Oceanogr Soc Japan 32:195-198
Servais P, Billen G, Rego JV (1985) Rate of bacterial mortality in aquatic environments. Appl Environ Microbiol 49:1448-1454
Sheldon RW, Prakash A, Sutcliff Jr WH (1972) Size distribution of particles in the ocean. Limnol Oceanogr 17: 327-340
Sheldon RW, Nival P, Rassoulzadegan F (1986) An experimental investigation of a flagellate-ciliate-copepod food chain with some observations relevant to the linear biomass hypothesis. Limnol Oceangr 31:184-189
Sherr BF, Sherr EB, Berman T (1982) Decomposition of organic detritus: a selective role for microflagellate protozoa. Limnol Oceanogr 27:765-769
Sherr BF, Sherr EB, Berman T (1983) Grazing, growth, and ammonium excretion rates of a heterotrophic microflagellate fed with four species of bacteria. Appl Environ Microbiol 45:1196-1201
Sherr BF, Sherr EB (1984) Role of heterotrophic protozoa in carbon and energy flow in aquatic ecosystems In: Klug M, Reddy CA (eds), Current Perspectives in Microbial Ecology. Amer Soc Microbiol Washington, p 412
Sherr BF, Sherr EB, Andrew TA, Fallon RD, Newell SY (1986a) Trophic interactions between heterotrophic protozoa and bacterioplankton in estuarine water analysed with selective metabolic inhibitors. Mar Ecol Prog Ser 32:169-180
Sherr EB, Sherr BF, Fallon RD, Newell SY (1986b) Small aloricate ciliates as a major component of the marine heterotrophic nanoplankton. Limnol Oceanogr 31:177-183
Sherr BF, Sherr EB, Fallon RD (1987) Use of monodispersed, fluorescently labeled bacteria to estimate *in situ* protozoan bacterivory. Appl Environ Microbiol 53:958-965
Sherr BF, Sherr EB, Hopkinson CS (1988) Trophic interactions within pelagic microbial communities: Indications of feedback regulation of carbon flow. Hydrobiologia 159:19-26
Sherr BF, Sherr EB (submitted) Distribution of numbers, biovolumes, and bacterivores within nanoplanktonic size spectra of apochlorotic nanoflagellates in several marine pelagic systems. Mar Microb Food Webs
Sherr BF, Sherr EB, Pedros-Alio C (submitted) Simultaneous measurement of bacterioplankton production and protozoan bacterivory in estuarine water. Mar Ecol Prog Ser
Sherr EB, Sherr BF (1983) Double staining epifluorescence

technique to assess frequency of dividing cells and bacterivory in natural populations of heterotrophic microprotozoa. Appl environ microbiol 46:1388-1393

Sherr EB (1988) Direct utilization of high molecular weight polysaccharide by heterotrophic flagellates. Nature 335:348-351

Sherr EB, Sherr BF, Paffenhofer G-A (1986c) Phagotrophic protozoa as food for metazoans: a 'missing' trophic link in marine pelagic food webs? Mar Microb Food Webs 1:61-80

Sherr EB, Sherr BF (1987) High rates of consumption of bacteria by pelagic ciliates. Nature 325:710-711

Sibbald MJ, Albright LJ (1988) Aggregated and free bacteria as food sources for heterotrophic microflagellates. Appl Environ microbiol 54:613-616

Sibbald MJ, Albright LJ, Sibbald PR (1987) Chemosensory responses of a heterotrophic microflagellate to bacteria and several nitrogen compounds. Mar Ecol Prog Ser 36:201-204

Sibbald MJ, Sibbald PR, Albright LJ (1988) How advantageous is a sensory prey detection mechanism to predatory microflagellates? J Plank Res 10:455-464

Sieburth JMcN (1979) Sea Microbes Oxford Univ Press

Sieburth JMcN (1984) Protozoan bacterivory in pelagic marine waters. In: Hobbie JE, Williams PJleB (eds), Heterotrophic activity in the sea Plenum Press, New York, p 405

Sieburth JMcN, Davis PG (1982) The role of heterotrophic nanoplankton in the grazing and nurturing of planktonic bacteria in the Sargasso and Caribbean Sea. Annls Inst Oceanogr 58:285-296

Sieracki ME, Haas LW, Caron DA, Lessard EJ (1987) The effect of fixation on particle retention by micro-flagellates: underestimation of grazing rates. Mar Ecol Prog Ser 38:251-258

Silver MW, Gowing MM, Brownlee DC, Corliss JO (1984) Ciliated protozoa associated with sinking oceanic detritus. Nature 309:246-248

Sleigh MA (1964) Flagella movement of the sessile flagellates *Actinomonas*, *Condonosiga*, *Monas* and *Proteriodendron*. Quart J Microsc Sci 105:405-414

Sleigh MA (1989) Protozoa and other protists. Edward Arnold, London

Small EB (1973) A study of ciliated protozoa from a small polluted stream in east-central Illinois. Amer Zool 13:225-230

Small EB (1984) An essay on the evolution of ciliophoran oral cytoarchitecture based on descent from within a karyorelictean ancestry. Origins of Life 13:217-228

Small EB, Lynn DH (1985) Phylum Ciliophora Doflein. In: Lee JJ, Hutner SH, Bovee EC (eds), Illustrated Guide to the Protozoa Allen Press and Soc of Protozoologists, Lawrence, Kansas, p 393

Smetacek VS (1981) The annual cycle of protozooplankton in the Kiel Bight. Mar Biol 63:1-11

Smith REH, Geider RJ, Platt T (1988) Microplankton productivity in the oligotrophic ocean. Nature 311:252-254

Sorokin YuI (1977) The heterotrophic phase of plankton succession in the Japan Sea. Mar Biol 41:107-117

Sorokin YuI (1981) Microheterotrophic organisms in marine ecosystems. In: Longhurst AR (ed), Analysis of Marine Ecosystems Academic Press, New York, p 293
Spindler M, Hemleben C, Salomons JB, Smit LP (1984) Feeding behavior of some planktonic foraminifers in laboratory cultures. J Foram Res 14:237
Spittler P (1973) Feeding experiments with tintinnids. Oikos 15:128-132
Spoon D (1986) Discovery of a predatory procaryote feeding on palmelloid and motile euglenids in the Georgetown coral reef microcosm Abstract 39th annual meeting, Society of Protozoologists
Stoecker D, Guillard RR, Kavee RM (1981) Selective predation by *Favella ehrenbergii* (Tintinnia) on and among dinoflagellates. Biol Bull 160:136-145
Stoecker DK, Davis LH, Anderson DM (1984) Fine scale spatial correlations between planktonic ciliates and dinoflagellates. J Plank Res 6:829-842
Stoecker DK, Cucci TL, Hulburt EM, Yentsch CM (1986) Selective feeding by *Balanion* sp (Ciliata: Balanionidae) on phytoplankton that best support its growth. J Exp Mar Biol Ecol 95:113-130
Stoecker DK, Michaels AE, Davis LH (1987) Large proportion of marine planktonic ciliates found to contain functional chloroplasts. Nature 326:790-792
Straskrabova-Prokesova V, Legner M (1966) Interrelations between bacteria and protozoa during glucose oxidation in water. Int Rev Gesamten Hydrobiol 51:279-293
Suttle CA, Chan AM, Taylor WD Harrison (1986) Grazing of planktonic diatoms by microflagellates. J Plank Res 8:393-398
Swanberg NR (1983) The trophic role of colonial radiolaria in oligotrophic oceanic environments. Limnol Oceanogr 28:655-666
Swanberg NR, Anderson OR (1985) The nutrition of radiolarians: trophic activity of some solitary Spumellaria. Limnol Oceanogr 30:646-652
Swanberg NR, Anderson OR, Lindsey JL, Bennett P (1986) The biology of *Physematium muelleri*: trophic activity. Deep-Sea Res 33:913-922
Swanberg NR, Harbison GR (1980) The ecology of *Collozoum longiforme*, sp nov, a new colonial radiolarian from the equatorial Atlantic Ocean. Deep-Sea Res 27A:715-732
Taniguchi A, Kawakami R (1983) Growth rates of ciliate *Eutintinnus lususundae* and *Favella taraikaensis* observed in the laboratory culture experiments. Bull Plank Soc Japan 30:33-40
Taylor DL, Seliger H (eds) (1979) Toxic dinoflagellate blooms Dev Mar Biol I. Elsevier/North Hollander, New York
Taylor GT, Pace ML (1987) Validity of eukaryotic inhibitors for assessing production and grazing mortality of marine bacterioplankton. Appl Environ Microbiol 53:119-128
Taylor GT, Karl DM, Pace ML (1986) Impact of bacteria and zooflagellates on the composition of sinking particles: an *in situ* experiment. Mar Ecol Prog Ser 29:141-155
Townsend DW, Cammen LM (1985) A deep protozoan maximum in the Gulf of Maine. Mar Ecol Prog Ser 24:177-182

Tremaine SC, Mills AL (1987a) Tests of the critical assumptions of the dilution method for estimating bacterivory by microeukaryotes. Appl Environ Microbiol 53:2914-2921
Tremaine SC, Mills AL (1987b) Inadequacy of the eucaryotic inhibitor cycloheximide in studies of protozoan grazing on bacteria at the freshwater-sediment interface. Appl Environ Microbiol 53:1969-1972
Triemer RE, Fritz L (1987) Structure and operation of the feeding apparatus in a colorless euglenoid, *Entosiphon sulcatum*. J Protozool 34:39-47
Turley CM, Newell RC, Robins DB (1986) Survival strategies of two small marine ciliates and their role in regulating bacterial community structure under experimental conditions. Mar Ecol Prog Ser 33:59-70
Turley CM, Lochte K, Patterson DJ (1988) A barophilic flagellate isolated from 4500 m in the mid-North Atlantic. Deep-Sea Res 35:1079-1092
Van Houten J (1988) Chemoresponse mechanisms: toward the molecular level. J Protozool 35:241-243
Verity PG (1985) Grazing, respiration, excretion, and growth rates of tintinnids. Limnol Oceanogr 30:1268-1282
Verity PG (1986a) Grazing of phototrophic nanoplankton by microzooplankton in Narragansett Bay. Mar Ecol Prog Ser 29:105-115
Verity PG (1986b) Growth rates of natural tintinnid populations in Narragansett Bay. Mar Ecol Prog Ser 29:117-126
Verity PG (1987) Abundance, community composition, size distribution and production rates of tintinnids in Narragansett Bay, Rhode Island. Est Coast Shelf Sci 24:671-690
Verity PG (1988) Chemosensory behavior in marine planktonic ciliates. Bull Mar Sci 43:772-782
Verity PG, Stoecker D (1982) Effects of *Olisthodiscus luteus* on the growth and abundance of tintinnids. Mar Biol 72:79-87
Verity PG, Villareal TA (1986) The relative food value of diatoms, dinoflagellates, flagellates, and cyanobacteria for tintinnid ciliates. Arch Protistenk 131:71-84
Vickerman K (1976) The diversity of kinetoplastid flagellates. In: Lumsden WHR, Evans DA (eds), Biology of the Kinetoplastida VI. Academic Press, London, p 5
Wailes GH (1927) Rhizopodia and Heliozoa from British Columbia. Ann Mag Nat Hist 9 Ser 20:153-156
Wailes GH (1937) Canadian Pacific fauna I. Protozoa (a, lobosa, b, reticulosa, c, heliozoa, d, radiolaria). Biol Bd Canada, Toronto. p 14
Wessenberg H, Antipa GA (1970) Capture and ingestion of *Paramecium* by *Didinium nasutum*. J Protozool 17:250-270
Wheeler PA, Kirchman DL (1986) Utilization of inorganic and organic nitrogen by bacteria in marine systems. Limnol Oceanogr 31:998-1009
Wikner J, Andersson A, Normark S, Hagstrom A (1986) Use of genetically marked minicells as a probe in measurement of predation on bacteria in aquatic environments. Appl Environ Microbiol 52:4-8
Wikner J, Hagstrom A (1988) Evidence for a tightly coupled

nanoplanktonic predator-prey link regulating the bacterivores in the Marine Environment. Mar Ecol Prog Ser 47:137-145
Wright RT, Coffin RB (1984) Measuring microzooplankton grazing on planktonic marine bacteria by its impact on bacterial production. Microb Ecol 10:137-149
Zumwalt GS, Delaca TE (1980) Utilization of brachiopod feeding currents by epizoic foraminifera. J Paleontol 54:477-484

PROTOZOAN ENERGETICS - SESSION SUMMARY

Johanna Laybourn-Parry
University of Lancaster
Department of Biological Sciences
Lancaster LA1 4YQ
UK

INTRODUCTION

The now widely accepted importance of protozoa in fundamental ecological processes such as energy flow and nutrient regeneration in aquatic environments, and in particular the pelagic zone, necessitates more detailed data on the energetics or physiological ecology of protistan groups. Such information is crucial to the construction of models of carbon cycling, energy flow and the remineralization of essential nutrients such as phosphorus and nitrogen, and to understanding the contribution made by specific trophic and taxonomic groups in community processes. It was Pomeroy (1974) who focussed attention on the significance of microorganisms as movers of energy and materials in marine waters, which is of course related to the small size of protists and their greater metabolic rate per unit weight. A later review by Williams (1981) elaborated the impact of microheterotrophic processes in more detail, leading to suggestions that at least half of primary production passes through planktonic microheterotrophs before mineralization, and that secondary production at the microbial level may be equal to or exceed that of larger zooplankton.

The complexity of the microbial food web was further extended by the 'microbial loop' hypothesis expounded by Azam *et al.* (1983) wherein bacteria exploiting dissolved organic matter (DOM), mainly of planktonic origin, have their population densities controlled by predatory heterotrophic flagellates, which are in turn exploited by microzooplankton (mainly ciliates) as a food source. By this means, energy released by primary producers as DOM is somewhat inefficiently returned to the major pelagic food chain via the 'loop'. It should be noted, however, that small pelagic ciliates can exploit

NATO ASI Series, Vol. G 25
Protozoa and Their Role in Marine Processes
Edited by P. C. Reid et al.

bacteria as an energy source at high rates of consumption, and thus by-pass the flagellate step in the microbial loop (Sherr and Sherr 1987). The significance of these energy flow pathways on nutrient cycling are considered elsewhere in this volume (Berman 1990, Caron 1990).

SPATIAL CHARACTERISTICS OF THE MICROBIAL FOOD WEB

An understanding of the functional and spatial processes in the microbial loop is fundamental to elucidating patterns of energy flow through bacteria, bacterivore flagellates (and ciliates) and their predators. The DOM exploited by bacteria is produced at discrete loci in seawater and can be expected in high concentrations around, for example, a phytoplankter which is exuding organic matter or autolysing. The concentrations of the majority of DOM components in seawater range from 10^{-12}M to 10^{-8}M (Ammerman and Azam 1981).

Farooq Azam further elaborated on a question put by Azam and Ammersham (1984) as to whether new dissolved organic matter is used by bacteria at high concentrations within the production

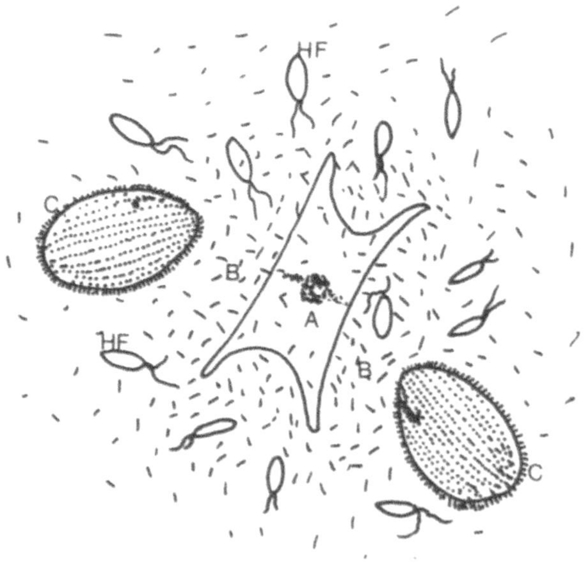

Fig. 1. Diagrammatic representation of a microbial loop consortium. A- algal cell, B- bacteria, HF- heterotrophic flagellate, C- ciliate, (After Azam)

(e.g. around an autolysing algal cell), or does much of it diffuse into a bulk phase in the water body before exploitation by structured nutrient field representing a region with sharp gradients of bacterial nutrients. In terms of efficiency of energy utilization and transfer it makes sense for the microbial loop organisms to be concentrated in such nutrient rich microzones. Thus such organisms may be expected to exist in temporal and spatial consortia, Figure 1. The hypothesis has yet to be proved empirically, but there is indirect evidence to support it. Most marine bacterial isolates have been shown to be motile (Azam and Ammerman 1984) and most are chemoattracted to algal exudates (Bell and Mitchell 1972), and are therefore capable of optimizing their position in a nutrient field. Investigations of the kinetics of nutrient uptake by pelagic bacterial assemblages indicate multiphasic kinetics of nutrient uptake, suggesting that some bacteria have high affinity systems, others low affinity systems and some may possess multiphasic abilities, thus leading to niche separation within structured nutrient microzones. (Azam and Hodson 1981). The concentration of nutrients within microzones may change quickly, particularly where exudation of organic matter is not sustained but released in pulses, such as in excretion by zooplankton. For example in a plume of released nutrient 100 µm across the concentration drops by three orders of magnitude at the pulse centre within 5 minutes (Jackson 1980).

HETEROTROPHIC FLAGELLATES

Detailed data on the energetics of protozoan microzooplankton species is still sparse, and inevitably studies have focussed on the more abundant groups, particularly ciliates like tintinnids. The abundance, occurrence and physiological ecology of heterotrophic bacterivore flagellates have been studied to some extent (Sorokin 1981, Fenchel 1982a, Davis and Sieburth 1982, Sherr and Sherr 1983) and there are also data on aspects of their physiological ecology (e.g. Fenchel 1982b, Sherr *et al.* 1983, Sherr *et al.* 1984). Other flagellates exploit phytoplankton and other flagellates as a food source, and among these are a poorly researched group, the heterotrophic dinoflagellates. This group is frequently common in oceanic waters, but is often overlooked because of their fragility and small size.

Greg Gaines described how heterotrophic dinoflagellates exploit a variety of food by various feeding mechanisms (Gaines and Elbrachter 1987). Studies on a range of species suggest that many possess high feeding rates comparable to loricate and non-loricate ciliates (Lessard and Swift 1985). Their success appears to lie in their ability to consume food items which are difficult for other microzooplankton to tackle, such as diatoms. Two species presented as examples, *Polykrikos kofoidii* and *Oxyrrhis marina*, appear to be able to feed profitably in high food concentrations. Increasing food concentration elicits a linear feeding response even at high food densities. In other words there is no evidence of feeding saturation. The reproductive rate, however, does not follow the same pattern, but reaches a peak at about one doubling per day at intermediate, but relatively high, food concentrations of 0.4-0.8 µg C ml^{-1} in *P. kofoidii* and 0.2-0.4 mg C ml^{-1} in *O. marina*. Presumably the digestive efficiency decreases at high levels of ingestion, thus limiting the energy available for production. Alternatively there may be greater energetic costs in feeding in high food densities which puts contraint on the energy which can be partitioned into production. Clearly the heterotrophic dinoflagellates, which can occur in densities equal to the ciliates, must be important grazers of elements of the phytoplankton and bacterioplankton and may make a significant contribution to energy flow in some oceanic waters during various phases in the annual cycle (Paasche and Kristiansen 1982, Lessard and Swift 1985).

CILIATE PRODUCTION

The main focus of this session was on the production of planktonic ciliates, and in particular tintinnids. Ciliates appear to be the dominant group of protists exploiting phytoplankton, heterotrophic flagellates and some elements of the bacterioplankton (Porter *et al.* 1985, Sherr *et al.* 1986, Smetacek 1981). There have been sufficient investigations of the group to allow some insight into the contribution made by ciliates to energy flow on a global scale in marine waters (Lynn and Montagnes 1989). However, there are innumerable problems associated with interpreting and quantifying the physiological ecology of these protists. From a trophic standpoint they are complex. While some species feed specifically on bacteria, or elements of the phytoplankton or heterotrophic flagellates, others are capable of exploiting a wide range of food items. Moreover, it is now clear that members of the suborder Oligotrichina, particularly the Stombidiidae, are mixotrophic by virtue of their ability to retain the plastids of their algal prey and employ them for autotrophic nutrition (Blackbourn *et al.* 1973, Laval-Peuto and Febvre 1986, Laval-Peuto and Rassoulzadegan 1988, Stoecker *et al.* 1987, Stoecker 1989). Plastidic ciliates are often an abundant component of the plankton (Rassoulzadegan 1977) and it is reasonable to suppose that mixotrophy may be a significant element in marine food webs. The degree to which mixotrophy occurs in waters of differing nutrient status has yet to be elucidated, but this is clearly an avenue which needs to be explored.

Other problems relate to the sporadic occurrence of many species in the annual cycle of the protistan community and the need for detailed and frequent sampling. There are also problems arising from the variety of techniques used by workers for sampling, fixing, counting and estimating biomass, which often renders comparisons of data on ecological energetics from different investigations difficult. Hopefully this latter problem will be resolved in the near future as a result of critical appraisal of the available techniques and their refinement (Rassoulzadegan 1990).

Tintinnid ciliates appear to be an important component in the planktonic food web of some marine waters. For example, in Narragansett Bay, Rhode Island they are estimated as grazing

26% of annual primary production (Verity 1987), in Long Island Sound 27% of annual primary production (Capriulo and Carpenter 1983) and in the Solent, UK as much as 60% of phytoplankton production may be removed by tintinnids (Burkill 1982).

Guy Gilron has shown that in other waters, such as the tropical seas of the Caribbean tintinnids may graze only 4% of primary production. Production by tintinnid ciliates off Jamaica was determined from *in situ* incubation experiments which excluded predators > 200 μm. The four major species showed no significant differences in their rates of production throughout the year, achieving a specific growth rate (see Table 1) of $r=0.033\ h^{-1}$, which equates to almost one doubling per day. Gilron reviewed the factors controlling growth by comparing his data with those for tintinnid production in Narragansett Bay (Verity, 1987). From Table 1 it can been seen that in terms of rates of growth and reproduction, the tintinnids in Verity's (1987) temperate study achieve twice those shown by tropical populations. Temperature appears not to be the dominating influence, since in the Caribbean temperature is normally constant around 27°C, whereas in Narragansett Bay the maximum temperature in August is 22°C. Food availability appeared to exert the major impact on the productivity of tintinnids. Tintinnid growth rate correlates positively with chlorophyll *a* concentration (Verity 1987). Thus the concentration of chlorophyll *a* in the size fraction of phytoplankton which can be exploited by tintinnids can be used as an index of food availability. In the Caribbean it was estimated that chlorophyll *a* in the tintinnid food size range was $0.8\ \mu g\ l^{-1}$, whereas in Narragansett Bay it was many times higher at $8.9\ \mu g\ l^{-1}$. The rates of production are considerably higher in the temperate site; $430\ J\ m^{-3}\ day^{-1}$ as opposed to $12\ J\ m^{-3}\ day^{-1}$ in tropical waters. However, such comparisons must be considered with care. When one considers ciliate production on a global scale it is clear that factors which control production are complex (Lynn and Montagnes 1989). This point is illustrated by Table 1, which presents specific growth rates and generation times from a selection of studies. The variation in rates is large even within fairly discrete zones. For example ciliate production in waters off Peru varied between 8.78-612.11 J m^{-3} day^{-1} along a 500 mile section from the coast seaward at 8°S (Tumantseva and Kopylov, 1985). Moreover there did not appear to be any clear correlation between the quantity of available food and the protistan biomass it supported.

Table 1. SPECIFIC GROWTH RATES $r = \ln N_t - \ln N_o$ WHERE N_o AND N_t ARE THE OBSERVED CONCENTRATIONS AT THE BEGINNING AND END OF THE TIME INTERVAL t, AND GENERATION TIMES T = 0.693/r FOR PLANKTONIC CILIATES FROM DIFFERENT LOCALITIES.

Species	T (h)	r (h^{-1})	Site
Urotricha marine[1]	7.3	0.090	Pacific off Peru
Strombidia sp[1]	18.73	0.040	"
Strombidium conicum[1]	14.77	0.056	"
S. acuminatum[1]	22.39	0.046	"
Tiarina fusus[1]	9.98	0.077	"
Tontonia gracillima[1]	12.38	0.067	"
Mesodinium sp[1]	9.13	0.080	"
Didinium nasutum[1]	12.40	0.061	"
Metastrombidium sp[1]	10.19	0.082	"
Tintinnids[1]	39.14	0.026	"
Tintinnopsis c.f. acuminata[2]	11.55	0.060	Laboratory
Eutintinnus pectinis[2]	11.55	0.060	"
Eutintinnus pectinis[3]	23.73	0.029	Falmouth, Mass
Balanion sp[3]	10.14	0.0683	USA
Tintinnopsis kofoiki[3]	10.16	0.0682	"
Favella sp[3]	21.65	0.032	"
naked ciliates[4]	72.18	0.0096	Solent, U.K.
Tintinnids[4]	39.00	0.0177	"
Tintinnopsis beroidea[5]	60-144	0.011-0.004	Laboratory
Tintinnids[6]	9.24	0.075	Narragansett, USA
Tintinnids[7]	21.0	0.033	Caribbean, Jamaica

[1]Tumantseva and Kopylov (1985), [2]Heinbokel (1978), [3]Stoecker *et al.* (1983), [4]Burkill (1982), [5]Gold (1971), [6]Verity (1987), [7]Gilron (pers. comm.)

Although the data are limited, attempts to draw together information in order to elucidate the factors which influence the productivity of protistan assemblages are essential. Such exercises need to be repeated at intervals as the pool of data swells, so that we can re-examine our ideas and highlight areas where we need to refine our techniques and approach. The global review of ciliate production offered by Lynn and Montagnes (1990) in this volume, indicates that estimates of ciliate production range from 5-500 J m^{-3} day^{-1}. These authors have critically considered the factors responsible for such wide variations with the available data and have suggested a number of directions for future research in protozoan production studies.

OTHER CONSIDERATIONS

On a smaller scale there are areas in the physiological ecology of protozoa which demand further investigation. As yet we have only sketchy information on the life-cycle 'strategies' of marine protozoa. We do not know how most species withstand periods when the physical chemical and biological conditions of their environment are unsuitable for their growth and reproduction, neither do we know how species disperse and recolonise environments. Many protozoans possess the ability to encyst and this probably plays a critical role in the seasonal life-cycle. There is some evidence to support this view. Encystment by tintinnids, an often abundant element in the microzooplankton, has been reviewed by Reid and John (1978). Tintinnids have five different cyst modes, some of which may have a reproductive function, while others are clearly resting or overwintering cysts. Such cysts have been seen attached to tintinnid loricae and in fine-grained marine sediments from around the British Isles. In one case a laboratory culture of the tintinnid *Helicostomella subulata*, which is a dominant member of the summer and autumn microzooplankton community off Nova Scotia, produced cysts under ideal growth conditions contemporaneously with the field population (Paranjape, 1980). Excystment was induced after simulating winter conditions, suggesting that in this species encystment is a timed event in the annual cycle. The oligotrich *Strombidium crassulum* has also been shown to undergo mass encystment following the main spring bloom of diatoms (Reid 1987). It is suggested that resuspension of bottom sediments by storms and tides ensures that cysts are always present in the water awaiting conditions suitable for excystment.

A second area which needs further detailed consideration is the way in which protozoans exploit their food resources. There is increasing evidence that some protozoa are selective feeders on elements in the plankton, which may in turn have consequences on the species composition of the phytoplankton or microzooplankton. Stoecker *et al.* (1986), for example have shown that the ciliate *Balanion* preferentially consumes the dinoflagellate *Heterocapsa* species in mixtures containing cryptophytes and green flagellates, even when the non-dinoflagellate algae are more abundant. Similarly, the tintinnid *Favella* is a selective predator of dinoflagellates (Stoecker *et al.* 1981). Sarcodines of various types also exhibit selective feeding (Mast and Hahnert 1935, Anderson 1980, Laybourn-Parry *et al.* 1987). Even the mixotrophic ciliates appear to exhibit prey preferences, presumably related to a liking for the plastids of particular species (Stoecker 1989). How protozoa recognise their preferred food is unclear.

Optimal foraging theory, which is concerned with understanding how organisms maximize their net energy gain from their feeding activities, has not yet been applied to protozoa. Selective feeding must confer some energetic advantage on those protozoa practising it. This may be a function of a variety of factors such as different digestive and production efficiencies achieved on a given diet, the energy costs involved in catching and handling the food item and the energy content of the prey. Perhaps protozoologists should attempt to apply optimality models to the acquisition of energy resources by protozoans. The approach provides a more detailed insight into the exploitation of available food resources by organisms and a clearer understanding of their physiological functioning.

The efficiency with which protozoa process and partition the energy gained from their feeding activities is critical to the study of energy flow and nutrient cycling. The so-called net and gross production efficiencies are widely used in the construction of models. There are only a limited number of such coefficients of efficiency in the literature and there is a need for more if we are to improve the accuracy of our models and predictions of energy flow and mineralization rates. Given the wide trophic diversity displayed by the different groups, it is not unreasonable to suppose that there is also a broad spectrum of net and gross production efficiencies. Moreover these efficiencies of energy partitioning are modified by environmental factors such as temperature (Laybourn and Stewart 1975, Laybourn 1976, Rogerson 1980, 1981), and food type (Rubin and Lee 1976).

REFERENCES

Ammerman JW, Azam F (1981) Dissolved cyclic adenosine monophosphate (cAMP) in the sea and uptake of cAMP by marine bacteria. Mar Ecol Prog Ser 5:85-91

Abdersib OR (1980) Radiolaria In: Levandowsky M, Huner SH (eds) Biochemistry and Physiology of Protozoa. Academic Press, New York

Azam F, Hodson RE (1981) Multiphasic kinetics for D-glucose uptake by assemblages of natural marine bacteria. Mar Ecol Prog Ser 6:213-220

Azam F, Fenchel T, Field JG, Gray JS, Meyer-Reil LA, Thingstad F (1983) The ecological role of water dolumn microbes in the sea. Mar Ecol Prog Ser 10:257-263

Azam F, Ammerman JW (1984) Cycling of organic matter by bacterioplankton in pelagic marine systems: microenvironmental considerations. In: Fasham MRJ (ed) Flows of Energy and Materials in Marine Ecosystems: Theory and Practise. NATO Conference Series 4 Marine Sciences Plenum Press, New York

Bell W, Mitchell R (1972) Chemotactic and growth responses of marine bacteria to algal extracellular products. Biol Bull Woods Hole 143:265

Berman T (1990) Protozoans as agents in planktonic nutrient cycling. In: Reid PC, Turley CM, Burkill PH (eds) Protozoa and their role in marine processes. Springer, Berlin Heidelberg New York

Blackbourn DJ, Taylor FJR, Blackbourn J (1973) Foreign organelle retention by ciliates. J Protozool 20:286-288

Burkill P (1982) Ciliates and other microplankton components of a nearshore foodweb: standing stocks and production processes. Ann Inst Oceanogr Paris 58:335-350

Capriulo GM, carpenter EJ (1983) Abundance, species composition

and feeding impact of tintinnid micro-zooplankton in central Long Island Sound. Mar Ecol Prog Ser 10:277-288
Caron DA (1990) Evolving role of protozoa in aquatic nutrient cycles. In: Reid PC, Turley CM, Burkill PH (eds) Protozoa and their role in marine processes. Springer, Berlin Heidelberg New York
Davis PG, Sieburth JMcN (1982) Differentiation of the photosynthetic and heterotrophic populations of nanoplankton by epifluorescence microscopy. Ann Inst Oceanogr Paris 58:249-260
Fenchel T (1982a) Ecology of heterotrophic microflagellates IV Quantitative occurrence and importance as bacterial consumers. Mar Ecol Prog Ser 9:35-42
Fenchel T (1982b) Ecology of heterotrophic microflagellates II Bioenergetics and growth. Mar Ecol Prog Ser 8:225-231
Gaines G, Elbrachter M (1987) Heterotrophic nutrition. In: Taylor FJR (ed) The Biology of Dinoflagellates. Blackwell, Oxford
Gold K (1971) Growth characteristics of the mass reared tintinnid Tintinnopsis beroidea. Mar Biol 8:105-108
Heinbokel JF (1978) Studies on the functional role of tintinnids in the Southern California Bight. I Grazing and growth rates in laboratory cultures. Mar Biol 47:177-189
Jackson GA 1980 Phytoplankton growth and zooplankton grazing in oligotrophic oceans. Nature 284:439-440
Laval-Peuto M, Febvre M (1986) On plastid symbiosis in *Pytontonia appendiculariformis* (Ciliophora, Oligotrichina). Biosystems 19:137-158
Laval-Peuto M, Rassoulzadegan F (1988) Autofluorescence of marine planktonic Oligotrichina and other ciliates. Hydrobiol 159:99-110
Laybourn J (1976) Energy budgets for Stentor coeruleus Ehrenberg (Ciliophors). Oecologia (Berl) 22:431-437
Laybourn J, Stewart JM (1975) Studies on consumption and growth in the ciliate Colpidium campylum Stokes. J Anim Ecol 44:165-174
Laybourn-Parry J, Jones K, Holdich JP (1987) Grazing by Mayorella sp (Protozoa:Sarcodina) on Cyanobacteria. Functional Ecol 1:99-104
Lessard EJ, Swift E (1985) Species-specific grazing rates of heterotrophic dinoflagellates in oceanic waters, measured with a dual-labelled radioisotope technique. Mar Biol 87:289-296
Lynn DH, Montagnes DJS (1990) Global production of heterotrophic marine planktonic ciliates. In: Reid PC, Turley CM, Burkill PH (eds) Protozoa and their role in marine processes. Springer, Berlin Heidelberg New York
Mast SO, Hahnert WF (1935) Feeding, digestion and starvation in Amoeba proteus Leidy. Physiol Zool 8:255-272
Paranjape MA (1980) Occurrence and significance of resting cysts in a hyaline tintinnid Helicostomella subulata (Ehre.) Jorgensen. J exp mar Biol Ecol 48:23-33
Paasche E, Kristiansen S (1982) Ammonium regeneration by micro-zooplankton in the Oslofjord. Mar Biol 69:55-63
Pomeroy LR (1974) The ocean's food web, a changing paradigm. Bioscience 24:499-504
Porter KG, Sherr EB, Sherr BF, Pace M, Sanders RW (1985)

Protozoa in planktonic food webs. J Protozool 32:409-415
Rassoulzadegan F (1977) Evolution anuelle des cilies pelagiques en Mediterranee nord-occidentale. I Cilies oligotriches 'non-tintinnides' (Oligotrichina). Ann Inst ocenaogr, Paris 53:125-134
Rassoulzadegan F (1990) Methods for the study of marine microzooplankton - Session summary. In: Reid PC, Turley CM, Burkill PH (eds) Protozoa and their role in marine processes. Springer, Berlin Heidelberg New York
Reid PC (1987) Mass encystment of a planktonic oligotrich ciliate. Mar Biol 95:221-230
Reid PC, John AWG (1978) Tintinnid cysts. J mar biol Ass, UK 58:551-557
Rogerson A (1980) Generation times and reproductive rates of Amoeba proteus (Leidy) as influenced by temperature and food concentration. Can J Zool 58:543-548
Rogerson A (1981) The ecological energetics of Amoeba proteus)Protozoa). Hydrobiologia 85:117-128
Sherr B, Sherr E (1983) Enumeration of heterotrophic microprotozoa by epifluorescence microscopy. Est Coast Shelf Sci 16:1-7
Sherr BF, Sherr EB, Berman T (1983) Grazing, growth and ammonium excretion rates of a heterotrophic microflagellate fed with four species of bacteria. Appl Environ Microbiol 45:1196-1201
Sherr BF, Sherr EB, Newell SY (1984) Abundance and productivity of heterotrophic nanoplankton in Georgia coastal waters. J Plankton Res 6:196-202
Sherr EB, Sherr BF (1987) High rates of consumption of bacteria by pelagic ciliates. Nature 325:710-711
Sherr EB, Sherr BF, Fallon RD, Newell SY (1986) Small aloricate ciliates as a major component of marine heterotrophic nanoplankton. Limnol Oceanogr 31:177-183
Smetacek V (1981) The annual cycle of protozooplankton in the Kiel Bight. Mar Biol 63:1-11
Sorokin Yu I (1981) Microheterotrophic organisms in marine ecosystems. In: Longhurst AR (ed) Analysis of Marine Ecosystems. Academic Press, New York
Stoecker D, Guillard RRL, Kavee RM (1981) Selective predation by Favella ehrenbergii (Tintinnina) on dinoflagellates. Biol Bull Woods Hole 160:136-145
Stoecker D, Davis LH, Provan A (1983) Growth of Favella (Ciliata: Tintinnina) and other microzooplankton in cages incubated in situ and comparison to growth *in vitro*. Mar Biol 75:293-302
Stoecker DK, Cucci TL, Hulbert EM, Yentsch CM (1986) Selective feeding by Balanion sp. (Ciliata:Balanionidae) on phytoplankton that best support its growth. J exp mar Biol Ecol 95:113-130
Stoecker DK, Michaels AE, Davies LH (1987) Large proportion of marine planktonic ciliates found to contain functional chloroplasts. Nature 326:790-792
Stoecker DK (1990) Mixotrophy in marine planktonic ciliates: physiological and ecological aspects of plastid-retention by oligotrichs. In: Reid PC, Turley CM, Burkill PH (eds) Protozoa and their role in marine processes. Springer, Berlin Heidelberg New York

Tumantseva NI, Kopylov AI (1985) Reproduction and production rates of planktonic infusoria in coastal waters of Peru. Oceanology 25:390-394

Verity PG (1987) Abundance, community composition, size distribution and production rates of tintinnids in Narragansett Rhode Island. Est Coast Shelf Sci 24:671-690

Williams PJ LeB (1981) Incorporation of microheterotrophic processes into the classical paradigm of the planktonic food web. Kieler Meerresforsch, Sonderh 5:1-28

GLOBAL PRODUCTION OF HETEROTROPHIC MARINE PLANKTONIC CILIATES

Denis H Lynn* and David J S Montagnes▪

*Department of Zoology
University of Guelph
Guelph
Ontario
CANADA
N1G 2W1

▪Department of Oceanography
University of British Columbia
Vancouver
BC V6T IN5

INTRODUCTION

The ecological role of marine microheterotrophs has recently received attention. Reviews by Sorokin (1981) and Porter *et al.* (1985) have emphasised that ciliates are important microheterotrophs in marine food webs and have provided some observations on the factors influencing ciliate production. Our goal is to up-date these reviews and concentrate on the production of heterotrophic planktonic marine ciliates.

Ciliates are often prominent members of planktonic food webs and can provide a link between smaller organisms and larger zooplankton (Porter *et al.* 1985, Sherr *et al.* 1986). Direct microscopic observations have indicated that they consume phytoplankton, heterotrophic flagellates, cyanobacteria, and heterotrophic bacterioplankton (Gifford, 1985, Johnson *et al.* 1982, Laval-Peuto *et al.* 1986 Rassoulzadegan *et al.* 1988, Sherr *et al* 1986, Smetacek, 1981). However, heterotrophic ciliates also act as facultative autotrophs (Laval-Peuto and Rassoulzadegan, 1988, McManus and Fuhrman, 1986, Stoecker *et al.* 1987a, 1988, 1989). Ciliates, like much of the plankton, are thought to be eaten by copepods. They are generally large enough for copepods to capture (Nival and Nival, 1976) and have been observed to be consumed by copepods in laboratory cultures (Berk *et al.* 1977) and in enclosures (Rassoulzadegan and Sheldon, 1986). Further, tintinnine loricae have been found in copepod faecal pellets (Turner and Anderson, 1983). Although other zooplankton (Stoecker *et al.* 1987b) and fish (e.g. Capriulo and Ninivaggi, 1982, references in Dale and Dahl, 1987a, Stoecker and Govoni, 1984) also eat ciliates, it has

NATO ASI Series, Vol. G 25
Protozoa and Their Role in Marine Processes
Edited by P. C. Reid et al.

been assumed that copepods are their dominant predator. In fact, copepods apparently prefer ciliates over other planktonic prey (Stoecker and Egloff 1987, Stoecker and Sanders 1985, Turner and Anderson, 1983).

In the past, quantifying the flow of energy through the 'ciliate compartment' in planktonic food webs has relied on estimates of their grazing impact, numerical abundance, biomass, and production. Of these production provides the most fundamental measure. Since definitions of production vary in the literature, production is defined here as the amount of biomass generated per unit time per unit space regardless of its fate (after Downing 1984). Several recent studies have adopted this general approach although not using the same formula that we use (e.g. Burkill 1982, Leppanen and Bruun 1986). Other studies on production (e.g. Verity 1987) have included a 'mortality' rate by assuming that it is an exponential function of population size. A loss by mortality is not included in our production estimates. A few estimates of ciliate production are available but the majority of studies report cell abundance or biomass. We have assembled estimates of biomass and production. Since cell size varies considerably we have not included studies that report abundance exclusively. The estimates span almost two decades of research, and the data are varied and often not comparable. They include values from single sites, single dates, annual averages, annual ranges, and regions, such as upwelling and convergence zones. Nevertheless, some trends are apparent and direction for further study can be realised from the present data.

METHODS

Data were collected from the literature and personal communications. Methods of presentation varied throughout, so data were converted to common units: J m^{-3} d^{-1}. Data that were originally graphically presented were interpreted from figures. Conversion factors were used when necessary and were as follows: 1) 0.17 g dry wt ml^{-1} protoplasm including vacuoles

(Laybourn in Finlay, 1978), 2) 20.15 J mg^{-1} dry wt (Laybourn and Stewart, 1975), 3) 0.071 g C ml^{-1} protoplasm (Fenchel and Finlay 1983, and 4) an average generation time of one day (see Discussion). When production was given in g C, volume was determined using the reported carbon:volume ratios or using 1), 2), and 3) above. Means reported in Table 1 were either taken from the original paper or calculated by integrating over vertical profiles or by obtaining a daily mean from reported monthly or annual estimates.

Methods of collection varied among studies. In many cases, abrasive and caustic techniques have undoubtedly underestimated ciliate biomass. However, it is impossible to correct for underestimation caused by these handling procedures since they are not consistent. Often only tintinnine biomass was reported. In these cases, tintinnine biomass is accompanied with an estimate of total ciliate biomass derived by assuming tintinnines to be 25% of the total (Sherr *et al.* 1986). This is only a rough estimate since several studies have reported tintinnines to be greater than 25% (Burkill 1982, Linley *et al.* 1983, Verity 1987) while others report them to be less (Montagnes 1987).

Ciliate vs. phytoplankton biomass and ciliate vs. bacterial biomass were correlated: > 170 data points for the former and > 80 for the latter. These data were recorded from a variety of places, including different depths at a single site, several sites in a region, many times during the year, and single and averaged observations (Beers and Stewart 1969, Beers *et al.* 1980, Berk *et al.* 1977, Landry *et al.* 1984, Leppanen and Bruun 1986, Linley *et al.* 1983, Middlebrook *et al.* 1987, Paranjape *et al.* 1985, Rassoulzadegan *et al.* 1988, Smetacek 1981, Sorokin 1977, Sorokin and Kogelschatz 1979).

PLANKTONIC CILIATE PRODUCTION

Biomass and production estimates for ciliates have been made in various locations, mainly in coastal areas (Figure 1). Ciliate production usually ranges from 5 - 500 J $m^{-3}d^{-1}$ with extremes

Table 1. ESTIMATES OF GLOBAL PRODUCTION OF CILIATES*

Lat	Long	Season	Distance from Shore (km)	Production Estimates ($J\ m^{-3} d^{-1}$) Mean	Range	Comments	Reference
32 N	117 W	Feb	50	8		Table 2, Fig. 1	Beers & Stewart 1969
31 N	117 W	Feb	90	7		integrated over water column	"
30 N	118 W	Feb	120	6		pump, formaldehyde, prefiltered	"
29 N	118 W	Feb	200	5			"
28 N	119 W	Feb	200	4			"
32 N	117 W	May	10	430		Table 2, pump, formaldehyde	Beers et al. 1980
32 N	117 W	Jun	10	51		integrated over water column	"
32 N	117 W	Jun	10	42			
38 N	77 W	Nov	0	9200	5700-12000	estuarine, 3 m, 10°/oo, mean of 3 sites, Table 2, bottle, Lugol	Berk et al. 1977
50 N	1 W	ann	1	73	16-180	Tables 2 and 5 pump, Lugol	Burkill 1982
58 N	9 E	May	5	120000		down-welling bucket, neutral formalin	Dale & Dahl 1987
18 N	77 W	ann	2	4**	0.04-19	shelf region, as production tintinnines only, protoplasmic volume, 20 µm net, Bouin	Gilron 1988
58 N	12 E	ann	5	2**	0.4-36	Fig. 6, net, formaldehyde tintinnines only 50% occupancy assumed	Hedin 1975
21 N	157 W	Sep	0.5	73	68, 78	small embayment Table 1, bottles, Lugol	Landrey et al. 1984
59 N	21 E	Apr/May	60	23	9-36	spring bloom, mean as mid-point, bottles, Lugol	Leppanen & Bruun 1986
48 N	5 W	Jul/Aug	100	130	68-179	frontal station	Linley et al. 1983
48 N	3 W	Jul/Aug	80	190	130-255	mixed waters	"
48 N	7 W	Jul/Aug	200	380	89-1433	stratified waters from Table 5, integrated pumped, prefiltered Lugol	"
60 N	160 E	Jun	500	106	5-340	bottle, live, integrated	Mamaeva 1983a
10 S	67 E	Aug/Oct	2000	6		Table 2	Mamaeva 1983b
12 S	70 E	Aug/Oct	2300	13		live, bottle, reverse filtrat^n	"

Table 1. ESTIMATES OF GLOBAL PRODUCTION OF CILIATES* (Continued)

Lat	Long	Season	Distance from Shore (km)	Production Estimates ($J\ m^{-3}d^{-1}$) Mean	Range	Comments	Reference
15 S	67 E	Aug/Oct	1700	0.6			Mamaeva 1983b
16 S	67 E	Aug/Oct	1700	17			"
22 S	70 E	Aug/Oct	2200	4			"
0 S	80 E	Aug/Oct	700	7			"
4 S	80 E	Aug/Oct	1000	7			
8 S	80 E	Aug/Oct	1300	6			"
10 S	80 E	Aug/Oct	1750	13			"
12 S	80 E	Aug/Oct	2000	20			"
15 S	80 E	Aug/Oct	2300	0.8			"
18 S	80 E	Aug/Oct	2600	0.3			"
20 S	80 E	Aug/Oct	2600	3			"
45 N	67 W	ann	3	27**		reported as production tintinnines only, enclosed bay, 50% occupancy assumed 30 µm net, bottles, Bouin	Middlebrook et al. 1987
42 N	70 W	ann	10	18	2.5-105	reported as production bottles, Bouin	Montagnes et al. 1988
45 N	64 W	ann	1	50**	1-460	enclosed basin, Fig. 1(b) tintinnines only, integrated 50% occupancy assumed bottles, formaldehyde	Paranjape 1987
74 N	75 W	Aug	<75	0.022		Table 1	Paranjape 1988
74 N	81 W	Aug	<75	0.023	0.016-0.029	pump, prefiltered, formaldehyde	"
74 N	85 W	Aug	<75	0.004		integrated	"
74 N	89 W	Aug	<75	0.005		50% occupancy assumed for	"
74 N	93 W	Aug	<75	0.003		tintinnines	"
43 N	62 W	Mar	100	80	48-140	prespring bloom Fig. 1, integrated over 32.5 m bottles, formaldehyde	Paranjape et al. 1985
40 N	10 E	ann	10	194		coastal, Table 1 aloricates only, bottle	Rassoulzadegan 1977

Table 1. ESTIMATES OF GLOBAL PRODUCTION OF CILIATES* (Continued)

Lat	Long	Season	Distance from Shore (km)	Production Estimates ($J\ m^{-3}d^{-1}$) Mean	Range	Comments	Reference
40 N	10 E	ann	10	15**		coastal, Table 2 tintinnines only total lorica volume	Rassoulzadegan 1979
43 N	8 E	ann	30	115	17-350	reported as production Table 2, bottle ?, reverse filtration	Rassoulzadegan et al. 1988
45 N	13 E	win	50	19	1-140	mixed waters	Revelante & Gilmartin 1983
45 N	13 E	sum	50	160	7-1 600	stratified waters Table 1, bottles, Lugol, integrated	"
64 N	92 W	Sep	1	20**	3-34	arctic estuary, Fig. 4 tintinnines only total lorica volume	Rogers et al. 1981
31 N	81 W	Aug	0	530		only small ciliates <20 µm	Sherr et al. 1986
31 N	81 W	Feb	0	120		from Table 1	"
31 N	81 W	Aug	1	77		estuarine	"
31 N	81 W	Feb	1	130		estuarine, hand pump, formaldehyde	"
32 N	80 W	Jun		4		shelf, bottles, formaldehyde	"
32 N	80 W	Jun	100	4		"	"
32 N	80 W	Sep	25	30		"	"
32 N	80 W	Sep	50	26		"	"
32 N	80 W	Sep	150	13		"	"
32 N	80 W	Sep	10	4		"	"
32 N	80 W	Apr		9			
19 S	148 E	Aug	30	1.3		Great Barrier Reef	"
54 N	10 E	ann	10	200	31-1 700	enclosed bight, 13-20^{0}/oo Fig. 1, bottle, Lugol	Smetacek 1981
40 N	134 E	Jun	500	100	10-350	heterotrophic phase from Table 2, integrated vertical profile, live counts	Sorokin 1977
0 N	97 W		1500	11		integrated 0-150 m	Sorokin 1981

Table 1. ESTIMATES OF GLOBAL PRODUCTION OF CILIATES* (Continued)

Lat	Long	Season	Distance from Shore (km)	Production Estimates ($J\ m^{-3} d^{-1}$) Mean	Range	Comments	Reference
0 N	122 W		3200	16		from Table 8, live counts	Sorokin 1981
0 N	139 W		4400	23		some reverse filtration used	"
0 N	154 W		4800	22		as well as bottles	"
15 S	75 W	Apr/May	10	3 131		an upwelling region	Sorokin & Kogelschatz
15 S	75 W	Apr/May	15	1 468		vertical profile	1979
15 S	75 W	Apr/May	20	1 468		with maximum biomass	"
15 S	75 W	Apr/May	15	2 055		from Table 1	"
15 S	75 W	Apr/May	10	195		live counts, bottles	"
15 S	75 W	Apr/May	25	1 419			"
15 S	75 W	Apr/May	10	98			"
15 S	75 W	Apr/May	65	343			"
15 S	75 W	Apr/May	150	1 860			"
15 S	75 W	Apr/May	10	2 349			"
15 S	75 W	Apr/May	25	4 698			"
15 S	75 W	Apr/May	35	97			"
9 S	68 E	Aug/Sep	2400	7		Table 6, Fig. 1	Sorokin et al. 1985
15 S	70 E	Aug/Sep	2200	17		live counts, reverse filtration	"
22 S	71 E	Aug/Sep	2900	5		bottles	"
24 S	61 E	Aug/Sep	1800	7		integrated 0-150 m	"
26 S	69 E	Aug/Sep	2900	7			"
43 N	144 E	ann	4	0.8**	0.2-12.7	coastal, Fig. 3	Taguchi 1976
						bottles	
						tintinnines only, total lorica	
48 N	123 W	win	0.5	26		deep fjord (0-5 m)	Takahashi & Hoskins
							1978
49 N	123 W	win	10	200		Strait of Georgia (total)	"
						Table 3, pump, Lugol	
8 S			<800	327		Table 4, reported as production	Tumantseva & Kopylov
8 S			<800	80		pail, bottles	1985
8 S			<800	52		live counts	"
8 S			<800	756		some reverse filtration	"
8 S			<800	809		from 80 W to 87 W	"
8 S			<800	336		map of stations not included	"

Table 1. ESTIMATES OF GLOBAL PRODUCTION OF CILIATES* (Continued)

Lat	Long	Season	Distance from Shore (km)	Production Estimates ($J\ m^{-3} d^{-1}$) Mean	Range	Comments	Reference
8 S			<800	93			Tumantseva & Kopylov 1985
8 S			<800	3 685			"
8 S			<800	778			"
8 S			<800	474			"
8 S			<800	1 542			"
8 S			<800	280			"
41 N	71 W	ann	2	430**		reported as production, tintinnines, bottle, formaldehyde	Verity 1987
45 N	33 E	sum	10	260		Table 3, live, bucket	Zaika & Averina 1968

* Biomass was multiplied by 1 generation per day to calculate daily production.

** These numbers should be multiplied by 4 (Sherr et al. 1986) to estimate total production of ciliates.

ranging from 0 - 10^5 J $m^{-3}d^{-1}$ (Table 1). Production levels may be higher near shore (Figure 1.) and in temperate regions (Figure 2). However, these observations may be misleading since there are few measurements from oceanic, tropical and circumpolar regions.

Ciliate and phytoplankton biomass are significantly positively correlated (r^2=0.41; $p < 0.05$) (Figure 3) while ciliate and bacterial biomass are poorly correlated (r^2=0.02) (Figure 4).

The estimates of production in Table 1, which are primarily derived from biomass estimates, range over five orders of magnitude. Ciliates in the plankton have generation times ranging from several hours to several days but typically are in the neighbourhood of one-half to two days (Gifford 1985,

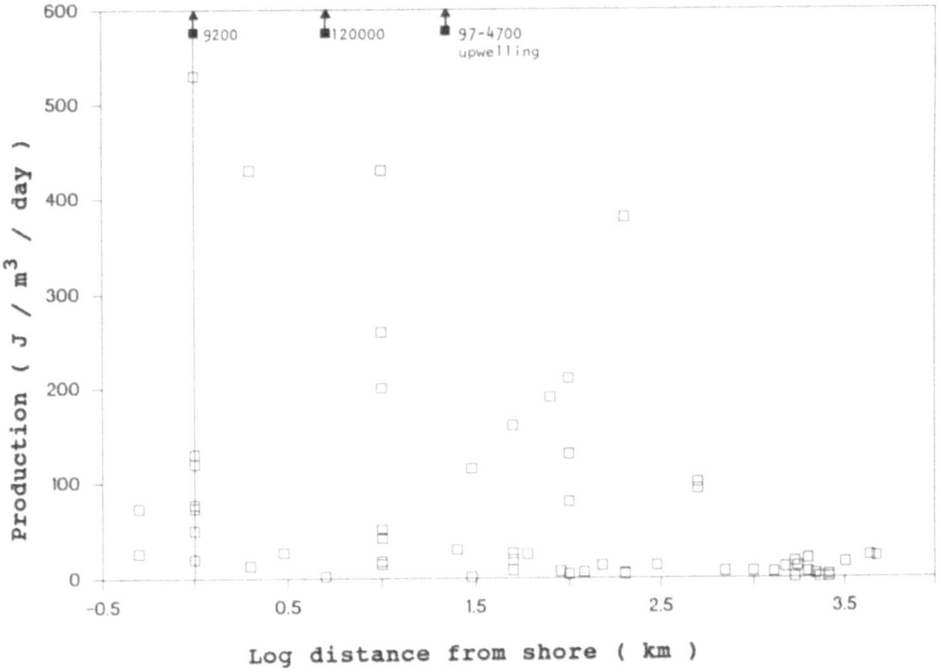

Fig. 1. The relation between ciliate production (J m^{-3} d^{-1}) and the logarithm of the distance (km) from shore that samples were collected at, from data presented in Table 1

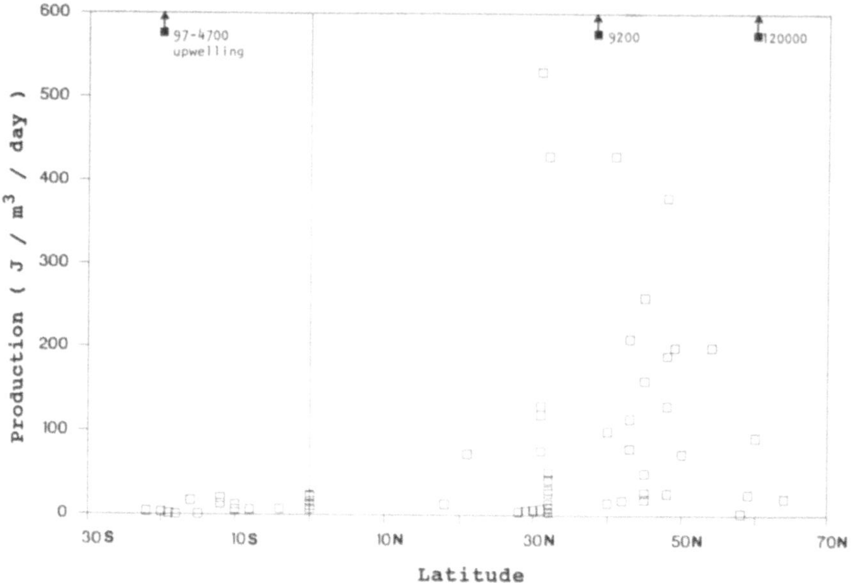

Fig. 2. The relation between ciliate production (J m^{-3} d^{-1}) and latitude (°N or °S) at which samples were collected, from data presented in Table 1

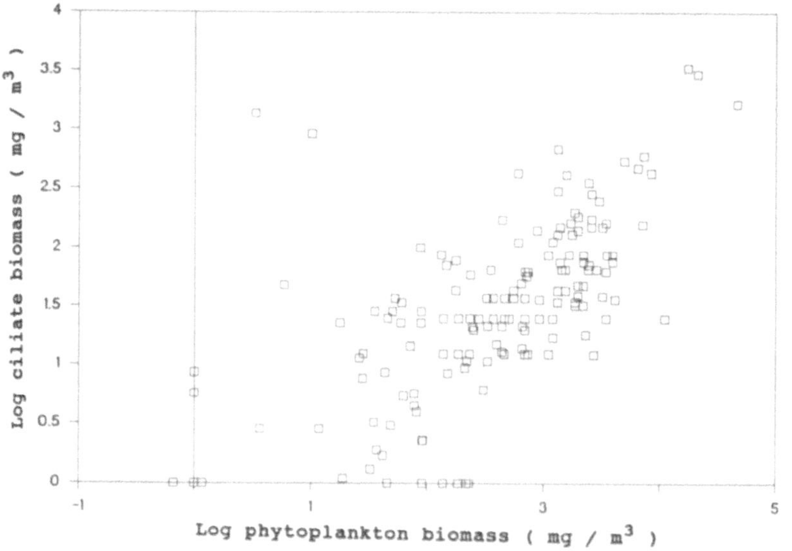

Fig. 3. The relation between the logarithm of ciliate biomass and the logarithm of phytoplankton biomass from data obtained from the literature (see Methods). $r^2 = 0.41$, d.f. = 169

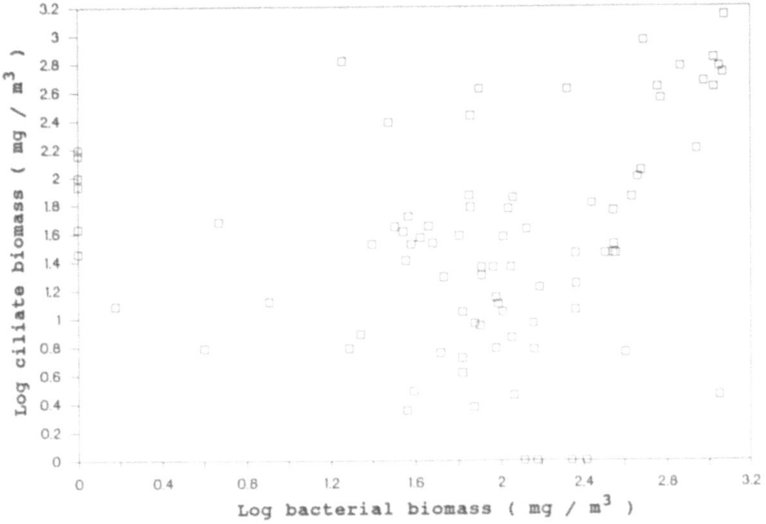

Fig. 4. The relation between the logarithm of ciliate biomass and the logarithm of bacterioplankton biomass from data obtained from the literature (see Methods). r^2 = 0.02, d.f. = 85

Gilron and Lynn 1989d, Heinbokel 1978, 1987, Jonsson 1986, Rassoulzadegan 1982, Rivier *et al.* 1985, Smetacek 1984, Tumantseva and Kopylov 1985, Turley *et al.* 1986 Verity 1986a). Thus, the assumption of one doubling per day may be imprecise but, in general, these calculations of production are likely to vary only by a factor of two. This is a small error considering other methodological and natural sources of variation.

SAMPLING ERRORS

Most planktonic ciliates are aloricate (Gilron 1988, Laval-Peuto *et al.* 1986, Montagnes *et al.* 1988, Sorokin 1981) and should be collected by bottle casts or other gentle methods to avoid cell loss and subsequent underestimation of biomass. It has been repeatedly noted that aloricate ciliates are lost by pre-filtration techniques (Gifford 1985, Laval-Peuto and Rassoulzadegan 1988, Sorokin 1977). Collecting by bottle and then filtering samples prior to fixation could result in an underestimate of aloricate ciliates by up to 95% (Gifford 1985, Sorokin 1977) while reverse filtration to concentrate live samples could result in losses of > 75% of the aloricate ciliates (Gifford, 1985, Laval-Peuto and Rassoulzadegan 1988). Sorokin (1981) suggested that even gentle concentration methods are "rather useless" and result in "loss averages of >95% of both naked and loricate forms".

Burkill (1982) documented how pumping systems influence estimates of ciliate biomass. He sampled using a pump and nets and found tintinnines to dominate. His pumping apparatus was checked for its effect on ciliates using *Uronema*, which is a relatively hardy ciliate, compared to the more delicate oligotrichs that dominate the marine plankton. Linley *et al.* (1983) and Paranjape (1988) using pumps and pre-filters of 200 and 35 µm respectively found tintinnines to be the dominant or only ciliates in their samples; it is possible that here also many of the more delicate naked ciliates were destroyed.

Poor fixation may also cause drastic cell loss. For this reason, a number of investigators prefer live counting (e.g. Dale and Burkill 1982, Mamaeva 1983a, Sorokin 1977, Tumantseva and Kopylov 1985, Zaika and Averina 1968). The major disadvantages of live counting are that sample temperatures must be kept close to the *in situ* temperature and samples must be counted usually within one hour after collection. This requires that observation equipment be available at or near the collecting site. Since most planktonic ciliates tend to move very fast, species identification is often inexact or impossible. Further, fast movements probably reduce counting accuracy and compound the difficulties already associated with measuring ciliates.

It is difficult to argue against the merits of live counting. However, given the logistical difficulties listed above, a good fixation protocol should produce abundance and biomass estimates equal in accuracy to live counting. In addition, staining of properly fixed material allows accurate species identification. Buffered formalin has been used extensively to fix ciliates. However, Revelante and Gilmartin (1983) observed that 30 to 70% of aloricate ciliates were lost upon formalin-fixation when compared to samples fixed with Rhode's acid version of Lugol's iodine (Lovegrove, 1960). Bouin's fluid is the recommended fixative for protargol silver-staining (Lee *et al.* 1985, Montagnes and Lynn 1987). Although we have not tested Bouin's fixed samples against live counting in the field, laboratory experiments indicate that this fixative performs well on a wide variety of taxa (Montagnes and Lynn 1987). However, Leakey *et al.* (1988) have reported that Bouin's fluid significantly underestimates abundance, especially of aloricate ciliates, when compared to acid Lugol's.

ERRORS IN BIOMASS ESTIMATION AND ENUMERATION

Variation in cell size of species occurs and should be recorded to accurately estimate biomass. It is inadvisable to use previously established sizes of described species since ciliate species may vary considerably in size depending upon their nutritional state and stage in their growth cycle (e.g. Lynn *et al.* 1987). Instead, organisms representing each species should be independently measured and that data used to estimate volume. Similarly, in the past, tintinnine cell volume has been assumed to be 50% of lorica volume (e.g. Beers and Stewart 1967, Capriulo and Carpenter 1983, Heinbokel 1978, Middlebrook *et al.* 1987), but this is not true for all tintinnines. Cell size ranges from 10-46% of the lorica volume (Gilron and Lynn 1989b), and assuming a 50% occupancy can significantly bias production estimates. Thus, cell dimensions should be directly measured on the samples in hand and not assumed.

Variation in ciliate biomass estimates occurs even when gentle methods of collection and appropriate fixatives are used. Although we found coefficients of variation (CV) of up to 50% when comparing: 1) replicate bottle casts; 2) methods of integrating a water column; 3) subsamples from a single bottle; and 4) measurements of cell volume, these sampling errors did not account for the large variation between sites approximately 1 km apart (CV>100%) or between weekly samples, which occasionally varied by almost an order of magnitude (Montagnes *et al.* 1988). Similar variation has been noted by others. Revelante and Gilmartin (1983, Table 1), sampling at several depths, recorded CV's (of biovolume) of 160% from mixed waters and 190% from stratified waters. Sherr *et al.* (1986, Table 1) examined several sites and found CV's (of biomass) ranging from 50-200% with a mean of 100% while data from Mamaeva (1983a, Table 1) gives CV's (of biomass) ranging from 36-105% with a mean of 68%. Natural factors and not sampling biases are likely to produce this variation.

DISTANCE FROM SHORE

Heinbokel and Beers (1979) showed that tintinnine abundance decreased with distance from shore and Porter *et al.* (1985) noted that ciliate production was higher near the shore and in estuarine regions. The data compiled here support this generalisation. Small flagellates and bacteria, which are the likely prey of ciliates, thrive in coastal regions and reach exceedingly high numbers (Fenchel 1980, Laval-Peuto *et al.* 1986, Rivier *et al.* 1985). Although some high estimates of production have been reported from the open ocean, these estimates are predominantly from the studies of Mamaeva (1983b), Tumantseva and Sorokin (1977 as Table 8 in Sorokin 1981), and Sorokin *et al.* (1985) and may be due to their use of live counting. Alternately, it is possible that in the open oceans where the dominant autotrophs are small (Beinfang and Szyper 1981, Herbland and Le Bouteiller 1981), ciliates are more abundant as grazers and to some extent replace the copepods.

LATITUDINAL AND TEMPORAL VARIABILITY

With the exception of the data of Sorokin and Kogelschatz (1979) from Peruvian upwelling (Table 1), production seems low in southern latitudes and highest in the mid-temperate region. However, since the data from the tropics are predominantly from the open ocean, abundances are generally relatively low. Comparing the ciliate production of tropical and temperate regions is unwarranted as yet. Similarly the polar regions are virtually unexplored. Globally, more work is needed before any general conclusions can be drawn as local physiography and climate may ultimately explain the variations more precisely than any broad parameter such as latitude Smetacek *et al.* (1984).

Within seasons, ciliate biomass and production can exhibit annual trends in which variations may occur of up to two orders of magnitude. Unlike small scale variations or short term blooms, these seasonal increases last several weeks to months and are predictable from year to year. They are likely to be influenced by local physiographic conditions and seasonal climatic changes that affect the food resources of ciliates (Sorokin, 1981, Smetacek *et al.* 1984).

The time of year at which seasonal peaks occur may be related to latitude. Smetacek (1981) noted March-April and September-November peaks in ciliate abundance in the Kiel Bight (54°N); the spring peak apparently grazed larger phytoplankton. In the Gulf of Maine (42°N), in spring, a peak of large-celled ciliates was suggested to feed on diatoms, other large phytoplankton, and ciliates; in summer, a small-celled assemblage supposedly ate small flagellates and bacteria-sized prey (Montagnes *et al.* 1988). In summer, Smetacek (1981) also observed an increase in the proportion of small ciliates. However, in the Japan Sea (40°N), ciliates contributed to a post-spring heterotrophic phase that 'used the energy from organic matter accumulated during the previous spring bloom' (Sorokin 1977). In the Mediterranean (43°N), only a single peak of both biomass and production occurred during July-August from 1973-1978 (Rassoulzadegan *et al.* 1988).

Several studies have concentrated only on the tintinnine ciliates. In Swedish waters (58°N), a variably sized, May-July peak occurred in tintinnine biomass from 1972-1975 (Hedin 1975). In coastal waters off the south coast of England (50°N) and in coastal waters off the coast of Rhode Island (41°N), peaks in biomass occurred in May and August-October (Burkill 1982) and May-June and July-September (Verity 1987). Two studies in large enclosed bays show differing annual trends: Middlebrook *et al.* (1987) observed a June-August peak (45°N) in 1983-1984 while Paranjape (1987) noted no conspicuous seasonal peak (45°N) in 1977-1978 and only a noticeable decrease in biomass during January-February. Tentatively, these results suggests that, in open coastal waters, two peaks in biomass occur between 40-50°N while north and south of this band of latitude there is only one seasonal peak.

There appear to be a number of times during the year when the biomass of individual species increases. If ciliates are 'tracking' food, it is predominantly on the basis of size (Burkill *et al.* 1987, Middlebrook *et al.* 1987, Rassoulzadegan and Sheldon 1986, Smetacek 1981, Takahashi and Hoskins 1978, Verity 1987). Thus, seasonal variation in the size of food sources will account for variation in the seasonal pattern of ciliates (Verity 1986a). Verity and Villareal (1986) have demonstrated that the *in vitro* growth rate of two *Tintinnopsis* species is negatively correlated with cross-sectional diameter of diatom prey. Undoubtedly, there are a number of planktonic ciliate assemblages, each one exploiting different-sized food resources that themselves vary seasonally. More research on the specific food requirements of different species will better establish the importance of this 'food tracking' hypothesis.

SMALL SCALE VARIATION

Small scale patches of protists may occur due to a number of physical factors (see Haury *et al.* 1978). This is indicated by coefficients of variation of mean biomass among sites about 1 km apart that may range up to 100% (Gilron 1988, Montagnes *et al.* 1988, Sherr *et al.* 1986). Stoecker *et al.* (1984) found ciliate densities in a tidal estuary to vary up to two orders of magnitude on a horizontal scale of 1 - 100 m, and they proposed that patches may have been caused by behavioural as well as physical factors. Such speculation is not unwarranted since planktonic ciliates are attracted to prey *in vitro* (Fenchel and Jonsson 1988, Verity 1988). Therefore, they may actively accumulate in areas of high abundance of prey in the field. Dale (1987) noted a 2 - 4 fold range in abundance in the vertical distribution of ciliates, and he suggested that, at least in part, this was due to vertical migration. Vertical migration has also been reported for other ciliates (Mamaeva 1983b, McManus and Fuhrman 1986). Small scale patchiness could also be due to the presence, abundance, and size of detrital aggregates. Beers *et al.* (1980) noted that these aggregates could support a series of trophic levels and indicated that sampling protocols would need to be modified to test this hypothesis. No attempts have yet been made although there is growing interest in this field (Lochte 1990). Downwelling may also contribute to increasing ciliate densities (Dale and Dahl 1987a, b). It appears that a combination of physical and behavioural factors could explain variation of an order of magnitude that is typical within and between similar regions.

BLOOMS

Blooms are defined as transient peaks in biomass and/or production, usually of a single species, that are considerably higher (e.g. > 3 fold) than background levels. Blooms can occur for extended periods of time in upwelling regions where ciliate production can exceed 4,000 J m^{-3} d^{-1} (Sorokin and Kogelschatz 1979) while in other regions, short term blooms can account for peaks in production. Montagnes *et al.* (1988) found the largest peak in the annual cycle of ciliate production was due to a bloom of a single species of *Strombidium*. Rassoulzadegan *et al.* (1988) also noted that a high production peak of 351 J m^{-3} d^{-1} was due to a single species of *Strombidium*.

The high reproductive rates of ciliates and, particularly, the ability to rapidly alter their numerical response to prey abundances (e.g. see Jonsson 1986) permit ciliates to rapidly exploit expanding food resources. This was illustrated in shipboard experiments in which *Strobilidium* (formerly *Lohmanniella*) *spiralis* increased in abundance by an order of magnitude over three days, grazing out the dinoflagellates on which it fed and dropping down to pre-bloom levels by the end of a week (Smetacek 1984). Heinbokel (1978) suggested that the high reproductive rates of tintinnines would permit rapid response to anomalous phytoplankton abundances. This was supported by Verity (1987) who correlated the abundance of a tintinnine species with its production; generally the more abundant a species was, the more productive it was.

During short blooms (1-7 days) ciliates may divide at least once a day. Smetacek (1984) calculated a generation time of 8 h for *Strobilidium spiralis*. Using this growth rate, if it is not grazed, an initial ciliate population of 10 J m^{-3} could reach > 40,000 J m^{-3} in four days. In fact, background levels of ciliate biomass may be maintained by growth limiting constraints, such as food abundance (Banse 1982), and not by predation. Thus, blooms occur when abundant resources are encountered and dissipate when the resources are exhausted.

The above 'model' assumes that a food resource patch is 'encountered' by the ciliates. Although this type of food patch exploitation is possible, it may be that the ciliates and prey co-exist. At some time, prey growth is stimulated, perhaps by changes in light, temperature, nutrients or bacterial abundance. The prey, such as autotrophic or heterotrophic nanoflagellates, would then be grazed by ciliates. Increased flagellate abundance would stimulate increased ciliate growth (Jonsson 1986), which, in turn, would increase the predation of ciliates on flagellates.

We have simulated a simple system where: 1) ciliate functional and numerical responses are related to prey abundance using equations derived from work on *Strobilidium spiralis* (Jonsson 1986), 2) initial ciliate abundance inputs of 1 to 4 ml^{-1} based on values from field studies (Montagnes 1987, Smetacek 1984), 3) flagellate generation time inputs of 12 to 24 hours based on *in situ* studies (Verity 1986b), 4) flagellate abundance input of 1000 ml^{-1} (Davis *et al*. 1985). We also assumed a loss rate of ciliates and flagellates based on: 5) a copepod functional response related to prey abundance using equations derived from graphs presented by Frost (1972), 6) an invariant copepod abundance of $1x10^{-3}$ ml^{-1} (Middlebrook and Roff 1986), and 7) an invariant net-phytoplankton standing stock available to copepods but not to ciliates of 10 ml^{-1} based on phytoplankton distributions reported by Marshall (1984).

We have run a number of simulations using this model and graphically present the results from one stable run that shows oscillations between prey and predator (Figure 5). Here, flagellate generation time is 24 hours and the starting

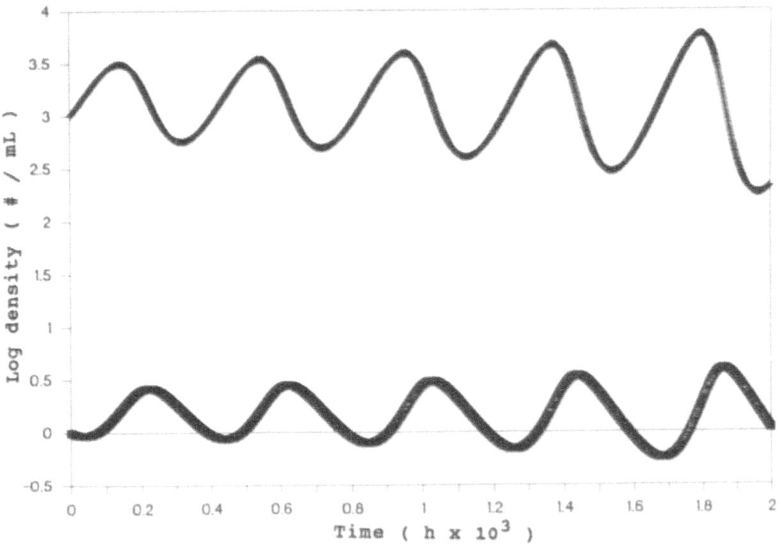

Fig. 5. Variation in density (logarithm) of flagellate prey (upper curve) and ciliates (lower curve) based on the model described in the text. Initial ciliate density = 1/ml; flagellate generation time = 24 h; initial flagellate density = 1000/ml

population of ciliates is 1 ml^{-1}. Although our model is simple, it produces a relatively steady state using densities and growth rates similar to those found in nature (Davis *et al.* 1985, Fenchel 1987, Gilron and Lynn 1988, Marshall 1984, Middlebrook and Roff 1986, Montagnes *et al.* 1988). However, a relatively steady state such as this is not likely to exist under natural conditions where all our parameters would be continuously changing. The most instructive observation we have made is that small changes in the parameters yield very different results. Most combinations of parameters other than those reported in Figure 5 yielded unstable systems with unrealistically large prey and predator blooms. A second observation is that both prey and predators can oscillate with a periodicity of days to weeks. Both these observations may offer some explanation for variation seen in natural samples. Population changes on a daily to weekly scale have been observed to be in the range of an order of magnitude, and the results of our modelling support this. Intensive sampling should be undertaken to explore the factors determining short term variation in the abundance of ciliates and their prey. As already noted by Sorokin and Kogelschatz (1979), 'the seasonal observations of the dynamics of natural plankton populations... are usually done with insufficient temporal resolution to catch and resolve the short bursts of growth...' (p 207).

CILIATE LIFE HISTORY AND PREY BIOMASS

Microorganisms, in general, and ciliates in particular, are considered to lead a 'feast and famine' existence due to the patchiness of their prey (Fenchel 1987, Lynn *et al.* 1987). Their life history strategies (e.g. see Jackson and Berger 1985) show traits adapted to this kind of existence: mixotrophy (Laval-Peuto and Rassoulzadegan 1988, Stoecker *et al.* 1987a) undoubtedly provides a source of energy during short 'famine' phases while encystment (e.g. see Paranjape 1980, Reid 1987) would permit survival over longer periods. When food is encountered, the functional and numerical responses of ciliates to increasing prey abundance permit a rapid reaction to these changing conditions. It is not surprising, therefore, to see positive correlations between the biomass of ciliates and their putative prey species (Figures 3, 4).

Porter *et al.* (1985) recorded a significant positive correlation between ciliate biomass and chlorophyll *a* in freshwater lakes. Previous studies on marine ciliates have demonstrated this relationship locally (Ibanez and Rassoulzadegan 1977, Smetacek 1981, Verity 1987). Our analysis of the data from the marine plankton corroborates this relation on a global scale, demonstrating that the direct observations of ingestion of phytoplankton by ciliates can be extended to the assemblage level.

Planktonic ciliates may also be major consumers of bacteria (Rivier *et al.* 1985, Sherr and Sherr 1987) and are often abundant in both coastal waters (Sherr *et al.* 1987) and in the deeper oceanic layers (Sorokin 1977, 1981, Townsend and Cammen 1985) where bacterial biomass is high. The data from the marine plankton suggest a possible correlation may exist between ciliate biomass and bacterial biomass (Figure 4), corroborating the trophic relation between these two groups. The absence of a strong positive correlation suggests, among other things, that a large portion of the heterotrophic bacterial production may go ungrazed (cf. Ducklow *et al.* 1986, Sherr *et al.* 1987). Perhaps this is because many bacterial species are 'unpalatable' to the ciliates. Alternately, only a subset of the ciliate biomass may be made up of bacterivores and it is this biomass that should be used in the analysis rather than the total ciliate biomass.

CONCLUSIONS

A review of the literature on marine planktonic ciliates has suggested a number of directions for future research. First, to ensure accurate estimates of ciliate abundance and biomass, sampling techniques should rely on bottle casts, followed by live counts and/or fixation with acid Lugol. Concentrated

Bouin's fixative is required for protargol staining ciliates. Protargol staining is useful for species identification, which should be an integral part of any study examining ciliate assemblages since ciliate species differ in their trophic status. Second, more information is needed globally, especially from the southern hemisphere, both in coastal and open ocean regions, to judge relative contributions to global production. Estimates should be established from longer term studies, at least annually, so that the influences of seasonality are normalised across latitudes. Third, although a pattern may emerge from this global approach, it is just as likely that local biological, physiographic, and climatic conditions may explain a good deal of the variation in production. To establish this, studies must be undertaken in which all planktonic trophic levels are considered simultaneously so that interactions between groups can be directly tested. Fourth, in addition to the global approach, more effort should be directed to examining the causes underlying smaller scale variations in biomass and production at small temporal (e.g. hours to days) and spatial scales (e.g. hundreds of microns to metres). Finally, more information on functional and numerical responses of a variety of species will permit dynamic modelling of shorter term interactions. Increasing our understanding of the causes of shorter term variations, such as blooms, may permit explanation of the longer term seasonal fluctuations in biomass and production.

ACKNOWLEDGEMENTS

This research was supported by NSERC of Canada Operating Grant A-6544 awarded to DHL. We are grateful to Colin Field for implementing in FORTRAN the set of equations used for the predator-prey interaction presented graphically in Figure 5. The equations are available from the authors on request. Guy Gilron, John Roff, and Peter Verity commented constructively on earlier drafts.

REFERENCES

Banse K (1982) Cell volumes, maximal growth rates of unicellular algae and ciliates, and the role of ciliates in the marine pelagial. Limnol Oceanogr 27:1059-1071

Beers JR, Stewart GL (1967) Micro-zooplankton in the euphotic zone at five locations across the California current. J Fish Res Bd Canada 24:2053-2068

Beers JR, Stewart G L (1969) Micro-zooplankton and its abundance relative to the larger zooplankton and other seston components. Mar Biol 4:182-189

Beers JR, Reid FMH, Stewart GL (1980) Microplankton population structure in southern California nearshore waters in later spring. Mar Biol 60:209-226

Beinfang PK, Szyper JP (1981) Phytoplankton dynamics in the subtropical Pacific Ocean off Hawaii. Deep Sea Res 28A:981-1000

Berk SG, Brownlee DC, Heinle DR, Kling JH, Colwell RR (1977) Ciliates as a food source for marine planktonic copepods. Microb Ecol 4:27-40

Burkill PH (1982) Ciliates and other microplankton components of a nearshore food web: standing stocks and production processes. Ann Inst Oceanogr Paris 58:335-350

Burkill PH, Mantoura RFC, Llewellyn CA, Owens NJP (1987) Microzooplankton grazing and selectivity of phytoplankton in coastal waters. Mar Biol 93:581-590

Capriulo GM, Carpenter EJ (1983) Abundance, species composition and feeding impact of tintinnid micro-zooplankton in central Long Island Sound. Mar Ecol Prog Ser 10:277-288

Capriulo GM, Ninivaggi DV (1982) A comparison of the feeding activities of field collected tintinnids and copepods fed identical natural particle assemblages. Ann Inst Oceanogr Paris 58:325-334

Dale T (1987) Diel vertical distribution of planktonic ciliates in Lindaspollene, western Norway. Mar Microb Food Webs 2:15-28

Dale T, Burkill PH (1982) 'Live Counting' - A quick and simple technique for enumerating pelagic ciliates. Ann Inst Oceanogr Paris 58:267-276

Dale T, Dahl E (1987a) Mass occurrence of planktonic oligotrichous ciliates in a bay in southern Norway. J Plank Res 9:871-879

Dale T, Dahl E (1987b) A red tide in southern Norway caused by mass occurrence of the planktonic ciliate *Tiarina fusus*. Fauna 40:98-103 (In Norwegian with English summary)

Davis PG, Caron DA, Johnson PW, Sieburth JMcN (1985) Phototrophic and apochlorotic components of picoplankton and nanoplankton in the North Atlantic: geographic, vertical, seasonal and diel distributions. Mar Ecol Prog Ser 21:15-26

Downing JA (1984) Assessment of secondary production: the first step. In: Downing JA, Rigler FH (eds) A Manual on Methods for the Assessment of Secondary Productivity in Fresh Waters. IBP Hand Book 17. Blackwell, London, p 1

Ducklow HW, Purdie DA, Williams PJLeB, Davies JM (1986) Bacterioplankton: A sink for carbon in a coastal marine plankton community. Science 232:865-867

Fenchel T (1980) Relation between particle size selection and clearance in suspension-feeding ciliates. Limnol Oceanogr 25:733-738

Fenchel T (1987) Ecology of Protozoa. The Biology of Free-living Phagotrophic Protists. Springer Verlag, New York

Fenchel T, Finlay B J(1983) Respiration rates in heterotrophic, free-living protozoa. Microb Ecol 9:99-122

Fenchel T, Jonsson PR (1988) The functional biology of *Strombidium sulcatum*, a marine oligotrich ciliate (Ciliophora, Oligotrichina). Mar Ecol Prog Ser 48:1-15

Finlay BJ (1978) Community production and respiration by ciliated protozoa in the benthos of a small eutrophic loch. Freshwater Biol 8:327-341

Frost BW (1972) Effects of size and concentration of food particles on the feeding behavior of the marine planktonic copepod *Calanus pacificus*. Limnol Oceanogr 17:805-815

Gifford D (1985) Laboratory culture of marine planktonic

oligotrichs (Ciliophora, Oligotrichida). Mar Ecol Prog Ser 23:257-267
Gilron GL (1988) Estimates of *in situ* growth rates and annual production of tintinnid ciliate populations near Kingston Harbour, Jamaica. MSc Thesis, University of Guelph, Guelph
Gilron GL, Lynn DH (1989a) Estimates of *in situ* population growth rates of four tintinnine ciliate species near Kingston Harbour, Jamaica. Est Coast Shelf Sci 29:1-10
Gilron GL, Lynn DH (1989b) Assuming a 50% cell occupancy of the lorica overestimates tintinnine ciliate biomass. Mar Biol 103:413-416
Haury LR, McGowan JA, Wiebe PH (1978) Patterns and processes in the time-space scales of plankton distributions. pp. 277-327. In: Steele JH (ed) Spatial Pattern in Plankton Communities. Plenum Press, New York
Hedin H (1975) On the ecology of tintinnids on the Swedish west coast. Zoon 3:125-140
Heinbokel JF (1978) Studies on the functional role of tintinnids in the southern California Bight. I. Grazing and growth rates in laboratory cultures. Mar Biol 47:177-189
Heinbokel JF (1987) Diel periodicities and rates of reproduction in natural populations of tintinnines in the oligotrophic waters off Hawaii, September 1982. Mar Microb Food Webs 2:1-14
Heinbokel JF, Beers JR (1979) Studies on the functional role of tintinnids in the southern California Bight. III. Grazing impact of natural assemblages. Mar Biol 52:23-32
Herbland A, Le Bouteiller A (1981) The size distribution of phytoplankton and particulate organic matter in the equatorial Atlantic Ocean: importance of ultraseston and consequences. J Plank Res 3:659-673
Ibanez F, Rassoulzadegan F (1977) A study of the relationships between pelagic ciliates (Oligotrichina) and planktonic nanoflagellates of the neritic ecosystem of the Bay of Villefranche-sur-Mer. Analysis of chronological series. Ann Inst Oceanogr Paris 53:17-30
Jackson KM, Berger J (1985) Life history attributes of some ciliated protozoa. Trans Amer Microsc Soc 104:52-63
Johnson PW, Xu H-S, Sieburth JMcN (1982) The utilization of chrococcoid cyanobacteria by marine protozooplankters but not by calanoid copepods. Ann Inst Oceanogr Paris 58:297-308
Jonsson PR (1986) Particle size selection, feeding rates and growth dynamics of marine planktonic oligotrichous ciliates (Ciliophora:Oligotrichina). Mar Ecol Progr Ser 33:265-277
Landry MR, Haas LW, Fagerness VL (1984) Dynamics of microbial plankton communities: experiments in Kaneohe Bay, Hawaii. Mar Ecol Progr Ser 16:127-133
Laval-Peuto M, Rassoulzadegan F (1988) Autofluorescence of marine planktonic Oligotrichina and other ciliates. Hydrobiologia 159:99-110
Laval-Peuto M, Heinbokel JF, Anderson OR, Rassoulzadegan F, Sherr BF (1986) Role of micro- and nanozooplankton in marine food webs. Insect Sci Appl 7:387-395
Laybourn J, Stewart JM (1975) Studies on consumption and growth in the ciliate *Colpidium campylum* Stokes. J Anim Ecol 44:165-174
Leakey RJG, Burkill PH, Sleigh M (1988) A comparison of fixatives for the quantification of pelagic ciliate populations. In: Protozoa and their role in marine processes. NATO ASI Abstracts
Lee JJ, Small EB, Lynn DH, Bovee EC (1985) Some techniques for collecting, cultivating and observing protozoa. pp. 1-7. In: Lee JJ, Hutner SH, Bovee EC (eds) An Illustrated Guide to the Protozoa. Society of Protozoologists, Lawrence, Kansas
Leppanen J-M, Bruun J-E (1986) The role of pelagic ciliates including the autotrophic *Mesodinium rubrum* during the spring bloom of 1982 in the open northern Baltic proper. Ophelia Suppl 4:147-157
Linley EAS, Newell RC, Lucas MI (1983) Quantitative relationships between phytoplankton, bacteria and heterotrophic microflagellates in shelf waters. Mar Ecol Progr Ser 12:77-89
Lochte K (1990) Protozoa as makers and breakers of marine aggregates. In: Reid PC, Turley CM, Burkill PH (eds)

Protozoa and their role in marine processes. Springer, Berlin Heidelberg New York
Lovegrove T (1960) An improved form of sedimentation apparatus for use with an inverted microscope. J de Conseil 25:279-284
Lynn DH, Montagnes DJS (1988) Taxonomic descriptions of some conspicuous species of strobilidiine ciliates (Ciliophora: Choreotrichida) from the Isles of Shoals, Gulf of Maine. J Mar Biol Ass UK 68:639-658
Lynn DH, Montagnes DJS, Riggs W (1987) Divider size and the cell cycle after prolonged starvation of *Tetrahymena corlissi*. Microb Ecol 13:115-127
Mamaeva NV (1983a) Planktonic infusorians in the Bering Sea. Soviet J Mar Biol 9:180-185
Mamaeva NV (1983b) Planktonic infusorians of the central part of the Indian Ocean. Soviet J Mar Biol 8:254-258
Marshall HG (1984) Phytoplankton distribution along the eastern coast of the USA. Part V. Seasonal density and cell volume patterns for the north-eastern continental shelf. J Plank Res 6:169-193
McManus GB, Fuhrman JA (1986) Photosynthetic pigments in the ciliate *Laboea strobila* from Long Island Sound, USA. J Plank Res 8:317-327
Middlebrook K, Roff JC (1986) Comparison of methods for estimating annual productivity of the copepods *Acartia hudsonica* and *Eurytemora herdmani* in Passamaquoddy Bay, New Brunswick. Can J Fish Aquat Sci 43:656-664
Middlebrook K, Emerson CW, Roff JC, Lynn DH (1987) Distribution and abundance of tintinnids in the Quoddy Region of the Bay of Fundy. Can J Zool 65:594-601
Montagnes DJS (1987) The annual cycle of planktonic ciliates in the waters surrounding the Isles of Shoals, Gulf of Maine: Estimates of biomass and production. MSc Thesis, University of Guelph, Guelph
Montagnes DJS, Lynn DH (1987) A quantitative protargol stain (QPS) for ciliates: method description and test of its quantitative nature. Mar Microb Food Webs 2:83-93
Montagnes DJS, Lynn DH, Roff JC, Taylor WD (1988) The annual cycle of heterotrophic planktonic ciliates in the waters surrounding the Isles of Shoals, Gulf of Maine: an assessment of their trophic role. Mar Biol 99:21-30
Nival P, Nival S (1976) Particle retention efficiencies of an herbivorous copepod, *Acartia clausi* (adult and copepodite stages): Effects on grazing. Limnol Oceanogr 21:24-38
Paranjape MA (1980) Occurrence and significance of resting cysts in a hyaline tintinnid, *Helicostomella subulata* (Ehre.) Jorgensen. J Exp Mar Biol Ecol 48:23-35
Paranjape MA (1987) The seasonal cycles and vertical distribution of tintinnines in Bedford Basin, Nova Scotia, Canada. Can J Zool 65:41-48
Paranjape MA (1988) Microzooplankton in Lancaster Sound (eastern Canadian Arctic) in summer: biomass and distribution. Deep-Sea Res 35:1547-1563
Paranjape MA, Conover RJ, Harding GC, Prouse NH (1985) Micro- and macrozooplankton on the Nova Scotian shelf in the prespring bloom period: A comparison of their potential resource utilisation. Can J Fish Aquat Sci 42:1484-1492
Porter KG, Sherr EB, Sherr BF, Pace M, Sanders RW (1985) Protozoa in planktonic food webs. J Protozool 32:409-415
Rassoulzadegan F (1977) Evolution annuelle des ciliés pélagiques en Mediterranée nord-occidentale. Ciliés oligotriches 'non-tintinnids' (Oligotrichina). Ann Inst Océanogr Paris 53:125-134
Rassoulzadegan F (1979) Evolution annuelle des ciliés pélagiques en Mediterranée nord-occidentale. II. Ciliés ologotriches. Tintinnides (Tintinnina). Invest Pesq 43:417-448
Rassoulzadegan F (1982) Dependence of grazing rate, gross growth efficiency and food size range on temperature in a pelagic oligotrichous ciliate *Lohmanniella spiralis* Leeg., fed on naturally occurring particulate matter. Ann Inst Oceanogr Paris 58:177-184
Rassoulzadegan F, Sheldon RW (1986) Predator-prey interactions of nanozooplankton and bacteria in an oligotrophic marine environment. Limnol Oceanogr 31:1010-1021
Rassoulzadegan F, Laval-Peuto M, Sheldon RW (1988)

Partitioning of the food ration of marine ciliates between pico-and nanoplankton. Hydrobiologia 159:75-88
Reid PC (1987) Mass encystment of a planktonic oligotrich ciliate. Mar Biol 95:221-230
Revelante N, Gilmartin M (1983) Microzooplankton distribution in the northern Adriatic Sea with emphasis on the relative abundance of ciliated protozoans. Oceanol Acta 6:407-415
Rivier A, Brownlee DC, Sheldon RW, Rassoulzadegan F (1985) Growth of microzooplankton: a comparative study of bactivorous zooflagellates and ciliates. Mar Microb Food Webs 1:51-60
Rogers GF, Roff JC, Lynn DH (1981) Tintinnids of Chesterfield Inlet, Northwest Territories. Can J Zool 59:2360-2364
Sherr EB, Sherr BF (1987) High rates of consumption of bacteria by pelagic ciliates. Nature 325:710-711
Sherr EB, Sherr BF, Fallon RD, Newell SY (1986) Small, aloricate ciliates as a major component of the marine heterotrophic nanoplankton. Limnol Oceanogr 31:177-183
Sherr EB, Sherr BF, Albright LJ (1987) Bacteria: Link or sink? Science 235:88-89
Smetacek V (1981) The annual cycle of protozooplankton in the Keil Bight. Mar Biol 63:1-11
Smetacek V (1984) Growth dynamics of a common Baltic protozooplankter: the ciliate genus *Lohmanniella*. Limnologica (Berlin) 15:371-376
Smetacek V, von Bodungen B, Knoppers B, Peinert R, Pollehne F, Stegmann P, Zeitzschel B (1984) Seasonal stages characterizing the annual cycle of an inshore pelagic system. Rapp P-v Reunion cons interntl Explor Mer 183:126-135
Sorokin YuI (1977) The heterotrophic phase of plankton succession in the Japan Sea. Mar Biol 41:107-117
Sorokin YuI (1981) Microheterotrophic organisms in marine ecosystems. p 293-342. In: Longhurst A T (ed) Analysis of Marine Ecosystems. Academic Press, New York
Sorokin YuI, Kogelschatz JE (1979) Analysis of heterotrophic microplankton in an upwelling area. Hydrobiologia 66:195-208
Sorokin YuI, Kopylov AI, Mamaeva NV (1985) Abundance and dynamics of microplankton in the central tropical Indian Ocean. Mar Ecol Progr Ser 24:27-41
Stoecker DK, Egloff DA (1987) Predation by *Acartia tonsa* Dana on planktonic ciliates and rotifers. J Exp Mar Biol Ecol 110:53-68
Stoecker DK, Govoni JJ (1984) Food selection by young larval gulf menhaden (*Brevoortia patronus*). Mar Biol 80:299-306
Stoecker DK, Sanders NK (1985) Differential grazing by *Acartia tonsa* on a dinoflagellate and a tintinnid. J Plank Res 7:85-100
Stoecker DK, Davis LH, Anderson DM (1984) Fine scale spatial correlations between planktonic ciliates and dinoflagellates. J Plank Res 6:829-842
Stoecker DK, Michaels AE, Davis LH (1987a) Large proportion of marine planktonic ciliates found to contain functional chloroplasts. Nature 326:790-792
Stoecker DK, Verity PG, Michaels AE, Davis LH (1987b) Feeding by larval and post-larval ctenophores on microzooplankton. J Plank Res 9:667-683
Stoecker DK, Silver MW, Michaels AE, Davis LH (1988) Obligate mixotrophy in *Laboea strobila*, a ciliate which retains chloroplasts. Mar Biol 99:415-423
Stoecker DK, Taniguchi A, Michaels AE (1989) Abundance of autotrophic, mixotrophic and heterotrophic planktonic ciliates in shelf and slope waters. Mar Ecol Prog Ser 50:241-254
Taguchi S (1976) Microzooplankton and seston in Akkeshi Bay, Japan. Hydrobiologia 50:195-204
Takahashi M, Hoskins KD (1978) Winter conditions of marine plankton populations in Saanich Inlet, B.C., Canada. II. Micro-zooplankton. J Exp Mar Biol Ecol 32:27-37
Townsend DW, Cammen LM (1985) A deep protozoan maximum in the Gulf of Maine. Mar Ecol Progr Ser 24:177-182
Tumantseva NI, Kopylov AI (1985) Reproduction and production rates of planktic infusoria in coastal waters of Peru. Oceanology 25:390-394
Turley CM, Newell RC, Robins DB (1986) Survival strategies of

two small marine ciliates and their role in regulating bacterial community structure under experimental conditions. Mar Ecol Prog Ser 33:59-70
Turner JT, Anderson DM (1983) Zooplankton grazing during dinoflagellate blooms in a Cape Cod embayment, with observations of predation upon tintinnids by copepods. Mar Ecol 4:359-374
Verity PG (1986a) Growth rates of natural tintinnid populations in Narragansett Bay. Mar Ecol Progr Ser 29:117-126
Verity PG (1986b) Grazing of phototrophic nanoplankton by microzooplankton in Narragansett Bay. Mar Ecol Progr Ser 29:105-115
Verity PG (1987) Abundance, community composition, size distribution, and production rates of tintinnids in Narragansett Bay, Rhode Island. Est Coast Shelf Sci 24:671-690
Verity PG (1988) Chemosensory behaviour in marine planktonic ciliates. Bull Mar Sci 43:772-782
Verity PG, Villareal TA (1986) The relative food value of diatoms, dinoflagellates, flagellates, and cyanobacteria for tintinnid ciliates. Arch Protistenk 131:71-84
Zaika VYe, Averina TYu (1968) Proportions of infusoria in the plankton of Sevastopol Bay, Black Sea. Oceanology 8:843-845

PROTOZOA ASSOCIATED WITH MARINE 'SNOW' AND 'FLUFF' - SESSION SUMMARY

Carol M Turley
Department of Zoology
University of Bristol
Woodland Road
Bristol BS8 1UG
UK

Present address:
Plymouth Marine Laboratory
Citadel Hill
Plymouth PL1 2PB
Devon
UK

INTRODUCTION

The session opened with a brief kaleidoscopic introduction to marine 'snow' and 'fluff' prior to the key note address by **Karin Lochte** (1990). Points of particular interest have been highlighted from the introduction, the contributed papers and the discussion forum.

Larry Pomeroy: Macroaggregates or marine 'snow' contain living phytoplankters (Alldredge and Silver 1988, Silver and Alldredge 1981) as well as degradation products of chlorophyll which indicate that there is an active microbial community and that the aggregates may have already passed through a gut. Large rod-shaped bacteria typify the active bacteria associated with macroaggregates. This is in contrast to the small cocci which are typical of free living oceanic bacteria. Bacterial biomass seems to make up much of the particulate biomass. Often extracellular material with bacteria around and within it occurs in early stages of aggregation. Protozoan populations are dominated numerically by heterotrophic flagellates although ciliates and amoeba have also been observed (Caron *et al.* 1982).

Aggregates can be made from dying marine phytoplankton cultures. Within 24 hr bacteria are on the surfaces and penetrating the material (Biddanda and Pomeroy 1988). After 48 hr, bacteria are not only on the surface of the aggregates but many are free in the water. Most of the bacterial production is associated with the bacteria in the water and not with bacteria on aggregates. When observed after 92 hr most of

NATO ASI Series, Vol. G 25
Protozoa and Their Role in Marine Processes
Edited by P. C. Reid et al.

the particulate material is degraded and the particles are formed largely of bacteria with numerous flagellates grazing on them. Macroaggregates may also be made from a mixture of natural biogenic material (Figure. 1A-D).

Carol Turley: When macroaggregates or marine 'snow' reach the sea bed they can incorporate some of the sediment and have been called 'fluff'. Mass occurrance of 'fluff' on the deep-sea bed was discovered in the Porcupine Sea Bight (SW of Ireland: 50°35'N, 13°W) in the early 1980's (Billett *et al.* 1983, Lampitt 1985, Rice *et al.* 1986). Mass sedimentation of phytoplankton blooms may also explain the origin of the organic material found on the surface of the sediment on the

Fig. 1. (A-D) Aggregates formed in the laboratory from natural planktonic material sampled, using a 150μm sized mesh net, during March at the onset of the spring bloom in the North Sea south of Helgoland. *Coscinodiscus wailesii*, an important mucus producer, and a filamentous cyanobacteria were the dominant phototrophs. Filaments of heterotrophic bacteria (A) and cyanobacteria (B) in a mucus matrix, seem to be important in initial aggregate formation. After 2 weeks large aggregates (>0.5 mm) form and incorporate larger diatoms (C,D) which are still active with intact nuclei (C) and autofluorescing chlorophyll (D). (A-C) are aggregates stained with the DNA-specific fluorochrome diamidinophenylindole (DAPI), mounted in glycerin and viewed under epifluorescence with UV excitation (340-380 nm) while (D) shows the red autofluorescing cells containing chlorophyll in the same field of view as (C) when seen under blue excitation (450-490 nm). (E-I) are macroaggregates sampled using a vertical net haul from 3,885 m to the surface at the BIOTRANS site, at the same time 'fluff' was discovered on the sea bed. (E-G) show the same field of view of mini pellets, of various sizes and contents lying in the phaeodium of *Phaeodaria* under (E) transmitted light (F) epifluorescence with UV excitation and (G) epifluorescence with green excitation (515-560 nm) making the orange-red autofluorescence of phycoerythrin containing cyanobacteria visible. Some pellets seem to contain little organic material. Two are almost totally comprised of the crushed lorica of *Dictyocysta elegans* (i), some contain many cyanobacteria (ii), while others have few or no cyanobacteria but contain a dense-looking, structureless material (iii). A number of dinoflagellate species (H) and acantharian needles of strontium sulphate in a matrix of organic material containing mini pellets (I) were not infrequent in these aggregates. Scale bars are 20 μm. Photographs taken by C Turley

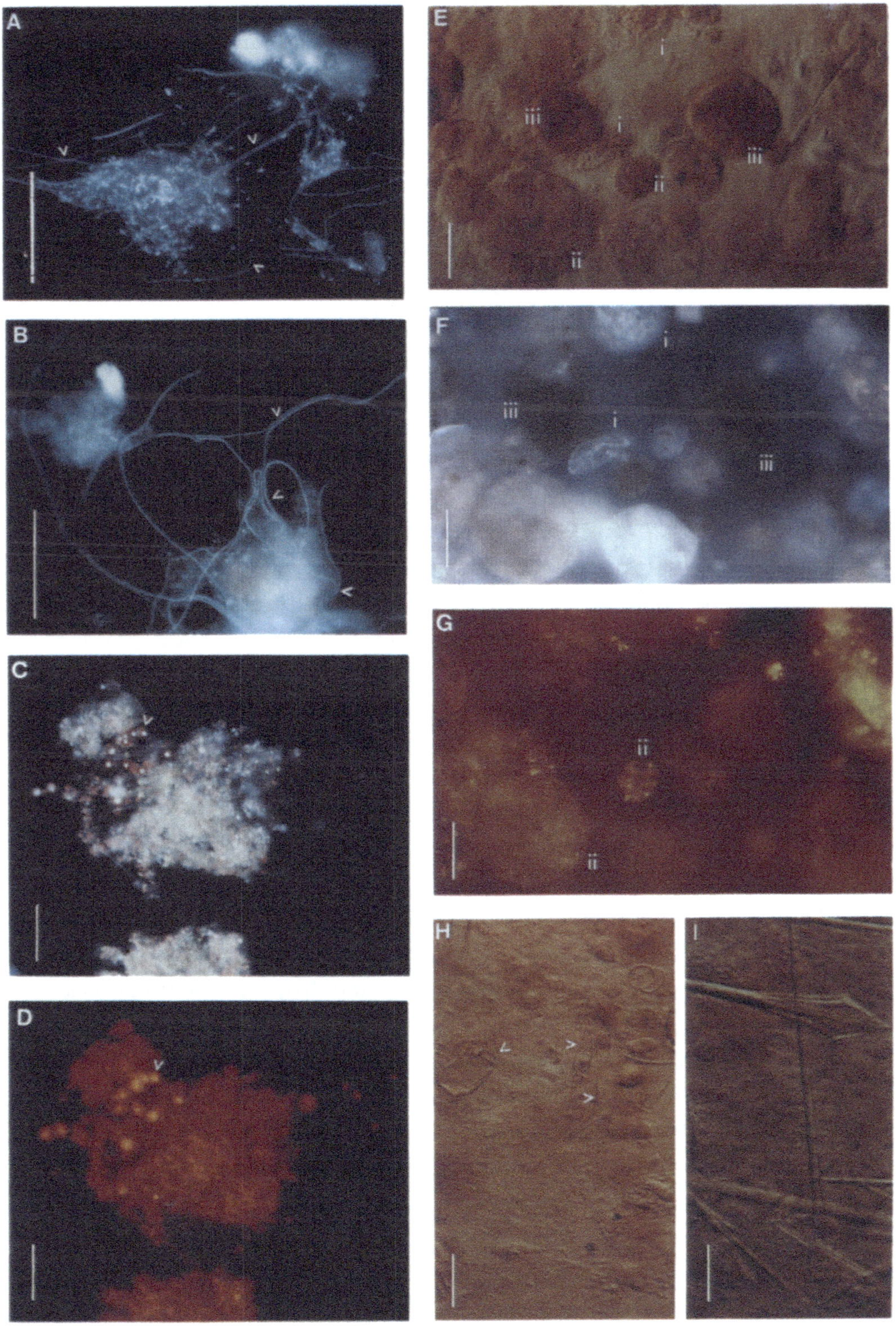
A
B
C
D
E
F
G
H
I
i
ii
iii

DOS 1 site (39° 46.7'N; 70° 40.8'W; 1,800 m, Grassle and Morse-Porteous 1987) and in relic tubes and burrows at the HEBBLE site (40° 26'N; 62° 20'W; 4, 800 m, Aller and Aller 1986) on the continental slope of the western N. Atlantic. Lampitt (1985) found seasonal deposition on the sea bed in the Porcupine Sea Bight at 4,025 m using a sequence of photographs from a bathysnap camera. Aggregates started to arrive at the beginning of May and by the end of June and throughout July there was nearly total coverage of the sea bed. However, it was not certain whether these deposits were typical of open oceanic waters or if it was a local phenomenon caused by its close proximity to the shelf break. In July and August 1986, scientists on the Meteor found similar deposits (Thiel *et al.* 1990) on the sea bed at 4,500 m in the NE Atlantic at the BIOTRANS site (47° 30'N, 20°W), far away from the influence of the shelf break. In 1987, a Norwegian cruise found layers of slimy, green, detrital coccolithophorid aggregates north of Iceland, at depths of 2,600 m (O Tendal pers. comm.). This material can be easily wafted away by the bow waves generated by box corers. However, the multiple corer (Barnett *et al.* 1984) took undisturbed samples of this light, flocculant, fluffy detritus. Pigment analysis of material recovered from the BIOTRANS area using this method contained 25% of the phytodetrital pigment as chlorophyll *a* indicating that this material was relatively undegraded, (Thiel *et al.* 1990).

Exuvia of mid water harpacticoids, empty tintinnid lorica of *Dictyocysta elegans*, empty diatom frustules (Figure 2A-B), coccolithospheres, 3 µm chlorophytes and high concentrations of large heterotrophic bacteria (Figure 2C) and *Synechococcus*-like cyanobacteria (Figure 2D) were characteristic of the 'fluff' on the BIOTRANS site (Lochte and Turley 1988, Riemann 1989, Thiel *et al.* 1990). Cyanobacteria and picoplankton algae have also been found to be concentrated on marine 'snow' down to 2,000 m (Silver and Alldredge 1981, Silver *et al.* 1986). Macroaggregates may, therefore, be a vehicle for the rapid transport of even small cells which would never reach those depths as single cells. If these depositions happen in the mid oceanic waters on a seasonal basis, as those observed in the Porcupine Sea Bight, the rapid transfer of particulate organic material from the euphotic zone to the deep-sea through the process of aggregate formation may be important in biogeochemical ocean flux. The degree of remineralisation of these aggregates by bacteria and protozoa and larger consumer organisms during and after their sedimentation to the sea bed, will determine the amount of organic carbon buried in the sediment and the importance of the oceans as a sink to atmospheric carbon dioxide (Turley and Lochte 1990).

CONTRIBUTIONS

Gabriel Gorsky presented a paper on the Appendicularian or Larvacean filter feeding tunicates which are capable of effectively filtering nanoplankton by means of a mucus 'house'. This is secreted by the organism and can be 0.1-50 cm wide. The larvaceans abandon their houses at intervals (10-12 times per day, Fenaux 1985). They are rich in non ingested particles and rapidly become colonised by bacteria and protozoa. Larvaceans grow rapidly, have a short generation time (14-20 d), fast metabolic rates (Gorsky *et al.* 1987) and 'blooms' are not infrequent (Seki 1973). Their abandoned mucus houses are

Fig. 2. Exuvia of mid water harpacticoid copepods, lorica of *Dictyocysta elgans*, empty centric and pennate diatom frustules (A and B), large numbers of heterotrophic bacteria (C), *Synechococcus*-like cyanobacteria (D) are common in the phytodetritus on the deep-sea bed at the BIOTRANS site. Opportunistic foraminifera, such as the allogromeiid *Tinogullmia* sp. nov. (E) and the browsing bacterivorous barophilic flagellate *Bodo* sp. (F-I), can also take advantage of this relatively rich organic food source. (A-C) are DAPI stained, (D) is viewed under green excitation (515-560 nm) for orange-red autofluorescence of phycoerythrin, common in *Synechococcus*, (E) is taken using phase contrast microscopy showing the nucleus and the protoplasm stained lightly with Rose Bengal. (F-I) is stained with the DNA fluorochrome Acridine Orange photographed under blue excitation (450-490 nm) showing the bright orange kinetoplast of *Bodo* sp., their association with detritus (G) and their two flagella in focus (H, I). Scale bars are 20µm. Photograph E taken by A Gooday, others taken by C Turley

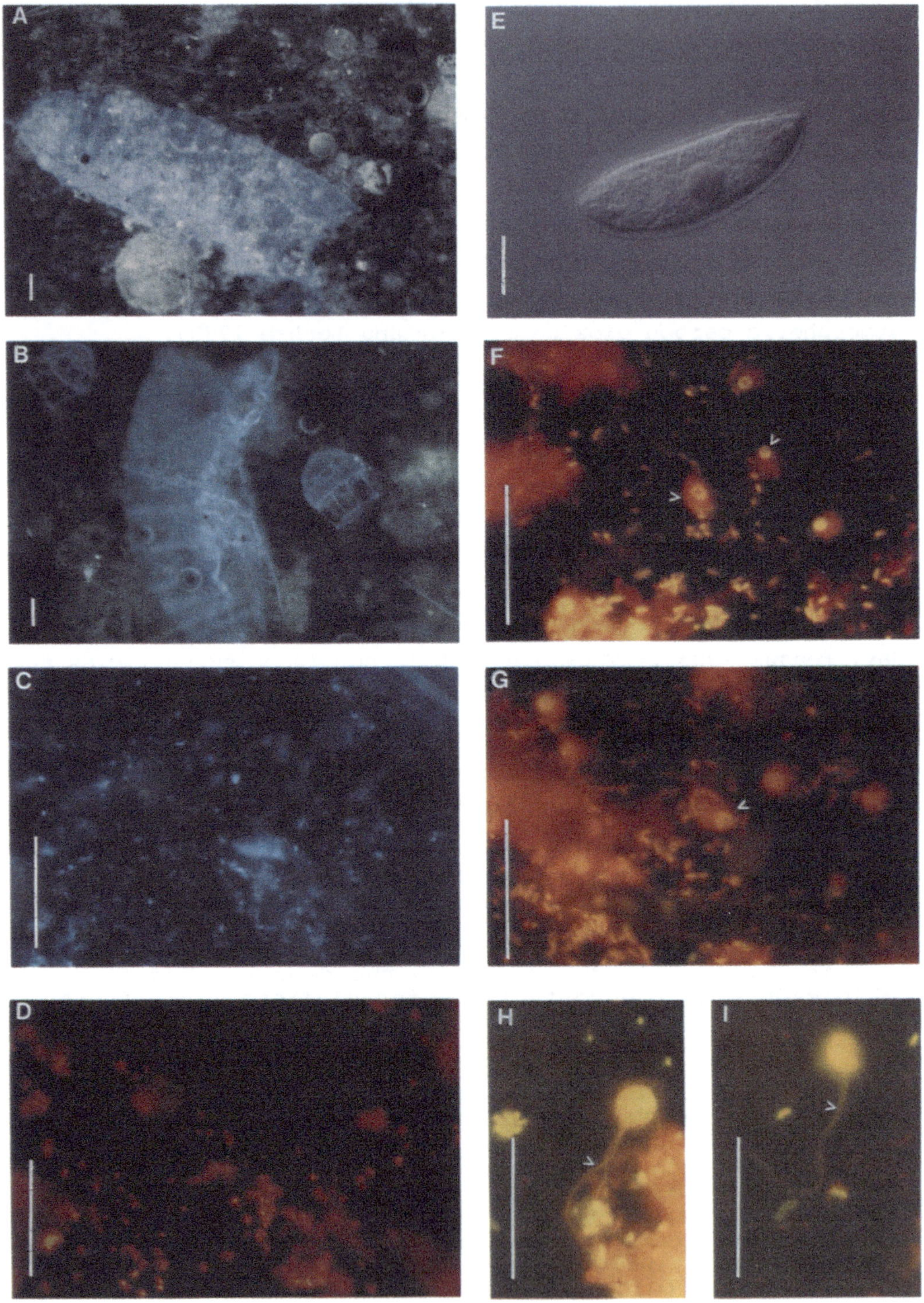
A
B
C
D
E
F
G
H
I

important in the formation of marine 'snow' and the vertical transport of organic matter (Alldredge 1976, Gorsky *et al.* 1984, Davoll and Silver 1986). Large populations were found at intermediate depths of 400-700 m and it was proposed that further aggregation may occur at this depth through the mediation of these Larvaceans.

Gaby Gorsky's unforgettable video of these beautiful creatures was an excellent method of demonstrating and understanding their structure, feeding mechanisms and commensal ciliate populations. It is a loss to science that there is not yet a method available of 'publishing' such recordings.

Ian Jenkinson focused attention on the mechanics of disaggregation. Both shearing forces in the seawater, generally turbulent but often laminar in density discontinuities, and drag resistance to sinking will break the aggregate when the combined stress (stress = force/area) exceeds the 'strength' or yield stress of the aggregates. From knowledge of the ranges of shear rates encountered in the sea, as well as of the sizes and sinking speeds of aggregates, he showed that disaggregation stress is dominated by seawater shearing in some cases and by drag in others.

Where seawater shearing dominates, a homogeneous aggregate will be 'smeared out', irrespective of size, when the shearing stress exceeds the aggregate yield stress; aggregates sheared in a rheometer were found to stretch elastically between 1- and 6-fold before yield stress was exceeded and they deformed permanently. Where drag dominates, however, the upper limits on size and sinking rate are determined by the excess density and yield stress of the aggregate material. However, as aggregates are often composed of different combinations of material (Figures 1 and 2), breaking would generally follow lines of weakness, resulting in smaller stronger aggregates.

A call was made for more characterisation of seawater flow properties. Aggregates, often transiently associating and separating, may be thought of as bits of seawater with enough yield stress to remain functionally solid. Despite greatly improved methods for seeing aggregates, we have practically no idea of the relative volume occupied by any which remain invisible. These might be detectable by high-resolution analysis of seawater flow (Jenkinson 1986).

Andy Gooday introduced the world of deep-sea benthic foraminifera and their response to sedimented phytodetritus or 'fluff' at two locations, the abyssal BIOTRANS area (4,500 m) and the Porcupine Sea Bight (1,300 m). Phytodetrital aggregates from the latter station sampled during July were dominated by *Alabaminella weddellensis*. The number of species (7-10) in the 'fluff' was lower than that in the sediment (80-90). Most of the species in the aggregates were also common in the sediment during July but were significantly less abundant in the sediments during April prior to 'fluff' deposition. Three of these species (*A. weddellensis*, *Epistominella exigua*, and *Tinogullmia* sp. nov. (see Figure 2E) dominated the large foraminiferal population inhabiting the 'fluff' at the deeper BIOTRANS site (Gooday 1988) but were rare in the underlying sediment. This and their size structure indicated that they appeared to be opportunist species multiplying in response to the deposition of the phytodetritus and that the distribution of some deep-sea benthic foraminifera may be controlled by food availability as well as bathymetry and water masses. It was proposed that the phytodetrital populations did not constitute a distinct foraminiferal community but rather that they are opportunists, drawn from the sediment to this nutritious food source (Lambshead and Gooday in prep.).

Kozo Takahashi described how PARFLUX sediment traps deployed at 1,000 and 3,800 m (water column depth 4,200 m) and sampled every 2 weeks at station PAPA in the eastern N. Pacific could be used to estimate sinking speeds of individual diatoms and radiolaria. Details of this and seasonal fluxes of radiolarians and silicon flagellates are in this volume (Takahashi 1990).

Evi Nöthig posed an interesting question as to the identity of ellipsoidal objects (50-300 µm in diameter) found in sediment traps deployed for short periods at 100 and 350 m in the Weddell Sea (Antarctica). These 'faecal pellets' comprised up to 80% of the POC in the traps and therefore were the major contributors to the daily particle flux (0.2 - 1.9% of the standing stock was found in the traps). Since all the 'faecal pellets' contained intact diatom frustules without any plasma content it is unlikely that they were produced by metazoans, but rather by organisms with a specialised feeding method such as that used by some heterotrophic dinoflagellates (described below by Elbraechter). The clue to the producers of these 'faecal pellets' may lie in the composition of organisms in the upper water column where metazoan numbers were low and the ciliates and heterotrophic dinoflagellates contributed between 16 - 42% of the combined phytoplankton and protozooplankton cell carbon. Radiolaria and foraminifera were also present in low numbers in most samples but could not be quantified. It may well be that protozoa can contribute significant amounts to particle flux if this faecal material originates from this source (Nöthig and von Bodungen 1989) These unidentified objects are substantially bigger than the faecal aggregates (20 - 30 µm diameter) produced in laboratory feeding experiments by planktonic marine ciliates (Stoecker 1984). Their small size has led to the current view of particulate flux, which is contrary to Nöthig's hypothesis, in which faecal pellets produced by protozoa are not considered to contribute to flux.

Through a series of slides and video recordings of live specimens **Malte Elbraechter** addressed the little known subject of faecal pellet production by dinoflagellates and radiolarians (Gaines and Elbraechter 1987). Small dinoflagellate species produce spherical (3-8 µm in diameter) or ovoid (5 x 3 µm) faecal pellets. Large specimens of *Gymnodinium gracile* can produce larger faecal pellets up to 42 x 25 x 18 µm about 2 hr after the ingestion of the food particle. The large unarmoured dinoflagellate *Gyrodinium spirale* can ingest up to four different sized food particles and produce several faecal pellets of different sizes in a short time. The organic contents of these pellets are variable indicating different efficiencies of digestion of the food bodies prior to egestion. However, some mixotrophic particle-ingesting dinoflagellates do not produce faecal pellets. For example, the peduncle of *Protoperidinium* envelopes a diatom and digests the cell extracellularly and while discarding the empty diatom frustule, produces no visible faecal material. Radiolaria also produce many small faecal pellets, some are green in colour; large faecal pellets can also be produced. After they are egested they eventually fall off the matrix of mucus around the radiolarian, which also traps the food particles, into the water column (Riemann 1989). The process of feeding and egestion in radiolaria and dinoflagellates needs further careful investigation.

DISCUSSION FORUM

This discussion was based on a series of questions which were identified prior to the meeting.

Q 1. **Are protozoa important aggregate formers and do they play a significant rôle in the rapid transportation of material to the deep-sea?**

Carol Turley: Phaeodaria approximately 3-6 mm in diameter were on occasion packed in the 'fluff' discovered at the BIOTRANS site. There were several species of dinoflagellates (Figure 1H) and considerable numbers of mini faecal pellets associated with phaeodaria in macroaggregates collected from the surface waters at the same time (Figure 1E-G). The mini pellets contained a variety of material; autofluorescing cyanobacteria (Figure 1G), folded exuvia of zooplankton, lorica of the tintinnid *Dictyocysta elegans* (Figure 1E-F) and some contained a darker mass of undistinguishable material (Figure 1E-G). Possibly the phaeodaria are important repackagers of small picoplankton in the upper mixed layer which results in their rapid transport to the deep-sea. Acantharia (Figure 1I) were also components of the aggregates in the surface waters. They have needles made of strontium sulphate which dissolve with depth. Thus, although they may be important in aggregate formation, because of dissolution during sedimentation, their role may become less obvious.

Kozo Takahashi: There are 30-60 species of phaeodaria depending on habitat. Flux data indicate that some of the populations are living in the deep water, around 400-5,000 m. Are the deep phaeodaria important in repackaging in the meso- and bathypelagic zone? The deep-dwelling phaeodaria are small, 30-200 µm, far smaller than the surface growing species.

David Caron: Cultures of flagellates, with no live bacteria, did not form conspicuous aggregation until stationary growth. Then flagellates embedded and swimming around the macroaggregates can be easily seen. Flagellates may, therefore, have something to do with aggregate formation.

Q 2. **Do protozoa make an important contribution to the breakdown of aggregates during their sedimentation and after their arrival on the sea bed?**

Carol Turley: A bacterivorous, barophilic microflagellate, *Bodo* sp. (Figure 2 F-I) was isolated from the 'fluff' and sediment-contact water at the BIOTRANS site on the same occasion (Turley *et al.* 1988, Lochte and Turley 1988) that Andy Gooday (1988) found the opportunistic 'fluff' foraminifera. No growth of *Bodo* sp. occurred at 1 atm. The bacterial loading in the deep-sea sediment and 'fluff' was sufficient to support a population of bacterivorous flagellates. It would seem that this browsing, bacterivorous flagellate is adapted to growth in the deep-sea and contributes to the breakdown of seasonal, sporadic inputs of organically rich phytodetritus on its arrival on the sea bed.

Andy Gooday: A footnote to the work mentioned earlier, is some current work, between Carol Turley and myself, investigating the diet of the three dominant deep-sea benthic foraminifera (in particular *Epistominella exigua* and *Tinogullmia* sp. nov.) which proliferated in the deep-sea 'fluff'. Both epifluorescent microscopy and TEM revealed that cyanobacteria, bacteria and small eucaryotic algae, which were typical of the 'fluff' microflora, were abundant inside these 'fluff' foraminifera. They may, therefore, play some role in the decomposition of the deep-sea phytodetritus although this is unlikely to be as important as bacteria and flagellates.

It should also be remembered that the megabenthos such as detrital feeding holothurians and echinoids also eat and reprocess the 'fluff' layers (Lampitt 1985, Billett *et al.* 1988) with its microbial and meiofaunal inhabitants. Furthermore, specialist foraminiferal feeding scaphopods can be abundant in the deep-sea. For a fuller review of the benthic response to organic input to the deep-sea see Gooday and Turley (1990).

Q 3. **Do 'snow' and 'fluff' have a unique, adapted protozoan population and if so, how and why?**

David Caron: Figure 3 is a theoretical model of what marine 'snow' may mean to protozoa in the open ocean (Caron 1984). Bacterial concentration is plotted against the distance between bacteria, calculated by the nearest neighbour model, assuming they were evenly distributed. The solid line shows where oceanic environments fall. Benthic environments have high bacterial concentrations around 10^{9-10} cells ml^{-1}, neritic environments have less, around 10^{6-7} cells ml^{-1} and in oceanic environments bacteria are even less abundant with around 10^5 cells ml^{-1}. The theoretical distance between bacteria for these environments is indicated by this line. Macroaggregates create unique habitats by presenting a microenvironment with high bacterial abundance in relatively oligotrophic waters. Macroaggregates would, therefore, allow the existence of different types of bacterivorous organisms, some of which are ordinarily more representative of neritic and benthic environments.

Gene Small: Some ciliates (e.g. *Mesolimbus fecalis*) were found to be living in a faecal pellet, feeding off the residual

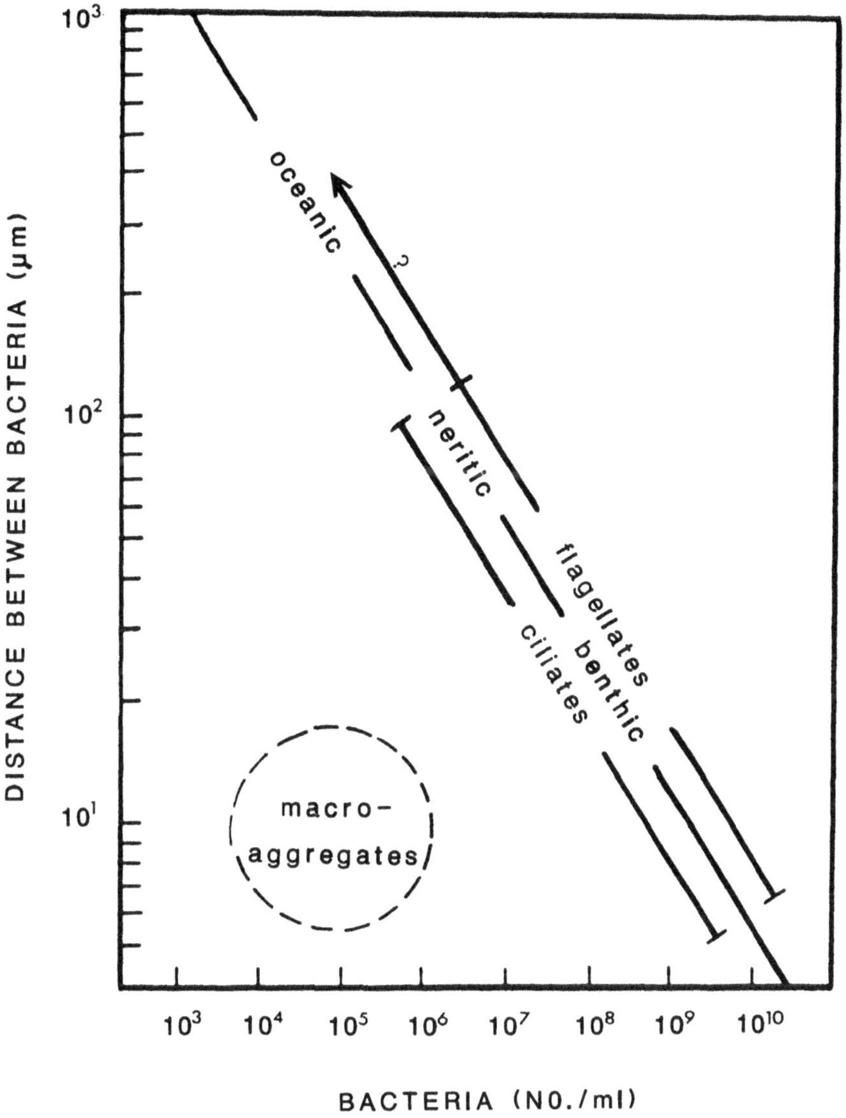

Fig. 3. Theoretical model of where macroaggregates lie in terms of bacterial concentration and distance between bacteria relative to other habitats and food concentration of bacterivorous protozoa (Caron 1984)

material or bacteria in the pellet and are unique to that microhabitat. The macroaggregates host a range of different and unique ciliates some of which have never been seen elsewhere (e.g. *Ovolimbus caronii*, Figure 4). Some are bacterivores

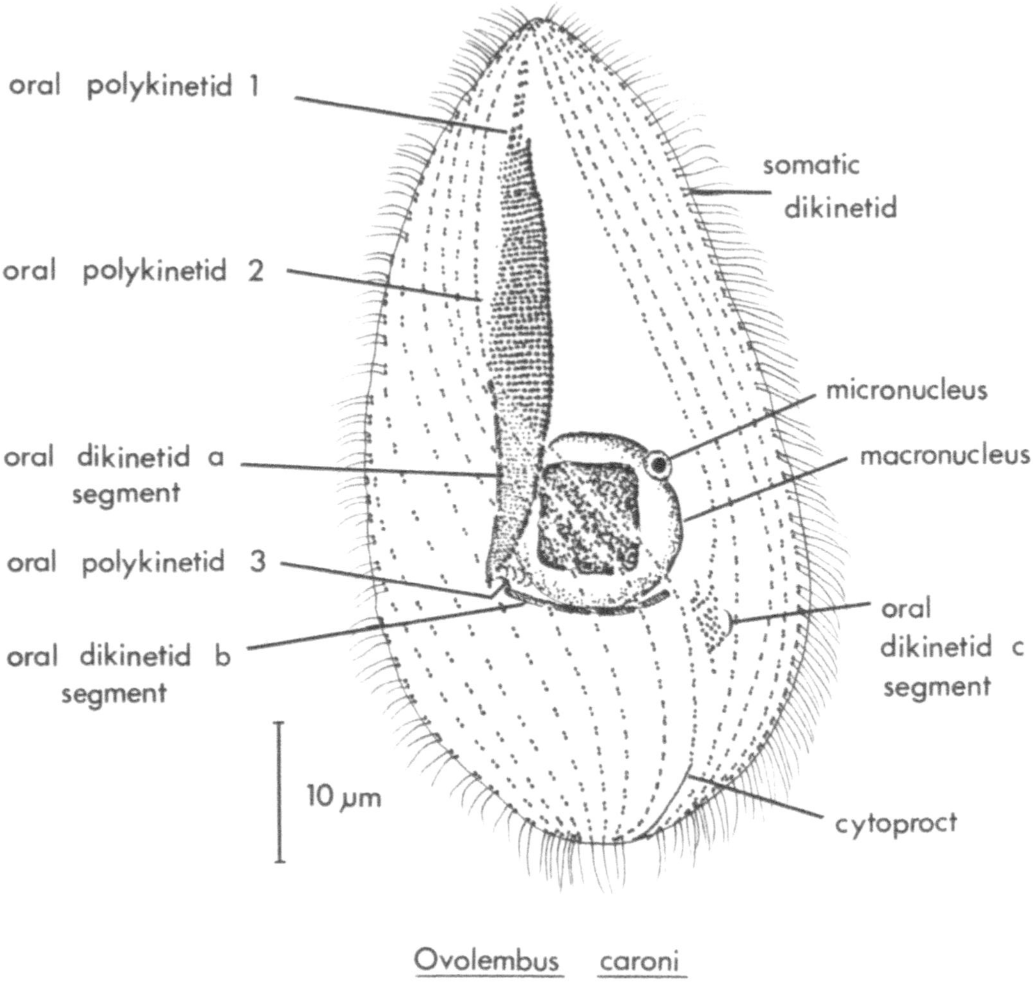

Fig. 4. A unique marine 'snow' ciliate, *Ovolembus caroni*, drawn from a Protargol stained specimen. All somatic kinotosomes bear cilia. However, for the sake of clarity only the marginal somatic paired kinetosome kinetids are illustrated with cilia. Original illustration by Small and Lynn in Lee *et al.* (1985)

(Caron *et al.* 1982), others appear to be detritivores, a few are feeding on diatoms and some are raptorial carnivores. This unique population of protozoa, associated with open ocean macroaggregates, does not include tintinnids or microalgal protozoan feeders, which are typical ciliates found in the euphotic zone. In contrast to marine 'snow', ciliates found in hydrothermal vents all have comparable forms in shallow water (Small and Gross 1985).

Karin Lochte: While it is probably understood how an adapted population can survive in the sediment and then colonise detrital aggregates, the question of how specialised protozoan populations can develop in pelagic particles is quite significant. How do they travel from particle to particle? How do they survive? Animal vectors, such as the appendicularia, may play a role.

David Caron: Many of the large mucus-producing zooplankton (pteropods, ctenophores, salps and appendicularians) are responsible for the production of these large particles or aggregates, particularly in open oceans. Many of them may have endemic populations of protozoa and may, therefore, be candidates for the role of animal vectors.

Carol Turley: Karin Lochte mentioned earlier that the discrepancy between the sinking rate of macroaggregates and the swimming speeds of protozoa and how difficult it would be for them to transfer between aggregates. We see from Gene Small that there are unique protozoa on marine 'snow' even at depths of 2,000 m (Silver *et al.* 1984). However, if macroaggregates sink at 100-300 m day^{-1} (Silver 1986) then attached protozoa have to cope with pressure changes of 10-30 atm day^{-1}. Do they remain associated with the macroaggregates throughout the journey to the deep-sea, or are they adapted to a range of pressures and detach when they reach the limit of their pressure adaptation? If so, there must be a method of maintaining them within a range of depths such as migrating animal vectors or upward fluxes of particles. Experiments looking at the barotolerances of protozoa isolated from macroaggregates from different depths may throw some light on this enigma.

Little is known about the mesopelagic zone and the organisms and processes in it. The thermocline at the bottom of the surface mixed layer can be a place of aggregate formation. A series of density discontinuities can occur in the mesopelagic zone and macroaggregates can accumulate on them (Lampitt pers. comm.). These 'mesopelagic traffic lights' may be areas where there is further mico- and macro- processing of the aggregates (which may cause disaggregation and/or reaggregation), where bacterial and protozoan attachment and detachment can occur and where protozoa can particle-hop or hitchhike on upward migrating animals and positively buoyant particles.

CONCLUSIONS

It was the general conclusion of the session, based on the information currently available in the literature (Lochte 1990), and the information presented by the participants during the session, that protozoa can play a significant role in the formation and transformation of macroaggregates, and that these microenvironments may be habitats with adapted endemic populations of protozoa. It was also obvious that there were great gaps in our knowledge with regard to the interplay between protozoa and macroaggregates and biological processes particularly in the mesopelagic zone. The biggest challenge may be to quantify and sample the entire spectrum of particles throughout the water column and determine the overall importance of the activity of the biota associated with these particles to the activity of the whole community.

ACKNOWLEDGEMENTS

My gratitude to Dr B Biddanda for assistance in making the aggregates photographed in Figure 1A-D, Dr F Riemann for allowing me to photograph aggregates he collected at the BIOTRANS site (Figure 1E-I) and Mrs A Smith for technical assistance. Finally, this paper would not exist but for the enthusiasm and quality of presentation and discussion of all the contributors mentioned in this session. Biogeochemical Ocean Flux Study (BOFS) contribution No 4.

REFERENCES

Alldredge AL (1976) Discarded appendicularian houses as sources of food, surface habitats and particulate organic matter in planktonic environments. Limnol Oceanogr 21:14-23
Alldredge AL, Silver MW (1988) Characteristics dynamics and significance of marine snow. Prog Oceanogr 20:41-82
Aller JY, Aller RC (1986) Evidence for localised enhancement of biological activity associated with tube and burrow structures in deep sea sediments at the HEBBLE site, western North Atlantic. Deep-Sea Res 33:755-790
Barnett PRO, Watson J, Conelly D (1984) A multiple corer for taking virtually undisturbed samples from shelf, bathyal and abyssal sediments. Oceanol Acta 7:399-408
Biddanda BA, Pomeroy LR (1988) Microbial aggregation and degradation of phytoplankton-derived detritus in seawater. I. Microbial succession. Mar Ecol Prog Ser 42:79-88
Billett SM, Lampitt RS, Rice AL, Mantoura RFC (1983) Seasonal sedimentation of phytoplankton to the deep-sea benthos. Nature 302:520-522
Billett DSM, Llewellyn C, Watson J (1988) Are deep-sea holothurians selective feeders? In: Burke RD, Mladenov PV, Lambert P, Parsley RL (eds) Echinoderm Biology. Proceedings of the Sixth International Echinoderm Conference Victoria, Balkema, Rotterdam, Brookfield, 421-429
Caron DA (1984) The role of heterotrophic microflagellates in plankton communities. PhD Thesis. Woods Hole Oceanographic Institution and Massachusetts USA, 1-268
Caron DA, Davis PG, Madin LP, Sieburth JMcN (1982) Heterotrophic bacteria and bacterivorous protozoa in oceanic macroaggregates. Science 218:795-797
Davoll PJ, Silver MW (1986) Marine snow aggregates: Life history sequence and microbial community of abandoned larvacean houses from Monterey Bay, California. Mar Ecol Prog Ser 33:111-120
Fenaux R (1985) Rhythm of secretion of Oikopleurid's houses. Bull Mar Sci 37:498-503
Gaines G, Elbraechter M (1987) Heterotrophic nutrition. In: Taylor FJR (ed) The Biology of Dinoflagellates, Blackwell, Oxford 224-268
Gorsky G, Fisher FS, Fowler SW (1984) Biogenic debris from the pelagic tunicate, *Oikopleura dioica* and its role in the vertical transport of a transuranium element. Est Coast Shelf Sci 18:13-23
Gorsky G, Palazzoli I, Fenaux R (1987) Influence of temperature changes on oxygen uptake and ammonia and phosphate excretion, in relation to body size and weight, in *Oikopleura dioica* (Appendicularia). Mar Biol 94:191-201
Gooday AJ (1988) A benthic foraminiferal response to the deposition of phytodetritus in the deep-sea. Nature 332:70-73
Gooday AJ, Turley CM (1990) Responses by benthic organisms to inputs of organic material to the ocean floor: a review. Phil Trans R Soc Lond, A 331:119-138
Grassle JF, Morse-Porteous LS (1987) Macrofaunal colonisation of disturbed deep-sea environments and the structure of deep-sea benthic communities. Deep-Sea Res 34:1911-1950
Jenkinson IR (1986) Oceanographic implications of non-newtonian properties found in phytoplankton cultures. Nature 323:435-437
Lambshead PJD, Gooday AJ In prep. The impact of seasonally deposited phytodetritus on benthic foraminiferal populations in the bathyal north- east Atlantic: the community response.
Lampitt RS (1985) Evidence for the seasonal deposition of detritus to the deep-sea floor (Porcupine Bight, NE Atlantic) and its subsequent resuspension. Deep-Sea Res 32:885-897
Lee JJ, Hutner SH, Bovee EC (eds) (1985) Illustrated guide to the protozoa. USA Society of Protozoologists. Allen Press, 1-534
Lochte K (1990) Protozoa as makers and breakers of marine aggregates. In: Reid PC, Turley CM, Burkill PH (eds)

Protozoa and their role in marine processes. Springer, Berlin Heidelberg New York
Lochte K, Turley CM (1988) Bacteria and cyanobacteria associated with phytodetritus in the deep-sea. Nature 333: 67-69
Nöthig E-M, von Bodungen B (1989) Occurrence and vertical flux of faecal pellets of probable protozoan origin in the southeastern Weddel Sea (Antractica). Mar Ecol Prog Ser 56:281-289
Rice AL, Billet DSM, Fry J, John AWG, Lampitt RS, Mantoura RFC, Morris RJ (1986) Seasonal deposition of phytodetritus to the deep-sea floor. Proc R Soc Edin 88B:265-279
Riemann F,(1989) Gelatinous phytoplankton detritus aggregates on the Atlantic deep-sea bed: structure and mode of formation. Mar Biol 100: 533-539
Seki H (1973) Red tide of *Oikopleura* in Saanlich Inlet. La Mer 11:152-158
Silver MW, Alldredge AL (1981) Bathypelagic marine snow: deep-sea algal and detrital community. J Mar Res 39:501-530
Silver MW, Gowing MM, Brownlee DC, Corliss JO (1984) Ciliated protozoa associated with oceanic sinking detritus. Nature 309:246-248
Silver MW, Gowing MM, Davoll PJ (1986) The association of photosynthetic picoplankton and ultraplankton with pelagic detritus through the water column (0 - 2000 m). Can Bull Fish Aqu Sci. In: Platt T and Li WKW (ed) Photosynthetic picoplankton 214:311-341
Silver MW (1986) Characteristics and distribution of marine snow. In: Alldredge AL, Hartwig EO (eds) Aggregate dynamics in the sea. Workshop report, Office of Naval Research, ASILOMAR Conference Center, Pacific Grove California Sept 22-24 1986, 60-89
Small EB, Gross ME (1985) Preliminary observations of protistan organisms, especially ciliates from the 21N hydrothermal vent site. Biol Soc Wash Bull 6:401-410
Stoecker DK (1984) Particle production by planktonic ciliates. Limnol Oceanogr 29: 930-940
Takahashi K (1990) Mineral flux and biogeochemical cycles of marine plantkonic protozoa - Session summary. In: Reid PC, Turley CM, Burkill PH (eds) Protozoa and their role in marine processes. Springer, Berlin Heidelberg New York
Thiel H, Pfannkuche O, Schriever G, Lochte K, Gooday AJ, Hemleben C, Mantoura RFC, Turley CM, Patching JW, Riemann F (1990) Phytodetritus on the deep-sea floor in a central oceanic region of the Northeast Atlantic. Biol Oceanogr 6:203-239
Turley CM, Lochte K, Patterson DJ (1988) A barophilic flagellate isolated from 4500 m in the mid-North Atlantic. Deep-Sea Res 35:1079-1092
Turley CM, Lochte K (1990) Microbial response to the input of fresh detritus to the deep-sea bed. In: Labeyrie L, Jeandel C (eds) Global and Planetary Change Section, Elsevier, Amsterdam, (in press)

PROTOZOA AS MAKERS AND BREAKERS OF MARINE AGGREGATES

Karin Lochte
Institut für Meereskunde
an der Universität Kiel
Düsternbrooker Weg 20
D-2300 Kiel
Federal Republic of Germany

INTRODUCTION

Aggregates have captured the attention of researchers, as sites of enhanced nutrient and biomass concentrations as well as increased microbial activity, for a long time (e.g. Suzuki and Kato 1953, Tsujita 1953, Riley 1963, Nishizawa 1966, Gordon 1970, Seki 1971, Wiebe and Pomeroy 1972, reviews by Fowler and Knauer 1986, Alldredge and Silver 1988). Microscopic observations of detrital aggregates usually reveal the presence of numerous bacteria and algal cells and often the occurrence of protozoa. Phototrophic and heterotrophic nanoflagellates are nearly always present; ciliates seem second in importance (Caron *et al.* 1982 and 1986, Silver *et al.* 1984); amoebae (Caron *et al.* 1982 and 1986, Paerl 1984), radiolaria (Silver and Alldredge 1981, Riemann 1989) and choanoflagellates (Silver and Alldredge 1981) have also been noted. Some of these organisms, especially naked amoebae, are probably highly underestimated because it is difficult to detect them in routine microscopic investigations.

Different types of organic particles can be distinguished by their size and visual appearance as marine snow, which are amorphous aggregates larger than 500 µm, and as microaggregates smaller than 500 µm (Alldredge and Hartwig 1986). Small aggregates are much more abundant than marine snow particles (Trent *et al.* 1978). Number and size distribution is difficult to determine, because large aggregates easily fragment when sampled. Using a camera system, Honjo *et al.* (1984) found highest concentrations in the surface mixed layer (2 - 7 particles ml^{-1}). They decreased below the thermocline to fairly constant numbers in the deeper water column (below 2 particles

NATO ASI Series, Vol. G 25
Protozoa and Their Role in Marine Processes
Edited by P. C. Reid et al.

ml^{-1}). Occasionally, localized peaks in particle numbers may occur at density gradients in deeper waters (Silver and Alldredge 1981, Alldredge and Youngbluth 1985). In the near bottom nepheloid layer numbers of particles increase again, which is caused by resuspension of sediment material (Nyffeler and Godet 1986, Smith *et al.* 1987).

Sedimentation of organic aggregates is responsible for most of the vertical flow of energy and matter within the water column and provides the deep oceans and the benthos with food (Angel 1984, Smetacek 1984). Formation, transformation and degradation of these sedimenting aggregates is, therefore, a question of consequence. Protozoa are likely to significantly influence the rate of aggregate remineralization (e.g. Caron *et al.* 1985). However, we know little about the process itself and the quantitative role of protozoa in it. At the same time, the high concentration of organic matter on aggregates may be of vital importance for protozoa in open ocean waters, where food particles are scarce.

In this article I will discuss the interaction between protozoa and detrital aggregates in the open ocean environment. The potential role of protozoa in the formation, transformation and breakdown of aggregates is described. Observed distribution of particles in the water column and their sedimentation rates are integrated with existing concepts of aggregates as microenvironments, which provide a survival mechanism for protozoa in the food limited oceanic environment. This combines the two apparently antagonistic processes of degradation of aggregates and of flux of organic matter into the deep ocean.

ROLE OF PROTOZOA IN AGGREGATE FORMATION

The characteristics and quantities of aggregates are largely dependent on the assemblage of planktonic organisms present in the upper mixed water layer. In many cases, the formation of aggregates is linked to phytoplankton bloom situations and exhibits a seasonal pattern (Tsujita 1953, Suzuki and Kato 1953, Nishizawa 1966, Smetacek *et al.* 1978, Cole *et al.* 1985). Often, the production of mucus by different organisms seems to initiate formation of marine snow. Ageing or nutrient depleted phytoplankton (Smetacek 1985), bacteria (Pearl 1978, Biddanda 1985), abandoned appendicularian houses (Alldredge 1976, Barham 1979, Alldredge and Youngbluth 1985, Caron *et al.* 1986, Davoll and Silver 1986), pteropod webs (Caron *et al.* 1982), mucus from reef corals (Johannes 1967), faecal pellets and mucus from tunicates and other gelatinuous zooplankton (Pomeroy and Deibel 1980, Bruland and Silver 1981, Pomeroy *et al.* 1984, Caron *et al.* 1986) have been reported or were suspected to provide the adhesive mucus matrix of detrital aggregates.

Little is known, so far, about the role of pelagic protozoa in aggregate formation (Table 1). For instance, mucus was observed associated with high densities of planktonic foraminifera by divers in the vicinity of the thermocline (Alldredge and Youngbluth 1985) and, thus, may be involved in aggregate formation. However, foraminiferan shells were rarely reported as being part of marine snow (Silver *et al.* 1978). Although sedimentation of foraminifera and radiolaria is of great importance in the vertical flux of organic carbon and various elements to the deep sea (Takahashi and Honjo 1981 and 1983, Gowing 1986), the data based on sediment trap material give no information on whether these organisms settle as individuals or are associated with marine snow. In some detrital aggregates radiolaria (Silver *et al.* 1978, Silver and Alldredge 1981) were found. Gowing (1986) and (Riemann 1989) describe phaeodaria and ancantharia. While the silicious tests and needles of phaeodaria are more easily preserved and may be observed also in material deposited on the sea bed, the acantharian spicules composed of strontium sulfate readily dissolve when released after the death of the organism. Acantharia may, therefore, be easily overlooked in microscopic examinations of aggregates. So-called 'minipellets', suspected to be faecal pellets of phaeodaria, spumellaria or small zooplankton (Gowing and Silver 1985), have been observed

Table 1 SUMMARY OF THE POTENTIAL ROLES OF PROTOZOA IN THE FORMATION AND DEGRADATION OF MARINE AGGREGATES

organism	potential interaction	reference
AGGREGATE FORMATION:		
radiolaria	present in aggregates	Silver *et al.* (1978) Silver & Alldredge (1981)
	possibly clumping of detrital particles production of 'minipellets' found in aggregates possibly increase of specific weight of aggregates through skeletal parts	Riemann (1989) Gowing & Silver (1985) Riemann (1989)
foraminifera	rarely reported as part of aggregates interaction with aggregates unknown	Silver *et al.* (1978)
tintinnids	present in aggregates possibly increase of specific weight of aggregates through loricae	Silver *et al.* (1978)
ciliates flagellates	flocculation of bacteria by mucus production (?)	review by Sieburth (1984)
TRANSFORMATION & DEGRADATION:		
radiolaria	feeding on detritus aggregates and repackaging into 'minipellets'	Gowing (1986)
heterotrophic flagellates	primary grazers of attached bacteria	Linley & Newell (1984) Biddanda & Pomeroy (1988) Turley *et al.* (1988)
	remineralization of nutrients	Pomeroy *et al.* (1984) Goldman *et al.* (1985) Taylor *et al.* (1986)
	mechanical disruption	pers. obs.
ciliates	grazing of attached microorganisms and control of mineralization rates	Silver *et al.* (1984) Caron *et al.* (1986)
amoebae	grazing of attached microorganisms mechanical disruption	Caron *et al.* (1982, 1986) Pearl (1984)

frequently in marine snow (Silver and Alldredge 1981, Riemann 1989).

The skeletal parts of radiolaria incorporated into detrital

aggregates may increase its specific weight and accelerate sinking speed. Similarly, tintinnid loricae frequently observed in marine snow (Silver *et al.* 1978, and pers. obs.) may have the same effect (Table 1).

Aggregations of prey bacteria may actively be caused by ciliates and flagellates. For instance, polysaccharide and mucoprotein can be excreted and flocculate bacteria in cultures of ciliates (reviewed by Sieburth 1984). How far this mechanism may be of importance to marine planktonic protozoa is unknown.

SUCCESSION, TRANSFORMATION AND DEGRADATION

In laboratory studies, the microbial succession on detrital particles is characterized by bacterial growth and detritus aggregation followed by development of a mixed assemblage of bacterivorous protozoa and eventual detritus disintegration (Biddanda and Pomeroy 1988). Similar processes were also observed on tunicate faecal pellets, which are important components of marine snow (Pomeroy and Deibel 1980, Pomeroy *et al.* 1984). If aggregates are already precharged with living bacteria and protozoa, as in abandoned larvacean houses, then a quicker development of bacterivorous protozoa ensues, which graze immediately upon the resident bacteria (Davoll and Silver 1986). Silver *et al.* (1984) suggest that ciliate feeding on bacterivorous flagellates as well as on bacteria themselves is the controlling agent of bacterial numbers and of particle degradation rate.

During microbial breakdown of natural tunicate faecal pellets (Pomeroy *et al.* 1984) organic carbon was channelled into bacteria in the initial 24 h and subsequently into protozoa. Most of the labile carbon was utilized within 2-3 d, after which fragmentation of the particles occurred. Similar time spans are found for degradation of faecal pellets of tunicates maintained in a laboratory (5 d, Pomeroy and Deibel 1980), abandoned larvacean houses (6 d, Davoll and Silver

1986) and phytoplankton aggregate disintegration (8-16 d, Biddanda and Pomeroy 1988). Paerl (1984) observed microflagellates and amoebae deeply embedded inside marine snow which appeared able to burrow through it. Similarly, in phytodetritus resting on the deep-sea floor bacterivorous microflagellates were found in great numbers inside the aggregates (Turley *et al.* 1988). Such burrowing activities may affect chemical microzones within the aggregate, loosen its structure and make it susceptible to fragmentation.

A defined 3-member food web consisting of one diatom species, an assemblage of bacteria and one microflagellate species showed an approximately 65% decrease in particulate organic carbon (POC) within 8 d (Caron *et al.* 1985). However, when only bacteria were present a much smaller percentage of POC was degraded while they were very efficient in the utilization of dissolved organic carbon. High remineralization rates of organic nitrogen to ammonia and release of phosphate occurred when protozoa developed on aggregates (Pomeroy *et al.* 1984, Goldman *et al.* 1985). Experiments carried out *in situ* on natural sedimenting matter in sediment traps showed that degradation of particulate organic matter and production of ammonia was enhanced when active zooflagellates were present (Taylor *et al.* 1986). These observations indicate that protozoa strongly influence the rate of POC remineralization within detrital aggregates. Their activity is largely responsible for the release of ammonia, which may create and maintain an enriched microzone around the particle. This is supported by the significantly higher concentrations of ammonia found in marine snow compared to the surrounding water (Shanks and Trent 1979) (Table 2).

The feeding vacuoles and faecal 'minipellets' of phaeodarians caught in sediment traps in the North Pacific central gyre contained detrital material similar to marine snow (Gowing 1986). It is likely that phaeodarians feed directly on sinking detrital aggregates taking advantage of the higher organic matter concentrations. Gowing (1986) suggests that, similarly to ciliates, phaeodarian feeding may regulate decomposition rates of sinking detritus and that they compete with ciliates for food.

Table 2. CONCENTRATION OF AMMONIA IN AGGREGATES AND THE SURROUNDING SEA WATER FROM THE SOUTHERN CALIFORNIA BIGHT (AFTER SHANKS AND TRENT 1979)

Depth (m)	Surrounding water ($\mu M\ l^{-1}$)	Aggregate min - max ($\mu M\ l^{-1}$)	Enrichment min - max
5	0.7	8.5 - 79	12.1 - 113
5	0.3	2.7 - 66	9.0 - 220
10	0.2	2.8 - 17	14.0 - 85
5	0.7	6.0 - 324	8.6 - 438
10	1.8	12.0 - 483	6.7 - 273

The role of protozoa in the transformation and breakdown of organic aggregates, therefore, seems to have three main aspects (Table 1). Firstly, bacterivorous protozoa (mainly flagellates, ciliates and also naked amoebae) settle on aggregates, feed on the attached bacteria and thereby control remineralization rates of nutrients. This has been observed in laboratory experiments and is substantiated by field observations. Secondly, mechanical disruption of aggregates by 'burrowing' protozoa is likely to occur towards the end of the microbial succession. The process itself has not yet been directly investigated but it can be implied from microscopic observations on natural detritus and from the final disintegration of aggregates in laboratory studies. Once aggregate binding strength has been weakened by microbial degradation and 'digging' activities of protozoa, turbulent and shear stresses more easily disrupt aggregates along lines of weakness. Thirdly, detritus may be consumed directly by protozoa, as described for phaeodarians (Gowing 1986).

Table 3. ABUNDANCE OF MARINE BACTERIA INHABITING MARINE SNOW IN COMPARISON TO SURROUNDING SEA WATER IN THE NORTH ATLANTIC. (AFTER ALLDREDGE ET AL. 1986)

Sampling Location	Abundance		% Bacteria on Aggregates
	10^5 cells agg.$^{-1}$	10^5 cells ml^{-1}	
Slope Water	18.3	9.13±1.58	0.2
Slope Water	67.9	2.92±0.71	2.3
Continental Shelf	154	5.51±0.94	2.8
Gulf Stream	64.6	5.32±0.01	0.1
Gulf Stream	2,780	6.4 ±1.0	4.4
Gulf Stream	890	10.4 ±0.9	0.9

MICROZONES

In comparison to the bulk of the water, aggregates are highly enriched in nutrients and microbial biomass (Tables 2 and 3) (Silver *et al.* 1978, Trent *et al.* 1978, Alldredge 1979, Shanks and Trent 1979, Alldredge *et al.* 1986, Beers *et al.* 1986, Caron *et al.* 1986). This enrichment factor is highly variable and depends on the state of degradation. However, the degree of enrichment was more noticeable in the open ocean than in near shore waters (Caron *et al.* 1986). Often phytoplankton cells were incorporated in marine snow (Trent *et al.* 1978, Beers *et al.* 1986) and showed active photosynthesis (Alldredge and Cox 1982, Prezelin and Alldredge 1983, Alldredge and Cohen 1987). Cyanobacteria, which as single cells would sink extremely slowly, were found in high numbers in detritus aggregates on the sea floor at 4500 m depth (Lochte and Turley 1988). It indicates that these photosynthetic cells must have been concentrated on marine snow particles in the euphotic zone and were transported to the deep sea on these rapidly sinking aggregates. Oxygen depletion in the dark due to microbial respiration created persistent oxygen and pH gradients around particles; inside marine snow oxygen concentrations were low but true anoxia was only found in large faecal pellets with intact peritrophic membranes (Alldredge and Cohen 1987, see also Paerl 1984). Hence, relatively stable concentration gradients of oxygen, pH, ammonia, DOC and other dissolved substances may be formed and maintained by microbial action around aggregates.

Phagotrophic protozoa living in open ocean waters face the problem of surviving in an environment with low bacterial concentrations of around 10^5 to 10^6 cells ml^{-1} (Ducklow 1983, Lochte and Pfannkuche 1987), while approximately 10^5 to 10^7 bacteria may inhabit a single aggregate (Alldredge and Silver 1988) (Table 3). Experimental evidence suggests that concentrations of food particles in the bulk of the water may be too low to support growth of bacterivorous ciliates (Fenchel 1980). Even the revised threshold levels of 8.5 - 96 x 10^5 cells ml^{-1}, published by Sherr and Sherr (1987) for ciliate grazing in coastal waters, are still insufficient to guarantee a survival of ciliates in open ocean waters. Nevertheless, bacterivorous ciliates are found there, often associated with aggregates (Silver *et al.* 1984, Caron *et al.* 1986, Small *et al.* 1983) or in sediment traps (Gowing 1986). Caron *et al.* (1986) suggested that feeding and growth of bacterivorous ciliates and amoebae may mainly take place on particles and that their occurrence in the surrounding water is only a transient state during their search for a new particle to invade. Although some species of microflagellates are probably capable of a free-living existence, many of the planktonic species tend to attach to detrital material (and to the walls of culture vessels) (Fenchel 1982 a and b).

Conceptual models of aggregates as microenvironments with enhanced primary production and nutrient recycling have been proposed to explain high metabolic rates despite apparently low concentrations of dissolved inorganic and organic nutrients in the bulk of the water. Goldman (1984) calculated that the swimming speed of flagellates is sufficient to overcome molecular diffusion of nutrients around a particulate source, which would enable the protozoa to locate nutrient rich patches by chemotaxis and to migrate between particles. A close coupling between phototrophic utilization and heterotrophic remineralization of inorganic nutrients would allow this microcosm to function at a high rate, which is determined by its slowest component. Development of redox gradients within the aggregate would cause each participant to occupy its specific niche and would facilitate oxygen-sensitive processes in oxygenated waters (Paerl 1984). Non-random distribution of bacteria around healthy algal cells was also proposed as a mechanisms to enhance protozoan predation (Azam and Ammerman 1984).

SEDIMENTATION

Marine snow was found to settle at rates of up to 350 m day^{-1} (which is equivalent to 4000 µm sec^{-1}), but more commonly around 100 m day^{-1} (equivalent to 1150 µm sec^{-1}) (Alldredge and Silver 1988). Swimming speeds of microflagellates only range up to 370 µm sec^{-1} (Throndsen 1973, Fenchel 1982a). Fast settling aggregates may, therefore, not be detected or reached by these organisms unless they are in the vicinity of the particle's path. This implies that the conceptual model suggested by Goldman (1984) may only be valid for slow sinking aggregates. Therefore, marine snow with little excess density may act as the proposed oases in the marine environment and will, as a consequence, be mineralized during the succession of microorganisms developing on it (Figure 1, Table 4). Asper (1987) reported very slow sinking rates of large aggregates (mean: 1 m day^{-1}) while smaller ones settled on average faster (mean: 36 m day^{-1}). Fast sinking particles, however, probably traverse the water column with less invasion by microorganisms and degradation (Figure 1, Table 4).

Apart from sinking rates, turbulent diffusion is an important factor determining the time spent by particles at various depths. Lande and Wood (1987) estimated that due to high turbulent diffusion in the upper mixed layer particles may reach the thermocline in a few days or even hours, but they make repeated excursions back and forth between the interior of the mixed layer and the top of the thermocline. Slow sinking particles are particularly affected and the time spent in the upper mixed layer can extend over a long period of time. This facilitates microbial breakdown of aggregates in this upper water mass. Fast sinking particles or particles which are produced or penetrate several meters below the top of the thermocline are unlikely to be mixed up again before sinking permanently out of the euphotic zone, because of low turbulent diffusion in the thermocline itself. Alldredge *et al.* (1987) found large macrocrustacean faecal pellets in the upper mixed water layer, which had partially or totally decayed peritrophic membranes. This indicated that the faeces were several days old despite their rapid sinking rates of 18 to 170 m day^{-1}. The authors hypothesized that turbulent mixing in the upper mixed layer and a potential reduction in specific density of the feacal pellets during their decay may cause the observed long retention time in the upper waters.

During settling of aggregates in the deeper part of the water column decreasing temperature and increasing hydrostatic pressure exert stress on the particle inhabitants. In the mesopelagic and bathypelagic zone the rate of particle degradation is reduced compared to surface waters (Taylor *et al.* 1986). This implies that once particles travel below the upper mixed water layer rates of biologically mediated exchange of matter become successively lower than in the surface water. Lal (1977) developed a conceptual model of particle sedimentation and dissolution based on Stoke's law and size dependent dissolution processes, which are constant throughout the water column.

However, biological action and, therefore, particle degradation must be considered variable and depth dependent. It should also be noted that aggregates may change size, shape and weight by incorporation of new particles or by biological breakdown and fragmentation which alters their settling characteristics.

Silver *et al.* (1984) observed that aggregates from different depths were characterized by ciliate populations, which seemed

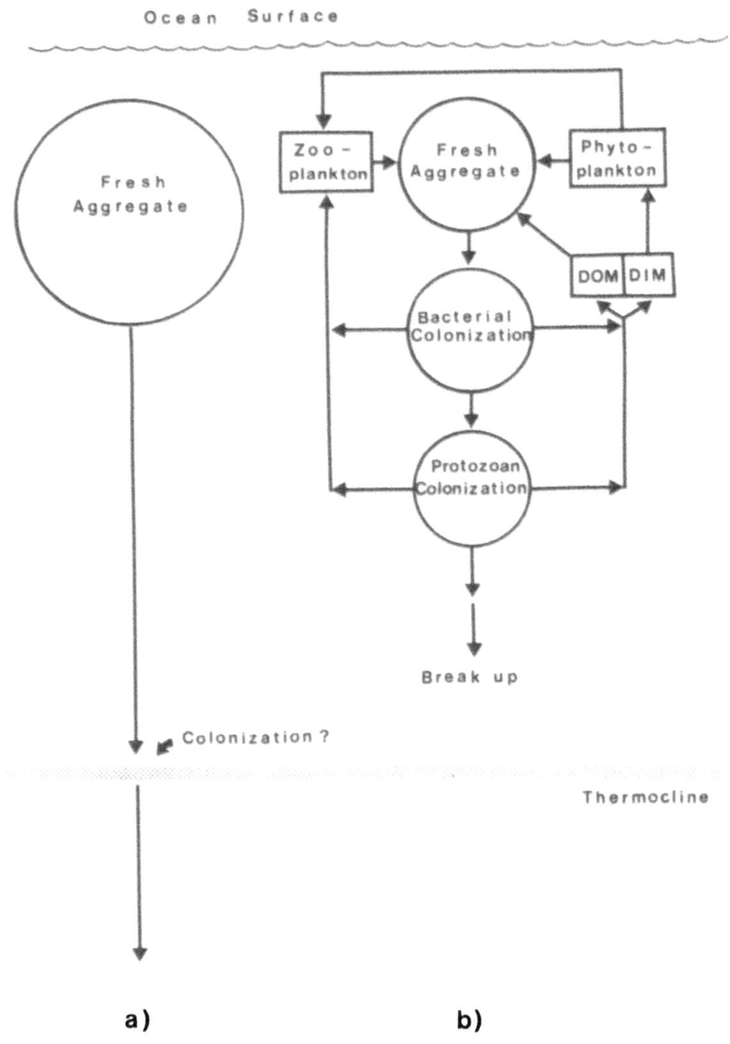

Fig. 1. Schematic representation of the fate of aggregates with different sedimentation rates in the upper mixed water layer. DOM = dissolved organic matter, DIM = dissolved inorganic matter

a) Fast sinking aggregates (around 100 m day^{-1}) leave the upper mixed water layer in less than a day and little of their organic content is being recycled within this water mass. These aggregates transport organic matter into the meso- and bathypelagic zone and the sediment.

b) Slowly sinking aggregates, which are kept in suspension in the upper mixed water layer by turbulent diffusion for several days and which may be within the reach of swimming protozoa at speeds of $\leq$ 30 m day^{-1}, are colonized by microorganisms and are degraded within this water mass. The remineralized nutrients become available for primary production

endemic for the depth of collection and not remnants of surface populations. This poses the question whether the protozoa only remain attached to a particle while it is within the specific tolerance zone for the respective organism. It is also unkown how, in the meso- and bathypelagic zone, where particles occur in low numbers (Honjo *et al.* 1984, Asper 1987), protozoa may encounter aggregates or interchange between them. There is evidence that repackaging of sedimenting aggregates by zooplankton feeding may influence particle dynamics in deep waters (Urrère and Knauer 1981). Feeding by phaeodarians on detritus also plays a role in this process (Gowing 1986). A balance of aggregation and disaggregation determines the chemical nature of different size particles in the deep water column (Wakeham and Canuel 1988). Therefore, it seems likely that aggregate-dwelling protozoa are carried between particles by zooplankton.

The occurrence of large amounts of phytodetrital matter on the sea floor (Billett *et al.* 1983, Rice *et al.* 1986, Thiel *et al.* in press) proves that sedimentation of fairly undegraded matter to the deep sea is possible. When natural deep-sea detrital matter ('fluff') was incubated it was degraded by microorganisms and a barophilic bodonid microflagellate developed (Lochte and Turley 1988, Turley *et al.* 1988). This organism was capable of rapid growth under deep-sea conditions and influenced the degradation of the detrital particulate matter. Opportunistic benthic foraminifera invaded and fed on phytodetritus lying on the sediment surface and, thus, contributed to its decomposition (Gooday 1988). These observations illustrate the labile nature of 'fluff' and show that it is a food source for deep-sea protozoa. It also demonstrates that this type of aggregate experienced little degradation in the water column, probably due to rapid fallout (Table 4).

SYNTHESIS

The fact that certain types of protozoa exist in open ocean waters, which are seemingly too barren to support them,

Table 4. SUMMARY OF FATE AND FUNCTION OF FAST AND SLOW SINKING AGGREGATES IN THE PELAGIC ECOSYSTEM

	Rapidly sinking aggregates	Slow sinking aggregates
Concentration in		
-surface water	low	high
-deep water	low	low
Colonization of aggregates by external microorganisms	inhibited	possible
Development of microbial Succession	incomplete	complete
Degradation of POM and mineralization	incomplete	extensive
Functional role in the Surface water	removal of material	zones of nutrient recycling,
Functional role in deep ocean	input of organic matter	unknown

challenges the current conceptual models of open ocean pelagic systems. The survival of these species depends on small scalefood enrichment as suggested by Azam and Ammerman (1984), Goldman (1984), Caron *et al.* (1986). However, these concepts have to be reconciled with the observed distribution and sinking speed of aggregates, and a more differentiated view of the interaction between protozoa and aggregates has to be adopted.

Sinking speed and nutrient concentration of aggregates determine their fate and both these factors vary greatly depending on their composition and the process of aggregation (Figure 1, Table 4). Slowly sinking particles would be within the range of swimming speeds of protozoa. When the abundance of this type of particle is high, so that the intermediate distances are short enough for protozoan migration between them, they probably enable survival of even those types of protozoa requiring higher food concentrations than provided in open ocean waters. These aggregates are likely to experience the full succession of microorganisms and may be mineralized to a large extent within one or two weeks. During this time a slow sinking particle would still be within the upper mixed water layer (Lande and Wood 1987). The remineralization products would become available to primary production in this water mass. Assuming swimming speeds of protozoa of up to 370 µm sec^{-1} particles with sinking rates of less than 30 m d^{-1} can theoretically be reached by these organisms. Of course, the slower a particle settles the higher is the chance of colonization and exhaustive particle degradation. Turbulent mixing also influences the encounter between protozoa and aggregates, but the magnitude of this process has not been estimated yet. By grazing and multiplying on slow settling aggregates the organisms also ensure that the population will be maintained within the water layer suitable for its survival.

Attachment to fast sinking particles or to aggregates produced within the thermocline itself, would result in loss of the attached organisms from the upper water mass. At sinking speeds of more than 100 m d^{-1} they would reside in the upper mixed water layer for less than a day (Lande and Wood 1987). The transfer to deep-sea conditions would be in most cases probably lethal. These aggregates, irrespective of their organic content, are likely to pass into deeper waters less degraded by microorganisms. The low numbers of particles in the deep water masses makes colonization by endemic species via chemotaxis unlikely. Other mechanisms are required, for instance protozoan transfer by zooplankton, to understand interaction between protozoa and aggregates in meso- and bathypelagic zones. Temporary concentrations of particles on pycnoclines may not only serve as zones of increased aggregate formation by particle collision, but may also enable renewed particle colonisation by motile organisms. The fast sinking aggregates probably transport their load of organic matter in a fairly undegraded state to the deep sea and provide an energy input for the benthos. Biological processes occurring in the meso- and bathypelagic water masses are largely unknown; many speculations and only few reliable data exist (Table 4).

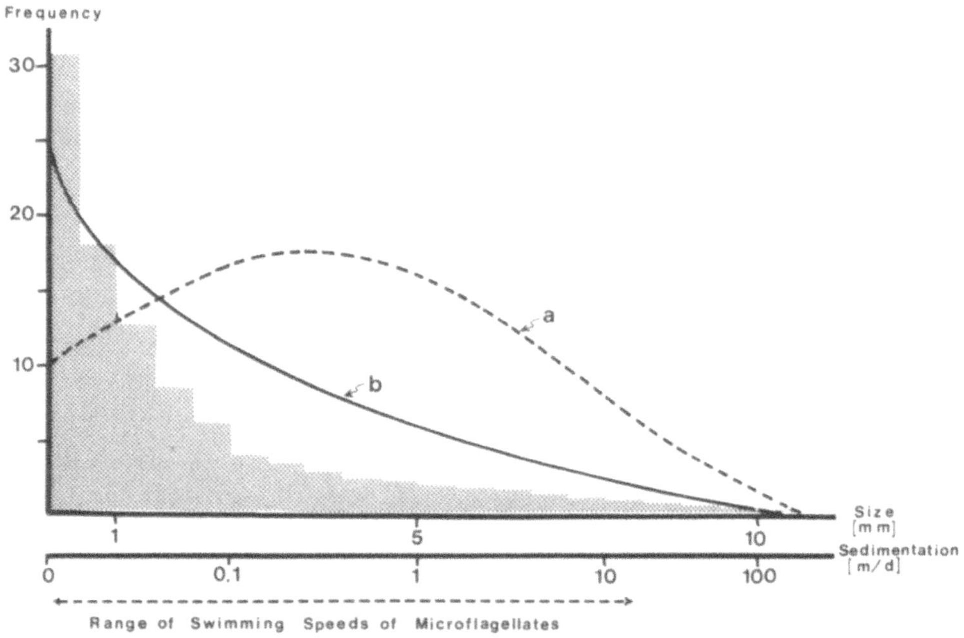

Fig. 2. Hypothetical frequency distributions of aggregate sizes and two different potential distributions of sedimentation rates. Small aggregates are generally more numerous than large ones. Depending on their excess density these aggregates may show different patterns of sedimentation rates:
a) when fast settling aggregates are more abundant, more organic matter will pass out of the upper mixed water column into the deep part of the water column ('leaky' system);
b) when the majority of aggregates settle slower than a few meters per day, most of the organic matter is likely to be recycled within the upper mixed water column ('retaining' system). The range of swimming speeds of microflagellates is shown in the same scale as the sedimentation rate

The above examples are certainly the extremes of a wide range of particle sizes and sinking speeds. The relative importance of the two opposing processes, degradation and sedimentation, is determined by the abundance of different sized particles having a certain distribution of sinking speeds. This is schematically expressed in figure 2. A certain size distribution of aggregates in the upper mixed water layer may have a different frequency distribution of sinking speeds. In case a) more rapidly settling particles are present compared to case b) which has a majority of slowly sinking ones. The former situation may be termed a 'leaky' system, where a certain fraction of particulate organic matter will leave the upper mixed zone, while the latter is a 'retaining' system with more of the particulate organic matter remaining suspended above the thermocline and being recycled in this water mass. Distance between particles and their residence times within a certain water depth determine the chances of protozoa to reach suitable particles and to take advantage of their nutrient contents.

In order to take advantage of the enriched microzones which the aggregates represent in the open ocean, protozoa have to possess the following characeristics, typical of opportunistic organisms: 1) rapid swimming; 2) chemotaxis with high sensitivity; 3) rapid food intake and proliferation; 4) tolerance towards starvation, e.g. encystment. However, we know very little about the taxonomy of the protozoa associated with particles, their feeding and response to times of famine, motility, chemotactic responses and their distribution within the water column. Knowledge of the interaction between protozoa and aggregates is the key to understanding their survival in the pelagic open ocean as well as important aspects of nutrient regeneration.

ACKNOWLEDGEMENTS

BIOTRANS-publication number 19. The work was supported by funds from the Minister für Forschung und Technologie der Bundesrepublik Deutschland (MFU 0572/1).

REFERENCES

Alldredge AL (1976) Discarded appendicularian houses as sources of food, surface habitats, and particulate organic matter

in planktonic environments. Limnol Oceanogr 21:14-23
Alldredge AL (1979) The chemical composition of macroscopic aggregates in two neritic seas. Limnol Oceanogr 24:855-866
Alldredge AL, Cohen Y (1987) Can microscale chemical patches persist in the sea? Microelectrode study of marine snow, fecal pellets. Science 235:689-691
Alldredge AL, Cole JJ, Caron DA (1986) Production of heterotrophic bacteria inhabiting macroscopic organic aggregates (marine snow) from surface waters. Limnol Oceanogr 31:68-78
Alldredge AL, Cox JL (1982) Primary productivity and chemical composition of marine snow in surface waters of the Southern California Bight. J Mar Res 40:517-527
Alldredge AL, Gotschalk CC, MacIntyre S (1987) Evidence for sustained residence of macrocrustacean fecal pellets in surface waters off Southern California. Deep-Sea Res 34:1641-1652
Alldredge AL, Hartwig EO (eds) (1986) Aggregate dynamics in the sea, workshop report. Office of Naval Research, Asilomar Conference Center, Pacific Grove, California, September 22-24, 1986. American Institute of Biological Sciences. p 211
Alldredge AL, Silver MW (1988) Characteristics, dynamics and significance of marine snow. Prog Oceanog 20:41-82
Alldredge AL, Youngbluth MJ (1985) The significance of macroscopic aggregates (marine snow) as sites for heterotrophic bacterial production in the mesopelagic zone of the subtropical Atlantic. Deep-Sea Res 32:1445-1456
Angel MV (1984) Detrital organic fluxes through pelagic ecosystems. In: Fasham MJR (ed)Flows of energy and materials in marine ecosystems, theory and practice. Plenum Press, New York, p 547
Asper VL (1987) Measuring the flux and sinking speed of marine snow aggregates. Deep-Sea Res 34:1-17
Azam F, Ammerman JW (1984) Cycling of organic matter by bacterioplankton in pelagic marine ecosystems: microenvironmental considerations. In: Fasham MJR (ed)Flows of energy and materials in marine ecosystems, theory and practice. Plenum Press, New York, p 345
Barham EG (1979) Giant larvacean houses: observations from deep submersibles. Science 205:1129-1131
Beers JR, Trent JD, Reid FMH, Shanks AL (1986) Macroaggregates and their phytoplanktonic components in the Southern California Bight. J Plankton Res 8:475-487
Biddanda BA (1985) Microbial synthesis of macroparticulate matter. Mar Ecol Prog Ser 20:241-251
Biddanda BA, Pomeroy LR (1988) Microbial aggregation and degradation of phytoplankton-derived detritus in seawater. I. Microbial succession. Mar Ecol Prog Ser 42:79-88
Billett DSM, Lampitt RS, Rice AL, Mantoura RFC (1983) Seasonal sedimentation of phytoplankton to the deep-sea benthos. Nature 302:520-522
Bruland KW, Silver MW (1981) Sinking rates of fecal pellets from gelatinous zooplankton (salps, pteropods, doliolids). Mar Biol 63:295-300
Caron DA, Davis PG, Madin LP, Sieburth JMcN (1982) Heterotrophic bacteria and bacterivorous protozoa in oceanic macroaggregates. Science 218:795-797
Caron DA, Davis PG, Madin LP, Sieburth JMcN (1986) Enrichment of microbial populations in macroaggregates (marine snow) from surface waters of the North Atlantic. J Mar Res 44:543-565
Caron DA, Goldman JC, Andersen OK, Dennett MR (1985) Nutrient cycling in a microflagellate food chain: II. Population dynamics and carbon cycling. Mar Ecol Prog Ser 24:243-254
Cole JJ, Honjo S, Caraco N (1985) Seasonal variation in the flux of algal pigments to a deep-water site in the Panama Basin. Hydrobiologia 122:193-197
Davoll PJ, Silver MW (1986) Marine snow aggregates: life history sequence and microbial community of abandoned larvacean houses from Monterey Bay, California. Mar Ecol Prog Ser 33:111-120
Ducklow HW (1983) Production and fate of bacteria in the oceans. BioScience 33:494-501
Fenchel T (1980) Relation between particle size selection and clearance in suspension feeding ciliates. Limnol Oceanogr 25:735-740

Fenchel T (1982a) Ecology of heterotrophic microflagellates. I. Some important forms and their functional morphology. Mar Ecol Prog Ser 8:211-223

Fenchel T (1982b) Ecology of heterotrophic microflagellates. IV. Quantitative occurrence and importance as bacterial consumers. Mar Ecol Prog Ser 9:35-42

Fowler SW, Knauer GA (1986) Role of large particles in the transport of elements and organic compounds through the oceanic water column. Prog Oceanogr 16:147-194

Goldman JC (1984) Conceptual role for microaggregates in pelagic waters. Bull Mar Sci 35:462-476

Goldman JC, Caron DA, Andersen OK, Dennett MR (1985) Nutrient cycling in a microflagellate food chain: I. Nitrogen dynamics. Mar Ecol Prog Ser 24:231-242

Gooday AJ (1988) A response by benthic Foraminifera to the deposition of phytodetritus in the deep-sea. Nature 332:70-73

Gordon DC (1970) A microscopic study of organic particles in the North Atlantic Ocean. Deep-Sea Res 17:175-185

Gowing MM (1986) Trophic biology of phaeodarian radiolarians and flux of living radiolarians in the upper 2000 m of the North Pacific central gyre. Deep-Sea Res 33:655-674

Gowing MM, Silver MW (1985) Minipellets: a new and abundant size class of marine fecal pellets. J Mar Res 43:395-418

Honjo S, Doherty KW, Agrawal YC, Asper VL (1984) Direct optical assessment of large amorphous aggregates (marine snow) in the deep ocean. Deep-Sea Res 31:67-76

Johannes RE (1967) Ecology of organic aggregates in the vicinity of a coral reef. Limnol Oceanogr 12:189-195

Lal D (1977) The oceanic microcosm of particles. Science 198:997- 1009

Lande R, Wood AM (1987) Suspension times of particles in the upper ocean. Deep-Sea Res 34:61-72

Linley EAS, Newell RC (1984) Estimates of bacterial growth yields based on plankton detritus. Bull Mar Sci 35:409-425

Lochte K, Pfannkuche O (1987) Cyclonic cold-core eddy in the eastern North Atlantic. II. Nutrients, phytoplankton and bacterioplankton. Mar Ecol Prog Ser 39:153-164

Lochte K, Turley CM (1988) Bacteria and cyanobacteria associated with phytodetritus in the deep sea. Nature 333:67-69

Nishizawa S (1966) Suspended material in the sea: from detritus to symbiotic microcosmos. Bull Plank Soc Japan 13:1-33

Nyffeler F, Godet CH (1986) The structural parameters of the benthic nepheloid layer in the northeast Atlantic. Deep-Sea Res 33:195-207

Paerl HW (1978) Microbial organic carbon recovery in aquatic ecosystems. Limnol Oceanogr 23:927-935

Paerl HW (1984) Alteration of microbial metabolic activities in association with detritus. Bull Mar Sci 35:393-408

Pomeroy LR, Deibel D (1980) Aggregation of organic matter by pelagic tunicates. Limnol Oceanogr 25:643-652

Pomeroy LR, Hanson RB, McGillivary PA, Sherr BF, Kirchman D, Deibel D (1984) Microbiology and chemistry of fecal products of pelagic tunicates: rates and fates. Bull Mar Sci 35:426-439

Prezelin BB, Alldredge AL (1983) Primary production of marine snow during and after an upwelling event. Limnol Oceanogr 28:156-167

Rice AL, Billett DSM, Fry J, John AWG, Lampitt RS, Mantoura RFC, Morris RJ (1986) Seasonal deposition of phytodetritus to the deep-sea floor. Proc R Soc Edinburgh 88B:265-279

Riemann F (1989) Gelatinous phytoplankton detritus aggregates on the Atlantic deep-sea bed. Structure and mode of formation. Mar Biol 100:533-539

Riley GA (1963) Organic aggregates in seawater and the dynamics of their formation and utilization. Limnol Oceanogr 8:372-381

Seki H (1971) Microbial clumps in seawater in the euphotic zone of Saanich Inlet (British Columbia). Mar Biol 9:4-8

Shanks AL, Trent JD (1979) Marine snow: Microscale nutrient patches. Limnol Oceanogr 24:850-854

Sherr EB, Sherr BF (1987) High rates of consumption of bacteria by pelagic ciliates. Nature 325:710-711

Sieburth JMcN (1984) Protozoan bacterivory in pelagic marine waters. In: Hobbie JE, Williams PJLeB (eds) Heterotrophic activity in the sea. Plenum Press, New York, p 405

Silver MW, Alldredge AL (1981) Bathypelagic marine snow: deep-sea algal and detrital community. J Mar Res 39:501-530

Siver MW, Gowing MM, Brownlee DC, Corliss JO (1984) Ciliated protozoa associated with oceanic sinking detritus. Nature 309:246-248

Silver MW, Shanks AL, Trent JD (1978) Marine snow: microplankton habitat and source of small-scale patchiness in pelagic populations. Science 201:371-373

Small EB, Neun B, Caron D, Davis P (1983) Ciliates associated with Gulf Stream 'snow'. J Protozool, Abstracts 30:15A-16A

Smetacek V (1984) The supply of food to the benthos. In: Fasham MJR (ed) Flows of energy and materials in marine ecosystems, theory and practice. Plenum Press, New York, p 517

Smetacek V (1985) Role of sinking diatom life-history cycles: ecological, evolutionary and geological significance. Mar Biol 84:239-251

Smetacek V, Bröckel K von, Zeitzschel B, Zenk W (1978) Sedimentation of particulate matter during a phytoplankton spring bloom in relation to the hydrographcal regime. Mar Biol 47:211-226

Smith KL, Carlucci AF, Jahnke RA, Craven DB (1987) Organic carbon mineralization in the Sánta Catalina Basin: benthic boundary layer metabolism. Deep-Sea Res 34:185-211

Suzuki N, Kato K (1953) Studies on suspended materials *marine snow* in the sea. Part I. Sources of *marine snow*. Bull Fac Fish Hokkaido Univ 4:132-135

Takahashi K, Honjo S (1981) Vertical flux of radiolaria: a taxon-quantitative sediment trap study from the western tropical Atlantic. Micropalaeontol 27:140-190

Takahashi K, Honjo S (1983) Radiolarian skeletons: size, weight, sinking speed and residence time in tropical pelagic oceans. Deep-Sea Res 30:543-568

Taylor GT, Karl DM, Pace ML (1986) Impact of bacteria and zooflagellates on the composition of sinking particles: an *in situ* experiment. Mar Ecol Prog Ser 29:141-155

Thiel H, Pfannkuche O, Schriever G, Lochte K, Gooday AJ, Hemleben C, Mantoura RFC, Turley CM, Patching JW, Riemann F. Phytodetritus on the deep-sea floor in a central oceanic region of the northeast Atlantic. Biol Oceanogr, 6:203-239

Throndsen J (1973) Motility in some marine nanoplankton flagellates. Norw J Zool 21:193-200

Trent JD, Shanks AL, Silver MW (1978) In situ and laboratory measurements on macroscopic aggregates in Monterey Bay, California. Limnol Oceanogr 23:626-635

Tsujita T (1953) A preliminary study on naturally occurring suspended organic matter in waters adjacent to Japan. J Oceanogr Soc Japan 8:113-125

Turley CM, Lochte K, Patterson DJ (1988) A barophilic flagellate isolated from 4500 m in the mid-North Atlanic. Deep-Sea Res 35:1079-1092

Urrére MA, Knauer GA (1981) Zooplankton fecal pellet fluxes and vertical transport of particulate organic material in the pelagic environment. J Plankon Res 3:369-387

Wakeham SG, Canuel EA (1988) Organic geochemistry of particulate matter in the eastern tropical North Pacific Ocean: implications for particle dynamics. J Mar Res 46:183-213

Wiebe WJ, Pomeroy LR (1972) Microorganisms and their association with aggregates and detritus in the sea: a microscopic study. Mem Ist Ital Idrobiol 29 Suppl:325-352

MINERAL FLUX AND BIOGEOCHEMICAL CYCLES OF MARINE PLANKTONIC PROTOZOA - SESSION SUMMARY

Kozo Takahashi
Woods Hole Oceanographic Institution
Woods Hole
MA 02543
USA

INTRODUCTION

Since the first NATO-ASI Workshop on the Ecology of Marine Planktonic Protozoa was held in Villefranch-sur-mer in 1981, a marked advance has been made in the study of shell-bearing marine planktonic protozoa. In particular, vertical flux measurements using sediment traps (e.g. Deuser *et al.* 1981, Takahashi and Honjo 1981, Reid 1982) have contributed to the understanding of marine ecosystems, and the material balance, and seasonal, interannual, and spatial distribution of plankton (e.g. Thunell *et al.* 1983, Bè *et al.* 1985, Smetacek 1985, Pisias *et al.* 1986, Takahashi 1986, 1987a b c, Thunell and Honjo 1987, Leventer and Dunbar 1987, Gersonde and Wefer 1987, Sancetta and Calver 1988.

Fluxes of biogenically precipitated minerals in the oceans, largely represented by biogenic silica and calcium carbonate, are important in global ecosystems. Both of the above minerals are produced in large quantities in marine environments and are partially entered into the geologic record. It should be noted that oceanic upper layer biological productivity has changed with time in the past: for example, significantly different productivity levels from those of the present day occurred in parts of the world oceans during the maximum extent of the last glacial age (Sundquist and Broecker 1985). Environmental shifts will likely cause changes not only in productivity but also in the preservation of sedimenting biogenic minerals to the sea-floor (Andersen and Malahoff 1977). Between production and preservation there is some remineralization of material into the water. The degree of remineralization will cause changes in productivity in a feed back manner. These

NATO ASI Series, Vol. G 25
Protozoa and Their Role in Marine Processes
Edited by P. C. Reid et al.

processes, and in partiuclar the fate of carbon dioxide produced by anthropogenic fossil fuel combustion and incorporated into the planktonic system, have great significance for global climate change.

SYNTHESIS OF SESSION PAPERS

Barry Leadbeater (1990) presented the keynote lecture of the mineral flux session. The focal points of this lecture were the silica and calcium carbonate cycles in the oceans and the fact that several planktonic protozoan groups play important roles in these cycles. Selected questions from the discussion were:

Q: Fecal pellets form a microenvironment within themselves. How does this affect dissolution of silica? (K Lochte)

A: Guts of copepods as well as fecal pellets themselves have been studied. Neither chemical or mechanical effects have been reported to be effective in the dissolution of silica. It seems that fecal pellets are protective environments for silica just as for carbonates (no etching or dissolution of coccoliths was reported from fresh fecal pellets with peritrophic membranes).

Q: Carbonate shell-bearing organisms are much more important in terms of transmitting carbon through the ocean, but how much do we know about factors governing production and fate of siliceous and carbonate organisms in the oceans? (H Ducklow)

A: We do not know much. One thing we know is that dissolution of silica will affect the dissolution of carbonate components. It should not be neglected that large amount of calcium carbonate are deposited by large-sized benthic foraminifera around the edges of the seas. (J Lee)

Kozo Takahashi presented results from the Pelagic Opal program in the subarctic Pacific which focused first on seasonal signals, and then on silica and organic carbon partition among siliceous plankton groups. Vertical fluxes of all the major siliceous microplankton particles were measured almost continuously for four years from 1982 through 1986 in the

Alaskan gyre system where high seasonal amplitudes of hydrographic parameters are known. Sediment traps were deployed at 3800 m for all of the sampling periods and at 1000 m for a total of one and half years in the 4200 m water column. The results clearly demonstrated a high amplitude of not only seasonal but also interannual flux variability, implying significant temporary changes in upper layer productivity. The sinking siliceous plankton particles include 18 diatom, 7 silicoflagellate, 1 dinoflagellate and 102 radiolarian taxa. The time-series fluxes of all of these organisms were enumerated and compiled. Seasonal and productivity indicator taxa were indentifed based on temporal flux signals. For example, *Lithomelissa setosa*, a nassellarian radiolarian, consistently increased its flux during spring time, around May, with considerable interannual flux differences in the four years of record (Figure 1). The maximum flux of *Rhizosolenia styliformis*, a large diatom, occurred during summer periods, indicating requirements for high temperature, high light intensity, and intermediate nutrient concentrations.

For global silica and carbon cycles it is important to understand the seasonal flux partition of diatoms, radiolarians and others since the production and fate of biogenic silica and organic carbon differ depending on organism groups. For example, dissolution rates of polycystine and phaeodarian Radiolaria are contrastingly different (Erez *et al.* 1982). High dissolution rates of phaeodarians cause complete dissolution of their skeletons during sinking in the water column or on the sea-floor whereas lower rates for polycystines result in partial preservation of polycystine skeletons in the fossil record (Takahashi *et al.* 1983). Previous studies on the partition of biogenic silica have emphasized the predominant contribution of diatoms e.g. 99.8 wt % in the Antarctic ocean based on discrete bottle castings (Lisitzin 1972). However, the present results from the subarctic Pacific clearly demonstrate a high seasonal variation of the partition. Diatoms are certainly the major contributors of biogenic silica flux in certain seasons such as early spring when their flux

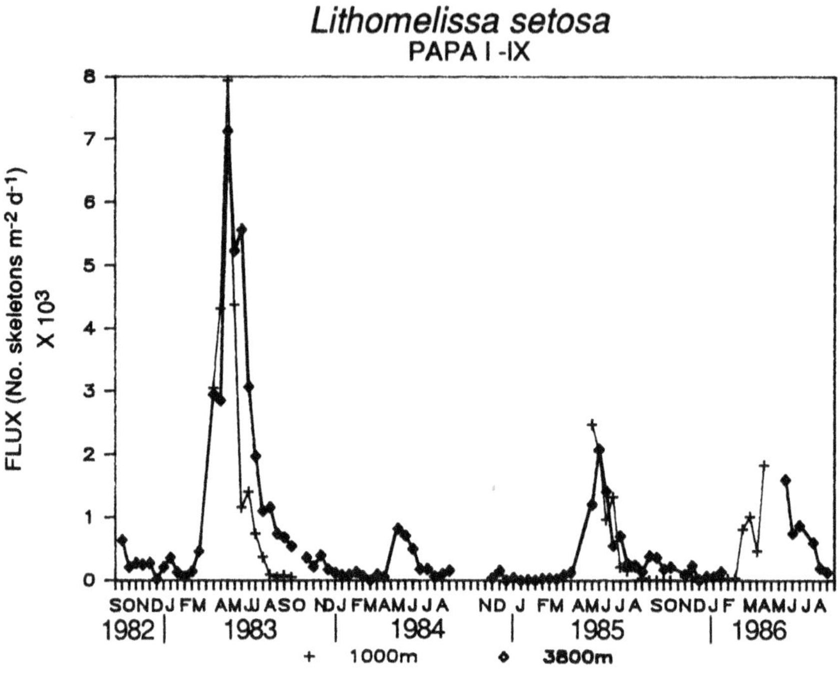

Fig. 1. Time-series flux of *Lithomelissa setosa*, a nassellarian Radiolarian, measured at 3800 m and 1000 m (Station Papa, 50°N 145°W, water depth 4200 m) in the eastern subarctic Pacific during 1982 through 1986. Note the clear spring seasonal signals exhibited by the flux of this taxon. The high variability of interannual flux observed here is due to significant change in hydrography (Takahashi 1987b,c)

forms approximately 90 wt % of total biogenic silica while radiolarians constitute the remaining 10 wt %. On the other hand, radiolarian contributions reach maximal values of approximately 50 wt % around late spring to early summer (Figure 2). Within the radiolarian flux fraction a significant portion (37 wt % on the average) was contributed by phaeodarian Radiolaria (Figure 3) whose dissolution rate is greater than the remaining polycystine Radiolaria and thus has a significant effect on the silica cycle. Dissolution effects during sinking have been taken into account in the calculated percent seasonal production of phaeodaria. When phaeodarians sink from the main production layer at the surface (i.e. 100 m) to 1000 m, approximately 18 wt % of their shells are lost; a total of 75 wt % is lost by the time they reach 3800 m depth. Large colonial phaeodarian species deserve particular attention because of their mass and consequent rapid vertical transport of readily soluble biogenic silica to the deep-sea (Takahashi 1987c).

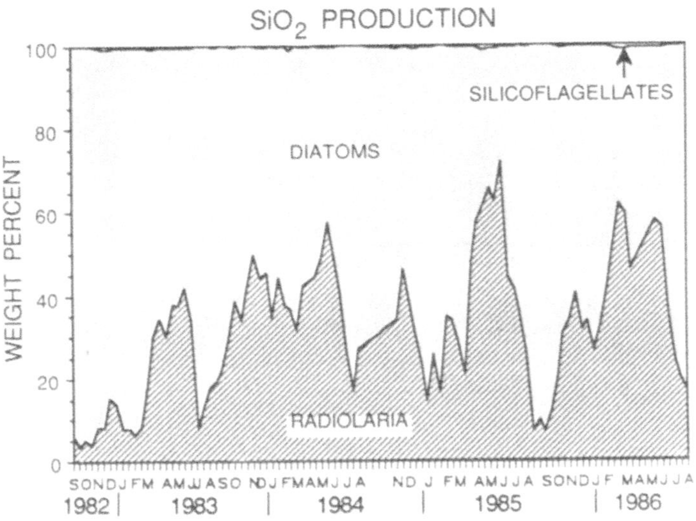

Fig. 2. Time-series partition (wt %) of biogenic silica production among diatoms, radiolarians and silicoflagellates. The percent values (most of them calculated from 3800 m) are shifted one sample toward earlier dates to reflect surface production, taking into account their sinking speeds. Dissolution effect of phaeodarian skeletons has been taken into account to reconstruct total radiolarian silica production in the upper layer (see Figure 3 caption for more details)

Silicoflagellates also have been studied for details of their seasonal fluxes, mass and carbon and silica partition. Among seven taxa, *Distephanus speculum* and *Dictyocha mandrai* are the two dominant components in the flux, representing greater than 75% of the total silicoflagellate shell numbers throughout the seasons. Due to its large size, *D. mandrai* contributes the major amount of carbon and silica among the total silicoflagellate silica flux. With respect to total biogenic silica the silicoflagellate silica contributes only 0.02 to 1.2 wt % with a mean value of 0.4 wt %. Furthermore, silicoflagellate organic carbon production is large enough to deserve attention: 0.5 to 10.5 wt % with a mean of 3.1 wt % of total carbon produced by the sum of all major siliceous plankton (diatoms, silicoflagellates and radiolarians) in the upper water layers.

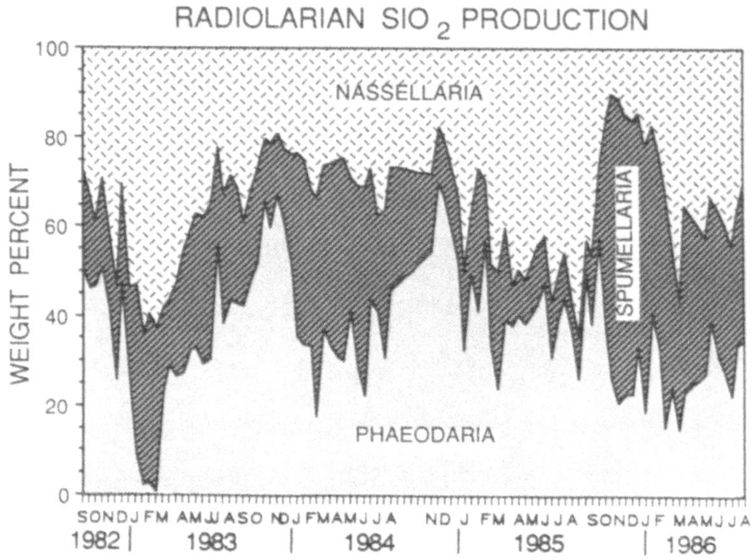

Fig. 3. Time-series partition (wt %) of radiolarian silica production among Spumellaria, Nassellaria, and Phaeodaria. Note that phaeodarian contributions are generally high (mean: 37 wt %) and that they significantly change with seasons. Dissolution loss on phaeodarians between upper layer (ie 100m) and 3800 m has been corrected for thier relative production in the upper water. The percent values (most of them calculated from 3800 m) are shifted one sample toward earlier dates to reflect surface production, taking into account their sinking speeds

The temporal variability of the partition has been clearly depicted from the above seasonal data. Partition among siliceous plankton fluxes would be spatially and temporally different elsewhere depending on environmental conditions and ecosystems, necessitating wide coverage by future studies in the world oceans.

Q: You have reported the fluxes of representative radiolarian species besides other siliceous organisms in terms of thousands of shells per square meter per day. What are the sizes of the radiolarians and how do they affect biomass transport? (D Caron)

A: Radiolarian skeletal sizes range from 30 µm for nassellarian species up to ~cm for colonial phaeodarians. Because of the large mass of carbon and silica per unit cell of large species they tend to play an important role in material transport depending on the flux values of shell numbers. As a result, mass flux of radiolarian silica is rather evenly distributed among several dozen taxa of mainly 50 µm to 500 µm in diameter.

Q: Why are dissolution rates of phaeodarian and polycystine radiolarians so different? (K Lochte)

A: There are two major reasons why that is the case. First, morphologic integrity of the basic structure of silica granules, which are on the order of 100 Angstroms in diameter, are markedly different between the two groups based on our previous TEM studies conducted on plankton tows, sediment trap and fossil samples. Second, the chemical nature of the silica such as water content may be different. Also, as **Barry Leadbeater** pointed out, incorporation of organic molecules in siliceous skeletons may play an important role in the dissolution process.

Jelle Bijma discussed the life cycle of *Globigerinoides sacculifer* (Brady), a spinose planktonic foraminifera, which has a profound effect on calcium carbonate flux. Because planktonic foraminifera are one of the major calcium carbonate producers they are an important component for carbonate flux to the sea-floor. *G. sacculifer* lives in the euphotic zone with symbiotic organisms. As it grows it adds calcite chambers of spherical shape. High potential growth and a short life span results in high abundance of this species in tropical and sub-tropical sediments. Within 24 to 48 hours following sac-like chamber formation, gametogenesis takes place. It is essential for this species to live in the euphotic zone and grow and then sink toward deeper water to reproduce. Reproduction depths and lengths of life cycle are species specific. It is known that *G. sacculifer, Hastigerina pelagica,* and *Orbulina universa* respond to the lunar cycle in terms of their reproduction

(Hemleben and Spindler 1983). In the study region Gulf of Elat/Aquaba, time-series plankton tows were taken every two days for the analysis of population and size spectra data on *G. sacculifer*. Soon after a full moon the population in the surface waters decreases drastically and increases again one week after the full moon. Size classes change with time also. Statistical residual analysis was performed to clarify the relative frequency of the populations. Maximum output values of empty shells just before the full moon can be implied from the data. Juvenile forms are found just after the full moon. Such a lunar cyclicity of *G. sacculifer* reproduction results in pulses of calcium carbonate output occurring around every full moon (Bijma *et al.* in press).

Q: Does this species' standing stock change annually? Does *G. sacculifer* appears only for part of the year in your study area? (V Smetacek)
A: At least in Elat *G. sacculifer* lives throughout the year. The standing stock changes seasonally only to a small degree. During winter months the standing stock becomes less and other species move in. Compared to other parts of the world oceans the seasonal changes are small.
Comment: Your study can be viewed as analogous to synchronized cultures of planktonic organisms in the laboratory. Where reproduction occurs just before the optimal condition and then later decreases again. (M Legner)
Q: The reason why they have such a lunar cycle is because of the maximum potential for sexual reproduction which is synchronized for the population. Whether it is indigenous or not can be discussed by comparing it with *H. pelagica*. (D Caron)
A: *H. pelagica* carries indigenous rhythms in culture, but *G. sacculifer* does not respond to the timing of the full moon in the laboratory. This species undergoes gametogenesis whenever it can accomplish it. This is partially related to sampling; depending on when you catch it, *G. sacculifer* will spend another 8 days in the laboratory before reproducing.

Eva Nöthig described particle sedimentation of protozoan minipellets (Gowing and Silver 1985) from the Weddell Sea region in the Antarctica. The minipellets enhance silica sedimentation but their organic carbon content is low since many diatoms contained in the pellets are devoid of protoplasm. The minipellets of unidentified protozoan origin are abundant in the vertical flux of materials collected from the Weddell Sea. Analogous to **Dean Jacobsen**'s observation of the extra-cellular feeding on diatoms by *Protoperidinium*, phytoplankton particles without protoplasm, but with intact shell structures may be common in the oceans. The contribution of minipellets to total flux is significant in the Weddell sea. Multisample sediment traps of the Honjo type (Honjo and Doherty 1988) were used as well as discrete water sampling in two different regions, one on the continental margin and the other in the central pelagic Weddell Sea with over 4 Km water depth. Two investigations at the margin were conducted 10 days apart. Drifting sediment traps at 100 m depth and moored traps were deployed for one half day to one day length. Ciliates which are important in the high southern latitudes formed a considerable part of the protozooplankton biomass. Minipellets of protozoan origin were normally between 50 µm and 100 µm in diameter. They consisted of diatom frustules and cell membranes. The pellets observed earlier in shallower depths were sampled 10 days later at 300 m to 500 m. This represents a sinking speed of 50 m d^{-1}. Comparisons of trap and suspended material show that their sinking speeds were between 10 to 50 m d^{-1}. Approximately 10 to 30% of the surface minipellet standing stock sinks from the euphotic layer daily making a large contribution to the vertical flux based on biovolume computations (Nöthig and Bodungen in prep). A multisampling sediment trap (Honjo and Doherty 1988) was moored at 860 m in the central Weddell Sea. The bulk of sedimenting materials was collected in March (Fisher *et al.* in press). Their constituents were similar to those of the trap deployed at the shallower site at 100 m, confirming the importance of minipellets in the Weddell Sea. The sedimentation rate observed in the central Weddell Sea is the lowest recorded in the world. The conclusion, therefore, is that these minipellets are important in vertical flux and that they are likely to exist elsewhere. Their contribution may be obscured by sinking aggregates such as marine snow composed of phytoplankton particles.

CONCLUSIONS AND AREAS FOR FUTURE RESEARCH

A comprehensive geographic coverage of studies of biogenic mineral flux will be required to enhance our understanding of the global cycles of silica and calcium carbonate. Priority areas for research which may have global significance include the following. Antarctic regions between 50° and 60°S siliceous belts should be thoroughly investigated since a large quantity of silica production and sedimentation takes place there. No seasonal data exist and winter data are virtually non-existence due to the logistical difficulties of operating in this area. Time-series measurements assisted by state of the art equipment will be needed in the future. The equatorial high productivity belt is globally important for both silica and carbonate systems because of the large area of oceans covered. Studies of the biogenic silica flux partition between diatoms and Radiolaria in the equatorial region, where fates may be different should be researched and modeling of the silica cycle is still to be achieved.

As **Barry Leadbeater** pointed out in the keynote lecture, and **Chris Reid** illustrated in his introductory presentation, the importance of the production and fate of coccolithophores must now be evident to marine scientists who are interested in material cycles in the oceans. Quantitatively, coccolithophores may contribute an order of magnitude more calcium carbonate production than planktonic foraminifera based on recent foraminiferal (Thunell and Honjo 1987) and coccolithophore flux studies (Steinmetz in press) which revise an earlier estimate of two orders of magnitude difference (Honjo 1977). Besides quantitative studies of calcium carbonate production and fate, studies of species fluxes on a seasonal and interannual basis need to be conducted in the future since ecological and climatic signals may be carried by them as shown in siliceous plankton (Takahashi 1986, 1987a, and c). Partitioning among coccolithophores, planktonic foraminifera and pteropods would greatly enhance our understanding of the calcium carbonate cycle since their respective dissolution rates are significantly different.

The modulation of particle sedimentation by specific organisms must be further clarified. One of the most important findings in the recent past is that sedimentation of particles is mainly mediated by large aggregates and the resulting accelerated sinking speeds are greater by orders of magnitude than small discrete particles. Faecal pellets, marine snow and other aggregate sinking will be a major focus for research in the next several years.

Many of the themes discussed in this session will be investigated by international teams of scientists participating in the Joint Global Ocean Flux Study (JGOFS) Program (e.g. Brewer *et al.* 1986). In JGOFS, pilot studies in the Atlantic will be conducted during 1989 through 1991 along 20°W between 33°N and 60°N on the relationships between hydrographic forcing such as deep winter mixing and the size of the spring plankton bloom. The fate of the spring bloom must be understood in order to decipher the material cycles of many important elements. There is no doubt that studies of protista, including planktonic protozoa such as the ones discussed in this session, will greatly contribute to our future understanding of a variety of ecosystems in the world oceans.

ACKNOWLEDGEMENTS

This work was funded by US National Science Foundation Grants OCE-86-08255. Funds to participate in the NATO-ASI Workshop were provided by NATO, US NSF and ONR. This is Woods Hole Oceanographic Institution Contribution 6878.

REFERENCES

Andersen NR, Malahoff A (1977) The Fate of Fossil Fuel CO_2. Plenum Press, New York

Bè AWH, Bishop JKB, Sverdlove MS, Gardner WD (1985) Standing stock, vertical distribution and flux of planktonic foraminifera in the Panama Basin. Marine Micropaleontology 9:307-333
Bijma J, Erez J, Hemleben C. Circalunar reproduction cycles in planktonic Foramininifera in the Gulf of Elat/Aquaba and off Barbados, in press
Brewer PG, Bruland RW, Eppley RW, McCarthy JJ (1986) The Global Ocean Flux Study (GOFS): Status of the US GOFS program EOS Trans AGU 67:827-832
Deuser WG, Ross EH, Hemleben C, Spindler M (1981) Seasonal changes in species composition, numbers mass, size, and isotopic composition, of planktonic foraminifera settling into the deep Sargasso Sea, Paleogeogr Paleoclimatol Paleoecol 33:103-127
Erez J, Takahashi K, Honjo S (1982) *In Situ* dissolution of Radiolaria in the central North Pacific Ocean, Earth Planet Sci Lett 59:245-254
Gersonde R, Wefer G (1987) Sedimentation of biogenic siliceous particles in Antarctic waters from the Atlantic sector, Mar Micropaleontol, in press
Gowing MM, Silver MW (1985) Minipellets: a new and abundant size class of marine fecal pellets. Journal of Marine Research 43:395-418
Fischer G, Futterer D, Gersonde R, Honjo D, Osterman D, Wefer G. Seasonal variability of particle flux in the Weddell Sea and its relation to ice cover. Nature, in press
Hemleben C, Spindler M (1983) Recent advances in research on living planktonic foraminifera. In: Meulenkamp JE (ed) Reconstruction of Marine Paleoenvironments Utrecht Micropaleontol Bull 30:141-170
Honjo S (1977) Biogenic carbonate particles in the oceans; do they dissolve in the water column? In: Andersen NR, Malahoff A (eds) Plenum Press, New York p 269-294
Honjo S (1984) Study of ocean fluxes in time and space by bottom-tethered sediment trap arrays: A recommendation, In: Global Ocean Flux Study: Proceedings of a Workshop, September 10-14, 1984 p 305-324 National Academy Press, Washington DC
Honjo S, Doherty KW (1988) Large aperture time series oceanic sediment traps; design objectives, construction, and application. Deep Sea Res 35:133-149
Leadbeater BSC (1990) Protozoa and mineral cycling in the sea. In: Reid PC, Turley CM, Burkill PH (eds) Protozoa and their role in marine processes. Springer, Berlin Heidelberg New York
Leventer A, Dunbar RB (1987) Diatom flux in McMurdo Sound, Antarctica, Mar Micropaleontol 12:49-64
Lisitzin AP (1972) Sedimentation in the World Oceans. Soc Econ Paleont Mineral Spec Publ No 17 p 218
Nöthig E-M, Bodungen BV, in prep. Sedimentation in the southeastern Weddell Sea (Antarctica) in January/February 1985
Pisias NG, Murray DW, Roelofs AK (1986) Radiolarian and silicoflagellate response to oceanographic changes associated with the 1983 El Nino. Nature 320:259-262
Reid PC (1982) Patterns of spatial and temporal variability of marine planktonic protozoa. Marine Pelagic Protozoa and Microzooplankton Ecology Ann Inst oceanogr Paris 58:(S) 179-190
Sancetta C, Calvert S E (1988) Vertical flux of diatom assemblages in Saanich Inlet, British Columbia, Deep-Sea Res 35:71-90
Smetacek V (1985) Role of sinking in diatom life-history cycles: ecological, evolutionary, and geographical significance. Marine Biology 84:239-251
Steinmetz JC in press. Calcareous nannoplankton biocoenosis: sediment trap studies in the Equatorial Atlantic, Central Pacific, and Panama Basin. In: Honjo S (ed) Ocean Biocoenosis: Micropaleontologically Relevant Plankton Species in Modern Oceans collected by Sediment Trap Experiments. Woods Hole Oceanographic Institution Press
Sundquist ET, Broecker WS (1985) The Carbon Cycle and Atmospheric CO_2: Natural Variation Archean to Present. Geophysical Monograph Series, 32 American Geophysical Union, Washington DC p 627

Takahashi K (1986) Seasonal fluxes of pelagic diatoms in the subarctic Pacific, 1982-1983, Deep Sea Res 33:1225-1251
Takahashi K (1987a) Seasonal fluxes of silicoflagellates and Actiniscus in the subarctic Pacific during 1982-1984 J Mar Res 45:397-425
Takahashi K (9187b) Response of subarctic Pacific diatom fluxes to the 1982-1983 El Nino disturbance J Geophys Res 92:14387-14392
Takahashi K (1987c) Radiolarian flux and seasonality: climatic and El Nino response in the subarctic Pacific, 1982-1984 Global Biogeochem Cycles 1:213-231
Takahashi K, Honjo S (1981) Vertical flux of Radiolaria: A taxon-quantitative sediment trap study from the western tropical Atlantic, Micropaleontology 22:140-190
Takahashi K, Hurd DC, Honjo S (1983) Phaeodarian skeletons: Their róle in silica transport to the deep-sea Science 222:616-618
Thunell RC, Curry WB, Honjo S (1983) Seasonal variation in the flux of planktonic foraminifera: time series sediment trap results from the Panama Basin, Earth Planet Science Lett 64:244-55
Thunel RC, Honjo S (1987) Seasonal and interannual changes in planktonic foraminiferal production in the North Pacific, Nature 328:335-337

PROTISTA AND MINERAL CYCLING IN THE SEA

Barry S C Leadbeater
School of Biological Sciences
University of Birmingham
PO Box 363
Birmingham B15 2TT
UK

INTRODUCTION

Biomineralization is the process by which living organisms assemble structures from naturally occurring inorganic compounds. The majority of groups of living organisms include representatives that deposit minerals and in most instances the mineralized structures provide skeletal support and protection for 'softer' organic parts (Leadbeater and Riding 1986). Although the range of inorganic compounds utilized by Protista includes most elements of the periodic table, the two most common and widespread minerals to be incorporated into structures are silica and calcium carbonate with strontium sulphate of relatively minor significance.

Within aquatic habitats Protista have considerable potential for modifying the direction and rate of elemental fluxes (Wollast 1974, Whitfield and Watson 1983, Reynolds 1986). This is particularly applicable to organisms that are abundant and utilise mineral compounds that are relatively stable and of limited availability. Protista are responsible for rapidly transporting minerals from surface waters to the depths of the oceans and they may remove them altogether from aquatic systems by depositing them in the sediments. The deposits of chalk, limestone and diatomaceous earths which form part of our geological landscape bear witness to the magnitude of the role of the Protista in mineral sedimentation.

This review is principally concerned with the role that Protista play in the flux of silica and calcium carbonate within the oceans. Whilst both minerals are used for skeletal purposes and both become involved in major oceanic biogenic

NATO ASI Series, Vol. G 25
Protozoa and Their Role in Marine Processes
Edited by P. C. Reid et al.

fluxes the solution chemistries of calcite and silica are quite different. In surface waters calcite is never limiting whereas silica is rarely if ever present in saturating quantities. Silica becomes less soluble at greater depths whereas calcite becomes more soluble. Biogenic siliceous deposits reflect biological activity in surface waters whereas biogenic calcite deposits reflect the preservation of calcite at depth. The differences in behaviour between calcite and silica in oceanic systems makes it more convenient to consider these minerals separately and briefly make comparisons at the end.

Table 1. MAJOR GROUPS OF PROTISTA THAT ASSEMBLE SILICA (Si) AND CALCITE (Ca) STRUCTURES. DETAILS OF MINERALIZED STRUCTURE, NATURE OF AQUATIC HABITAT: F FRESHWATER, M MARINE. MODE OF NUTRITION: A AUTOTROPHIC, P PHAGOTROPY, S CONTAINING PHOTOAUTROPHIC SYMBIONTS. DISTRIBUTION IN PLANKTON AND SEDIMENTS: U UNIVERSAL, X AT TIMES ABUNDANT, T ABUNDANT IN TROPICS, NR NOT REPORTED, FWC FRESHWATER CYSTS.

PROTOZOAN	MINERAL	STRUCTURE	HABITAT	NUTRITIONAL MODE	PLANKTON DISTRIB^{n}	SEDIMENT DISTRIB^{n}
Choanoflagellates	Si	lorica	M	A	U,X	NR
Chrysophytes	Si	scales, cysts	F,M	A	U	FWC
Coccolithophorids	Ca	coccoliths	F,M	A	U,X,T	U,X
Diatoms	Si	frustule	F,M	A	U,X	U
Foraminiferans	Ca	skeleton	M	P,S	U,T	U,X
Heliozoa	Si	scales	F,M	P	U	NR
Radiolaria	Si	skeleton	M	P,S	U,T	X
Silicoflagellates	Si	skeleton	M	A	U,X	X
Testate amoebae	Si	scales	F	P	U	NR

MINERAL DEPOSITING PROTISTA

Table 1 lists the major groups of Protista that assemble silica and calcium carbonate structures. Of the Protista which utilize silica, diatoms are the most abundant in both the plankton and in siliceous sediments of biological origin (Whitfield and Watson 1983). Diatoms are universally present in the oceans with the highest levels of production in temperate latitudes and where there is major upwelling such as the Antarctic convergence, west coasts of South Africa and South America, Gulf Stream, subarctic Pacific and in the equatorial zones (Lisitzin 1972). Together planktonic and epipelic (benthic) diatoms may account for as much as 25% of the world's net primary production (Volcani 1981).

Radiolaria are holoplanktonic Protista widely distributed in the oceans although they are generally more abundant in warmer waters. Concentrations vary greatly but in surface waters of the central Pacific Ocean at temperatures between 23-28°C they range from 5-15 x 10^3 specimens m^{-3} (Figure 3a) (Anderson 1983, Takahashi 1983). In the tropical Pacific, Radiolaria may account for 62-99% of the weight of biogenic opal (Lisitzin 1972). Radiolaria are subdivided into Phaeodaria and Polycystina on the basis of skeletal morphology and capsular substructure (Lee *et al.* 1985).

Most marine planktonic silica-depositing chrysophytes are phagotrophic and belong to the genus *Paraphysomonas*. Together with loricate choanoflagellates (Acanthoecidae) they are universally distributed in marine and brackish water environments (Leadbeater 1981, 1986). They may be locally abundant and in some waters, especially Arctic and Antarctic, they may be the predominant group of plankton.

Of the calcium carbonate depositing Protista only coccolithophorids and Foraminifera are of major importance. Coccolithophorids are autotrophic monads bearing calcified scales, discs, spines or other structures and form one order within the algal class Prymnesiophyceae (=Haptophyceae) (Green 1986). Coccolithophorids live almost exclusively in the upper 350 m in the open ocean. Maximum densities have been observed in the 40-80 m layer in the Central Gyre, Transitional and Equatorial zones of the Pacific and Atlantic Oceans (Honjo 1977). They can also occur in 'bloom' proportions, at concentrations up to 115 x 10^6 cells l^{-1}, in northern temperate waters (Holligan *et al.* 1983).

The Foraminifera are a diverse group of single-celled Protista with shells or tests of calcite that display immense architectural variety (Lee *et al.* 1985). Most Foraminifera are benthic: they are found at all latitudes and highest diversities are found in tropical areas. Distribution of planktonic taxa is largely temperature controlled with a greater abundance and diversity in the euphotic zone of warmer waters (Cifelli and Sachs 1966) where maximal densities occur at 6-30 m. Many planktonic species have complex relationships with symbiotic algae which play an important role in calcification.

Closely related to the Radiolaria are the Acantharia which are generally characterized by skeletons composed of strontium sulphate spicules (Lee *et al.* 1985). Although the fate of strontium sulphate is not considered further the Acantharia can occur in densities as high as 30 cells l^{-1} (Michaels 1988) and integrated abundances of 1.58 to 5.34 x 10^5 cells m^{-2} have been recorded in the Equatorial Pacific Ocean. Up to 90% were concentrated near the surface, their abundance declined sharply below 20 m. The whole water column is undersaturated with respect to strontium sulphate (celestite) and the degree of undersaturation increases with depth (Whitfield and Watson 1983). The flux of acantharian spicules from surface waters carries with them 30 µg Sr cm^{-2} a^{-1} equivalent to a whole ocean flux of 1.1 x 10^{14} g Sr a^{-1}. The redissolution of celestite spicules at depth is more or less complete.

SILICA AND ITS UTILIZATION BY PROTISTA

Silcon is present in natural waters in the form of monomeric monosilicic acid $Si(OH)_4$, virtually all of it dissociated except in extremely alkaline waters (Stumm and Morgan 1970). Monosilicic acid or soluble reactive silicate probably accounts for all dissolved silica that can be used by silica depositing protozoa (Burton and Leatherland 1970).

Although silicon is the second most abundant element in the earth's crust, seawater is substantially undersaturated with respect to this element (Whitfield and Watson 1983). Weathering of rocks and fresh water run off from land is a major source of silicic acid for the sea. Levels of reactive silicate in fresh water range from 10-180 µm and are usually at least one order of magnitude greater than those found in surface seawater where levels between <1-20 µm are not uncommon (Liss and Spencer 1970, Paasche 1973). In the open ocean the concentration of dissolved silica increases by about two orders of magnitude in deep water (Grill 1970, Edmond 1974). Characteristic vertical profiles of soluble silicate in the Atlantic and Pacific oceans are illustrated in Figure 1.

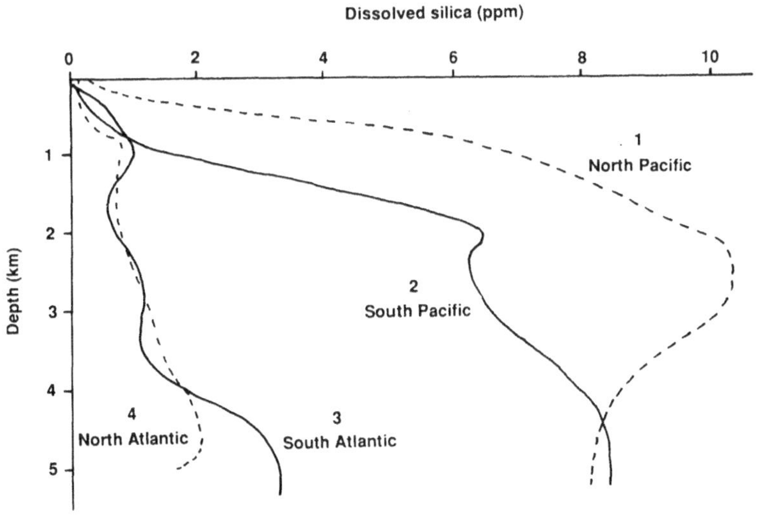

Fig. 1. Vertical profiles of dissolved silica in the Atlantic and Pacific Oceans (Redrawn after Heath 1974)

Detailed information on the physiology and biochemistry of silica metabolism by planktonic cells is limited to studies on diatoms (Volcani 1981, Sullivan 1986) with some ecophysiological data on choanoflagellates (Leadbeater 1986). Raven (1980, 1983) has reviewed the evidence for active transport of $Si(OH)_4$ into diatoms. Based on published data, Raven (1983) estimated that only 2% of the energy budget of a diatom cell is used in the production of a frustule which, with a whole cell Si/C atomic ratio of 0.25, comprises some 20% of the cell dry weight. Most species of diatoms are able to adjust to reduced levels of silicate in the medium by decreasing the amount of silica accumulated in their frustules (Sicko-Goad 1984). However, growth is halted by the complete absence of soluble silicate in the medium (Busby and Lewin 1967). In contrast to diatoms, choanoflagellates do not have a facultative requirement for silicon and grow unhindered in silica depleted medium (Leadbeater 1985). Chrysophytes and some heliozoa can also be grown in silica impoverished medium (Klaveness and Guillard 1975, Patterson and Durrschmidt 1988).

Despite the low concentrations of reactive silicate in the sea, silicon is frequently present in excess of nitrogen and phosphorus even in extremely nutrient poor waters such as the Central Pacific Ocean (Thomas and Dodson 1975). However, in upwelling regions which are noted for the intensities of their diatom blooms silicon may be exhausted from the euphotic zone before either nitrogen or phosphorus and therefore may be an important reason for the termination of a diatom bloom (Paasche 1973).

FLUXES OF MINERAL PARTICLES

A sedimentary mineral flux is taken to mean the vertical movement of mineralized particles from surface waters to the benthos. The overall distance travelled ranges from tens or hundreds of metres in coastal waters and on continental shelves to many kilometres in mid-ocean.

Based on Stokes's law the terminal velocity of a sinking sphere is related to the radius of the particle, its excess density over seawater and factors influencing drag. It is represented by the equation:

$$v = \frac{2gr^2 (p^1 - p)}{9n} \qquad (1)$$

Where v terminal velocity of sinking particle, g gravitational acceleration, r radius of particle, p density of medium, p^1 density of particle, n coefficient of viscosity of fluid medium. Active vertical mixing and density discontinuities may slow or even reverse sinking rates under certain conditions. Alternatively particles may be accelerated through the water column by migrating organisms or within faecal pellets. Although most mineral depositing Protista are not spherical and the movement of water past the sinking particle may not be laminar, as indicated by Reynolds numbers that exceed 0.5 (Takahashi and Honjo 1983, Takahashi and Bé 1984), nevertheless stokesian principles have formed the basis of studies on plankton sinking rates (Smayda 1970, Reynolds 1984).

According to Stokes's equation the terminal sinking velocity of a sphere increases with an increase in radius provided there is no change in density. In the case of diatoms there is a straight line relationship between the logarithm of mean sinking velocity and the logarithm of mean cell diameter (Smayda 1970, Reynolds 1984). However, the slope of the regression is nearer to 1 than to 2 as would be expected from Stokes's equation ($v \alpha r^2$) if r were the only variable. This can probably be explained by the cell compensating for increased size by a reduction in density or an increase in form resistance (Reynolds 1984).

Mineralization of cells adds considerably to their density. Opaline silica deposits in diatom cell walls have a density of 2.0-2.6 g cm^{-3} and overall densities of diatoms range from 1.078-1.263 g cm^{-3} (Reynolds 1984). It therefore follows that mineralized cells will sink in water. However, in nature the majority of planktonic cells remain suspended in surface water (0-100 m in depth) and this is essential for photoautotrophs which must remain in the euphotic zone. Water movement, and in particular vertical turbulence, is of paramount importance in maintaining planktonic cells in suspension (Smayda 1970). In addition 'form resistance', as manifested by distortions in cell shape and major morphological protuberances, leads to reduced sinking rates (Smayda 1970, Reynolds 1984). Many planktonic diatoms possess spines and may occur in chains; both of these phenomena reduce rates of sinking (Walsby and Xypolyta 1977). Likewise spinose shells of Foraminifera sink more slowly than non-spinose shells (Takahashi and Bé 1984). Dead diatoms sink 3-5 times more rapidly than living cells for reasons that are not fully understood (Reynolds 1984). Many planktonic Radiolaria and Foraminifera contain fluid filled alveoli which aid buoyancy and some Radiolaria can apparently alter their position in the water column by shedding much of their alveolated extracapsulum (Anderson 1983).

Fluxes of the mineralized remains of protista in the oceans have been monitored with the aid of sediment traps deployed at selected depths for several months at a time. Throughout each experiment a time series of samples, each sample comprising sedimented particles harvested over a predetermined period of between 4-16 d, was collected automatically and preserved in fixative. From samples collected in this way, Takahashi (1986, 1987a, b) has been able to analyse in detail vertical fluxes of all major classes of siliceous particles during a four year period, 1982-1986, at station PAPA in the subarctic Pacific. Sediment traps were deployed at 1000 m and 3800 m in an overall column depth of 4200 m. Other objectives of this work included: 1) the identification of 'marker' species that have well defined maxima and minima which are clearly correlated to specific seasonal and climatic events 2) determination of the effects of dissolution and mechanical damage on mineralised structures during their descent through the water column. Altogether 122 taxa were recorded comprising 18 diatom, 7 silicoflagellate and 97 radiolarian species from station PAPA. The published results demonstrate high amplitude fluxes for all siliceous plankton on a seasonal basis (Figure 2). For diatoms there was variation in the seasonal timing of maxima and minima for individual species although all taxa could be allocated to one of five different patterns of seasonal flux (Takahashi 1986). Based on correlation analyses three of these patterns are considered to be indicators of high total mass, silica and organic carbon fluxes at this station. Amongst the diatoms fluxes of *Denticulopsis seminae*, *Chaetoceros atlanticum* and *Asteromphalus robustus* are the best indicators of production. Within the silicoflagellates, *Distephanus speculum* is a good productivity indicator (Takahashi 1987a). Although silicoflagellates contribute only 0.02-1.2% (mean 0.4%) of total weight of biogenic silica to the total flux their contribution of carbon is proportionately higher; 3.1% of total weight of carbon produced by the sum of all major siliceous plankton in upper water layers (Takahashi 1990). The influence of seasonal and other temporal variations on productivity and the flux of biogenic particles in the Atlantic Ocean and Panama Basin has been investigated by Deuser *et al.* (1988) and Honjo (1982).

Partitioning of silica between taxonomic groups varies from one season to another. For instance in spring at station PAPA 90% by weight of biogenic silica particles are diatoms and only 10% Radiolarians whereas in late spring and early summer Radiolaria account for 50% of total weight. Of the Radiolaria about 37% are Phaeodaria and the remainder Polycystina (Takahashi 1990).

Aggregation of biogenic silica particles into clumps with other organic matter as marine snow is not uncommon in nature (Angel 1984). By reference to Stokes's equation noted above, the increased size of particles, in this case aggregates, as compared with solitary cells will inevitably result in an increased sinking rate. At times of high productivity with increased densities of plankton in surface waters aggregation of biogenic material to form marine snow is common. Detailed comparison of the counts of mineralized particles, Radiolaria, diatoms and silicoflagellates, collected in sediment traps deployed at 1000 m and 3800 m in the same water column generally shows that maxima and minima for individual taxa are of similar magnitude but are offset from each other by a period of two weeks (Figure 2). This translates into a sinking speed of 175 m d^{-1} or faster (Takahashi 1986, 1987a, c). Such a rate is significantly higher, by as much as one order of magnitude, than the rates obtained in laboratory experiments for unialgal cultures (Smayda 1970) and heterogeneous plankton populations (Bienfang 1981). Takahashi (1987c) concludes that in the case of radiolarians there are two independent sinking mechanisms in operation. One, an accelerated rate (175 m d^{-1}), for skeletons incorporated into aggregates or marine snow, and the second, a slower rate (c 58 m d^{-1}), for solitary particles. Published sinking rates for marine snow range from 43-95 m d^{-1} (Shanks and Trent 1980) or higher 20-200 m d^{-1} (Takahashi 1986).

Faecal pellets are now regarded as being an important form in which organic and inorganic material is transported down the water column. Assimilation efficiencies of herbivores grazing on plankton vary considerably according to the food source but a value of 70% can be considered reasonable (Angel 1984). If this is so then at least 30% of primary production is almost immediately converted into detrital form. In a study on the faecal pellets produced by copepods in Lake Michigan, fragments of diatom frustules constituted a major fraction of the pellet contents (Ferrante and Parker 1977). The incorporation of diatom frustules and other mineralized structures into faecal pellets has two important consequences for mineral recycling. Firstly, sinking rates of pellets are considerably higher than those of solitary frustules and therefore the transfer of particulate silica to the meso- and bathypelagic regions is accelerated. Secondly, silica particles within a fecal pellet are protected from dissolution by the surrounding peritrophic membrane, a coating secreted onto the pellet during its passage

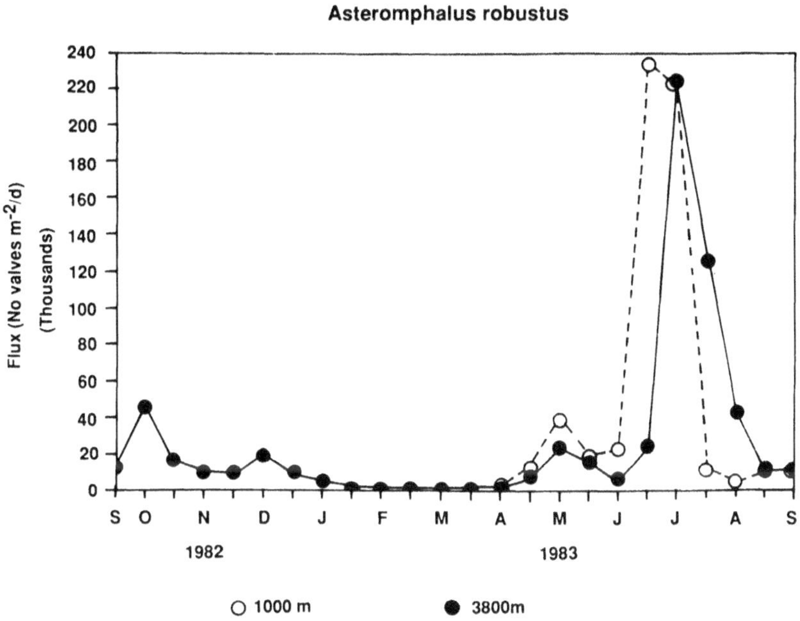

Fig. 2. Seasonal fluxes of the diatom taxa *Asteromphalus robustus* at the subarctic Pacific Station PAPA during 1982 and 1983 (redrawn after Takahashi 1986)

through the copepod gut. Sinking rates of faecal pellets vary according to size, shape and composition but range from 20-220 m d^{-1} for pellets of marine copepods (Angel 1984). Once exposed to the water column the peritrophic membrane is subject to degradation by microorganisms and mechanical forces. However, pellets may stay intact long enough for them to provide a mechanism for the rapid transport of particulate silica from surface waters without dissolution.

DISSOLUTION OF BIOGENIC SILICA

Once in contact with water silica structures will dissolve and thereby return reactive silicate to the medium. The rate of dissolution of soluble particles in water is dependent on: 1) the absolute dissolution rate of each mineral phase 2) the surface area in contact with water 3) the concentration of soluble silicate present in the water. Protistan siliceous structures are heterogeneous in composition and do not dissolve uniformly (Lawson *et al.* 1978, Hurd *et al.* 1981, Leadbeater and Davies 1984, Takahashi *et al.* 1983). Organic material within or on the surface of siliceous structures has a major effect on their solubility. Diatoms, for instance, have an organic layer on the frustule surface which protects the underlying silica from dissolution. Removal of this layer by microbial degradation or treatment with acid exposes the silica and only then does significant dissolution occur (Lewin 1961). The highly porous substructure of many protistan mineralized structures means that the surface area in contact with the water is very large and is continually changing with time as dissolution proceeds (Hurd *et al.* 1981, Hurd and Birdwhistell 1983, Takahashi *et al.* 1983, Leadbeater and Davies 1984).

Conditions in surface waters are particularly favourable for silica dissolution since they are usually warm, turbulent and highly undersaturated with respect to silica. However, a large proportion of mineralized plankton in this layer is living (Figure 3) and the mineralized structures are protected by organic matter from dissolution. Nevertheless, once dead the organic layer is degraded by microbial action, which in diatoms occurs within a few days of cell death (Grill and Richards 1964), and substantial recycling of silica occurs although it may not be apparent since it can be immediately reused by other organisms. The introduction of radioactive silicon tracers into field work has lead to a realization that the intensive production of biogenic silica in surface waters in upwelling regions is more or less balanced by the high rates of silica dissolution (Nelson and Goering 1977).

Once out of surface water and into the colder meso- and bathypelagic strata rates of silica dissolution are slower although increased pressure to some extent compensates for the effects of the reduced temperature. The fate of silica particles will depend on the period of time they remain surrounded by water. At these depths diatom frustules are usually fragmented either as a result of grazing or mechanical damage. Taking into account rates of sinking and dissolution

Wollast (1974) has computed the weights of diatom debris particles that remain at particular depths for various initial radii of particles. Considering that the mean Stokes's radius of diatom debris particles in nature is 7 mm (Wollast 1974) then according to his calculations few particles of this dimension should remain below 1000 m. However, if diatom frustules or silicoflagellate skeletons are transported into the meso- and bathypelagic regions either in marine snow or faecal pellets then there seems to be little subsequent dissolution during their passage through the water column. This is evident from Takahashi's (1986, 1987a) data on the patterns of deep sea fluxes of diatoms and silicoflagellates in the subarctic Pacific.

Erez *et al.* (1982) carried out *in situ* experiments designed to measure the dissolution rates of monospecific phaeodarian radiolarian assemblages of *Castanidium longispinum* at depths between 378 and 5582 m in the central North Pacific. Highest weight losses of about 90% were observed at 378 m where water temperature was 8.62°C. Below the thermocline at a temperature of 1.4°C weight loss was about 40%. Kinetic considerations show that temperature is the major factor that controls dissolution of the skeletons. Under natural conditions large phaeodarian radiolarians appear to reach the water-sediment interface relatively quickly and ultimately dissolve whereas small specimens dissolve *en-route* through the water column.

Phaeodarians are poorly represented in the fossil record. In contrast, polycystine Radiolaria are more resistant to dissolution and are, as a result, well preserved in the fossil record. Takahashi *et al.* (1983) speculate that phaeodarian silica may be more hydrated than that of polycystines and as a result phaeodarian skeletons quickly become porous thereby accelerating dissolution.

THE FATE OF BIOGENIC SILICA SEDIMENTS

Of the siliceous remains of Protista that reach the sediment surface only about 2% or less are eventually preserved in fossil deposits (Hurd 1973, Heath 1974). Takahashi (1986, 1987a) estimates that 2% and 1% respectively of diatom and silicoflagellate fluxes at station PAPA in the subarctic Pacific are preserved in the fossil record. Within the sediment diagenetic chemical reactions take place. These may be biogenic, which occur in the presence of organic matter and involve bacteria, or abiogenic which are not biologically controlled (Berner 1974a). The actual processes whereby silica interacts with pore waters via dissolution, precipitation or ion exchange are abiogenic whether or not they are ultimately brought about by biogenic reactions. Soluble silica within pores of the sediment is returned to the water column by a combination of diffusion, advection and chemical reaction. Compaction of fine-grained sediments may result in silicate-rich pore water being squeezed out of the sediments. In shallow waters turbulent action by bioturbation, waves and currents is important in circulating soluble silicate (Siever 1957, Berner 1974a).

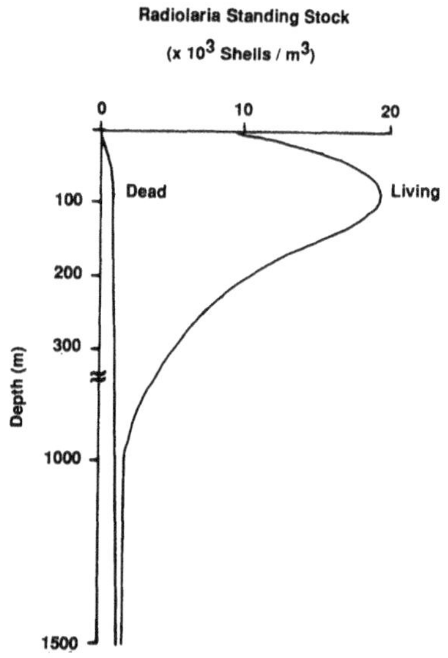

Fig. 3. Characteristic profiles for living and dead radiolarian standing stock in a water column (Redrawn after Takahashi 1983)

The distribution of sediments of biogenic silica is closely correlated with distribution of plankton species in the overlying surface waters. This is demonstrated by the close correlation between distinctive diatom assemblages in different water masses in the northernmost Pacific and in the underlying surface sediments (Sancetta 1981). Principal sites of biogenic silica deposition are in high latitudes of the Northern Hemisphere and South of the Antarctic Convergence around Antarctica. The latter circum-global deposit is 900-2000 km wide and contains 75% of all oceanic silica deposits of which diatom frustules form up to 70% by weight. Biogenic sediments near the polar front of the South Atlantic Antarctic have an average accumulation rate of 0.17 mm a^{-1} (DeMaster 1981).

SILICA MASS BALANCES IN THE OCEAN

The impact of mineralizing Protista on the geochemical cycling of silica is substantial (Figure 4). Considering the geochemical balance alone, inputs of silica from rivers, submersive weathering and volcanism are estimated to be 4.30×10^{14} g SiO_2 a^{-1} (Wollast 1974). This is balanced by geochemical removal processes. The total weight of dissolved silica in the oceans is estimated as 5.5×10^{18} g. Residence time of silica in the oceans, obtained by dividing concentration by rate of input or output, is 1.3×10^4 years. Estimates of the annual uptake of silica by organisms range from $0.77\text{-}16 \times 10^{16}$ g a^{-1} (Wollast 1974) which is at least one order of magnitude greater than the annual input. Based on these figures the biological impact on geochemical cycling is to reduce the mean residence time of silica in seawater from 1.3×10^4 years to 200-300 years (Heath 1974).

CALCIUM CARBONATE AND ITS UTILIZATION BY PROTISTA

It is necessary to consider fluxes of calcium carbonate in seawater independently since the solution chemistry of calcite is quite different from that of silica. All surface waters are supersaturated with respect to calcite and aragonite (Figure 5) (Whitfield and Watson 1983).

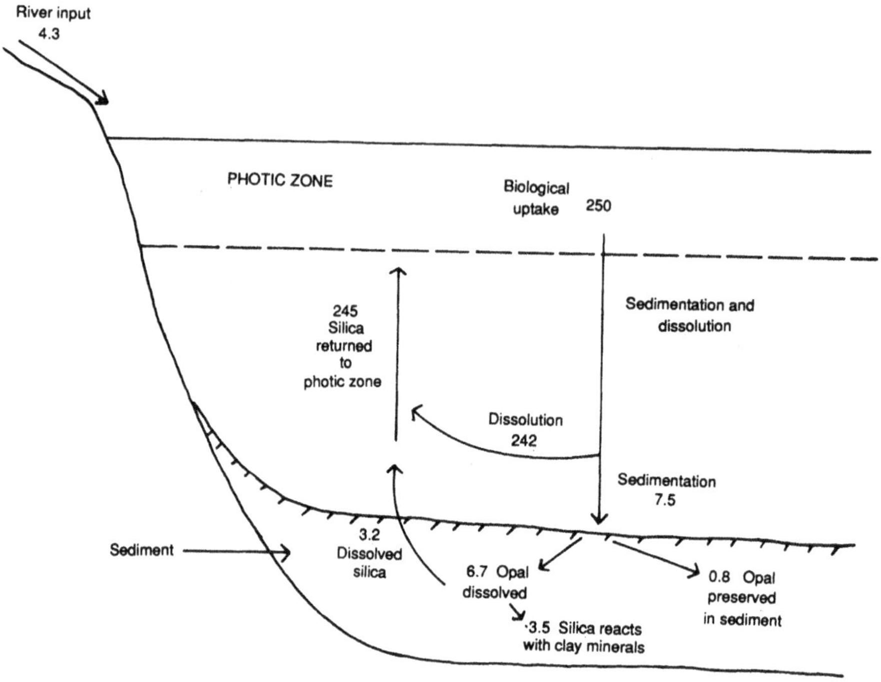

Fig. 4. Summary of the overall fluxes of silicon ($\times 10^{14}$ g a^{-1}) along various parts of the oceanic cycle. Values for annual fluxes are taken from Wollast (1974) and are included to show the magnitude and relative importance of the component processes

Since calcium carbonate is more soluble in cold water than warm the extent of supersaturation depends on latitude and varies from 2.5-5.5 fold for calcite and 1.7-3.7 for aragonite. Since the two major groups of calcium carbonate secreting Protista, the coccolithophorids and the Foraminifera, deposit calcite structures it is only this form of calcium carbonate that we will be concerned with here. The solubility of calcite increases with depth because of a combination of increasing pressure, decreasing temperature and pH (Edmond 1974, Whitfield and Watson 1983). The level at which water becomes just saturated with respect to calcium carbonate is known as the lysocline, below this depth dissolution of calcite particles will occur (Whitfield and Watson 1983). The lysocline is deeper in the Atlantic Ocean than in the Pacific and is deeper at the Equator than at higher latitudes. Some authors define the lysocline as being the depth at which dissolution of sedimentary calcium carbonate becomes apparent (Broecker and Broecker 1974). Thus in the Pacific Ocean only the surface 122 m of water is supersaturated with respect of calcite (Berger 1970). However, in spite of the remainder of the water column being undersaturated it is not until a depth of 3800 m is reached in the central Pacific that calcite dissolution increases drastically. At a specific depth below the lysocline the downward flux of calcite particles is equal to the rate at which they are dissolved; this depth is known as the carbonate compensation depth (Figure 5). Calcite oozes and sediments would only be expected in shallower waters where dissolution is limited or cannot take place.

Based on data collected from the Pacific Ocean, Honjo (1977) has calculated that the standing crop of calcite as coccoliths on living coccolithophorids in the upper 200 m photic zone is about 350 mg m^{-2} in the Equatorial region, 100 mg m^{-2} in the Central Gyre and as high as 1000 mg m^{-2} at a station in the North Pacific. Annual calcite productivity can be roughly estimated in tropical and subtropical waters where seasonal fluctuations of biomass are relatively small. In the Equatorial region of the Pacific Ocean annual calcite production by coccoliths is of the order of 10 g m^{-2} a^{-1}, production in Gyre water is lower and in Transition zones higher (Honjo 1977).

Estimates of calcite production by foraminiferans are less easily made since there is less information on their concentrations and turnover rates (Honjo 1977). Records of abundance range from 1 specimen m^{-3} in the Pacific Central Gyre to 4-5.6 specimens m^{-3} in the Atlantic Central Gyre. In some regions 10^2-10^3 specimens m^{-3} is more common. The approximate world average concentration is of the order of 10 specimens m^{-3} (Honjo 1977). Total abundances in the Atlantic are about half those in the Pacific and yet carbonate accumulation rates at the sediment surface are less than half those in the Pacific demonstrating the importance of preservation in controlling carbonate distribution. Based on Berger and Soutar's (1967) data Honjo (1977) estimates that the annual production of calcite by Foraminifera in the Santa Barbara basin is of the order of 0.1 g m^{-2} a^{-1}. These figures demonstrate that calcite production by coccolithophorids is several orders of magnitude greater than that of Foraminifera.

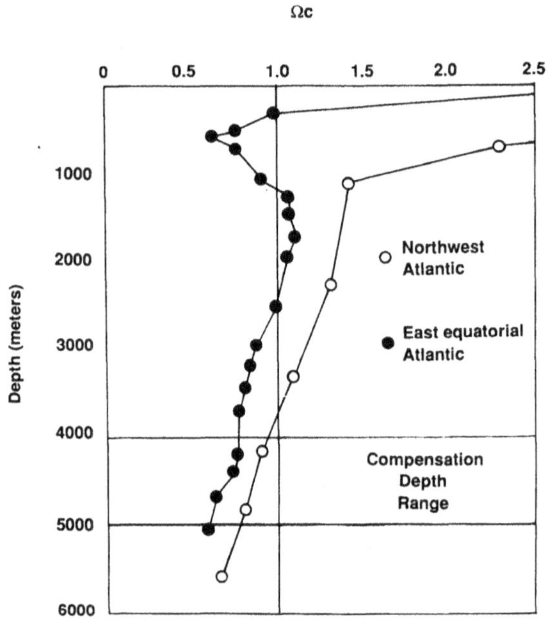

Fig. 5. Degree of saturation of seawater with respect to calcite (Ωc) as a function of depth in the East Equatorial Atlantic and the Northwest Atlantic (redrawn after Berner 1974b)

FLUXES OF CALCITE PARTICLES

Sinking rates for individual coccoliths, which weigh between 5-400 pg, are very slow. Honjo (1977) has calculated that the coccoliths of *Emiliania huxleyi*, *Coccolithus neohelis* and *C. leptopora* have terminal sinking velocities of 0.138 m d^{-1}, 0.156 m d^{-1} and 1.3 m d^{-1} respectively. At these velocities a typical residence time in an ocean water column would be 100 years. Sinking speeds of foraminiferan shells are much faster ranging from 30-1000 m d^{-1}. Sinking speed is governed primarily by shell weight and the presence or absence of spines

(Takahashi and Bé 1984). Based on data from plankton tows and sediment traps most planktonic foraminifera >150 µm reach the mean ocean depth of 3800 m within 3-12 d. Calcite oozes generally are deposited at rates of 0.1-0.3 mm a^{-1}.

Faecal pellets provide a means of accelerated sinking for coccoliths and the covering peritrophic membrane inhibits calcite dissolution. Honjo (1976, 1977) concludes that faecal pellets produced by grazers in cold water below the thermocline must play an important role in transporting coccoliths to the sediments in deep undersaturated waters. At an Equatorial Pacific station it was estimated that 92% of coccoliths produced in the euphotic layer were being transported to the deep-sea bottom in this way. Coccoliths, like diatoms, are subject to mechanical damage when eaten by grazers but the coccoliths of some species such as *Emiliania huxleyi* are not seriously damaged either mechanically or chemically after passage through a copepod gut (Honjo 1977).

DISSOLUTION OF CALCITE PARTICLES

In supersaturated waters undamaged calcite particles are usually well preserved and where the sea floor is above the level of the lysocline coccolith and foraminiferan oozes accumulate. However, in deeper waters below the lysocline rapid dissolution takes place and unless the rate of supply of calcite particles is in excess of their dissolution no permanent calcite sediment will be formed.

Skeletons of different taxa dissolve at different rates with the result that during sedimentation below the lysocline planktonic coccolith and foraminiferan assemblages are modified in species composition, size distribution, content of damaged particles and average particle weight (Berger 1967). *In situ* and laboratory experiments have been carried out to determine solubility, as measured by weight loss, of different foraminiferan taxa under different environmental conditions. From these experiments taxa have been ranked according to their susceptibilty to dissolution or alternatively to their resistance to solution (Berger 1967, Berger and Piper 1972). Ranking data is of particular assistance to geologists when attempting to reconstruct past climates and oceanic conditions that might have prevailed when fossil assemblages of coccoliths and foraminiferans were deposited. The percentage of total calcium carbonate flux that redissolves ranges from 69% in the Atlantic to 83% in the Pacific. This is sufficient to produce a 1.5% enrichment of Ca^{2+} and 15% of HCO_3^- in the deep sea (Whitfield and Watson 1983).

THE FATE OF BIOGENIC CALCITE SEDIMENTS

Once on the ocean floor calcite deposits, in the form of coccolith and foraminiferan oozes, are still subject to dissolution if the sediment is below the lysocline. Dissolution is further aided in aerobic conditions where there is organic matter undergoing microbial degradation with the resulting production of CO_2. Different taxa dissolve at different rates so that as a sediment ages the species composition changes. In this way sediments become enriched with solution-resistant species (Berger 1970). Mixing and horizontal displacement by water currents leads to uneven deposition with the result that sediments are usually thicker in depressions than on hills (Berger and Piper 1972). Bottom currents may also serve to redeposit taxa in a graded series so that enrichment of species in a sediment is related to settling velocity. Of the original populations of Foraminifera in surface waters of the Pacific only about 15% of foraminiferan calcite is ultimately preserved in the deep sea record. Preservation of calcium carbonate, notably in the Atlantic Ocean, is sufficient to provide a major flux of both Ca^{2+} and C from the oceans to the sea floor. Calcite oozes are present over about half the ocean floor and are more extensive in the Atlantic Ocean than in the Pacific. Oozes are generally restricted to lower latitudes although they are extensive in the North Atlantic. They rarely occur at depths greater than 4500 m. Since coccolith oozes are slightly more resistant to dissolution than foraminiferan oozes they are often found at slightly greater depths immediately above the carbonate compensation depth.

CONCLUDING COMMENTS

Superficially there are many similarities in the role that Protista play in the cycling of calcite and silica in the oceans. Both minerals are utilized by planktonic Protista for the production of skeletons or cell coverings and eventually these inorganic structures form vertical fluxes transporting mineral matter to the sediment surface. Here they may dissolve or become lithified to form rock.

However, closer examination of the solution chemistries of the two minerals and the distribution of soluble silicate and calcite in the sea reveals major differences that have important consequences for productivity and sediment deposition. On the one hand silica can be a limiting nutrient for phytoplankton growth and most silica dissolution, particularly of diatom frustules, takes place in surface strata. Without this recycling diatom growth would be severely limited. Biogenic siliceous deposits reflect biological activity in surface waters. On the other hand soluble calcite is never limiting to phytoplankton growth and recycling does not occur in surface waters. Biogenic calcite deposits reflect the preservation of calcite at depth and whilst productivity must provide the organisms in the first place the distribution of sediments is largely independent of the amount of biological activity occurring in the surface waters.

Whilst we are slowly building up a comprehensive picture of the role of Protista in mineral cycling in the oceans there still remains the unresolved question of why planktonic cells require mineral structures in the first place. Unfortunately we are still some way from providing a convincing answer to this question.

REFERENCES

Anderson OR (1983) Radiolaria, Springer, New York Berlin Heidelberg Tokyo
Angel MV (1984) Detrital organic fluxes through pelagic ecosystems. In: Fasham MJR (ed) Flows of energy and materials in marine ecosystems: theory and practice. Plenum, New York, p 475
Berger WH (1967) Foraminiferal ooze: solution at depths. Science 156:383-385
Berger WH (1970) Planktonic foraminifera: selective solution and the lysocline. Marine Geol 8:111-138
Berger WH, Piper DJW (1972) Planktonic foraminifera: differential settling, dissolution, and redisposition. Limnol Oceanogr 17:275-287
Berger WH, Soutar A (1967) Planktonic foraminifera: field experiment on production rate. Science 156:1495-1497
Berner RA (1974a) The benthic boundary layer from the viewpoint of a geochemist. In: McCave IN (ed) The benthic boundary layer. Plenum, New York, p 33
Berner RA (1974b) Physical chemistry of carbonate in the oceans. Soc Econ Palaeo Min Special Publ 20:37-43
Bienfang PK (1981) Sinking rates of heterogeneous, temperate phytoplankton populations. J Plank Res 3:235-255
Broecker WS, Broecker S (1974) Carbonate dissolution on the western flank of the East Pacific Rise. Soc Econ Palaeo Min Special Publ 20:44-57
Burton JD, Leatherland TM (1970) The reactivity of dissolved silicon in some natural waters. Limnol Oceanogr 15:473-476
Busby WF, Lewin J (1967) Silicate uptake and silica shell formation by synchronously dividing cells of the diatom *Navicula pelliculosa* (Breb) Hilse. J Phycol 3:127-131
Cifelli R, Sachs KN (1966) Abundance relationships of planktonic Foraminifera and Radiolaria. Deep Sea Res 13:751-753
DeMaster DJ (1981) The supply and accumulation of silica in the marine environment. Geochim et Cosmochim Acta 45:1715-1732
Deuser WG, Muller-Karger FE, Hemleben C (1988) Temporal variations of particle fluxes in the deep subtropical and tropical North Atlantic: eulerian versus lagrangian effects. J Geophys Res 93:6857-6862
Edmond JM (1974) On the dissolution of carbonate and silicate in the deep ocean. Deep Sea Res 21:455-480
Erez J, Takahashi K, Honjo S (1982) In-situ dissolution experiment of Radiolaria in the central North Pacific Ocean, Earth Planet Lett 59:245-254
Ferrante JG, Parker JI (1977) Transport of diatom frustules by copepod fecal pellets to the sediments of Lake Michigan. Limnol Oceanogr 22:92-98
Green JC (1986) Biomineralization in the algal class Prymnesiophyceae In: Leadbeater BSC, Riding R (eds) Biomineralization in lower plants and animals. Clarendon, Oxford p 173
Grill EV (1970) A mathematical model for the marine dissolved silicate cycle. Deep Sea Res 17:245-266
Grill EV, Richards FA (1964) Nutrient regeneration from phytoplankton decomposing in seawater. J Mar Res 22:51-71

Heath GR (1974) Dissolved silica and deep sea sediments. Soc Econ Palaeo Min Special Publ 20:77-93

Holligan PM, Viollier M, Harbour DS, Camus P, Champagne Phillipe M (1983) Satellite and ship studies on coccolithophore production along a continental shelf edge. Nature London 304:339-342

Honjo S (1976) Coccolith production, transportation and sedimentation. Mar Micro 1:65-79

Honjo S (1977) Biogenic carbonate particles in the ocean; do they dissolve in the water column? In: Andersen NR, Malahoff A (eds) The fate of fossil fuel CO_2 in the oceans. Plenum, New York, p 269

Honjo S (1982) Seasonality and interaction of biogenic and lithogenic particulate flux at the Panama Basin. Science 218:883-884.

Hurd DC (1973) Interactions of biogenic opal, sediment and seawater in the Central Equatorial Pacific. Geochim Cosmochim Acta 37:2257-2282

Hurd DC, Birdwhistell S (1983) On producing a more general model for biogenic silica dissolution. Am J Sci 283:1-28

Hurd DC, Pankratz HS, Asper V, Fugate J, Morrow H (1981) Changes in the physical and chemical properties of biogenic silica from the central equatorial Pacific: Part III, specific pore volume, mean pore size, and skeletal ultrastructure of acid-cleaned samples. Am J Sci 281:833-895

Klaveness D, Guillard RRL (1975) The requirement for silicon in *Synura petersenii* (Chrysophyceae). J Phycol 13:349-355

Lawson DS, Hurd DC, Pankratz HS (1978) Silica dissolution rates of decomposing phytoplankton assemblages at various temperatures. Am J Sci 278:1373-1393.

Leadbeater BSC (1981) Ultrastructure and deposition of silica in loricate choanoflagellates. In: Simpson TL, Volcani BE (eds) Silicon and siliceous structures in biological systems. Springer, New York, p 295

Leadbeater BSC (1985) Developmental studies on the loricate choanoflagellate *Stephanoeca diplocostata* Ellis. IV Effects of silica deprivation on growth and lorica production. Protoplasma 127:171-179

Leadbeater BSC (1986) Silica deposition and lorica assembly in choanoflagellates. In: Leadbeater BSC, Riding R (eds) Biomineralization in lower plants and animals. Clarendon, Oxford p 345

Leadbeater BSC, Davies ME (1984) Developmental studies on the loricate choanoflagellate *Stephanoeca diplocostata* Ellis. III. Growth and turnover of silica, preliminary observations. J exp mar Biol Ecol 81:251-268

Leadbeater BSC, Riding R (1986) Biomineralization in lower plants and animals. Clarendon, Oxford

Lee JJ, Hutner SH, Bovee EC (1985) An illustrated guide to the Protozoa. Society of Protozoologists, Kansas

Lewin JC (1961) The dissolution of silica from diatom walls. Geochim et Cosmochim Acta 21:182-196

Lisitzin AP (1972) Sedimentation in the world oceans. Soc Econ Paleont Min Special Publ 17

Liss PS, Spencer CP (1970) Abiological processes in the removal of silicate from seawater. Geochim et Cosmochim Acta

34:1073-1088
Michaels AF (1988) Vertical distribution and abundance of Acantharia and their symbionts. Mar Biol 97:559-569.
Morse JW, Berner RA (1972) Dissolution kinetics of calcium carbonate in sea water: II. A kinetic origin for the lysocline. Am J Sci 272:840-851
Nelson DM, Goering JJ (1977) Near-surface silica dissolution in the upwelling region off northwest Africa. Deep Sea Res 24:65-73
Paasche E (1973) Silicon. In: Morris I (ed) The physiological ecology of phytoplankton. Blackwell, Oxford, p 259
Patterson DJ, Durrschmidt M (1988) The formation of siliceous scales by *Raphidiophrys ambigua* (Protista, Centroheliozoa). J Cell Sci 91:33-39
Raven JA (1980) Nutrient transport in microalgae. Adv Mic Physiol 21:47-226
Raven JA (1983) The transport and function of silicon in plants. Biol Rev 58:179-207
Reynolds CS (1984) The ecology of freshwater phytoplankton, University Press, Cambridge
Reynolds CS (1986) Diatoms and the geochemical cycling of silicon. In: Leadbeater BSC, Riding R (eds) Biomineralization in lower plants and animals. University Press, Oxford, p 269
Sancetta C (1981) Oceanographic and ecologic significance of diatoms in surface sediments of the Bering and Okhotsk seas. Deep Sea Res 28:789-817
Shanks AL, Trent JD (1980) Marine snow: sinking rates and potential role in vertical flux. Deep Sea Res 27:137-143
Sicko-Goad LM, Schelske CL, Stoermer EF (1984) Estimation of intracellular carbon and silica content of diatoms from natural assemblages using morphometric techniques. Limnol Oceanogr 29:1170-1178
Siever R (1957) The silica budget in the sedimentary cycle. Amer Mineralog 42:821-841
Smayda TJ (1970) The suspension and sinking of phytoplankton in the sea. Oceanogr Mar Biol Ann Rev 8:353-414
Stumm W, Morgan JJ (1970) Aquatic chemistry. Wiley, New York
Sullivan CW (1986) Silicification in diatoms. In: Evered D, O'Connor M (eds) Silicon biochemistry. Ciba Foundation Symposium Volume 121, Wiley, Chichester p 59
Takahashi K (1983) Radiolaria: sinking population, standing stock, and production rate. Mar Micropalaeontol 8:171-181
Takahashi K (1986) Seasonal fluxes of pelagic diatoms in the subarctic Pacific, 1982-1983. Deep Sea Res 33:1225-1251
Takahashi K (1987a) Seasonal fluxes of silicoflagellates and *Actiniscus* in the subarctic Pacific during 1982-1984. J Mar Res 45:397-425
Takahashi K (1987b) Response of subarctic Pacific diatom fluxes to the 1982-1983 El Niño disturbance. J Geophys Res 92:14387- 14392
Takahashi K (1987c) Radiolarian flux and seasonality: climatic and El Niño response in the subarctic Pacific, 1982-1984. Global Biogeochem Cycles 1:213-231
Takahashi K (1990) Mineral flux and biogeochemical cycles of marine plantkonic protozoa - Session summary In: Reid PC, Turley CM, Burkill PH (eds) Protozoa and their role in

marine processes. Springer, Berlin Heidelberg New York

Takahashi K, Bé AWH (1984) Planktonic foraminifera: factors controlling sinking speeds. Deep Sea Res 31:1477-1500

Takahashi K, Honjo S (1983) Radiolarian skeletons: size, weight, sinking speed, and residence time in tropical pelagic oceans. Deep Sea Res 30:543-568

Takahashi K, Hurd DC, Honjo S (1983) Phaeodarian skeletons: their role in silica transport to the deep sea. Science 222: 616-618

Thomas WH, Dodson AN (1975) On silicic acid limitation of diatoms in near-surface waters of the eastern tropical Pacific Ocean. Deep Sea Res 22:671-677

Volcani BE (1981) Cell wall formation in diatoms: morphogenesis and biochemistry. In: Simpson TL, Volcani BE (eds) Silicon and siliceous structures in biological systems. Springer, New York Heidelberg Berlin p 157

Walsby AE, Xypolyta A (1977) The form resistance of chitan fibres attached to the cells of *Thalassiosira fluviatilis* Hustedt. Br Phycol J 12:215-223

Whitfield M, Watson AJ (1983) The influence of biomineralisation on the composition of seawater. In: Westbroek P, Jong EW de (eds) Biomineralization and biological metal accumulation. Riedel, Dordrecht p 57

Wollast R (1974) The silica problem. In: Goldberg ED (ed) The sea. Wiley, New York, p 359

EVOLVING ROLE OF PROTOZOA IN AQUATIC NUTRIENT CYCLES

David A Caron
Biology Department
Woods Hole Oceanographic Institution
Woods Hole, Massachusetts 02543
USA

OVERVIEW OF PROTOZOAN NUTRIENT CYCLING

There has been considerable discussion over the last 10 years on the role that protozoa play in the cycling of major nutrients (nitrogen and phosphorus) in the marine environment. This discussion has been fostered by recent studies which have clearly demonstrated the importance of these heterotrophic protists in the flow of energy and materials in the ocean, and by recent efforts to model their activities in marine ecosystems (Pomeroy 1974, Sieburth 1976, Anderson 1980, Fenchel 1967, 1980, 1986, Azam *et al.* 1983, Laacke *et al.* 1983, Sherr and Sherr 1984, Sherr *et al.* 1986, 1988, Goldman and Caron 1985, Stoecker and Evans 1985).

Given this newfound importance of protozoa in aquatic nutrient cycling, there still remains a considerable controversy over the relative importance of protozoa as nutrient assimilators or nutrient regenerators in aquatic communities. There have been strong arguments on the one hand that protozoa constitute an important trophic **link** for the transfer of vital nutrients from small phototrophic or heterotrophic microorganisms to higher organisms, and on the other hand that protozoa are a major source of remineralized nutrients (i.e. a trophic **sink**).

This debate is, of course, closely related to the 'link or sink' controversy concerning the fate of organic material in the microbial loop of aquatic communities (Ducklow *et al.* 1986, Sherr *et al.* 1987). The primary role of the heterotrophic bacteria in aquatic ecosystems has traditionally been considered to be that of remineralizers of organic carbon and nutrients (Strickland 1970, Fenchel and Jorgensen 1977, Fenchel and Blackburn 1979). As decomposers bacteria respire dissolved

NATO ASI Series, Vol. G 25
Protozoa and Their Role in Marine Processes
Edited by P.C. Reid et al.

and non-living particulate organic material and release the contained carbon and nutrients primarily as dissolved inorganic nutrients and carbon dioxide. Accordingly, the primary role of the protozoa has been considered to be that of maintaining the bacteria in a state of physiological youth by their grazing activities on this population (Fenchel and Harrison 1976, Barsdate *et al.* 1977, Fenchel 1977). These pools of remineralized elements are then once again available to the primary producers for uptake and incorporation into plant biomass. Although the present paradigm still recognizes an important role for bacteria as remineralizers of organic material in aquatic environments, we now know that this view of the ecology of the heterotrophic bacteria is clearly an oversimplification of the activities of this microbial assemblage.

Attention has recently been drawn to the potentially large amount of bacterial productivity in the ocean, and the importance of consumers (Sieburth 1976, Fuhrman and Azam 1980, 1982, Azam and Fuhrman 1984, Ducklow and Hill 1985a,b). In particular, a number of studies conducted within the last decade have focused on the hypothesis that a significant fraction of the dissolved organic material and non-living particulate organic material acted on by the bacteria is converted into living bacterial biomass, thus becoming available to bacterial consumers and, subsequently, to higher organisms that consume these bacterivores. The conversion of non-living organic matter into bacterial biomass presents the possibility that, rather than acting solely as a nutritional 'dead end' in aquatic food webs, the bacteria and the organisms that consume them constitute a potentially important mechanism for the return of some of the organic material entering the detrital food web back to the classical phytoplankton-copepod-fish food chain (Azam *et al.* 1983). A growing consensus of opinion would indicate that the protozoa fill the role of primary consumers of bacterial biomass in most aquatic environments (Fenchel 1980, 1982b,d, Sieburth and Davis 1982,

Linley *et al.* 1983, Sherr *et al.* 1986, Wright and Coffin 1984, Andersen and Fenchel 1985, Hobbie and Helfrich 1988).

In addition to their important role as bacterial consumers, many of the protozoa are, of course, consumers of primary producers. The importance of grazing by ciliated protozoa on phytoplankton has been clearly demonstrated in a number of field and laboratory studies (Heinbokel and Beers 1979, Stoecker *et al.* 1981, Capriulo and Carpenter 1980, 1983, Capriulo 1982, Gifford 1985, Verity 1986). Herbivory by the larger sarcodines also is well-known (Lipps and Valentine 1970, Lee 1980), as is their ability to feed on much larger metazoan prey (Anderson 1980, Anderson *et al.* 1979, Swanberg *et al.* 1986, Caron and Bé 1984, Spindler *et al.* 1984). More recently, the potential importance of microflagellate grazing on small phytoplankton and cyanobacteria has been demonstrated (Johnson *et al.* 1982, Goldman *et al.* 1985, 1987b, Caron *et al.* 1985, Andersen *et al.* 1986, Campbell and Carpenter 1986, Parslow *et al.* 1986, Nicholls 1987). It now appears certain that an important fraction of the total living biomass in marine communities passes through the protozoa in these environments. These 'new' ideas concerning the function of heterotrophic aquatic microorganisms, and in particular the role of protozoa in these processes, are not new discoveries; their ecological significance has been largely overlooked until relatively recently. They present the possibility of two potentially conflicting views of nutrient flow through the microbial loop, as summarized in Figure 1. The arrows in this figure indicate pathways of nutrient flow among the microbial groups. Thicker arrows indicate pathways of greater importance. This flow diagram is not meant to be a complete representation of our present knowledge of microbial trophic relationships, but rather to emphasize major differences in nutrient and energy flow according to the 'link or sink' hypotheses.

In the first view (Figure 1A) nutrients contained in the organic substrates consumed by bacteria are largely incorporated into bacterial biomass (arrow 1) and subsequently into their consumers (arrows 2,3), presenting the possibility

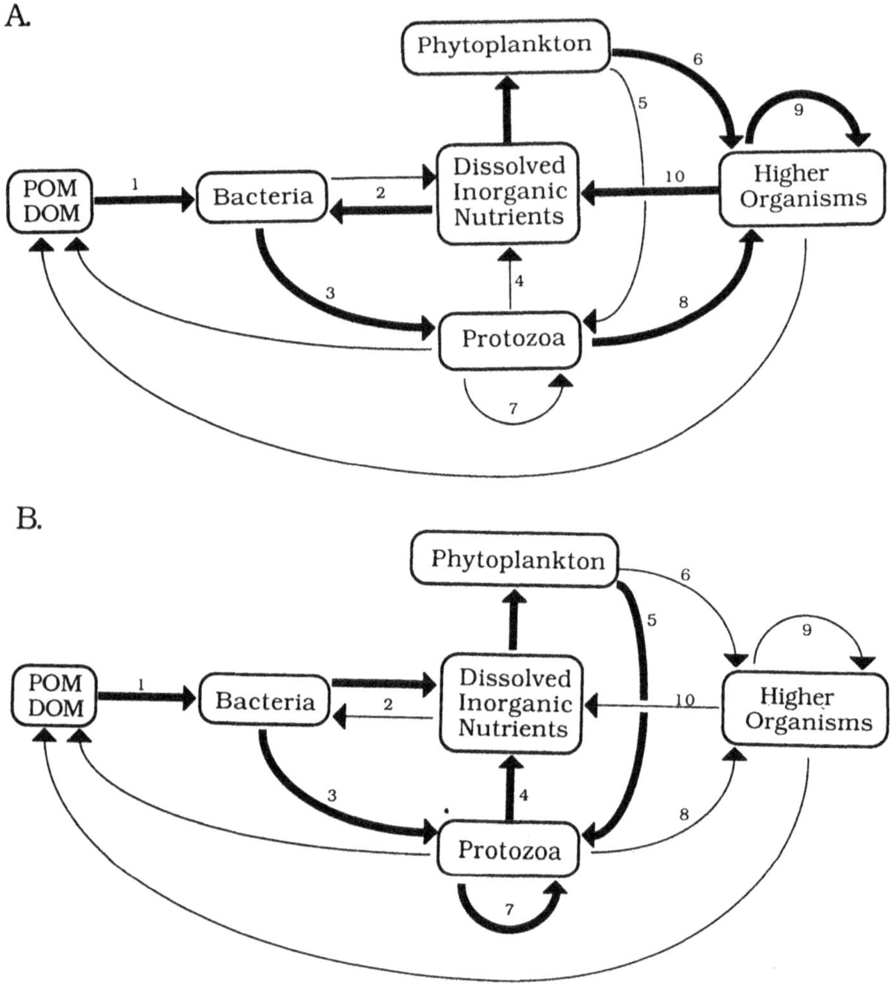

Fig. 1. Idealized models for nutrient cycling through the 'microbial loop'of aquatic communities. The thickness of the arrows indicates the hypothesized relative importance of the pathway for nutrient flow. A). A microbial loop characterized by a high efficiency of transfer of nutrients to higher organisms. B). A microbial loop characterized by a high efficiency of nutrient remineralization

of a significant trophic 'link' between detrital material and higher organisms (arrow 8). Nutrient remineralization in this scenario is dominated by the larger zooplankton (arrow 10), and the efficiencies of transfer through the various microbial groups are high. Accordingly, nutrient remineralization in the microbial loop component of the community is small (arrow 4).

In the opposing view (Figure 1B), it is contended that most of the nutrients contained in the bacterial substrates are remineralized directly by the bacteria (arrow 2), or by their primary consumers (arrow 4). In this case, very little nutrient is available to higher organisms as living biomass, and thus the microbial loop functions as a nutrient 'sink'. The fate of the nutrients taken up by the phytoplankton must also be considered in this scheme. To some extent the size structure, species composition, and perhaps rate of production of the phytoplankton community will determine the percentage of this assemblage that is grazed directly by the larger zooplankton (arrow 6), and the percentage that is available to the protozoa. If trophic efficiency is <100%, a protozoan trophic link would decrease the efficiency of nutrient transfer to the zooplankton and support the supposition that protozoa are primarily remineralizers in aquatic food webs. Additionally, a high degree of trophic complexity within the protozoan assemblage (arrow 7) would decrease the efficiency of transfer of nutrients to higher organisms.

FACTORS CONTROLLING PROTOZOAN NUTRIENT CYCLING

In the real world, it is probable that the extreme cases depicted in Figure 1 are rare, and the role aquatic microbes play is between these extremes. The importance of protozoa as nutrient assimilators and nutrient regenerators, and their contribution to nutrient cycling in the ocean is determined by a number of factors. These are 1) the percentage of organic matter that is consumed by protozoa relative to the amount consumed by other organisms (which in turn is controlled by protozoan biomass and feeding rates relative to other organisms), 2) the nutrient content of their food, and 3) the efficiency of incorporation of food constituents into protozoan biomass.

Ultimately, the magnitude of protozoan grazing pressure will dictate the overall importance of these organisms in oceanic nutrient flow. However, other considerations control the degree to which protozoa incorporate or remineralize the nutrients that they ingest. The remainder of this paper will focus on the factors controlling the fate of ingested nutrients, and discuss the importance of various protozoan and other microbial groups with respect to these considerations.

CELL SIZE CONSIDERATIONS

The nutrient excretion rates of a large number of marine metazoa have been measured and reported in the literature (Hargrave and Geen 1968, Ikeda 1974, Mullin *et al.* 1975, Ikeda *et al.* 1982, Vidal and Whitledge 1982, Verity 1985). These data display a positive correlation with animal body mass. When expressed on a weight-specific basis, however, nutrient excretion rates of marine metazoa (and animals in general) are seen to decrease with increasing body size. That is, as animal size increases, nutrient excretion rate **per unit body weight** decreases. This well-known relationship is intuitively obvious to most biologists, but has rarely been extrapolated to single-celled organisms.

The potential implications of this relationship for protozoan nutrient cycling are expressed in the hypothetical data of Figure 2. The absolute body weights of a bacterivorous microflagellate of 5μm diameter (specific gravity = 1.0) and a 70 kg human being, and the biomass that they might consume in a day are portrayed in Figure 2A. These values assume that the microflagellate eats 100 bacteria hr^{-1} (each bacterium = 0.3 μm^3; specific gravity = 1.0) and the human eats 1 kg of food day^{-1}. As expected from their vast differences in body size (>14 orders of magnitude), the absolute amount of food consumed per individual is much greater (≈ 12 orders of magnitude) for the human. However, when the biomass of the microflagellate and the human are normalized to 1 g (Figure 2B), the rate of food consumption for the microflagellate **per unit biomass** is approximately 3 orders or magnitude greater

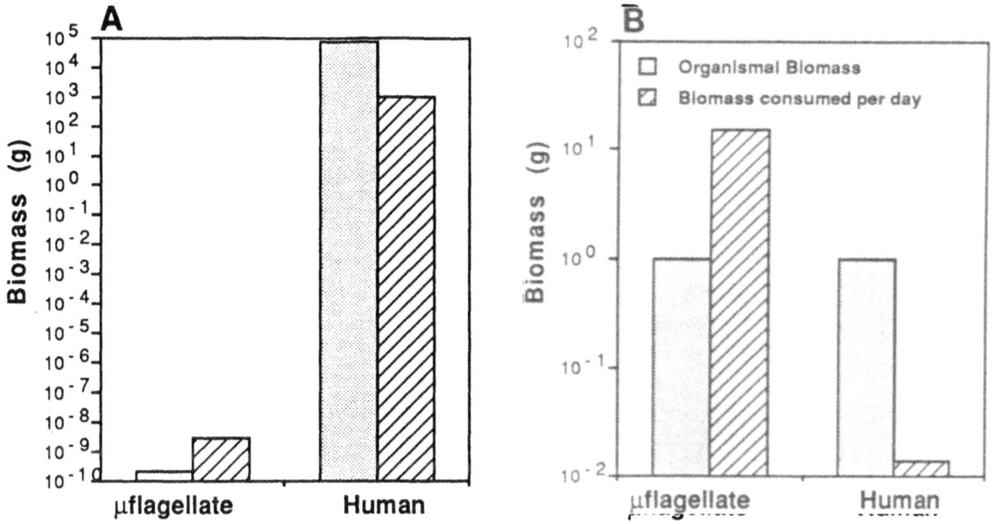

Fig. 2. Comparison of the approximate amount of biomass consumed per individual (A) and per gram of body weight (B) for a 70 kg human being and a 5µm diameter bacterivorous microflagellate. The calculated body weight turnover rates for these organisms would be approximately 0.07 days and 70 days for the microflagellate and human, respectively

than the rate for the human. This result of the relationship between size and weight-specific metabolic rate (i.e. the more rapid rate of turnover of the smaller organism), although seemingly obvious, is often overlooked in conceptualizations of the flow of energy and nutrient in aquatic communities.

Unfortunately, at present there are only a handful of nutrient excretion rates for protozoa to compare with the substantial data set on metazoan excretion rates. It is therefore difficult to determine the validity of extrapolations of the weight-specific excretion rates of marine zooplankton to protozoan-sized organisms. However, the few protozoan excretion rate measurements that are available do not contradict the possibility that, in a general sense, this extrapolation may be valid (Figure 3, summarized from Caron and Goldman, in press). A similar comparison using a larger data set on protozoan respiration rate has given relatively similar results (Fenchel and Finlay 1983, Caron *et al.* in press). If the relationship described in Figure 3 continues to remain true as more data are amassed, then an important role for protozoa in nutrient cycling in the marine environment is indicated. At the very least, it is clear that the absolute weight-specific excretion rates of protozoa that have been determined to date are generally much greater than the weight-specific rates of most metazoa.

It should be noted that the relationship between body size and organismal metabolic rate does not necessarily address the 'link or sink' question concerning the fate of nutrient ingested by protozoa. The nitrogen excretion data of Figure 3 have been used to demonstrate weight-specific nutrient excretion rate, but it is possible that a similar relationship exists between body size and weight-specific nutrient incorporation rate. For example, given the ingestion rate information hypothesized in Figure 2, and assuming a gross growth efficiency for the microflagellate of 50%, then this protozoan would incorporate and excrete > five times its own weight in bacterial biomass per day. Reducing the growth efficiency of the microflagellate to 10% would still result in the incorporation by the protozoan of an amount of food that would be greater than its own weight per day. This rate is clearly far in excess of the possible weight-specific excretion rate and weight-specific incorporation rate of a human.

FOOD QUALITY AND NUTRIENT CYCLING

The example provided in Figure 2 and discussed above points out two important features of the relationship between an organism's size and its weight-specific metabolic rate. First, it is possible that small organisms may contribute more to nutrient incorporation and remineralization than larger organisms because, on a weight-specific basis they simply process much more food per unit time than larger organisms. Using this reasoning, it is predicted that protozoa may

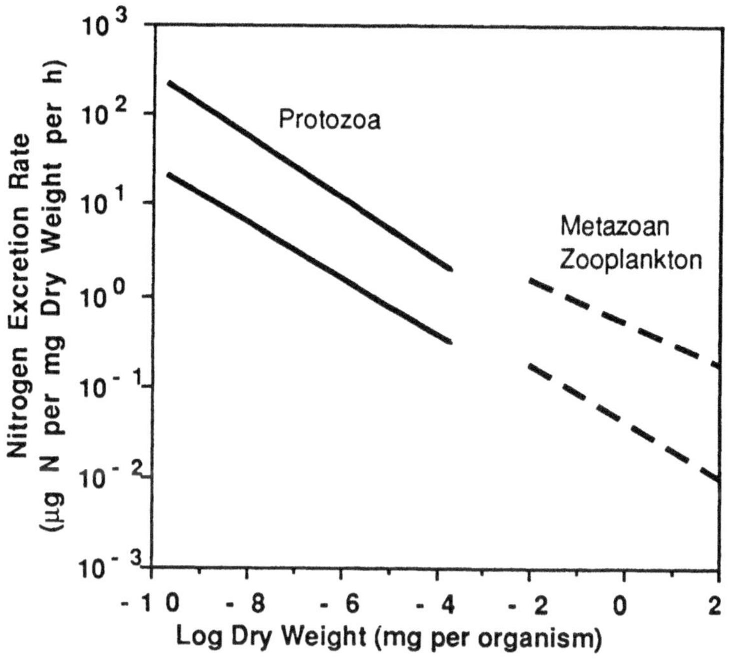

Fig. 3. Summary of the weight-specific ammonium excretion rates of metazoan zooplankton and protozoa. The lines encompass the range of values provided in individual studies. Metazoan data are a summary of several hundred individual measurements from five studies and the data on protozoa are a summary of approximately a dozen studies, from Caron and Goldman in press

contribute more significantly to nutrient cycling in aquatic communities than metazoan zooplankton. Second, the example indicates the effect of incorporation efficiency as a major factor controlling the relative contribution of the organism to nutrient incorporation or nutrient remineralization (the link or sink aspect).

It is also clear from this example that the 'link ' and 'sink' hypotheses are two sides of the same coin. Ingested nutrients that are not incorporated by a microbial population (the link aspect) are largely remineralized and returned to the pools of inorganic nutrients for subsequent uptake and incorporation by the primary producers (the sink aspect). By the same reasoning, nutrients that are remineralized cannot, at the same time, be passed up the food chain. Therefore, we are not dealing with two distinct processes, but with complementary facets of the same process.

What factors control the proportions of ingested nutrient that are remineralized or incorporated? From a theoretical viewpoint, the efficiency with which a protozoan will incorporate an ingested nutrient into its own biomass is determined by the amount of the nutrient contained in an ingested food item, and the amount of nutrient required to produce protozoan biomass from that amount of food. The former term is, of course, determined by the chemical nature of the protozoan prey, while the latter term is dependent upon the unique stoichiometry of the protozoan (i.e. the proportions of carbon, nitrogen and phosphorus in the cell) and the gross growth efficiency of the organism (i.e. how much of the carbon from the food is converted to protozoan biomass).

There have been several formal descriptions of this relationship for microbial populations (Fenchel and Blackburn 1979, Blackburn 1983, Lancelot and Billen 1985, Caron and Goldman 1988, and in press). Caron and Goldman (1988) have expressed the nutrient excretion rate of a protozoan as:

$$E = \frac{R}{1 - GGE} \times \left(\frac{1}{C{:}Nu_{prey}} - \frac{GGE}{C{:}Nu_{pro}} \right)$$

where E, R, GGE and $C{:}Nu_{pro}$ are the nutrient excretion rate, respiration rate, gross growth efficiency and carbon:nutrient ratio of the protozoan, respectively, and $C{:}Nu_{prey}$ is the carbon:nutrient ratio of the prey organism. Gross growth efficiency and respiration are both expressed in units of carbon in this equation. Note that this formalization assumes that all ingested carbon and nutrients are either incorporated into consumer biomass or released in remineralized form. This assumption has been shown to be reasonable for some populations (e.g. bacteria, Lancelot and Billen 1985), but its applicability for other organisms remains to be demonstrated. This relationship recognizes the complementary nature of nutrient incorporation and remineralization.

The elemental composition of the food of a protozoan ($C{:}Nu_{prey}$) will depend upon the nutrional mode of the species. A useful relationship for comparing the nutrient content of different organisms has been the Redfield ratio of 106C:16N:1P atoms (Redfield *et al.* 1963). Although there are extremely few data available, it appears that at least some protozoa produce biomass that is at times close to this relationship (Goldman *et al.* 1987b). In contrast, the prey of some protozoa may be quite different from this relationship (Table 1).

On the one hand, the C:N and C:P ratios of bacterial biomass typically are less than the proportions of the Redfield ratio. Therefore, bacterivorous protozoa generally would consume biomass that contains an excess of major nutrients (relative to carbon) compared to the protozoan biomass that is being produced from this food. On the other hand, the C:N and C:P ratios of phytoplankton biomass can be high compared to the Redfield ratio due to their ability to store carbon-rich compounds when nutrients are not available, and then convert these carbon skeletons into protein and other cellular constituents as nutrients become available (Cuhel *et al.* 1984). This is particularly true for phytoplankton grown under severe nutrient limitation. Therefore, herbivorous protozoa might consume biomass that contains a nutrient deficiency (relative to carbon) compared to the protozoan biomass that is being produced. These contrasting situations may have highly significant, and different, consequences for the percentages of nutrient incorporated or remineralized by bacterivorous and herbivorous protozoa. Based on these considerations of the nutritional mode of marine protozoa, these microorganisms may play a role as nutrient assimilators or remineralizers in the ocean.

Table 1. APPROXIMATE RANGES OF CARBON:NITROGEN (C:N) AND CARBON:PHOSPHORUS (C:P) RATIOS (BY ATOMS) FOR PHYTOPLANKTON AND BACTERIA FROM SOME RECENT STUDIES

Population	C:N Ratio	C:P Ratio	Reference
Phytoplankton			
Dunaliella tertiolecta	7.8-22.7		Caperon and Meyer (1972)
Coccochloris stagnina	7.9-13.3		
Cyclotella nana	4.3-20		
Monochrysis lutheri	10.9-19.6		
M. lutheri	7.1-11.3	106-1300	Goldman *et al.* (1979)
D. tertiolecta	7.0-20.0	35-600	
Thalassiosira pseudonana	7.1-12.6	63-106	
D. tertiolecta	6-21		Goldman and Peavey (1979)
Thalassiosira fluviatilis	5.1-25		Laws and Bannister (1980)
Phaeodactylum tricornutum	6.1-18	95-500	Terry *et al.* (1985)
Hymenomonas elongata	7.9-12.8		Claustre and Gostan (1987)
P. tricornutum	6.8-11.5	89-359	Goldman *et al.* (1987b)
D. tertiolecta	7.8-16.5	72-521	
Bacteria			
Arthrobacter globiformis	5.1-6.9	46-83	vanVeen and Paul (1979)
Enterobacter aerogenes	3.8-6.2	48-110	
Klebsiella pneumoniae	4.0-4.9		Esener *et al.* (1982)
Pseudomonas putida	4.5-5.6	16.1-500	Bratbak (1985)
Mixed assemblage	4.8-6.7	7.7-56	
Mixed assemblage	3.8-8.6		Nagata (1986)
Mixed assemblage	5.3		Berman *et al.* (1987)
Mixed assemblage	4.2-8.7	39-131	Goldman *et al.* (1987a)
Mixed assemblage	2.9-5.0		Lee and Fuhrman (1987)
Pasteurella sp.	4.6	63.3	Caron (unpubl.)

The gross growth efficiencies of protozoa that have been reported also do not clearly indicate whether these organisms will function primarily as nutrient incorporators or remineralizers. Growth efficiencies of 2 to 82% have been reported in the literature (summarized in Caron and Goldman in press). However, the median values for this growth parameter tend to be high compared to larger organisms (~ 40-50%). These values would tend to indicate a significant but not overwhelming role for protozoa as 'links' in the movement of nutrients through food webs. Their importance for the transfer of nutrients to higher organisms should become greater if protozoa are feeding on prey with high C:N and C:P ratios (which would promote efficient retention of the growth-limiting nutrient). Unfortunately, it has proven to be exceedingly difficult to obtain experimental information on microbial populations to confirm the theoretical considerations discussed above. Thus far this goal has been attained with limited success for bacterial assemblages (reviewed by Lancelot and Billen 1985). This success is due primarily to two simplifying features of bacterial growth: bacteria produce biomass of relatively constant stoichiometry and, for the most part, they release carbon and nutrients only in remineralized form.

The validity of these simplifying assumptions are more questionable for protozoa. For example, it is unclear to what degree the C:N:P ratio of protozoan biomass can vary. Similarly, it is not yet known which element or compound exerts the most control in determining the utilization of prey biomass, and whether or not this feature is the same for all protozoan species. The extent to which nutrient and carbon utilization can be uncoupled to allow more efficient extraction and retention of a growth-limiting element from prey of varying stoichiometry is also unknown.

Another potentially important consideration for nutrient incorporation and/or remineralization by protozoa is the effect of starvation on nutrient release. Rapid temporal oscillations in population abundance have been documented for natural protozoan assemblages (Fenchel 1982d, Andersen and Fenchel 1985, Bjørnsen *et al.* 1988), and a 'boom and bust' strategy for population growth has been proposed for these populations. Although protozoan metabolic rates decrease rapidly during starvation, metabolism continues at a significant but reduced rate in non-encysting species. Nutrient excretion during this period apparently reflects the nutrients that are associated with the cellular constituents that are autophagocytized at this time (Fenchel 1982c). Based on recent studies with phagotrophic protists, it appears that nitrogen-rich compounds are metabolized more rapidly than phosphorus-rich compounds during starvation (Goldman *et al.* 1985, 1987b, Andersen *et al.*

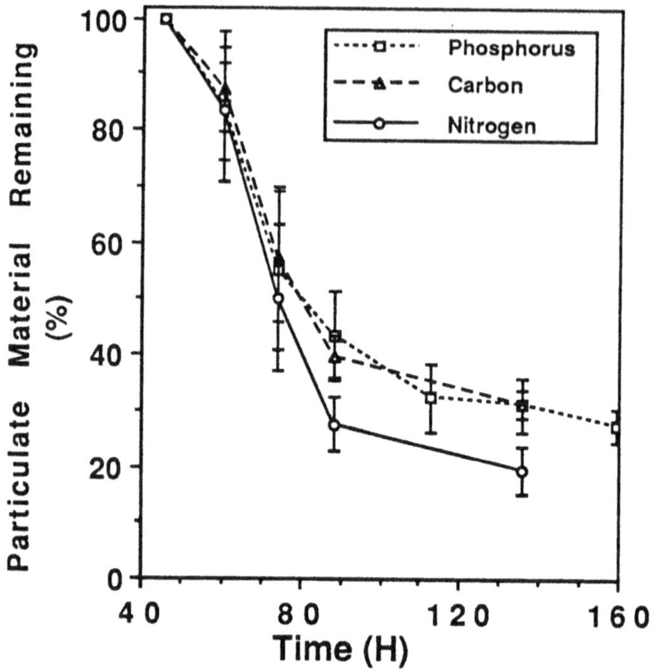

Fig. 4. Losses in particulate carbon, nitrogen and phosphorus (± 1 standard deviation) from twelve cultures of the freshwater mixotrophic phytoflagellate *Poterioochromonas malhamensis* growing heterotrophically on a heat-killed bacterium. Time = 0 hr was the time of phytoflagellate inoculation. Summarized from Caron, Porter and Sanders (in press)

1986), as evidenced by the observation that phosphorus is retained more efficiently than nitrogen by the protists during this growth stage (Figure 4). These results indicate that the fate of a nutrient ingested by a protozoan is dependent not only on the growth efficiency and elemental composition of the protozoan, but also on the growth stage of the organism.

BACTERIAL NUTRIENT CYCLING

A pivotal piece of information in deciphering the role of marine protozoa in oceanic nutrient cycles is, unquestionably, the activity of the bacterial assemblage. Bacteria directly affect the contribution of protozoa to nutrient cycling in the ocean by serving as the main food source for many of these protists. In addition, however, bacterial nutrient incorporation and/or remineralization theoretically can proceed faster than protist nutrient cycling because of the generally more rapid metabolic rates for these heterotrophic prokaryotes (this can be predicted from extrapolation of Figure 3 to bacteria-sized organisms).

Bacterial nutrient incorporation and remineralization is, however, clearly controlled by the chemical composition of the substrate. This effect has been demonstrated with natural assemblages of bacteria in the field (Hollibaugh 1978, Lancelot and Billen 1985), and with laboratory cultures (Goldman *et al.* 1987a, Caron *et al.* 1988). Increasing the carbon:nutrient ratio of the bacterial substrate results in the remineralization and release of a smaller percentage of the nutrient contained in the substrate (Goldman *et al.* 1987a). Furthermore, at least some bacteria possess high affinity uptake systems for inorganic nitrogen and phosphorus. Therefore, in extreme cases where a major nutrient is present in the bacterial substrate in quantities that are insufficient for growth, bacteria may compete with the primary producers for the pool of dissolved inorganic nutrient in the surrounding water (Currie and Kalff 1984a,b,c, Berman 1985, Wheeler and Kirchman 1986).

The ecological implications of this behavior of bacteria, and the role of protozoa in this scenario, has been demonstrated using a model system of a diatom, a natural bacterial assemblage, and a bacterivorous protozoan (Figure 5, information summarized from Caron *et al.* 1988). In this system, growth of the diatom population (which was not preyed upon by the protozoan) was used as an indicator of nitrogen availability. The results of this experiment clearly showed that when the bacteria were provided with a low carbon:nitrogen substrate (glycine), nitrogen was readily remineralized and released by the bacteria (as evidenced by a higher algal

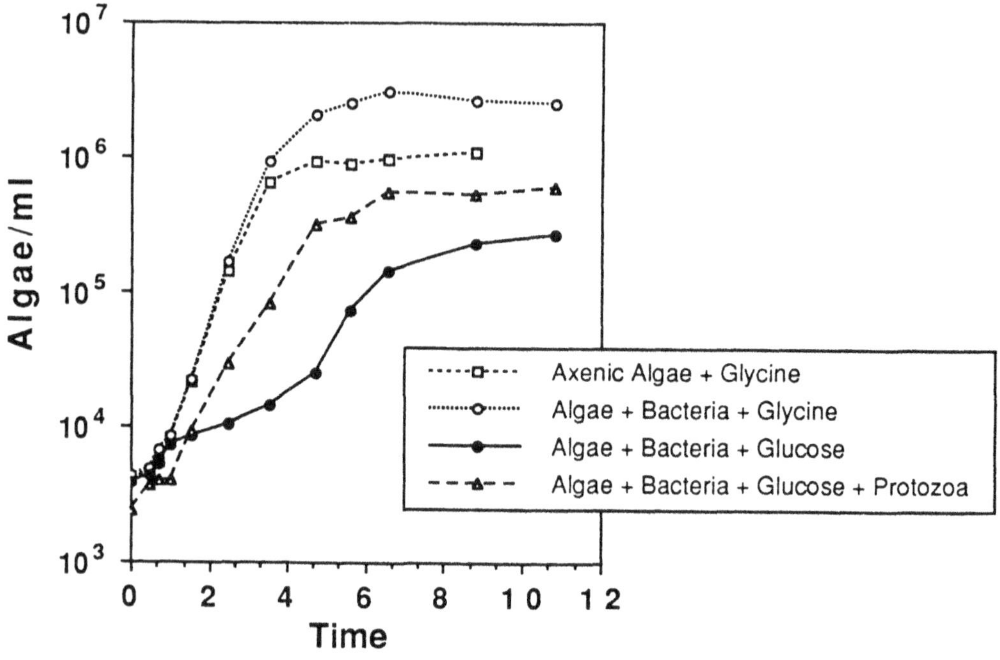

Fig. 5. Growth of the diatom *Phaeodactylum tricornutum* in four nitrogen-limited culture treatments, summarized from Caron *et al.* (1988). The treatments were an artificial seawater medium with the following constituents and microbial components: an axenic culture of the diatom with 50 µM of nitrogen as NH_4^+ and 100 µM of nitrogen as glycine ('Alage + Glycine'), a bacterized culture of the diatom with the same nitrogen sources ('Algae + Bacteria + Glycine'), a bacterized culture of the diatom with 50 µM of nitrogen as NH_4^+ and 1000 µM of carbon as glucose ('Algae + Bacteria + Glucose'), and the same chemical and microbial components as in the latter treatment but with a bacterivorous microflagellate added ('Algae + Bacteria + Glucose + Protozoa')

biomass relative to the axenic culture with glycine). When protozoa were inoculated into this bacterized culture, diatom growth was unaffected. In contrast, when a nitrogen-free substrate (glucose) was provided for the bacterial assemblage, the bacteria competed for the pool of inorganic nitrogen in the culture, and diatom growth was stunted. However, when bacterivorous protozoa were inoculated into a replicate flask of this treatment, nitrogen-limitation of the diatom was partially relieved due to consumption of the bacteria by the protozoa and remineralization and release of some of the nitrogen contained in the bacterial biomass. These results indicate the pivotal role that the carbon:nutrient ratio of the bacterial substrate plays in determining the changing role of the bacteria in nutrient remineralization or incorporation, and the cascading effect that this behavior has on the relative contribution of protozoa to nutrient regeneration.

IMPLICATIONS FOR NUTRIENT CYCLING AND REGENERATION

It will be some time before a reliable base of information exists from which to estimate the overall importance of protozoa to oceanic nutrient remineralization. However, based on the size and food quality considerations discussed above, some generalities concerning the relative importance of various protozoan groups in this process can be predicted (Figure 6).

Based purely on the relationship between weight-specific metabolic rate and the size of an organism, as described in Figures 2 and 3, it is predicted that protozoa play a greater role in nutrient cycling than larger zooplankton of similar nutritional mode for equivalent biomasses of these populations. This relationship does not predict whether the nutrient passing into the protozoa will be incorporated or remineralized, but only that the protozoa are capable of processing more food per unit time than an equivalent biomass of zooplankton.

Similarly, it is predicted that, for equivalent biomasses of small and large protozoa of similar nutritional mode, small protozoa play a greater role in nutrient cycling because of their higher weight-specific metabolic rates.

Exclusive of size considerations, food quality for similarly-sized organisms must be a major consideration for determining the percentages of an ingested nutrient that are incorporated or regenerated. Based on the theoretical considerations given above, it is concluded that bacterivorous protozoa should be

Protozoa $>$ Large Zooplankton

Small Protozoa $>$ Large Protozoa

Bacterivorous Protozoa $>$ Herbivorous Protozoa

Bacteria $\lessgtr$ Bacterivorous Protozoa

Fig. 6. Estimated importance in aquatic nutrient remineralization of different protozoan groups based on size and nutritional mode

more efficient remineralizers of major nutrients than herbivorous protozoa because of the large disparities that often occur in the C:N and C:P ratios of bacterial and phytoplankton biomass (Table 1). The large 'excess' of nitrogen and phosphorus contained in bacterial biomass will be released largely in remineralized form when these organisms are consumed by bacterivores (Goldman *et al.* 1985, Gude 1985, Andersen *et al.* 1986). In contrast, little remineralized nutrient should be released during the active growth phase of a protozoan consuming phytoplankton that have been grown under severe nutrient-limitation. The extrapolation of weight-specific metabolic rates to the bacteria would indicate that bacteria must be more important than the protozoan population in nutrient cycling. However, several factors complicate this generalization. Bacteria appear to be more capable of remaining dormant but viable in aquatic systems than higher organisms, and much of the bacterial biomass in an ecosystem may be metabolically inactive at any particular time (Meyer-Reil 1978, Stevenson 1978, Kogure *et al.* 1979, Roszak and Colwell 1987). Therefore, the biomass of bacteria in an ecosystem might be a significant overestimate of the 'active' biomass of this assemblage. In addition, it must be remembered that, although bacteria constitute an important food source for many protozoa, they are not the sole food source available to these organisms. Because food sources other than the bacteria exist for many protozoa (such as cyanobacteria and algae), the overall importance of these protists for nutrient and energy cycling in the marine environment is not totally dictated by bacterial biomass and activity.

The relative importance of bacteria and protozoa in the process of nutrient remineralization also is not settled. As described above, the chemical composition of the substrate available to the bacteria will strongly affect the importance of this assemblage in nutrient regeneration (Hollibaugh 1978, Lancelot and Billen 1985, Goldman *et al.* 1987a). For example, based on the available literature, it appears that the flux into the bacterioplankton of substrates with high C:N ratios exceeds the flux of substrates with low C:N ratios (Burney 1986a,b, Laacke *et al.* 1983), and that amino acids and inorganic nitrogen may be taken up by bacteria as a source of nitrogen (Ferguson and Sunda 1984, Fuhrman and Ferguson 1986, Wheeler and Kirchman 1986). This consideration complicates any generalization that might be attempted on the importance of bacteria as nutrient regenerators because it presents the possibility that bacteria may compete for, rather than contribute to, the pool of regenerated nutrients. It does not, however, diminish their important role in nutrient cycling.

The hypothesized relationships described above for the importance of various microbial groups in nutrient cycling are gross generalizations for equivalent weights of the different groups of microorganisms. The overall contribution of these various populations in the environment is, of course, dictated by the amount of biomass of each group present and the rate of food supply. A clearer understanding of the role that protozoa play in nutrient flow in the ocean will evolve as these latter features of the microbial community are established.

FUTURE GOALS AND DIRECTIONS

1). As a result of a concerted effort by a large scientific community over the past 20 years, the basic framework for conceptualizing the role of aquatic microbial populations, including protozoa, has begun to take shape (Pomeroy 1974, Sieburth 1976, Fenchel 1969, 1980, 1982a-d, Azam *et al.* 1983, Goldman and Caron 1985). However, less than a few dozen measurements exist on the nutrient remineralization rates of protozoa (Caron and Goldman in press), and several of these values are questionable due to the unknown contribution of living bacteria to nutrient cycling in these cultures. Much more data on nutrient assimilation and remineralization by individual species of protozoa will be required to construct accurate models and to enable a clearer understanding of protozoan nutrient cycling.

It is imperative that these future studies avoid, or at least determine the extent of, the perplexing problems of deciphering nutrient cycling in multi-species assemblages. Important considerations for future studies must be the presence and activity of prey in the culture vessels, and the elimination of this complication if at all possible. In addition, special attention must be given to the possibility of cell damage to the protozoa during experimental manipulation of cultured species. These potential artifacts are not new considerations, but they still present formidable impediments to experimental design and interpretation.

2). A determination of the extent to which bacteria assimilate or regenerate nutrients in aquatic environments is central to the question of protozoan nutrient cycling. To a large extent the activity of the bacteria dictates the relative importance of protozoan nutrient cycling (Figure 5). The potential for uptake of dissolved inorganic nutrients by the bacteria is significant, and the relative importance of nutrient release and uptake by this assemblage *in situ* is presently unknown. Because of the large contribution of the bacteria to the metabolic activity of aquatic communities (Sorokin 1981, Williams 1981), a major factor in determining the overall importance of nutrient regeneration by the protozoa is determining the activity of the bacteria. To a large extent, the activity of the bacteria will be controlled by the chemical composition (C:N:P) of algal exudates, and dissolved or particulate material released by heterotrophs. In general, we require better characterization of the composition of bacterial substrates in order to settle the question of bacterial nutrient cycling.

3). Protozoan biogeography also must be fostered in order to determine the overall importance of protozoa to oceanic nutrient cycles. As described above, the relationship between organismal size and weight-specific metabolic rate favors the contention that protozoa play an active and important role in nutrient cycling in aquatic ecosystems. The overall importance of the protozoan community is, however, dependent upon the abundance and biomass of these organisms relative to other populations. The biogeography of most protozoan species is still lacking, or in a very rudimentary stage. The exceptions to this rule are many of the planktonic sarcodines for which global patterns of abundance have been established (Bé 1977, Dworetzky and Morley 1987), and a few 'landmark' studies conducted in the oceanic plankton (Beers *et al.* 1980, 1982, studies summarized in Sorokin 1981) and in the benthos (Fenchel 1967, 1969) that have included other major taxa. Few other studies have attempted the prodigious task of characterizing the prokaryote, protistan and metazoan population abundances and biomass in an environment. However, it is imperative that the quantitative importance of protozoa (numbers, but more importantly, biomass) be established **relative to other plankton groups**, in order to be able to apply metabolic information on these protists to estimate their overall contribution to ecosystem activity.

4). Two groups for which nutrient studies may prove to be interesting are protozoan symbioses and mixotrophic protozoa. Protozoan symbioses have been documented in the marine environment for many years (e.g. Taylor *et al.* 1971, Stoecker 1990) and laboratory and field studies on chloroplast-retaining ciliates and mixotrophic algae are appearing at a rapid rate (Estep *et al.* 1986, Stoecker *et al.* 1987, Porter 1988, Laval-Peuto and Rassoulzadegan 1988, Boraas *et al.* 1988, Sanders and Porter 1988). These systems are potentially ideal models for studies on nutrient transfer (and perhaps conservation) between heterotrophs and phototrophs. However, the importance of nutrient transfer between host and symbiont remains almost completely speculative at the present (Jorgensen *et al.* 1985), and studies on nutrient cycling among mixotrophic algae are only now beginning.

5). Finally, it is generally accepted that the structure of the food web must certainly affect the efficiency of nutrient transfer to higher organisms in the ecosystem, but there are few data on which to base this generalization. To some extent the size structure of the phytoplankton community will determine whether the grazers of this biomass are primarily larger zooplankton or protozoa and micrometazoa (arrows, 5, 6 in Figure 1). Clearly, the percentage of the primary productivity that passes through the bacterial assemblage will affect the amount of nutrient flowing through the protozoa if these protists are the dominant consumers of bacterial biomass. Much more work will be required before we will obtain a clear understanding of the factors that control the importance of the microbial loop relative to the 'classical' plankton food chain. Our understanding of the importance of protozoa in oceanic nutrient cycles will increase, and our present conceptualization of this process will undoubtedly evolve, as information on these aspects of carbon and nutrient flow through the microbial loop becomes available.

REFERENCES

Andersen OK, Goldman JC, Caron DA, Dennett MR (1986) Nutrient cycling in a microflagellate food chain: III. Phosphorous dynamics. Mar Ecol Prog Ser 31:47-55

Andersen P, Fenchel T (1985) Bacterivory by microheterotrophic flagellates in seawater samples. Limnol Oceanogr 30:198-202

Anderson, OR (1980) Radiolaria. In: Levandowsky M, Hutner S (eds) Biochemistry and physiology of protozoa. Academic Press, New York, Vol 3, p 1

Anderson OR, Spindler M, Bé AWH, Hemleben C (1979) Trophic activity of planktonic foraminifera. J Mar Biol Ass UK 59:791-799

Azam F, Fuhrman JA (1984) Measurement of bacterioplankton growth in the sea and its regulation by environmental conditions. In: Hobbie JE, Williams PJL (eds) Heterotrophic activity in the sea. Plenum Publishing Corp, New York, p

179
Azam F, Fenchel T, Field JG, Gray JS, Meyer-Reil LA, Thingstad F (1983) The ecological role of water-column microbes in the sea. Mar Ecol Prog Ser 10:257-263
Barsdate RJ, Fenchel T, Prentki RT (1977) Phosphorus cycle of model ecosystems: significance for decomposer food chains and effect of bacterial grazers. Oikos 25 :239-251
Bé AWH (1977) An ecological, zoogeographic and taxonomic review of recent planktonic foraminifera. In: Ramsay ATS (ed) Oceanic micropaleontology. Academic Press, London, p 1
Beers, JR, Reid FMH, Stewart GL (1980) Microplankton population structure in southern California nearshore waters in late spring. Mar Biol 60:209-226
Beers JR, Reid FMH, Stewart GL (1982) Seasonal abundance of the microplankton population in the north Pacific central gyre. Deep-Sea Res 29:227-245
Berman T (1985) Uptake of [^{32}P] orthophosphate by algae and bacteria in Lake Kinneret. J Plank Res 7 (1):71-84
Berman T, Nawrocki M, Taylor GT, Karl DM (1987) Nutrient flux between bacteria, bacterivorous nanoplanktonic protists and algae. Mar Ecol Prog Ser 2:69-82
Bjørnsen PK, Riemann B, Horsted SJ, Nielsen TG, Pock-Sten J (1988) Trophic interactions between heterotrophic nanoflagellates and bacterioplankton in manipulated seawater enclosures. Limnol Oceanogr 33:409-420
Blackburn TH (1983) The microbial nitrogen cycle. In: Krumbein WE (ed) Microbial geochemistry. Blackwell Scientific Publications, Oxford, p 63
Boraas ME, Estep KW, Johnson PW, Sieburth JMcN (1988) Phagotrophic phototrophs: the ecological significance of mixotrophy. J Protozool 35:249-252
Bratbak G (1985) Bacterial biovolume and biomass estimation. Appl Environ Microbiol 49:1488-1493
Burney CM (1986a) Bacterial utilization of *in situ* dissolved carbohydrate in offshore waters. Limnol Oceanogr 31:427-431
Burney CM (1986b) Diel dissolved carbohydrate accumulation in coastal water of South Florida, Bermuda and Oahu. Estuarine Coast Shelf Sci 23:197-203
Campbell L, Carpenter EJ (1986) Estimating the grazing pressure of heterotrophic nanoplankton on *Synechococcus* spp. using the sea water dilution and selective inhibitor techniques. Mar Ecol Prog Ser 33:121-129
Caperon J, Meyer J (1972) Nitrogen-limited growth of marine phytoplankton. I. Changes in population characteristics with steady-state growth rate Deep-Sea Res 19:601-618
Capriulo GM (1982) Feeding of field collected tintinnid microzooplankton on natural food. Mar Biol 71:73-86
Capriulo GM, Carpenter EJ (1980) Grazing by 35 to 202 µm microzooplankton in Long Island Sound. Mar Biol 56:319-326
Capriulo GM, Carpenter EJ (1983) Abundance, species composition and feeding impact of tintinnid micro-zooplankton in central Long Island Sound. Mar Ecol Prog Ser 10:277-288
Caron DA, Bé AWH (1984) Predicted and observed feeding rates of the spinose planktonic foraminifer *Globigerinoides sacculifer*. Bull Mar Sci 35:1-10
Caron DA, Goldman JC (1988) Dynamics of protistan carbon and

nutrient cycling. J Protozool 35:247-249
Caron DA, Goldman JC (in press) Nutrient regeneration. In: Capriulo GM (ed) The ecology of marine protozoa. Oxford, New York
Caron DA, Goldman JC, Andersen OK, Dennett MR (1985) Nutrient cycling in a microflagellate food chain: II. Population dynamics and carbon cycling. Mar Ecol Prog Ser 24:243-254
Caron DA, Goldman JC, Fenchel T (In press) Protozoan respiration and metabolism. In: Capriulo GM (ed) The ecology of marine protozoa. Oxford, New York
Caron DA, Goldman JC, Dennett MR (1988) Experimental demonstration of the roles of bacteria and bacterivorous protozoa in plankton nutrient cycles. Hydrobiologia 159:27-40
Claustre H, Gostan J (1987) Adaptation of biochemical composition and cell size to irradiance in two microalgae: possible ecological implications. Mar Ecol Prog Ser 40:167-174
Cuhel RS, Ortner PB, Lean DRS (1984) Night synthesis of protein by algae. Limnol Oceanogr 29:731-734
Currie DJ, Kalff J (1984a) A comparison of the abilities of freshwater algae and bacteria to acquire and retain phosphorus. Limnol Oceanogr 29:298-310
Currie DJ, Kalff J (1984b) The relative importance of phytoplankton and bacterioplankton in phosphorus uptake in freshwater. Limnol Oceanogr 29:311-321
Currie DJ, Kalff J (1984c) Can bacteria outcompete phytoplankton for phosphorus? A chemostat test. Microb Ecol 10:205-216
Ducklow HW, Hill SM (1985) The growth of heterotrophic bacteria in the surface waters of warm core rings. Limnol Oceanogr 30:239-259
Ducklow HW, Hill SM (1985) Tritiated thymidine incorporation and the growth of heterotrophic bacteria in warm core rings. Limnol Oceanogr 30:260-272
Ducklow HW, Purdie DA, Williams PJleB, Davies JM (1986) Bacterioplankton: a sink for carbon in a coastal marine plankton community. Science 232:865-867
Dworetzky BA, Morley, JJ (1987) Vertical distribution of radiolaria in the eastern equatorial Atlantic: analysis of a multiple series of closely-spaced plankton tows. Mar Micropaleontol 12:1-19
Esener AA, Roels JA, Kossen NWF (1982) Dependence of the elemental composition of *K. pneumoniae* on the steady-state specific growth rate. Biotechnol Bioeng 24:1445-1449
Estep KW, Davis PG, Keller MD, Sieburth JMcN (1986) How important are oceanic algal nanoflagellates in bacterivory. Limnol Oceanogr 31:646-650
Fenchel T (1967) The ecology of marine microbenthos. I. The quantitative importance of ciliates as compared with metazoans in various types of sediments. Ophelia 4:121-137
Fenchel T (1969) The ecology of marine microbenthos. IV. Structure and function of the benthic ecosystem, its chemical and physical factors and the microfauna communities with special reference to the ciliated protozoa. Ophelia 6:1-182
Fenchel T (1977) The significance of bactivorous protozoa in

the microbial community of detrital particles. In: Cairns JJr (ed) Aquatic microbial communities. Garland Publishing, New York, p 529
Fenchel T (1980) Suspension feeding in ciliated protozoa: feeding rates and their ecological significance. Microb Ecol 6:13-25
Fenchel T (1982a) Ecology of heterotrophic microflagellates. I. Some important forms and their functional morphology. Mar Ecol Prog Ser 8:211-223
Fenchel T (1982b) Ecology of heterotrophic microflagellates. II. Bioenergetics and growth. Mar Ecol Prog Ser 8:225-231
Fenchel T (1982c) Ecology of heterotrophic microflagellates. III. Adaptations to heterogeneous environments. Mar Ecol Prog Ser 9:25-33
Fenchel T (1982d) Ecology of heterotrophic microflagellates. IV. Quantitative occurrence and importance as bacterial consumers. Mar Ecol Prog Ser 9:35-42
Fenchel T (1986) The ecology of heterotrophic microflagellates. Adv Microb Ecol 9:57-97
Fenchel T, Blackburn TH (1979) Bacteria and mineral cycling. Academic Press, London p 225
Fenchel T, Finlay BJ (1983) Respiration rates in heterotrophic free-living protozoa. Microb Ecol 9:99-122
Fenchel T, Harrison P (1976) The significance of bacterial grazing and mineral cycling for the decomposition of particulate detritus. In: Anderson JM (ed) The role of terrestrial and aquatic organisms in decomposition processes. Blackwell Scientific, Oxford, p 285
Fenchel TM, Jørgensen BB (1977) Detritus food chains of aquatic ecosystems: The role of bacteria. Adv Microb Ecol 1:1-58
Ferguson RL, Sunda WG (1984) Utilization of amino acids by planktonic marine bacteria: Importance of clean technique and low substrate additions. Limnol Oceanogr 29:258-274
Fuhrman JA, Azam F (1980) Bacterioplankton secondary production estimates for coastal waters of British Columbia, Antarctica, and California. Appl Environ Microbial 39:1085-1095
Fuhrman JA, Azam F (1982) Thymidine incorporation as a measure of heterotrophic bacterioplankton production in marine surface waters: evaluation and field results. Mar Biol 66:109-120
Fuhrman JA, Ferguson RL (1986) Nanomolar concentrations and rapid turnover of dissolved free amino acids in seawater: agreement between chemical and microbiological measurements. Mar Ecol Prog Ser 33:237-242
Gifford DJ (1985) Laboratory culture of marine planktonic oligotrichs (Ciliophora, Oligotrichida). Mar Ecol Prog Ser 23:257-267
Goldman JC, Caron DA (1985) Experimental studies on an omnivorous microflagellate: implications for grazing and nutrient regeneration in the marine microbial food chain. Deep-Sea Res 32:899-915
Goldman JC, Caron DA, Andersen OK, Dennett MR (1985) Nutrient cycling in a microflagellate food chain: I. Nitrogen dynamics. Mar Ecol Prog Ser 24:231-242
Goldman JC, Caron DA, Dennett MR (1987a) Regulation of gross growth efficiency and ammonium regeneration in bacteria by

substrate C:N ratio. Limnol Oceanogr 32:1239-1252

Goldman, JC, Caron DA, Dennett MR (1987b) Nutrient cycling in a microflagellate food chain: IV. Phytoplankton-microflagellate interactions. Mar Ecol Prog Ser 38:75-87

Goldman JC, McCarthy JJ, Peavey DG (1979) Growth rate influence on the chemical composition of phytoplankton in oceanic waters. Nature 279:210-215

Goldman JC, Peavey GD (1979) Steady-state growth and chemical composition of the marine chlorophyte *Dunaliella tertiolecta* in nitrogen-limited continuous culture. Appl Environ Microbiol 38:894-901

Gude H (1985) Influence of phagotrophic processes on the regeneration of nutrients in two-stage continuous culture systems. Microb Ecol 11:193-204

Hargrave BT, Green GH (1968) Phosphorus excretion by zooplankton. Limnol Oceanogr 13:332-342

Heinbokel JF, Beers JR (1979) Studies on the functional role of tintinnids in the Southern California Bight. III. Grazing impact of natural assemblages. Mar Biol 52:23-32

Hobbie JE, Helfrich JVK (1988) The effect of grazing by microprotozoans on production of bacteria. Arch Hydrobiol Beih Ergebn Limnol 31:281-288

Hollibaugh JT (1978) Nitrogen regeneration during the degradation of several amino acids by plankton communities collected near Halifax, Nova Scotia, Canada. Mar Biol 45:191-201

Ikeda T (1974) Nutritional ecology of marine zooplankton. Mem Fac Fish Hokkaido Univ 22:1-97

Ikeda T, Hing Fay E, Hutchinson SA, Boto GM (1982) Ammonia and inorganic phosphorus excretion by zooplankton from inshore waters of the Great Barrier Reef, Queensland. I. Relationship between excretion rates and body size. Aust J Mar Freshwater Res 33:55-70

Johnson PW, Xu H, Sieburth JMcN (1982) The utilization of chroococcoid cyanobacteria by marine protozooplankters but not by calanoid copepods. Ann Inst oceanogr Paris 58:297-308

Jørgensen BB, Erez J, Revsbech NP, Cohen Y (1985) Symbiotic photosynthesis in a planktonic foraminiferan, *Globigerinoides sacculifer* (Brady), studied with microelectrodes. Limnol Oceanogr 30:1253-1267

Kogure K, Simidu U, Taga N (1979) A tentative direct microscopic method for counting living marine bacteria. Can J Microbiol 25:415-420

Laacke M, Dahle AB, Eberlein K, Rein K (1983) A modelling approach to the interplay of carbohydrates, bacteria and non-pigmented flagellates in a controlled ecosystem experiment with *Skeletonema costatum*. Mar Ecol Prog Ser 14:71-79

Lancelot C, Billen G (1985) Carbon-nitrogen relationships in nutrient metabolism of coastal marine ecosystems. Adv Aquatic Microbiol 3:263-321

Laval-Peuto M, Rassoulzadegan F (1988) Autofluorescence of marine planktonic Oligotrichina and other ciliates. Hydrobiologia 159:99-110

Laws EA, Bannister TT (1980) Nutrient- and light-limited growth of *Thalassiosira fluviatilis* in continuous culture, with

implications for phytoplankton growth in the ocean. Limnol Oceanogr 25:457-473
Lee JJ (1980) Nutrition and physiology of the foraminifera. In: Levandowsky M, Hutner S (eds) Biochemistry and physiology of protozoa. Academic Press, New York, vol 3, p 43
Lee S, Fuhrman JA (1987) Relationships between biovolume and biomass of naturally derived marine bacterioplankton. Appl Environ Microbiol 53:1298-1303
Linley EAS, Newell RC, Lucas MI (1983) Quantitative relationships between phytoplankton, bacteria and heterotrophic microflagellates in shelf waters. Mar Ecol Prog Ser 12:77-89
Lipps JH, Valentine JW (1970) The role of foraminifera in the trophic structure of marine communities. Lethaia 3:279-286
Meyer-Reil L-A (1978) Autoradiography and epifluorescence microscopy combined for the determination of number and spectrum of actively metabolizing bacteria in natural waters. Appl Environ Microbiol 36:506-512
Mullin MM, Perry MJ, Renger EH, Evans PM (1975) Nutrient regeneration by oceanic zooplankton: A comparison of methods. Mar Sci Comm 1:1-13
Nagata T (1986) Carbon and nitrogen content of natural planktonic bacteria. Appl Environ Microbiol 52:28-32
Nicholls KH (1987) Predation on *Synura* spp. (Chrysophyceae) by *Bodo crassus* (Bodonaceae). Trans Amer Microsc Soc 106:359-363
Parslow JS, Doucette GJ, Taylor FJR, Harrison PJ (1986) Feeding by the zooflagellate *Pseudobodo* sp. on the picoplanktonic prasinomonad *Micromonas pusilla*. Mar Ecol Prog Ser 29:237-246
Pomeroy LR (1974) The ocean's food web, a changing paradigm. Bioscience 24:499-504
Porter, KG (1988) Phagotrophic phytoflagellates in microbial food webs. Hydrobiologia 159:89-97
Redfield, AC, Ketchum BH, Richards FA (1963) The influence of organisms on the composition of sea water. In: Hill MN (ed) The sea. Wiley, New York, vol 2, p 26
Roszak DB, Colwell RR (1987) Survival strategies of bacteria in the natural environment. Microbiol Rev 51:365-379
Sanders RW, Porter KG (1988) Phagotrophic phytoflagellates. Adv Microb Ecol 10:167-192
Sherr BF, Sherr EB (1984) Role of heterotrophic protozoa in carbon and energy flow in aquatic ecosystems. In: Klug M, Reddy CA (eds) Current perspectives in microbial ecology. ASM, Washington, DC, p 412
Sherr, EB, Sherr BF, Paffenhofer G-A (1986) Phagotrophic protozoa as food for metazoans: a 'missing' trophic link in marine pelagic food webs. Mar Microb Food Webs 1:61-80
Sherr EB, Sherr BF, Albright LJ (1987) Bacteria: Link or sink? Science 235:88
Sherr BF, Sherr EB, Hopkinson CS (1988) Trophic interactions within pelagic microbial communities: Indications of feedback regulation of carbon flow. Hydrobiologia 159:19-26
Sieburth JMcN (1976) Bacterial substrates and productivity in marine ecosystems. Ann Rev Ecol Syst 7:259-285
Sieburth JMcN, Davis PG (1982) The role of heterotrophic nanoplankton in the grazing and nurturing of planktonic

bacteria in the Sargasso and Caribbean Sea. Ann Inst oceanogr Paris 58:285-296
Sorokin YI (1981) Microheterotrophic organisms in marine ecosystems. In: Longhurst AR (ed) Analysis of marine ecosystems. Academic Press, London, p 293
Spindler M, Hemleben C, Salomons JB, Smit LP (1984) Feeding behavior of some planktonic foraminifers in laboratory cultures. J Foram Res 14:237-249
Stevenson LH (1978) A case for bacterial dormancy in aquatic systems. Microb Ecol 4:127-133
Stoecker DK, Michaels AE, Davis LH (1987) Large proportion of marine planktonic ciliates found to contain functional chloroplasts. Nature 326:790-792
Stoecker DK, Evans FT (1985) Effects of protozoan herbivory and carnivory in a microplankton food web. Mar Ecol Prog Ser 25:159-167
Stoecker DK, Guillard RRL, Kauvee RM (1981) Selective predation by *Favella ehrenbergii* (Tintinnia) on and among dinoflagellates. Biol Bull 160:136-145
Stoecker DK (1990) Mixotrophy in marine planktonic ciliates: physiological and ecological aspects of plastid-retention by oligotrichs. In: Reid PC, Turley CM, Burkill PH (eds) Protozoa and their role in marine processes. Springer, Berlin Heidelberg New York
Strickland JDH (1970) Introduction to recycling of organic matter. In: Steele JH (ed) Marine food chains. University of California Press, Berkeley, p 3
Swanberg NR, Bennett P, Lindsey JL Anderson OR (1986) A comparative study of predation in two Caribbean radiolarian populations. Mar Microb Food Webs 1:105-118
Taylor DL, Blackbourn DJ, Blackbourn J (1971) The redwater ciliate *Mesodinium rubrum* and its 'incomplete symbionts': a review including new ultrastructural observations. J Fish Res Bd Can 28:391-407
Terry KL, Hirata J, Laws EA (1985) Light-, nitrogen-, and phosphorus-limited growth of *Phaeodactylum tricornutum* Bohlin strain TFX-1: chemical composition, carbon partitioning, and the diel periodicity of physiological processes. J Exp Mar Biol Ecol 86:85-100
vanVeer JA, Paul EA (1979) Conversion of biovolume measurements of soil organisms, grown under various moisture tensions, to biomass and their nutrient content. Appl Environ Microbiol 37:686-692
Verity PG (1985) Ammonia excretion rates of oceanic copepods and implications for estimates of primary production in the Sargasso Sea. Biol Oceanogr 3:249-283
Verity PG (1986) Grazing of phototrophic nanoplankton by microzooplankton in Narragansett Bay. Mar Ecol Prog Ser 29:105-115
Vidal J, Whitledge TE (1982) Rates of metabolism of planktonic crustaceans as related to body weight and temperature of habitat. J Plankton Res 4:77-84
Wheeler PA, Kirchman DL (1986) Utilization of inorganic and organic nitrogen by bacteria in marine systems. Limnol Oceanogr 31:998-1009
Williams PJleB (1981) Microbial contribution to overall marine plankton metabolism: direct measurements of respiration.

Oceanol Acta 4:359-364
Wright RT, Coffin RB (1984) Measuring microzooplankton grazing on planktonic marine bacteria by its impact on bacterial production. Microb Ecol 10:137-149

PROTOZOANS AS AGENTS IN PLANKTONIC NUTRIENT CYCLING

Tom Berman
The Yigal Allon Kinneret Limnological Laboratory (IOLR)
PO Box 345, Tiberias 14102
Israel

INTRODUCTION

"I find" said 'e, "things very much
as 'ow I've always found,
For mostly they goes up or down
or else goes round and round"
(Roundabouts and Swings - PR Chalmers)

It is becoming evident that, at least from the ecosystem point of view, a major 'raison d'etre' for planktonic protozoans is their function as agents of nutrient transfer and recycling. The term 'nutrients' is used here in the broadest sense and applies to carbon, nitrogen, phosphorus and perhaps to other elements despite their different potential pathways. As yet we know almost nothing about the biological cycling in aquatic systems of micronutrients such as iron, magnesium, zinc, cobalt or growth factors such as vitamins (Phillips, 1984). This paper first addresses the important role bacteria play in converting dissolved nutrients to particulates before discussing remineralization by protists and priorities for further research.

THE ROLE OF BACTERIA

In an earlier and simpler world, our understanding was that soluble nutrients were first neatly packaged into phytoplankton biomass by the primary producers and then taken up by grazing metazoans and eventually passed on to fish, with inevitable losses in the form of respired carbon and released inanimate particulate or dissolved nutrients en route (Cushing and Walsh 1976). The assumed function of bacteria (the biomass of which was assumed to be relatively minor in comparison to the primary producers) was to degrade particulate organic matter and thereby to remineralise the nutrients therein (Steele and Frost 1977). We now know that this picture is an inadequate

NATO ASI Series, Vol. G 25
Protozoa and Their Role in Marine Processes
Edited by P. C. Reid et al.

description of nutrient flux in many aquatic situations (Sherr and Sherr 1984, Sieburth 1984, Berman 1988a). Primary producers, using their unique capability for photosynthesis, sequester nutrients in the form of particulate biomass. In some cases chemoautotrophic bacteria may also be contributors to primary production (Sieburth *et al.* 1987). Often a sizeable proportion of the primary produced organic carbon (in the form of dead, moribund or fragmented cells or derived from algal extracellular release) ends up as bacterial biomass rather than in the biomass of metazoan grazers (Pace *et al.* 1984, Williams 1984). Consequently the flux of primary produced materials through a bacterial pathway is now known to be of considerable significance (Azam *et al.* 1983). In an extensive review Cole *et al.* (1988) have shown that the values determined for bacterial production correlated well with planktonic primary production or with chlorophyll *a* for a large number of results reported from both marine and freshwater environments. Overall, bacterial productivity averaged about 20% or 30% of primary production based on a volumetric or areal measure respectively.

Within the plankton, the bacteria are generally the most efficient converters of dissolved nutrients into particulate biomass (Azam *et al.* 1989) and especially in oligotrophic situations, bacteria tend to retain rather than release nitrogen and phosphorus. Subsequently intensive grazing, mainly by flagellates and ciliates, on these bacteria is necessary to mobilise nutrients and to keep the ecosystem turning over (Goldman 1984, Sieburth 1984). Indeed one suspects, that in the absence of the phagotrophic nano-protists, we might be walking on seas solid, or at least undulating, gel-like with bacteria!

Before considering the impact of protozoan grazing it is worth noting some characteristics of planktonic bacteria. By far the majority of bacteria in the oceans and in many lakes appear to be free-floating, unattached cells (Derenbach and Williams 1974, Berman 1975). Of course, there are also active bacteria associated with particulate aggregates (marine snow and fluff, larvacean 'houses', dinoflagellate and copepod faeces, etc) which may carry out specialised chemical transformations only possible in such micro-environments (Aldredge and Cohen 1987). Azam and Ammerman (1984) and Azam (this Workshop), expanding on the ideas of Bell and Mitchell (1972) have suggested that even free-floating bacteria may be organised in loose consortia together with their erstwhile protozoan grazers within the vicinity of phytoplankton cells or other potential point sources of dissolved organic substrates. However, Mitchell *et al.* (1985) indicate that such microzone communities could only develop under specially favourable physical conditions, such as might exist at thermoclines.

In nutrient limited situations, planktonic bacteria can usually out-compete phytoplankton for both dissolved inorganic or organic phosphorus (Currie and Kalff 1984, Berman 1985, 1988b) and probably also for fixed inorganic or organic nitrogen. One should note, however, that the nutrient uptake characteristics of the pico-phytoplankton, which constitute a very large fraction of total phytoplankton in the open ocean and in some lakes (Joint 1986, Stockner and Antia 1986) are still largely undetermined and may be expected to approach those of the heterotrophic bacteria. At any rate, the importance of planktonic bacteria as agents for scavaging and converting dissolved inorganic and organic nutrients into particulate organic biomass is now well recognised.

Recent work has indicated that the relative amount of bacterial to phytoplankton biomass in euphotic zones increases with increasing oligotrophy of aquatic systems and that, in some circumstances, the standing stock biomass of bacteria can approach or exceed that of phytoplankton (Simon and Tilzer 1987, Azam *et al.* 1990). A similar trend also holds in respect to the ratio of pico-phytoplankton biomass to that of nano- and micro-phytoplankton in algal assemblages as oligotrophy increases. Therefore, the more oligotrophic an aquatic system, the greater relative amounts of nutrients may be expected to flow into and through the microbial loop, with bacteria and pico-phytoplankton cells being utilised as nutrient sources for protozoans which, in turn, may be grazed by metazoan consumers (Hagstrom *et al.* 1988).

REMINERALISATION BY PROTISTS

The realisation that, in many aquatic environments, relatively large amounts of the primary produced carbon are incorporated into a bacterial biomass which also ties up phosphorus and nitrogen has focussed attention on the central role of bacteriophagous protozoa in the processes of remineralization and nutrient recycling. For example, Gude (1985) showed that phagotrophic nanoflagellates were necessary for high mineralisation rates of N and P in a freshwater ecosystem. Protozoans are effective processors of particulate organic matter because of their high weight-specific clearance rates and rapid digestion of bacteria and other food (Fenchel 1982, Sherr and Sherr 1984, Gast and Horstman 1983, Caron 1989). Thus, flagellates are typically capable of clearing food particles from a water volume about 10^5 fold their own cell volume per hour (i.e. 2×10^{-6} to 2×10^{-5} ml h^{-1}) depending on size (Fenchel 1987). Larger protozoans, for example, dinoflagellates (Kimor 1981) tintinnids (Heinbokel and Beers 1979, Verity 1985, Lessard and Swift 1985), as well as metazoan zooplankters (Gophen *et al.* 1974, Peterson *et al.* 1978) also graze bacteria but presumably do not usually have as significant an impact on the ecosystem as nanoprotists.

The presence of phagotrophic protists has been shown to accelerate the breakdown of organic detritus, at least partly because protozoans speed up the rate of nutrient mineralisation which, in turn, stimulates bacterial growth and activity (Barsdate *et al.* 1974, Fenchel and Harrison 1976). The bacterivorous protozoans may maintain populations of detritus-degrading bacteria in a state of physiological 'youth' by preventing the accumulation of senescent bacterial cells (Sieburth and Davis 1982). Protozoans may also have an indirect qualitative effect upon the chemical patterns of detrital breakdown by selecting somehow for specific groups of bacteria adapted to particular substrates such as those with high carbohydrate content (Sherr *et al.* 1982). Sherr *et al.* (1988) have proposed that nanoprotozoans may have a major cybernetic function in material flux in aquatic environments because of their rapid growth and grazing rates, and their high release rates of ammonia and phosphorus (see below).

The impact of nanoprotists as remineralisers in a given aquatic system will depend on the degree to which N and P are incorporated into the cell biomass or excreted. The higher the efficiency of utilisation of nutrients by the protozoans, the lower the amounts of demineralised nutrients. Here one should distinguish between the pathways of carbon which, when respired, is essentially lost to the aquatic biota and that of mineralised N and P which are presumably recycled by being reassimilated by bacteria and algae.

Measured NH_4 excretion rates for protozoans range from 2.4 to 17 µg N mg^{-1} dry weight h^{-1} (Sherr *et al.* 1983, Verity 1985, Goldman *et al.* 1985, Berman *et al.* 1987, Lucas *et al.* 1987). By comparison, ammonia mineralisation rates for marine metazoan zooplankters range from 0.8-4.9 µg N mg^{-1} dry weight h^{-1} (listed by Goldman *et al.* 1985), average 0.13 µg mg^{-1} dry weight h^{-1} for marine microplankton-micronecton (Roget 1988) and range from 0.3-1.3 µg N mg^{-1} dry weight h^{-1} for freshwater copepods and cladocerans (Gophen 1978).

Probably because of the preponderance of studies on marine environments, less information is available about phosphorus excretion, although here too most, but not all (Taylor and Lean 1981), evidence indicates that protozoans have much higher weight-specific release rates than metazoans. Maximum phosphorus release rates of 2 and 3.8 µg P mg^{-1} dry weight h^{-1} were observed when *Paraphysomonas* were feeding on diatoms or bacteria respectively (Anderson *et al.* 1986). Excretion rates of soluble reactive phosphorus excretion ranging from 0.4-

15.1 µg mg^{-1} dry weight h^{-1} for a chrysomonad flagellate and a ciliate (*Cyclidium*) were measured by Berman *et al.* (1987).

Our limited knowledge of the grazing efficiencies of different protist species usually makes it difficult to estimate the amounts of carbon, nitrogen or phosphorus which may be recycled or transmitted to metazoans or fish via protozoans. The impact of nanoprotists as remineralisers in a given aquatic system will depend upon the degree to which N and P are incorporated into their biomass or excreted (Fenchel 1986) and this is determined to some extent by the chemical composition of their particulate food sources (Caron *et al.* 1988) as well as by the physiological state or growth stage of the grazers (Caron 1989). In an elegant series of studies using batch and continuous cultures of bacteria, cyanobacteria and a nanoprotist (Goldman and Caron 1985, Goldman *et al.* 1985, 1987, Anderson *et al.* 1986) have shown that the extent of recycling of NH_4 and orthophosphate by nanoflagellates was greatly dependent on the C:N:P ratios of the bacterial source. Release of reasonably high amounts of NH_4 and orthophosphate during exponential growth of the nanoprotists was observed only when bacteria or cyanobacteria with relatively low C:N or C:P ratios were provided as food for nanoprotists. Borsheim (pers. comm.) suggests that a bacterial C:N ratio less than 24 is a prerequisite for release of NH_4 by bacteriophagous protozoans. Moreover, even when recycling efficiency of protozoans is low, the fact that N and P are nicely packaged into particulate biomass which can be conveniently grazed by metazoan zooplankters may be an effective way of passing these nutrients on to the larger members of the food web (Hamilton and Taylor 1987, Turner *et al.* 1988). At least for oligotrophic systems, Berman *et al.* (1987) have demonstrated that nanoprotists could enhance the growth of photosynthetic microalgae by grazing on intact bacterial cells which comprised the initial input of N and P nutrients.

Although there has been a tendency to consider nutrient cycling only in terms of inorganic nutrient flux, I submit that the potential of planktonic protists to generate dissolved organic compounds as a result of their grazing or excretion functions may also have an important impact in aquatic ecosystems. This aspect of nutrient flux has not been examined closely, but there are some indications that at certain stages in their growth cycle, bacterivorous protozoans may release organic molecules such as amino acids, hypoxanthine, guanine, ATP, and urea (Antia *et al.* 1980, Goldman *et al.* 1985, Taylor 1982, Taylor *et al.* 1985, Andersson *et al.* 1985, Taylor 1986, Berman *et al.* 1987). Perhaps phagotrophic protists generate at least some of the relatively high (0-10 µg l^{-1}) amounts of dissolved DNA and RNA measured in marine waters (Paul and Carlson 1984, Karl pers. comm.). Such dissolved organic substrates may serve as sources of both nutrients and energy for planktonic microorganisms. Furthermore, some organic molecules such as DNA, RNA and cyclic-AMP, may be important as 'agents of evolution' by acting as carriers of genetic information or may serve as molecular messengers to activate biochemical responses in the plankton (Franko and Wetzel 1981).

Not much is known about nutrient cycling by mixotrophic protozoans. The phenomenon of mixotrophy may be more widespread and of greater significance than previously reckoned (Estep *et al.* 1986) especially in lakes and coastal waters (Bird and Kalff 1986, Porter 1988). With the increasing use of epifluorescent microscopy, greater numbers of chloroplastidic flagellates and ciliates (many of which are phagotrophic) are being described in plankton assemblages (Laval-Peuto and Rassoulzadegan 1988). In addition to photosynthesising, mixotrophs can also consume bacteria, pico-phytoplankton and even nanophytoplankton. (Perhaps some of these creatures should be classed as 'cannibalistic plants'?). The capacity of some phagotrophic protozoans to consume prey much larger than themselves (Goldman and Caron 1985, Goldman *et al.* 1987, Jacobson 1989) also complicates hypotheses of size-dependent, grazing relationships although probably this behaviour is more the exception than the rule (Sherr and Sherr submitted). Furthermore, E Sherr (1988) has recently reported that heterotrophic flagellates are also capable of ingesting and growing on high molecular weight polysaccharides, thus suggesting that some protists may use such dissolved compounds as alternative food sources.

It is impossible to discuss the topic of nutrient (in particular, carbon) cycling within aquatic microbial communities without mentioning the vexed 'sink' or 'source' question (Ducklow *et al.* 1986, Sherr *et al.* 1987). Judging from presentations at this Workshop, it now appears that this may be a 'non-issue' (Sherr and Sherr 1988). Nutrients following 'sink' or 'link' pathways (roundabouts or swings!) should be viewed as two concomitant aspects of the pattern of carbon flow within the microbial loop, with the ratio between these flux patterns varying under different environmental conditions (Caron 1989). In most situations, neither mode will operate exclusively. The emerging consensus would suggest that the 'microbial loop' when considered as a single ecosystem compartment, has two functions: 1) to mineralise nutrients which originate as organic detritus or as particulate cells of bacteria, pico- or nano-phytoplankton, and 2) to transfer nutrients in the form of cellular (mainly protozoan) biomass directly to mesozooplankton and perhaps to larval fish (Sherr *et al.* 1986, Turner *et al.* 1988). Various ecological forcing factors (as yet not clearly defined) will determine the overall balance between these sink or link modes.

Presently we have only meagre information about the autecology of planktonic protozoan species and about the environmental factors which determine the composition of protozoan populations. This also makes it difficult to quantify the contribution of phagotrophic protists in nutrient transfer or cycling within the microbial loop in nature. Both temporal and spatial considerations are likely to be critical in this respect. For example, nutrient flux within waters of the euphotic zone is almost certainly different at night than during the day (Sieburth 1984). The impact of stochastic events, such as internal waves or storm-induced mixing and turbulence, must obviously be important. Modelling exercises (Ducklow 1989) should prove useful in assessing the potential effects of such phenomena on the functioning of the microbial loop. Nutrient cycling will also obviously be influenced by the spatial relationships between micro-organisms (Bell and Mitchell 1972, Goldman 1984, Azam and Ammerman 1984). Current studies of protozoan and microbial activity associated with marine snow and fluff particles may give insights on chemical transformations which may be limited to such environments (Aldredge and Cohen 1987). Experimental approaches to verify the existence of consortia of unattached microorganisms in microzones are problematic and, as yet, rare (but see Lehman and Scavia 1982, Lehman 1987).

FUTURE RESEARCH

I will end this brief overview of the role of protozoans in planktonic nutrient transfer and cycling by noting a few important objectives relating to this topic which, in my opinion, are worthy of future research attention.

- Improving methodologies for preserving, identifying, counting and observing protozoans in natural aquatic assemblages. Epifluorescent microscopy has revolutionised our capabilities for observing many of these organisms and further advances are to be expected with the wider application of the techniques of flow cytometry and image analysis. However, there are still many protozoan groups which remain problematic. For example, as noted at this Workshop, our knowledge about the abundance and activity of marine amoebae is almost nil.
- Developing better techniques for measuring the growth, respiration and grazing rates of planktonic protists, especially *in situ*. Scanning transmission microscopy with the capability of providing chemical analysis of single cells (Heldal *et al.* 1985, Borsheim 1988) has potential to supply information on the nutritional and physiological state of planktonic microorganisms.
- Determining how environmental variables influence the functioning of the microbial loop and especially in what manner

they may affect the balance between the 'sink' or 'link' modes of nutrient flux.

- Defining the effects of short-term (diurnal) temporal fluctuations or long-term (seasonal) changes on the patterns of nutrient flux via phytoplankton, bacteria and protozoans.
- Evaluating how stochastic events (storms, seiches) may affect the structure and functions of aquatic microbial assemblages.
- Determining what, if any, specialised microbial transformations occur within, or on, aggregates such as marine snow and fluff (or their freshwater equivalents). Are the processes associated with such particles of overall ecosystem importance?
- Assessing the extent to which unattached planktonic microorganisms can form loose consortia in microzones centred around point sources of nutrition and exploring the implications of such associations for nutrient flux patterns.
- Determining and quantifying the links between the bacterial and protozoan components of the microbial loop and their mesoplanktonic and larval fish consumers.

It should be obvious that the above list is by no means comprehensive or in any particular order of priority. Only now are we beginning to comprehend the complex web of relationships and interactions subsumed under the term 'protozoan nutrient cycling'. Nevertheless, all the evidence points to the pivotal status of the phagotrophic protists as determinants of nutrient flux in oceans and lakes.

REFERENCES

Aldredge AL, Cohen Y (1987) Can Microscale chemical patches persist in the sea? Microelectrode Study of Marine Snow, Faecal Pellets. Science 235:689-691
Anderson OK, Goldman, DA, Caron A, Dennett MR (1986) Nutrient cycling in a microflagellate food chain: III. Phosphorus dynamics. Mar Ecol Progr Ser 31:47-55
Andersson A, Lee C, Azam F, Hogstrom A (1985) Release of amino acids and inorganic nutrients by heterotrophic marine microflagellates. Mar Ecol Prog Ser 23:99-106
Antia NJ, Berland BR, Bonin DJ (1980) Proposal for an abridged nitrogen turnover cycle in certain marine planktonic systems involving hypoxanthine-guanine excretion by ciliates and their reutilisation by phytoplankton. Mar Ecol Prog Ser 2:97-103
Azam F, Fenchel T, Field JG, Meyer-Reil LA and Thingstad T (1983) The Ecological Role of Watercolumn Microbes in the Sea. Mar Ecol Prog Ser 10:257-263
Azam F, Cho BC, Simon M (1990) Bacterial Cycling of Matter in

the Pelagic Zone. In: Tilzer MM, Serruya S, Imboden D (eds) Structural and Functional Properties of Large Lakes. (in press)
Azam F, Ammerman JW (1984) Cycling of Organic Matter by Bacterioplankton in Pelagic Marine Ecosystems: Microenvironmental Considerations. In: Fasham MJ (ed) Flows of energy and Materials in Marine Ecosystems: Theory and Practice. Plenum Press, London. p 345
Barsdate, RJ, Prentki RT, Fenchel T (1974) Phosphorus Cycle of Model Ecosystems: Significance for Decomposer Food Chains and Effect of Bacterial Grazers. Oikos 23:239-251
Bell WH, Mitchell R (1972) Chemotactic and Growth Responses of Marine Bacteria to Algal Extracellular Products. Biol Bull 143:265-277
Berman T (1975) Size Fractionation of Natural Aquatic Populations Associated with Autotrophic and Heterotrophic Carbon Uptake. Mar Biol 33:215-220
Berman T (1985) Uptake of 32p orthophosphate by algae and bacteria in Lake Kinneret. J Plank Res 7:71-84
Berman T (1988a) Microbes in a Watery World. Hydrobiol 159:5-6
Berman T (1988b) Differential uptake of orthophosphate and organic phosphorus substrates by bacteria and algae in Lake Kinneret. J Plank Res 10:1239-1249
Berman T (1990) Microbial Food Webs and Nutrient Cycling in Lakes: Changing Perspectives. In: Tilzer MM, Serruya S, Imboden D (eds) Structural and Functional Properties of Large Lakes. (in press)
Berman T, Nawrocki M, Taylor GT, Karl DM (1987) Nutrient Flux between bacteria, Bacterivorous Nanoplanktonic Protists and Algae. Mar Microbial Food Webs. 2:69-82
Bird DF, Kalff (1987) Algal Phagotrophy: Regulating factors and importance relative to photosynthesis in *Dinobryon* (Chrysophyceae). Limnol Oceanogr 32:277-284
Buechler, DG, Dillon RD (1974) Phosphorus regeneration in fresh-water paramecia. J Protozool 21:339-343
Caron DA, Goldman JG, Dennett MR (1988) Experimental demonstration of the roles of bacteria and bacterivorous protozoa in plankton nutrient cycles. Hydrobiol 159:27-40
Cole JJ Pace ML, Findley S (1988) Bacterial production in fresh and saltwater ecosystems : A cross-system overview. Mar Ecol. Progr Ser 43:1-10
Currie DJ, Kalff J (1984) The relative importance of bacterioplankton and phytoplankton in phosphorus uptake in freshwater. Limnol Oceanogr 29:311-321
Cushing DH, Walsh J (1976) The ecology of Seas. Blackwell Sci Publ, Oxford p 574
Derenbach JB, Williams PL LeB (1980) Autotrophic and bacterial production: Fractionation of plankton populations by differential filtration of samples from the English Channel. Mar Biol 25:263-269
Ducklow H (1988) Modelling and the structure of microbial foodwebs. Abstract. In: Reid PC, Turley CM, Burkill PH (eds) Protozoa and their role in marine processes. Springer, Berlin Heidelberg New York
Ducklow HW, Purdie DA, Williams PJ LeB, Davies JM (1986) Bacterioplankton: A sink for carbon in a coastal marine plankton community. Science 232:865-867
Estep KW, Davis PG, Keller MD, Sieburth JMcN (1986) How important are Oceanic algal nanoflagellates in bacterivory? Limnol Oceanogr 31:646-650
Fenchel T (1982) Ecology of heterotrophic microflagellates. II. Bioenergetics and growth. Mar Ecol Prog Ser 8: 225-231
Fenchel T (1986) The ecology of heterotrophic microflagellates. Quantitative occurrence and importance as bacterial consumers. Mar Ecol Prog Ser 9:35-42
Fenchel T (1987) Ecology of protozoa: The biology of free living phagotrophic protists. Science Tech. Publishers. Madison, Wisc. USA
Fenchel T, Harrison P (1976) The significance of bacterial grazing and mineral cycling for the decomposition of particulate detritus. In: Anderson JM, Macfadyen A (eds) The role of terrestrial and aquatic organisms in decomposition processes. Blackwell Scientific Publications, Oxford p 285
Franko DA, Wetzel RG (1981) Dynamics of cellular and extracellular cAMP in *Anabaena flos-aquae* (Cyanophyta):

Intrinsic culture variability and correlation with metabolic variables. J Phycol 17:129-134
Gast V, Horstmann (1983) N-remineralization of phyto- and bacterioplankton by the marine ciliate *Euplotes vannus* Mar Ecol Prog Ser 13:55-60
Goldman JC (1984) Conceptual role for microaggregates in pelagic waters. Bull Mar Sci 35:462-476
Goldman JC, Caron DA (1985) Experimental studies on an omnivorous microflagellate: Implications for grazing and nutrient regeneration in the marine microbial food chain. Deep Sea Res. 32: 899-915
Goldman JC, Caron DA, Anderson OK, Dennett MR (1985) Nutrient cycling in a microflagellate food chain : I. Nitrogen dynamics. Mar Ecol Prog Ser 24:231-242
Goldman JC, Caron DA, Dennett MR (1987) Nutrient cycling in a microflagellate food chain: IV. Phytoplankton-microflagellate interactions. Mar Ecol Prog Ser 38:75-87
Gophen M (1976) Temperature dependence of food intake, ammonia excretion and respiration in *Ceriodaphnia reticulata* (Jurine) (Lake Kinneret, Israel). Freshwat Biol 6:451-455
Gophen M, Cavari BZ, Berman T (1974) Zooplankton feeding on differentially labelled algae and bacteria. Nature 247: 393-394
Gude, H (1985) Influence of phagotrophic processes on the regeneration of nutrients in two-stage continuous culture systems. Microb Ecol 11:193-204
Hagstrom A, Azam F, Andersson A, Wikner J, Rassoulzadegan F (1988) Microbial loop in an oligotrophic pelagic marine ecosystem: possible roles of cyanobacteria and nanoflagellates in the organic fluxes. Mar Ecol Prog Ser 49:171-178
Hamilton DT, Taylor WD (1987) Short-term effects of zooplankton manipulations on phosphate uptake. Can J Fish Aquat Sci 44: 1038-1044
Heinbokel JF, Beers JR (1979) Studies on the functional role of tintinnids in the Southern California Bight. III. Grazing impact on natural assemblages. Mar Biol 52:23-32
Heldal M, Norland S, Tumyr O (1985) X-ray microanalytical method for measurement of dry matter and elemental content of individual bacteria. Appl Environ Micriobiol 50:1251-1257
Jacobsen DM (1988) Trophic behaviour of *protoperidinium* and related thecate heterotrophic dinoflagellates. In: Burkill PH, Turley CM and Reid PC eds. Abstracts of NATO-ASI Workshop, Plymouth, UK
Joint IR (1986) Physiological ecology of picopankton in various oceanographic provinces. Can J Fish Aqu Sci 214:289-309
Kimor B (1981) The role of phagotrophic dinoflagellates in marine ecosystems. Kieler Meeresforsch. Sonderh. 5:164-173
Laval-Peuto M, Rassoulzadegan F (1988) Autofluorescence of marine planktonic oligotrichina and other ciliates. Hydrobiol 159: 99-110
Lehman JT, Scavia D (1982) Microscale patchiness of nutrients in plankton communities. Science 216:729-730
Lehman JT (1987) Microscale nutrient patchiness in plankton communities. ASLO Abstracts p 46
Lessard EJ, Swift E (1985) Species-specific grazing rates of heterotrophic dinoflagellates in oceanic waters, measured with a dual-label radioisotope technique. Mar Biol 87: 289-296
Lucas MI, Probyn TA, Painting SJ (1987) An experimental study of microflagellate bacterivory: further evidence for the importance and complexity of microplanktonic interactions. S Afr J Mar Sci 5:791-808
Mitchell JG, Okubo A, Fuhrman JA (1985) Microzones surrounding phytoplankton form the basis for a stratified marine microbial ecosystem. Nature 316(4):58-59
Pace M, Glasser J, Pomeroy L (1984) a simulation analysis of continental shelf food webs. Marine Biology 31:47-63
Paul JH, Carlson DJ (1984) Genetic materials in the marine environment: Implications for bacterial DNA. Limnol Oceanogr 29:1091-1097
Peterson BJ, Hobbie JE, Haney JF (1978) *Daphnia* grazing on natural bacteria. Limnol Oceanogr 23:1039-1044
Phillips NW (1984) Role of different microbes and substrates as potential suppliers of specific, essential nutrients to marine detritivores. Bull Mar Sci 35:283-298

Porter KD (1988) Phagotrophic phytoflagellates in microbial food webs. Hydrobiol 159:89-98
Roget C (1988) Recyclage des sels nutritifs par le macroplancton-micronecton dans le Pacifique tropical Sud-Ouest. Oceanol Acta 11:107-116
Sherr, EB (1988) Direct use of high molecular weight polysaccharide by heterotrophic flagellates. Nature 335:348-351
Sherr, BF, Sherr EB, Berman T (1982) Decomposition of organic detritus: A selective role for microflagellate protozoa. Limnol Oceanogr 27:765-769
Sherr BF, Sherr EB, Berman T (1983) Grazing, growth, and ammonium excretion rates of a heterotrophic microflagellate fed with four species of bacteria. Apl environ Microbiol 45:1196-1201
Sherr BF, Sherr EB (1984) Role of heterotrophic protozoa in carbon and energy flow in aquatic ecosystems. In: Klug M, Redd CA (eds) Current perspectives in microbial ecology.
Sherr EB, Sherr BF, Paffenhofer G-A (1986) Phagotrophic protozoa as food for metazoans: a 'missing' trophic link in marine pelagic food webs? Mar Microb Food Webs 1:61-80
Sherr EB,Sherr BF, Albright LJ (1987) Bacteria: Link or sink. Science 235:88
Sherr BF, Sherr EB (1988) Role of microbes in pelagic food webs: a revised concept. Limnol Oceanogr 33:1225-1227
Sherr B, Sherr E, Hopkinson CS (1988) Trophic interactions within pelagic microbial communities: Indications of feedback regulation of carbon flow. Hydrobiologia 159:19-26
Sherr BF, Sherr EB Distribution of numbers, biovolumes, and bacterivores within nanoplanktonic size spectra of apochlorotic nanoflagellates in several marine pelagic systems. Mar Microb Food Webs, (submitted)
Sieburth JMcN (1984) Protozoan bacterivory in pelagic marine waters. In: Hobbie JE, Williams PJ LeB (eds) Heterotrophic activity in the sea. Plenum Press, New York p 405
Sieburth J McN, Davis PG (1982) The role of heterotrophic nano-plankton in the grazing and nurturing of planktonic bacteria in the Sargasso and Caribbean Sea. Ann Inst Oceanogr (Paris) 58 (Suppl):285-296
Sieburth J McN, Johnson PW, Eberhardt MA, Sieracki ME, Lidstrom M, Laux D (1987) The first methane-oxidizing bacterium from the upper mixing layer of the Deep Ocean: *Methylomonas pelagica* sp. nov. Curr Microbiol 14:285-293
Simon N, Tilzer MM (1987) Bacteral response to seasonal changes in primary production and phytoplankton biomass in Lake Constance. J Plank Res 9:535-552
Steele J, Frost B (1977) The structure of plankton communities. Philos Trans R Soc London, ser B 280:485-534
Stockner JG, Antia NJ (1986) Algal picoplankton from marine and freshwater ecosystems: a multidisciplinary perspective. Can J Fish Aquat Sci 43:2472-2503
Taylor GT (1982) The role of pelagic heterotrophic protozoa in nutrient cycling: a review. Ann Inst Oceanogr Paris 58:227-241
Taylor GT, Iturriaga R, Sullivan CW (1985) Interactions of bactivorous grazers and heterotrophic bacteria with dissolved organic matter. Mar Ecol Prog Ser 23:129-141
Taylor WD (1984) Phosphorus flux through epilimnetic zooplankton from Lake Ontario: relationship with body size and significance of phytoplankton. Can J Fish Aquat Sci 41:1702-1712
Taylor WD (1986) The effect of grazing by a ciliated protozoan on phosphorus limitation of heterotrophic bacteria in batch culture. J Protozool 33:47-52
Taylor WD, Lean DRS (1981) Radiotracer experiments on phosphorus uptake and release by limnetic microzooplankton. Can J Fish Aquat Sci 38:1316-1321
Turner JT, Tester PA, Ferguson RL (1988) The marine Cladoceran *Penila avirostris* and the "Microbial Loop" of pelagic food webs. Limnol Oceanogr 33:245-255
Verity PG (1985) Grazing, respiration and growth rates of tintinnids. Limnol Oceanogr 30:1268-1282
Williams PJ LeB (1984) Bacterial production in the marine food chain: the emperor's new suit of clothes? p 271-299. In: Fasham MR (ed) Flow of energy and materials in marine ecosystems - theory and practice. Plenum Press, Lond, p 733

MODELLING - SESSION SUMMARY

Hugh W Ducklow* and Arnold H Taylor▪

*Horn Point Laboratory
Centre for Environmental Estuarine Studies
University of Maryland
PO Box 775
Cambridge Maryland 21613
USA

▪Plymouth Marine Laboratory
Prospect Place
West Hoe
Plymouth
Devon PL1 3DH
UK

INTRODUCTION

The papers in this volume and in a companion collection (Marine Microbial Foodwebs) attest to the rapid growth and progress of marine microbiology and protozoology since the mid 1970's. An increasing number of articles dealing with models of the role of microbial systems and protozoa in plankton systems demonstrates the growth of this field. To reflect these trends the organizing committee for NATO ASI 604/87 decided to include practical and lecture sessions on modelling in the workshop programme, with the aim of a) assessing the need for and progress in modelling protozoan processes in the sea, b) introducing these models to the larger community of protozoologists, and c) identifying the kinds of information needed for better models. We review the main aspects of these sessions in this chapter.

SUMMARY OF PROGRAM SESSIONS

A session at the beginning of the ASI included a keynote presentation by **Mike Fasham** and papers by **Hugh Ducklow** and **Benno Ter Kuile** (Ter Kuile and Lee 1990). Fasham discussed both flow analysis (Fasham 1984) of steady state foodwebs with protozoan compartments and simulation models with and without protozoans. This latter approach is useful for identifying which ecosystem processes are most sensitive to the functional

NATO ASI Series, Vol. G 25
Protozoa and Their Role in Marine Processes
Edited by P.C. Reid et al.

description of protozoan activity (Frost 1984, 1987). Ducklow's presentation on the 'Link-Sink' question (Ducklow *et al.* 1986, Sherr *et al.*, 1987) stimulated a spirited discussion on the validity of this concept, which continued throughout the meeting (*cf* Sherr and Sherr 1988, Pomeroy 1990). Ter Kuile discussed his work with John Lee, which used physiological models of carbon flow in benthic foraminifera to interpret experimental data. **Arnold Taylor** concluded the session with a brief description of practical modelling tools available to ASI participants.

IBM PC computers were available throughout the meeting for demonstration of flow analysis and ecosystem simulation software. Programs included NETWORK 3, used for determining the structural properties of foodweb models constructed from observational data or from simulation models (Ulanowicz 1984, 1986, McManus 1990), a two-layer version of the mixed layer simulation model, MULES, described in Taylor (1988), and SYSL, a versatile ecosystem model demonstrated by **Philip Radford** of Plymouth Marine Laboratory. Among the more interesting activities of the practical modelling sessions was the construction of an *ad hoc* steady-state, flow network containing most of the protozoan groups discussed at the meeting (see below). This exercise highlighted perhaps the most important finding of the modelling sessions, that while protozoologists and modelers all recognize a need to include more protozoan groups in plankton models, there is a great scarcity of the sorts of data needed to write model equations and validate model output. This point has been made recently by Frost (1987) and will be discussed further below.

Modelling was readdressed near the close of the ASI when **John Field** gave a keynote lecture which again combined simulation and flow analysis of plankton systems with protozoan components (Moloney and Field 1990). Finally, **Arnold Taylor** and **Phillip Boyd**, of Queens University Ireland, presented observations on the annual cycles of three size classes of phytoplankton in the Irish sea, and the results of an attempt to simulate the data using three size classes of herbivores. The limited success of this model to date was attributed to the complexity of the biology and the hydrographic regime and to our overall ignorance of the correct parameter values describing microbial, protozoan and zooplankton dynamics (Taylor and Joint 1980). The calculations indicated only a minor role for bacteria in this region where nutrients were never limiting. In general, all the modelling talks and discussion during the practical sessions echoed this same message. Protozoan processes should be included in models of plankton dynamics, but there is still a severe lack of data on parameter values and annual cycles of stocks, which are needed to write the governing equations and calibrate the model results. This is in spite of the rapid growth of marine protozoology.

The reason why we need to model protozoan processes was questioned. Are, for example, protozoan components necessary for accurate description of plankton processes like nitrogen cycling or zooplankton feeding? These questions were addressed in the penultimate session of the ASI, a group discussion on understanding and modelling the role of protozoans and the microbial foodweb in marine carbon and nitrogen cycles. As a prelude to this discussion, **Hugh Ducklow** distributed to all ASI participants a questionnaire for a simple model of a microbial foodweb. It was hoped that the answers to the questionnaire would form the basis of the discussion, and of the chapter describing the 'consensus' view of microbial dynamics in the sea. It soon became apparent that no consensus was possible, given the lack of data on different protozoan groups and the great diversity of views on the importance of these groups. However, the following important points were made. First, while participants called for the inclusion of many different protozoan compartments (autotrophic, mixotrophic and heterotrophic microflagellates, ciliates, dinoflagellates, amoebae, etc) in models of plankton systems, the resulting models would be so complex as to render validation and interpretation practically impossible. Second, that we do not have enough information on the annual cycles, growth rates, physiology and feeding behaviour of most protozoans, especially mixotrophs, to develop such inclusive models, nor even to frame the correct questions we wish to ask of them. The discussions strongly endorsed the view that even simple models of protozoan processes will be useful for testing our knowledge of the processes and depicting ecosystem-scale functions like nutrient cycling more accurately.

MODELLING BENTHIC MICROBIAL FOODWEBS

An attempt was made through discussion of the questionnaires to identify the information available to construct simple models of carbon and nitrogen fluxes in benthic and planktonic foodwebs. The roles of protozoans in benthic foodwebs have been little investigated and remain poorly defined, even though the behaviour of several groups of benthic protozoans have been well studied (e.g. foraminifera) and even modelled (Lee 1990). Discussion focussed on a generic model for a benthic system in a shelf or oceanic region, which is shown in Figure 1. It was agreed that the taxonomic composition of most of the compartments was variable or even bizarre (Gooday 1986) and poorly known. Parameters governing rates of exchange between compartments remain almost totally unknown. But some important points were emphasized: First, the main input to benthic foodwebs appeared to be sedimentation of particulate detrital material (bold arrows in Figure 1), including marine snow, which may also serve to colonize the benthos with some protozoan groups (Lochte 1990). Second, much of the detrital input is thought to enter the benthic foodweb via bacterial metabolism or ingestion by protozoans. Also, the significance of the fine, unconsolidated layer of 'fluff' on the seabed was recognized as a unique habitat for microbes. Thus an important priority for research in this area is investigation of the utilization of sedimenting and recently sedimented particles by benthic bacteria and bacteriovores (Cho and Azam 1988, Lochte and Turley 1988).

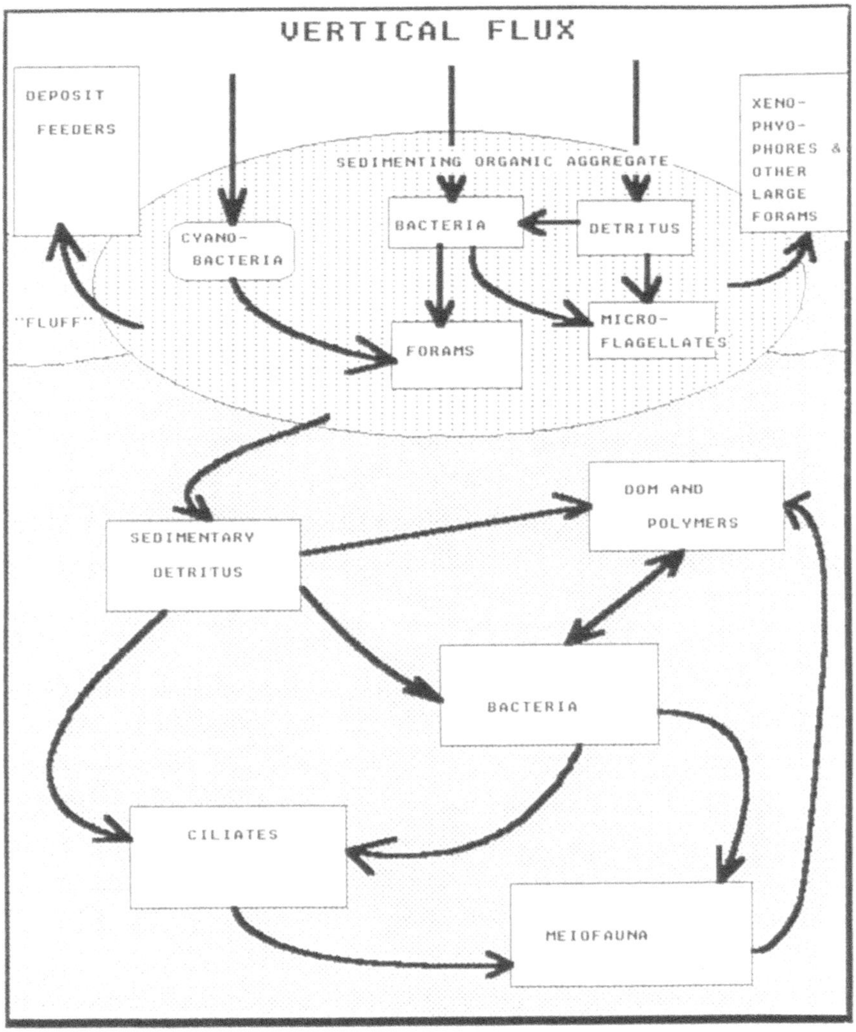

Fig. 1. Conceptual model of carbon flows in a shelf or open sea benthic ecosystem, emphasizing principal inputs and microbial components. This model reflects views presented during discussions on modelling microbial foodwebs. Three distinct habitats are shown: sedimenting organic aggregates (marine snow), 'fluff' or marine snow recently deposited on the sea floor, and the sediment proper

MODELLING PLANKTONIC MICROBIAL FOODWEBS

Most of the ASI participants had expertise in planktonic rather than benthic systems, and the questionnaire directly addressed plankton foodwebs. A variety of replies were received. They fell into three main categories:
1) Replies which claimed insufficient familiarity with modeling and no direct knowledge of the kinds of data requested; 2) others which had no data to offer but presented views on what a detailed model of a general microbial foodweb should contain; 3) a number which offered data answering the questionnaire. Here we present a brief synthesis of replies and subsequent discussion in the form of a simple model of a microbial foodweb and a more detailed model which contains a greater variety of protozoan components. A similar exercise is described in McManus (1990).

Figure 2 summarizes the answers from those who offered information on parameter values which govern C and N transfers in the plankton. This is offered as a simple representation of a stratified summer condition in a coastal or open sea. This model could serve as a subcomponent of a full plankton system, as shown in Ducklow (1990). It is important to stress that this model emphasizes flows and transformations within the microbial portion of the plankton system without specifying in detail the connections between microbes and other trophic levels or subsystems. Some of these connections are discussed in Ducklow (1990). Several ASI participants stated that the microbial foodweb could not be understood in isolation and that inputs from external sources of energy and matter (e.g. detritus) must be addressed (e.g. Sorokin 1981). This is implicit in the overall modelling approach.

Perhaps the most important point to be made from this exercise is the large range of estimates (and in some cases, small ranges) for many of the parameters. For example, the proportions of total primary or bacterial production which are removed by ciliates and heterotrophic flagellates are believed

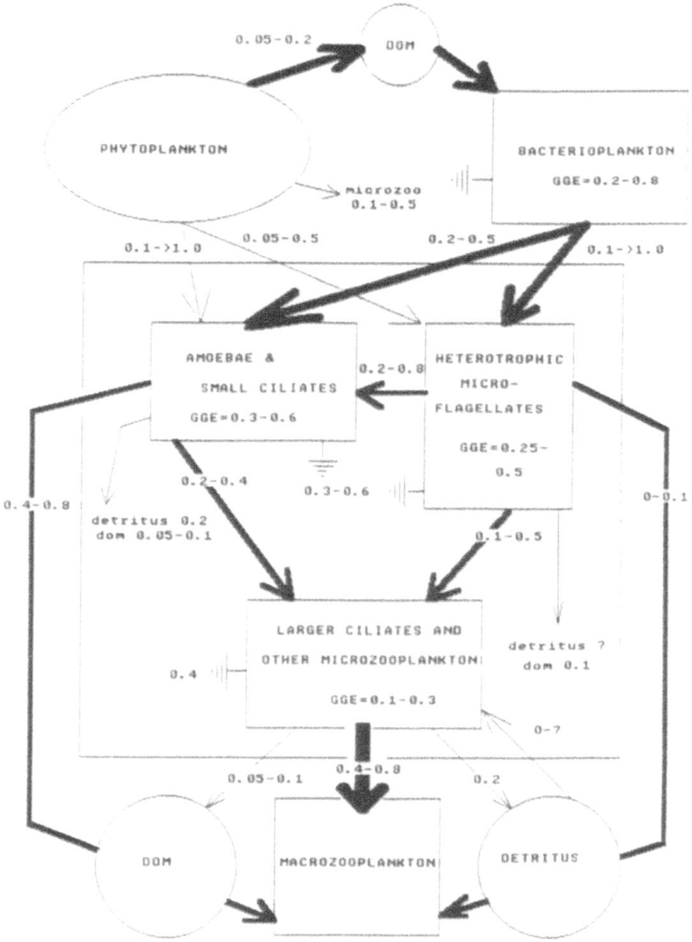

Fig. 2. Simple model of carbon flows in an aggregated microbial foodweb summarizing responses to a questionnaire distributed to ASI participants. This view represents a foodweb from temperate or subtropical, stratified shelf or open ocean regions. In this figure, outputs from each compartment to respiration and excretion are expressed as fractions of the total input. Thus for example, ciliates respire 25-60% of their input, etc. Grazing losses are represented as a fraction of the total production lost to each predator, e.g. 10-100% of the phytoplankton production is removed by ciliates, and 5-50% by heterotrophic microflagellates

to range over an order of magnitude. Based on these estimates, it seems impossible to conclude *a priori* that either ciliates or micro-flagellates are more important as bacterivores in any given circumstance. This view of the microbial foodweb also suggests considerable uncertainty over the trophic roles of microzooplankton and small metazoa: are they mostly herbivores removing up to 50% of the primary production, or secondary consumers eating mostly protozoans (Roman *et al.* 1988, Sherr and Sherr 1988)? These large ranges certainly reflect our current lack of observational data on feeding behaviour of microbial grazers from natural systems. Perhaps more interestingly, they also suggest the considerable variability and flexibility inherent in these groups of organisms.

Given the large uncertainties in the feeding preferences and diet composition indicated, it was surprising to see the narrow range of estimates for most of the physiological parameters. For example, estimates of gross growth efficiencies for the three groups of microbial grazers varied only a factor of 2-3. Whether this truly represents the real variability (or lack thereof) remains to be determined. Pomeroy (1990) suggests that we may still be "on the crest of the wave of optimism" regarding growth and conversion efficiencies for bacteriovores, and that more research on natural bacteriovores eating natural bacterioplankton may yield lower and more variable estimates. Similar narrow ranges were suggested for the other physiological parameters (DON and NH_4^+ excretion and egestion). Moloney and Field (1990) make the case for biomass-specific parameters to govern these processes, but the input-output approach was used here for simplicity's sake. Full dynamic models will need a more detailed approach.

This discussion of a simple, aggregated model suggests the great need we have for more data on the trophodynamics of protozoan communities in nature. Even simple models cannot yet be specified accurately and the data required for calibration do not yet exist.

Two of the more interesting findings reported at the ASI were the apparent importance in many situations of mixotrophic and symbiotic organisms (Stoecker 1990, Laval-Peuto 1990) and

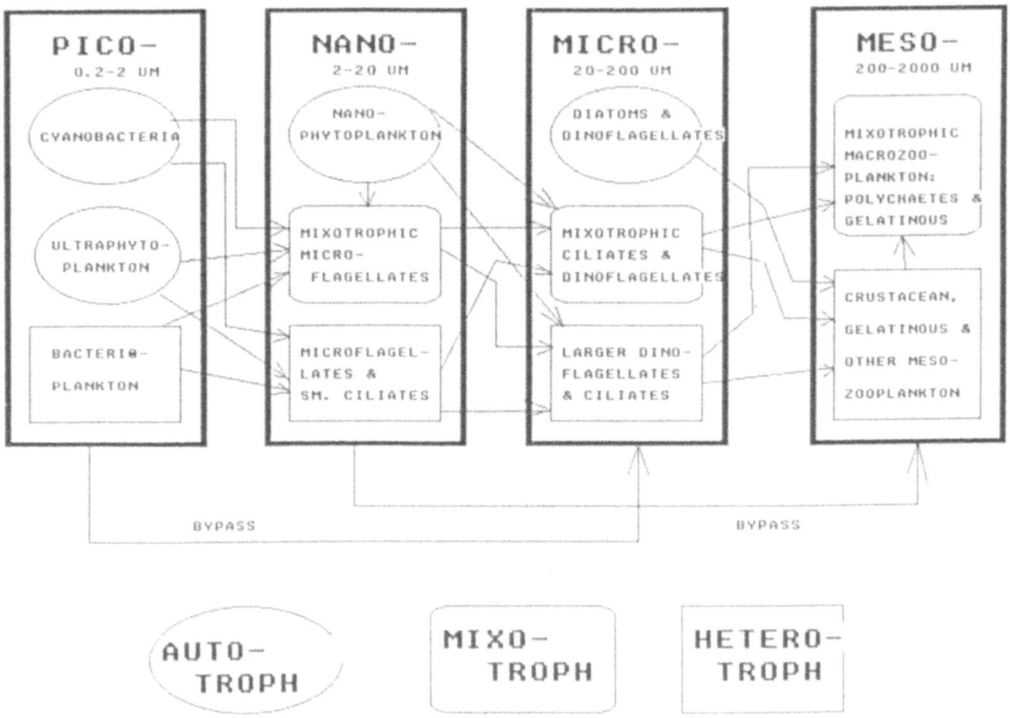

Fig. 3. Conceptual, size-based model of a microbial foodweb emphasizing the continuum of trophic modes (autotrophic---->mixotrophic---->heterotrophic) which characterize all size classes of plankton. This model includes some of the components discussed in the ASI which have not previously been included in plankton flow models. Circles: autotrophs; Rounded boxes: mixotrophs; Boxes: heterotrophs

heterotrophic dinoflagellates (Lessard 1990) in the plankton. Even rather ambitious models of plankton dynamics (e.g. Pace *et al.* 1984) have not yet addressed these groups which remain largely unknown to most non-specialists. Preliminary attempts to include symbiotic associations in simple models suggest that patterns of vertical carbon fluxes can be influenced strongly by the short-circuiting between production and heterotrophic utilization which symbiosis permits (A Michaels, pers. comm.).

Figure 3 presents a conceptual model of a more inclusive foodweb. This view summarizes the contributions from many participants, especially Diane Stoecker, Alan Longhurst and Peter Bjornsen. It was felt that future models should be organized on the basis of size (see also Sieburth *et al.* 1978, Moloney and Field 1990, Taylor 1988, Taylor and Joint 1990); thus the microbial foodweb is presented here as a hierarchy of size groupings each containing autotrophic, strictly heterotrphic, and mixotrophic compartments. Furthermore each compartment may be further subdivided into taxonomic groupings (e.g. mixotrophic ciliates and dinoflagellates; heterotrophic microflagellates and small ciliates, etc). In this way, a continuum of trophic capability from autotrophic to heterotrophic may be recognized. Finally, the variability, diversity and flexibility of protozoans must be analyzed in future models. The arrows connecting the nano- and meso-compartments, and the pico- and micro- compartments, respectively, suggest the possible by-passes of the simple pico-nano-micro-meso sequence which may occur under conditions of enrichment (Peter Bjornsen pers. comm.). Others noted that size is not always a good predictor of trophic relationships, since 'little things can eat big things' (e.g. Goldman and Caron 1985) thus many (or most??) of the arrows shown should be pointing in both directions. This feature was not included out of consideration for the reader! Finally, only the phagotrophic (grazing) flows are shown; the flows to and from dissolved and detrital pools were left out. The great complexity implied by this model is obvious. Flow analysis of quantitative versions of such a model is feasible and was attempted in the practical sessions using NETWORK 3.

CONCLUSIONS

As many ASI participants (including the modellers) stated, models are not ends in themselves. The modelling exercise, however, is always valuable in that it offers a formal way of stating our knowledge in rigorous, quantitative terms. Furthermore it offers a way of organizing our ignorance: What are the most important things we don't know? What are the most crucial measurements to make? As our discussions pointed out, there is no shortage of ways to model microbial systems, but few data exist to point the way, to provide parameter values, and to calibrate results. The most salient need is for data on the annual cycles of biomass or abundance and production for some of the major groups (Taylor and Joint 1990, Lynn and Montagnes 1990, Montagnes and Lynn 1990), with which models can be validated.

ACKNOWLEDGEMENTS

This work was sponsored in part by NSF Grant OCE-871690 and Office Naval Research Grant N-00014-88-J-1149 to HWD.

REFERENCES

Cho BC and Azam F (1988) Major role of bacteria in biogeochemical Fluxes in the ocean's interior. Nature 332:441-43
Duckow HW (1990) The passage of carbon through microbial foodwebs: results from flow network models. Mar Microb Food Webs, (submitted)
Ducklow HW, Purdie DA, Williams PJLeB, Davies JM (1986) Bacterioplankton: A sink for carbon in a coastal plankton community, Science 232:865-67
Fasham MJR (1984) Flows and Energy and Materials In marine Ecosystems; Theory and Practice, New York: Plenum p 733
Frost BW (1984) Utilization of phytoplankton production in the surface layer. Washington DC: Natl Academy Press. In: Global Ocean Flux Study: Proceedings of a workshop. p 125
Frost BW (1987) Grazing control of phytoplankton stock in the open subartic pacific ocean: a model assessing the role of mesozooplankton particularly the large calanoid copepods *Neocalanus* spp. Mar Ecol Prog Ser 39:49-68
Goldman JC, Caron DA (1985) Experimental studies on an omnivorous microflagellate: implications for grazing and nutrient regeneration in the marine microbial food chain. Deep Sea Res 32:899-915
Gooday AJ (1986) Meiofaunal foraminiferans from the bathyal

Porcupine Seabight (northeast Atlantic): size structure, standing stock, taxonomic composition, species diversity and vertical distribution in the sediment. Deep Sea Res 33:1345-1373
Laval-Peuto M (1990) Endosymbiosis in the protozoa - Session summary. In: Reid PC, Turley CM, Burkill PH (eds) Protozoa and their role in marine processes. Springer, Berlin Heidelberg New York
Lee JJ (1990) Brief perspective on the autecology of marine protozoa. In: Reid PC, Turley CM, Burkill PH (eds) Protozoa and their role in marine processes. Springer, Berlin Heidelberg New York
Lessard EJ (1990) The trophic role of heterotrophic dinoflagellates in diverse marine environments. Mar Microb Food Webs, (submitted)
Lochte K (1990) Protozoa as makers and breakers of marine aggregates. In: Reid PC, Turley CM, Burkill PH (eds) Protozoa and their role in marine processes. Springer, Berlin Heidelberg New York
Lochte K, Turley CM (1988) Bacteria and cyanobacteria associated with phytodetritus in the deep sea, Nature 333:67-69
Lynn D, Montagnes DJS (1990) Global production of heterotrophic marine planktonic ciliates. In: Reid PC, Turley CM, Burkill PH (eds) Protozoa and their role in marine processes. Springer, Berlin Heidelberg New York
Maloney CL, Field JG (1990) Modelling carbon and nitrogen flows in a microbial plankton community. In: Reid PC, Turley CM, Burkill PH (eds) Protozoa and their role in marine processes. Springer, Berlin Heidelberg New York
McManus GB (1990) Flow analysis of a planktonic microbial food web model. Mar Microb Food Webs, (submitted)
Montagnes DJS, Lynn DH (1990) Taxonomy of choreotrichs, the major planktonic ciliates. Mar Microb Food Webs, (submitted)
Pace ML, Glasser JE, Pomeroy LR (1984) A simulation analysis of continental shelf food webs. Mar Biol 82:47-63
Pomeroy LR (1990) Status and future needs in protozoan ecology. In: Reid PC, Turley CM, Burkill PH (eds) Protozoa and their role in marine processes. Springer, Berlin Heidelberg New York
Roman MR, Ducklow HW, Fuhrman JA, Garside C, Glibert PM, Malone TC, McManus GB (1988) Production, consumption and nutrient recycling in a laboratory mesocosm. Mar Ecol Prog Ser 42:39-52
Sherr EB, Sherr BJ, Albright LJ (1987) Bacteria: Link or Sink? Science 235:88
Sherr EB, Sherr BJ (1988) Role of microbes in pelagic food webs: A revised comment. Limnol Oceanogr 33:1225-1227
Sieburth JMcN, Smetacek V, Lenz J (1978) Pelagic ecosystem structure: Heterotrophic compartments of plankton and their relationship to plankton size fractions. Limnol Oceanogr 23:1256-1263
Sorokin YuI (1981) Microheterotrophic organisms in marine ecosystems, In: Longhurst (ed) Analysis of marine Ecosystems, Academic Press, New York p 293
Stoecker DK (1990) Mixotrophy in marine planktonic ciliates: physiological and ecological aspects of plastid-retention by oligotrichs. In: Reid PC, Turley CM, Burkill PH (eds) Protozoa and their role in marine processes. Springer, Berlin Heidelberg New York
Taylor AH (1988) Characteristic properties of models for the vertical distribution of phytopankton under stratification. Ecological modelling 40:175-199
Taylor AH, Joint I (1990) A steady-state analysis of the 'microbial loop' in stratified systems. Mar Ecol Prog Ser 59:1-17
Ter Kuile B, Lee JJ (1990) Mar Microb Food Webs, submitted
Ulanowicz RE (1984) Community measures of food networks and their possible applications, In: Fasham M (ed) Flows of energy and materials in marine ecosystems: Theory and practice. Plenum Press, p 23
Ulanowicz RE (1986) Growth and development: Ecosystems Phenomenology, Springer-Verlag, New York

MODELLING CARBON AND NITROGEN FLOWS IN A MICROBIAL PLANKTON COMMUNITY

Colleen L Moloney and John G Field

Marine Biology Research Institute
Zoology Department
University of Cape Town
7700 Rondebosch
South Africa

INTRODUCTION

The role of Protozoa in marine ecosystems can be studied from many aspects. In this paper we focus on the rates of material flow and nutrient cycling and how these rates change as a microplankton community develops in a succession after a disturbance.

In practice it is nearly impossible to directly measure all the flows of even one substance (such as nitrogen) amongst all the components in a simple community. The approach we have therefore adopted has been to reduce a hypothetical community to its essential components, and to develop a simulation model based on first principles. If the model behaves in a manner consistent with experimental and field observations, then we infer that the principles used to build the model are likely to be approximately correct and we can use the model to estimate the flows of carbon and nitrogen which are so difficult to measure simultaneously. These flows are necessary to cause the changes in biomass which are observed. The model is validated with observations in the Benguela upwelling region of the southeast Atlantic Ocean, and microcosm experiments of community succession using freshly upwelled water from this same region.

AN ALLOMETRIC BASIS FOR MODELLING

The size of organisms is important in influencing the rates of processes in, and interactions among, planktonic (and other) organisms (Peters 1983, Calder 1985, Dickie *et al.* 1987). This

NATO ASI Series, Vol. G 25
Protozoa and Their Role in Marine Processes
Edited by P. C. Reid et al.

has long been recognised by physiologists, but has more recently been absorbed into ecological thinking (Sheldon *et al.* 1972, 1977, Silvert and Platt 1980, Peters 1983). The structure of pelagic food webs is likewise largely dependent upon organism size (Sheldon *et al.* 1972, 1977, Platt and Denman 1978, Silvert and Platt 1980, Cousins 1985, Platt 1985). Table 1 reviews some general allometric equations for the mass-specific daily rates of growth, ingestion and respiration of living organisms of widely differing sizes. The daily rates are a power function of body size (M) so that in general the rates (R) may be expressed by the equation $R=aM^b$. This relationship has often been estimated by regression, in which both the exponent (b) and the constant (a) are calculated using standard regression techniques (see Table 1 for examples). However, it is very difficult to make comparisons when they are based on two values (a and b) simultaneously. If

Table 1. GENERAL ALLOMETRIC EQUATIONS OF THE FORM RATE = $a*(Mass)^b$ FOR RATES OF METABOLIC PROCESSES IN RELATION TO BODY SIZE FOR DIFFERENT GROUPS OF ORGANISMS

Process	Organisms	a	b	Source
Growth	Virus - mammals	20	-.25	Fenchel 1974
Growth	Virus - mammals	16.5	-.26	Blueweiss *et al.* 1978
Growth	Ciliates and amoebae	9.45	-.311	Baldock *et al.* 1980
Growth	Ciliates	12	-.247	Taylor and Shuter 1981
Growth	Copepod	9.2	-.15	Ross 1982
Ingestion	Marine amphipod	68	-.25	Dagg 1976
Ingestion	Detritivores	76	-.258	Cammen 1980
Ingestion	Invertebrates	54	-.306	Capriulo 1982
Respiration	Rat - steer	-	-.25	Kleiber 1932
Respiration	Mouse - elephant	-	-.266	Brody et al. 1934
Respiration	Bacteria - mammals	-	-.25	Hemmingsen 1960
Respiration	Marine plankton	33	-.309	Ikeda 1970
Respiration	Unicellular algae	0.4	-.10	Banse 1976
Respiration	Marine amphipod	11.6	-.225	Dagg 1976
Respiration	Daphnia	10	-.15	Lampert 1977
Respiration	Zooplankton	15.4	-.312	Ikeda and Motoda 1978
Respiration	Crustacea	16	-.268	Ivleva 1980
Respiration	Copepod	17	-.25	Ross 1982

(Parameters converted to correspond with carbon masses (pg C) and mass-specific rates (d^{-1}) at 20°C. (from Moloney and Field 1989, with permission from the publishers of Limnology and Oceanography)

there were grounds for assuming that one of the parameters were fixed, then one could fit the other and make comparisons based on only the fitted parameter.

There has been much debate about the true value of the exponent (b). Platt and Silvert (1981) for example, proposed on theoretical grounds that for terrestrial organisms b = -.25, whereas for aquatic organisms b = -.33. Inspection of Table 1 suggests that most of the values of (b) lie close to -.25 and it appears, on purely empirical grounds, that it might not be unreasonable to assume that the size-exponent for these processes is -.25 as suggested by many authors (e.g. Brody *et al*. 1934, Kleiber 1947, Hemmingsen 1960, Fenchel 1974, Blueweiss *et al*. 1978, Cammen 1980). It would then be possible to derive allometric equations with the same exponent (b = -0.25) for a number of different processes. This would give a common basis for comparison of size relationships simply by comparing the rate constants (a).

Consequently, we have calculated regressions which relate rates of nutrient uptake, ingestion and respiration to organism size, assuming b = -.25 [see Moloney and Field (1989) for details]. In order to estimate the rate constants (a) of the power functions, the data from many of the aquatic sources listed in Table 1 were graphed on log-log axes. For each graph, a regression line was drawn through the log-transformed centroid of the data points (i.e. the geometric means of the x- and y-values) and rotated about the centroid to have slope -.25. The regressions are thus forced to have slope b = -.25. The rate coefficients (a) were calculated from the intercepts (log a) of the lines when log M = 0 (See Moloney and Field 1989, summarised here in Figure 1).

Regressions were calculated for nutrient uptake, ingestion and respiration rates; they are compared in Table 2. Maximum specific nutrient uptake rates (V_{max}) of phytoplankton and bacteria were treated together, because both take up nutrients from solution (Azam *et al.* 1983) and thus similar kinetics are likely to apply. Maximum specific ingestion rates of particle-feeders have been shown to decrease with increasing body size both within-species (Dagg 1976) and between-species (Paffenhöfer 1971, Ikeda 1977, Fenchel 1980, Capriulo 1982,

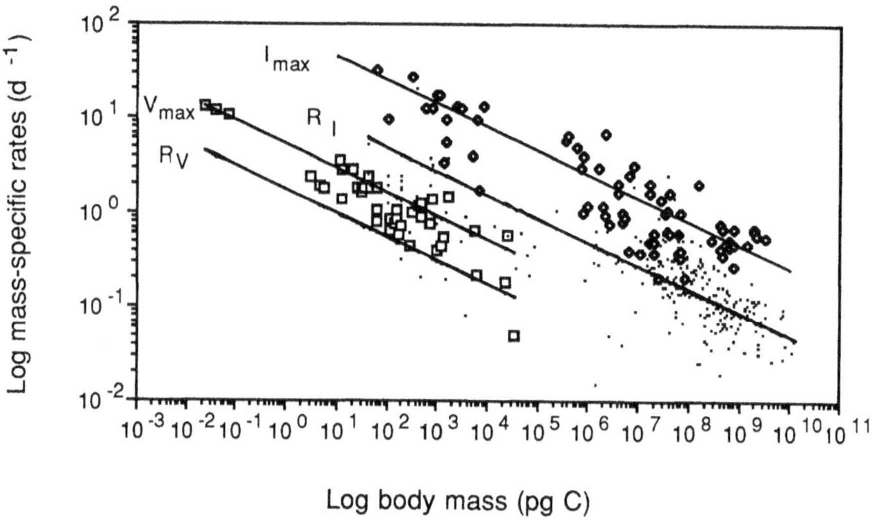

Fig. 1. Relationships of maximum nutrient uptake rates (V_{max}), maximum ingestion rates (I_{max}), and respiration rates of particle-feeders (R_I) to body mass, plotted on logarithmic axes. For details see Moloney and Field (1989). Regression equations are presented in Table 2. Data are plotted as squares for V_{max}, diamonds for I_{max}, and dots for R_I. The regression for respiration of phytoplankton and bacteria (R_V) was derived from equations presented by Banse (1982) and is plotted for comparison

Ross 1982). Data for aquatic particle-feeders ranging from ciliates to euphausiids were analysed by Moloney and Field (1989) to estimate the feeding coefficient when the exponent b = -.25. For the respiration coefficients, two equations were calculated; one for the respiration of phytoplankton and bacteria (R_V) and another for that of particle feeders (R_I).

The four regressions are based upon a wide range of data collected for different purposes under very different conditions. In spite of this, correlation coefficients of the regressions are surprisingly high for size- specific uptake and ingestion rates (Table 2). The correlation for respiration rates of particle feeders (R_I) is much lower, reflecting much greater scatter in the wide range of respiration data available. Individual data points were not published for size-specific respiration rates of prokaryotes and eukaryotes reviewed by Banse (1982), from whose equations the rate coefficient for R_V was derived (Moloney and Field 1989), hence tests of significance do not apply to those data.

Table 2. ALLOMETRIC EQUATIONS FOR MAXIMUM NUTRIENT UPTAKE RATES (V_{MAX}) OF PHYTOPLANKTON AND BACTERIA, AND MAXIMUM INGESTION RATES (I_{MAX}) AND RESPIRATION RATES (R_I) OF BOTH UNICELLULAR AND MULTICELLULAR PARTICLE-FEEDING HETEROTROPHS. THE EQUATION FOR RESPIRATION RATES OF PHYTOPLANKTON AND BACTERIA (R_V), MODIFIED FROM BANSE (1982), IS ALSO PRESENTED

Equation	N	r^2
V_{max} (d^{-1}) = 3.6 M (pg C)$^{-.25}$	44	0.77
I_{max} (d^{-1}) = 63 M (pg C)$^{-.25}$	74	0.85
R_I(d^{-1}) = 14 M (pg C)$^{-.25}$	284	0.43
R_V(d^{-1}) = 1.7 M (pg C)$^{-.25}$	-	-

Units of the rate coefficient a are pg C 0.25.d^{-1}. r^2 values were calculated for the variation of data points about the line of slope - 0.25. (from Moloney and Field 1989, with permission from the publishers of Limnology and Oceanography)

When the allometric relationships are assumed to have an exponent of = -.25, comparisons of the coeffcients (a) are simple and one can easily visualise the parallel lines on log-log axes, separated by the differences between the coefficients (a) (Figure 1). It is noteworthy (Table 2) that the ingestion rate coefficient of particle feeders is an order of magnitude larger than the uptake coefficient for bacteria and phytoplankton, and similarly the corresponding respiration rate coefficients differ by an order of magnitude. Thus particle-feeders (whether uni- or multi-cellular) appear to sustain mass-specific metabolic rates which are an order of magnitude faster than organisms which absorb nutrients through their body walls. This observation corroborates the finding of Fenchel and Finlay (1983) who noted that the size- specific respiration rates of ciliates (particle-feeders) were higher than the allometric regression for protozoa in general.

MODEL STRUCTURE

The purpose of the model described here is to study the role of different components of the microplankton community in carbon flow and nitrogen cycling. We attempted to develop the simplest possible model for this purpose (the rule of parsimony), and thus tried to build the model from first principles. Problems of parameter estimation are to a large extent obviated by the use of an independent criterion (body size) to estimate nearly all parameters. Organism size thus serves as a convenient theoretical and practical basis for developing a system model of a microplankton community.

If one assumes that the same form of allometric relationship holds for nutrient uptake, ingestion, and for respiration, one is able to make simple comparisons of the size-specific rates of processes over a wide range of body sizes. In graphical terms, this amounts to plotting the relationships on double-logarithmic axes, drawing the best-fit line of slope -.25 through the data, and comparing the intercepts of parallel lines. In modelling terms, this results in scaling various non-linear processes for organisms of widely different sizes with the same exponent. This is likely to give model output which is more stable than models in which there are different exponents for different processes over a wide range of body sizes.

The model is conceptually based upon the trophic continuum model of Cousins (1980, 1985), in which ecosystems are divided into three fundamental components: autotrophs, heterotrophs and detritus. Within each component there is a size continuum from small to large particles. Unlike Cousins' energy flow model, this model subdivides the size continuum into discrete size-classes for ease of computation, and it includes the flows of both carbon and nitrogen. Flows of carbon and nitrogen are modelled because it is believed that nitrogen often limits the growth of marine phytoplankton (Dugdale and Goering 1967, Eppley and Peterson 1979), whereas the supply of carbon may also limit bacterioplankton growth (Azam *et al.* 1983). The model is described in detail by Moloney and Field (1990). The present version has size categories arranged logarithmically to the base 10. This gives six living components, two of autotrophs in the 2- 200 µm size range and four of heterotrophs from 0.2 - 2000 µm (Table 3). Table 3 gives the names most commonly identified with each size-class, the geometric mean equivalent spherical diameter, the assumed C:N ratio and the proportion of phytoplankton sinking out of the euphotic zone. Sinking rates of phytoplankton were calculated from the data of Burns and Rosa (1980), Bienfang and Harrison (1984) and Bienfang (1985) using a power-curve sinking rate (S, $m.d^{-1}$) to size (M, pg C) relationship for live phytoplankton: $S = 0.029\ M^{0.42}$ ($n = 36$, $r^2 = 0.74$) (Moloney 1988, Moloney and Field 1990). Phytoplankton sinking was only included in the model after the bloom peak, to simulate the effects of senescence in stratified water.

Figure 2 depicts the model structure and the flows of nitrogen in the model system. One of the main features is the non-living nitrogen pool which consists initially of newly-upwelled nitrate-nitrogen, which can be taken up by bacteria, nano-phytoplankton and net-phytoplankton (oblique solid arrows). Carbon flows are not depicted, but are linked to nitrogen flows for sinking particles and predator-prey interactions. Carbon flows differ from nitrogen flows in that the supply of inorganic carbon to phytoplankton is assumed to be open and unlimited, each living component respires carbon (the carbon system is open), and dissolved organic carbon is secreted by phytoplankton and utilized by bacteria.

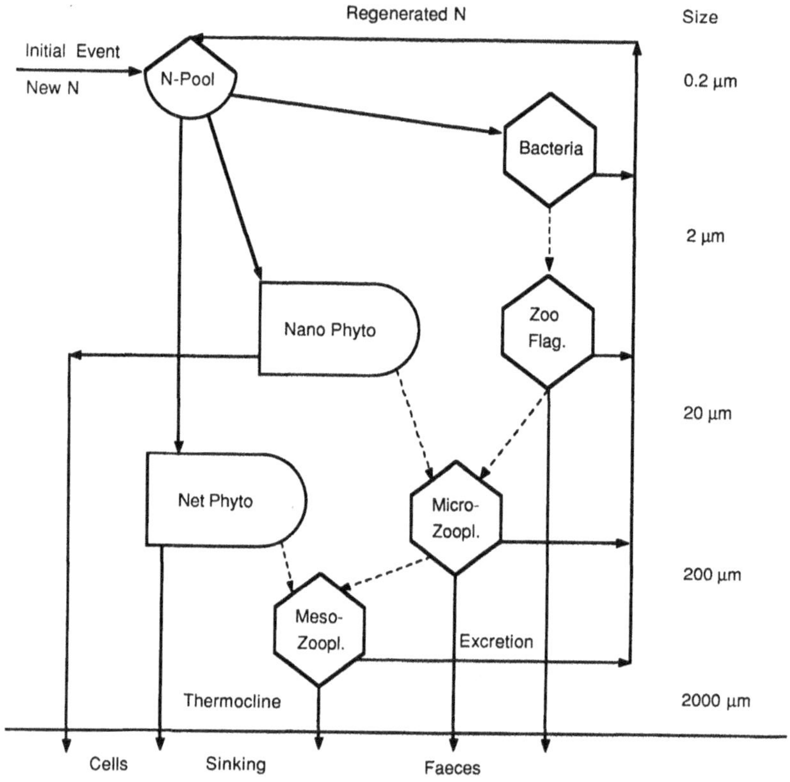

Fig. 2. Diagrammatic representation of nitrogen flows in the model microplankton community. Organisms are arranged by size: there are two size classes of phytoplankton (bullets) and four of heterotrophs (hexagons). An initial upwelling event adds new nitrogen (nitrate) to the nitrogen pool. Arrows represent nitrogen flows: vertical downward arrows represent sinking out of the surface layer, other solid arrows depict uptake or excretion of dissolved nitrogen, broken arrows show predator-prey interactions. Carbon flows are not illustrated but are modelled

Since the model uses dual currencies of carbon and nitrogen, it is necessary to relate the fluxes of the two to each other. This is done by assuming constant C:N ratios for each category (Table 3). For example, if the rate of bacterial production is limited by dissolved organic carbon availability, nitrogen uptake is modelled to give the appropriate C:N ratio in the bacteria. The same technique was used for losses of DOM, faeces etc., so that the C:N ratios of all components remained

Table 3. LIVING COMPONENTS OF THE MODEL

Component	esd Range µm	Mean esd µm	C:N ratio	Sinking Rate % d^{-1}
Nano-Phytoplankton	2-20	6.32	6.0	0.1
Net-Phytoplankton	20-200	63.2	6.0	15.1
Bacteria	0.2-2	.632	4.0	-
H-Flagellates	2-20	6.32	4.5	-
Micro-Zooplankton	20-200	63.2	4.5	-
Meso-Zooplankton	200-2000	632.	4.5	-

Equivalent spherical diameters (esd), assumed carbon to nitrogen ratios, and percent of biomass sinking out of 20 m euphotic zone per day.

constant. This may not always be realistic but mechanisms causing different rates of uptake or release of carbon and nitrogen are complex and poorly understood (Syrett 1981, Terry 1982). The assumption of constant C:N ratios is commonly made by ecologists converting carbon estimates to nitrogen and vice-versa using Redfield ratios (e.g. Probyn and Lucas 1987).

The model is described by a series of differential equations given in Table 4. These are solved numerically using a second-order Runge-Kutta method (Lapidus and Seinfeld 1971). Rates of change of standing stocks are determined by the input and output of carbon and nitrogen to and from each compartment. Diel effects are not included. Since the focus of this model is on the effects of body size, the physical and chemical environments were kept as simple as possible. The main assumptions are as follows:

1) The mixed-layer and euphotic zone are the same and constant at 20 m depth (i.e. the water column stratifies immediately after upwelling)

2) The temperature is constant at 10°C (newly upwelled water)
3) Light is constant and not limiting
4) There is one initial pulse of new nitrogen (one upwelling event) and no mixing or diffusion thereafter
5) Detritus is assumed to sink out of the euphotic zone in less than 12 h, hence bacterial degradation of detritus is not modelled here.

Table 4. EQUATIONS DESCRIBING CHANGES IN THE MAIN STATE VARIABLES OF THE SIMULATION MODEL OF A MICROPLANKTON COMMUNITY

dPC_j/dt	=	Carbon fixation-PDOC production-Respiration-Grazing-Sinking
dPN_j/dt	=	Nitrogen uptake-Grazing-Sinking
DHC_j/dt	=	Uptake/Ingestion-Egestion-Respiration-Predation
DHN_j/dt	=	Uptake/Ingestion-Egestion-Excretion-Predation
dDET/dt	=	Faeces egested - Sinking
dPDOC/dt	=	PDOC Production - Uptake
dNEWN/dt	=	(Initial upwelling) - Uptake
dREGN/dt	=	Excretion - Uptake

PC and PN represent phyto-plankton carbon and nitrogen respectively, HC and HN represent heterotroph carbon and nitrogen, DET represents detritus, PDOC represents photoassimilated dissolved organic carbon, NEWN and REGN represent new and regenerated nitrogen, respectively, and J denotes the size class of living components.

MODELLING OF PRODUCTION, UPTAKE, INGESTION AND EGESTION

Primary production is modelled in the simplest possible way. It is assumed that only nitrogen limits phytoplankton production and that light and other factors are not limiting. These and other simplifying assumptions need to be borne in mind in interpreting the model output. Primary production rates are assumed to be governed by Michaelis-Menten N-uptake kinetics (MacIsaac and Dugdale 1969):

$$V_j(d^{-1}) = V_{maxj}(d^{-1}) \frac{N(mg.m^{-3})}{K_{sj}(mg.m^{-3}) + N(mg.m^{-3})} \qquad (1)$$

where V_j is the mass-specific uptake rate for size-class j, V_{maxj} is the maximum mass-specific uptake rate for size-class j, K_{sj} is the half saturation constant for size-class j, and N is the ambient nitrogen concentration. V_j and N are variables, whereas V_{maxj} and K_{sj} are size dependent parameters whose values are given in Table 5. The size dependence of the half-saturation constants is modelled as a power function obtained by regression using published data and detailed by Moloney and Field (1990). Carbon fixation rates are calculated from the specific uptake rate (V_j) calculated in equation (1), multiplied by the carbon standing stock (B_j).

Table 5. LIVING COMPONENTS AND EQUATION PARAMETERS

	V_{max} (d^{-1})	K_s ($mgN.m^{-3}$)	I_{max} (d^{-1})	PER (%)	AE (%)	R (d^{-1})
Component						
Nano Phytoplankton	1.49	.713	-	30	-	.487
Net Phytoplankton	.265	66.28	-	30	-	.087
Bacteria	8.38	.00072	-	-	100	2.74
H-Flagellates	-	-	22.3	-	90	4.30
Microzooplankton	-	-	3.97	-	90	.764
Mesozooplankton	-	-	.707	-	90	.136

Nitrogen uptake (V_{max} and K_s), ingestion (I_{max}), percent nitrogen absorption efficiency (AE), percent extracellular carbon release (PER), and proportion of biomass respired daily (R), based on size-specific processes at 10°C. V_{max} and K_s are used in equations (1) and (2), I_{max} in equation (3).

A proportion of primary production is exuded by phytoplankton as dissolved organic matter (Berman and Holm-Hansen 1974), but the fraction released is difficult to measure because of rapid bacterial uptake, and varies from small amounts to up to 70% of net production (Lancelot 1983). The percentage of extracellular release (PER) varies inversely with ambient nitrogen concentration (Azam *et al.* 1983, Lancelot 1983), but is also affected by light levels, species and age of cells etc. (Lancelot 1983). For simplicity we have assumed a constant value of 30% in the present model. This refers only to the labile portion of photo-assimilated dissolved organic carbon (PDOC). The refractory portion has long residence times beyond the time scale of the model. The dissolved organic nitrogen component of dissolved organic matter is not modelled explicitly, but can be visualised as returning to the nitrogen pool.

Bacterioplankton take up nitrogen (organic and inorganic) from solution in much the same way as do phytoplankton (Azam *et al.* 1983). The same Michaelis-Menten kinetics [equation (1)] are assumed to apply, with the maximum and half-saturation parameters modified by bacterial body size (see Table 5 for values). Bacterioplankton are assumed to take up carbon solely from the PDOC pool. Detritus, such as the senescent phytoplankton cells and animal faeces that comprise the particulate organic carbon (POC), sinks much faster than living phytoplankton (Bruland and Silver 1981) and is assumed to be broken down below the euphotic zone. Thus bacterial utilization of POC is not included in this model.

Bacteria are assumed to take up PDOC according to Michaelis-Menten kinetics in a similar fashion to nitrogen when carbon is the limiting nutrient for bacterial growth. Thus carbon uptake is modelled as:

$$V_j(d^{-1}) = V_{maxj}(d^{-1}) \frac{PDOC(mg.m^{-3})}{K_{sj}(mg.m^{-3}) + PDOC(mg.m^{-3})} \qquad (2)$$

where V_j is the mass-specific uptake rate for bacteria as determined by substrate availability, V_{maxj} is the maximum mass-specific uptake rate, PDOC is the ambient dissolved carbon concentration, and K_{sj} is the mass-specific half saturation constant for PDOC uptake (assumed to equal K_{sj} for N-uptake multiplied by the C:N ratio for bacteria). Bacterioplankton growth is thus limited by either nitrogen or carbon, depending upon which uptake rate V_j [equation (1) or (2)] is slower. Bacterial growth in terms of the limiting nutrient is calculated by multiplying the specific uptake rate by the bacterial standing stock (B_j). The amount of the other, non-limiting nutrient taken up is calculated using the C:N ratio for bacteria. When nitrogen is limiting, carbon uptake may be underestimated because bacteria may take up carbon in excess of their biosynthetic and bioenergetic demands (Tempest and Neijssel 1978).

Particle-feeding heterotrophs obtain carbon and nitrogen by ingestion of appropriate sized food, either autotroph or heterotroph. The predator-prey model is a saturation function:

$$\mathrm{Ingestion}_{jk}(d^{-1}) = I_{maxj}(d^{-1}) \frac{B'_k}{K_{sk} + \Sigma B'_r} \qquad (3)$$

where the specific ingestion rate of size class k by class j is determined by the maximum mass-specific size- dependent ingestion rate of size class j (I_{maxj}), the standing stock (B'_k) of size class k available to size class j, the total standing stock ($\Sigma B'_r$) of all prey size classes available to size-class j, where r represents the range of size-classes, and the half saturation constant, K_{sk}, which depends upon the size of the prey according to a power function (Moloney 1988, Moloney and Field 1990). Thus the maximum mass-specific ingestion rate (I_{maxj}) is modelled as a function of *predator* body size, decreasing with increasing size (Table 5), and the half saturation constant (K_{sk}) is a function of *prey* body size, increasing with increasing body size of prey (Table 6).

Heterotrophs do not assimilate all that they ingest, a proportion of ingested material is released as faeces which contribute to the detritus and sink from the euphotic zone in the present model. Hall *et al.* (1976) were not able to show any size-dependence of absorption efficiencies, and the nitrogen absorption efficiencies are assumed to be 90% for all particle-feeders (Dagg 1976, Barthel 1983, Miller and Landry 1984). The N- absorption efficiency for bacteria is assumed to be 100%. After absorption, nitrogen is metabolized and a portion may be excreted, so that the net nitrogen uptake will be less than that absorbed (see 'Excretion' below).

Table 6. MATRIX OF PREDATOR-PREY INTERACTIONS, WITH K_{sk} VALUES (HALF-SATURATION CONCENTRATION, (mg $C.m^{-3}$) FOR PREY OF A PARTICULAR MEAN SIZE [SEE TEXT, EQUATION (3)]

	PREDATOR		
PREY	H-Flagellates	Microzooplankton	Mesozooplankton
Nano-Phytopl.	-	64.5	-
Net-Phytopl.	-	-	112
Bacteria	37.1	-	-
H-Flagellates	-	64.5	-
Microzooplankton	-	-	112

MODELLING, RESPIRATION AND EXCRETION

Respiratory carbon losses are modelled as a constant fraction of the standing stock of each size class:

$$\text{Respiration } (mg.C.m^{-3}.d^{-1}) = R_j(d^{-1}) * B_j(mg\ C.m^{-3}) \qquad (4)$$

where R_j is the mass-specific size-dependent respiration rate, and B_j the carbon standing stock of size-class j. Respiration rates vary during feeding and other activities, but these effects are not included in the model. This may cause underestimation of respiratory losses, but is probably only significant for large heterotrophs, because the motility of protozoans uses an insignificant fraction of their energy budgets (Fenchel and Finlay 1983). The values of R_j are derived from the size regressions discussed above (Figure 1, Table 2) and are given in Table 5.

Metabolic carbon losses by heterotrophs are matched by equivalent nitrogen losses in order to maintain constant C:N ratios in each heterotroph size class. Excretion was calculated as:

$$\text{Excretion (mg N.m}^{-3}\text{.d}^{-1}\text{)} = R_j * B_j \text{(mg N.m}^{-3}\text{)} \qquad (5)$$

where R_j is the same as in equation (4), and B_j is the nitrogen standing stock of size-class j. For bacteria this may not be realistic. Bacteria can take up nitrogen and carbon separately (Azam *et al.* 1983), because the dissolved pools consist of both inorganic and organic material and a variety of chemical species. Bacterial carbon and nitrogen uptake rates are therefore not necessarily coupled, as is assumed here. When nitrogen is limiting, bacteria may conserve this nutrient. An alternative model would be to consider only the net uptake of nitrogen, assuming that bacteria take up only sufficient nitrogen to meet their needs, and excrete none. This approach makes the *a priori* assumption that bacteria excrete no nitrogen at all. The real situation is probably intermediate between these two extremes, and the present model [equation (5)] will probably overestimate the amount of nitrogen excreted by bacteria.

Tables 5 and 6 show the values used in the main interactions of the model. Values for maximum uptake rates, maximum ingestion rates and respiration rates for each size class were obtained from the allometric rate-size regressions for the geometric mean mass of each size class (see above). Half-saturation parameter values were derived from regressions detailed in Moloney (1988) and Moloney and Field (1990). All values were standardized to 10°C, a reasonable value for newly upwelled water in the southern Benguela region (Shannon 1985, Brown and Field 1986, Brown and Hutchings 1987). Maximum ingestion rates are a function of predator body-size, and half-saturation constants for the predator-prey interactions are a function of prey body-size.

The parameter values of the model were not tuned, they were used as derived from the size-regressions. The model contains no time-lag functions. Any lags that occur do so as a result of the size-based rate parameters. Starting values of the variables were taken from published and unpublished results of fieldwork in the Benguela Ecology Programme. Sensitivity analyses have been conducted on both the structure of the model, and the parameter and variable values used. These are reported fully in Moloney (1988) and Moloney and Field (1990). In brief, altering the value of the allometric exponent from the assumed value of -.25 had a small effect on the standing stocks. When *b* was increased to -.17, small organisms were favoured because their rate constants were faster relative to those of larger organisms. When b was decreased to -.33, larger organisms were favoured, since they then had the faster relative rates. Varying the half-saturation constants for ingestion affected predators and their prey considerably, but similar sensitivity analysis on the nitrogen uptake constants had little effect. This suggests that grazing is an important factor in limiting net phytoplankton growth in the model. Halving or doubling the value of the percentage of extracellular release (PER) from the assumed value of 30% affected bacterial biomass by a similar factor, but had only a small effect upon other components. The model is sensitive to the initial values of some of the standing stocks at the start of each simulation. Net phytoplankton were the most sensitive to starting values, and took a long time to peak (18 or 28 days) when initiated with a small standing stock (10 or 1 mgC m^{-3}, respectively) (Moloney 1988).

MODEL RESULTS

The model was run to simulate the effects of a single upwelling event in the southern Benguela system, in which cool, nitrate rich water (25 µgat. NO_3-N.l^{-1}) is upwelled into the euphotic zone. The surface water is assumed to stratify immediately, and the development of the microplankton community is followed for 40 days after the disturbance caused by upwelling.

Figure 3 shows the changing standing stocks of the nitrogen pool and six living components after upwelling. Newly upwelled water in the southern Benguela region, coming from depths of 200-300 m which are well below the euphotic zone, has very little living biomass (Andrews and Hutchings 1980). Nevertheless, it is to some extent seeded with net phytoplankton (G Pitcher, Sea Fisheries Research Institute, *pers. comm.*) and this is reflected in the higher starting value on day 0. The model phytoplankton blooms take some 7 days to peak, both nano- and net phytoplankton peaks coinciding. The nitrogen pool declines slowly at first, then rapidly as exponential phytoplankton growth occurs. There is a rapid succession in the microplankton from days 6-9. Bacteria peak simultaneously with the phytoplankton, succeeded by heterotrophic (H-) flagellates and micro-zooplankton. There is some delay before the meso-zooplankton populations build up from days 14-20. All components gradually decline as nitrogen sinks out of the surface layer.

Model results are unlikely to occur in nature beyond days 15-20, because it is most unusual for the pulsing wind system to remain calm for such a long period in the southern Benguela region. Normally, there are upwelling events at 5-15 day intervals during the 8-month summer upwelling season (Andrews and Hutchings 1980, Wulff and Field 1983) which would cause mixing and/or upwelling into the surface layer. Nevertheless, the model does give results consistent with observations in the southern Benguela system, and meso-zooplankton populations do build up in less pulsed regions of the Benguela system.

It is seldom possible to follow a time series of observations on an identified parcel of water in the open ocean for weeks, but standing stocks of phytoplankton, bacteria, heterotrophic flagellates and meso-zooplankton are of the same order as those sampled from nature (Probyn 1985, Lucas 1986, Lucas *et al.* 1986, 1987, Brown and Hutchings 1987). Table 7 shows peak model

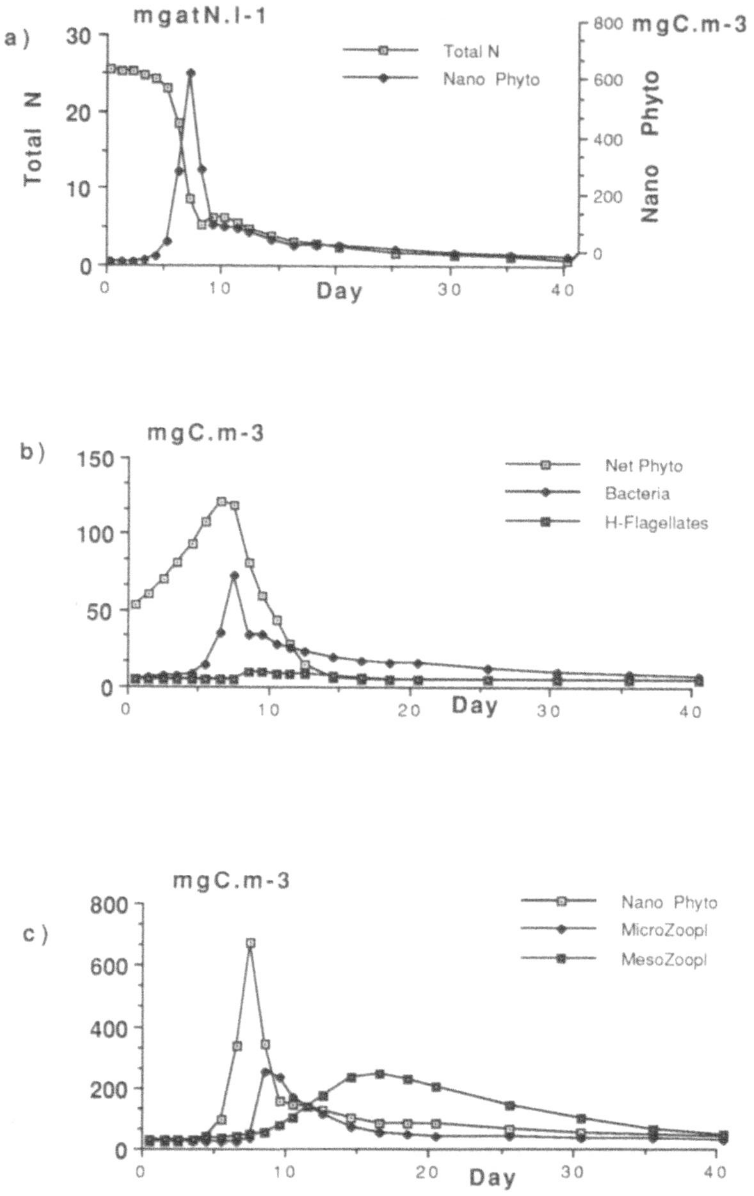

Fig. 3. Model output of carbon standing stocks and the nitrogen pool for a 40-day period after simulated upwelling. (a) Nitrogen pool (mgat-N.m^{-3}) and nanophytoplankton standing stock (mgC.m^{-3}); (b) standing stocks of net phytoplankton, bacteria and heterotrophic flagellates (mgC.m^{-3}); (c) standing stocks of microzooplankton and meso-zooplankton (mgC. m^{-3})

biomass values in relation to some comparable observations made in the southern Benguela system. For example, peak net phytoplankton (>20 µm) values of some 130 mgC m^{-3} in the model correspond to maximum field measurements of 300 mgC m^{-3} for total particulate organic carbon in the >10 µm size fraction on the southern Benguela shelf. Similarly, peak model nanoplankton (<20 µm) values of 600 mgC m^{-3} are roughly matched by maximum field measurements of some 400 mgC m^{-3} in the <10 µm

Table 7. MAXIMUM BIOMASS VALUES PREDICTED IN MODEL RESULTS AND OBSERVED IN THE SOUTHERN BENGUELA SYSTEM

Category	Predicted	Observed	Source of observation
Net phyto. >20 µm	130		
POC >10µm		300	Probyn 1985
Nano phyto. <20µm	600		
POC <10 µm		400	Probyn 1985
Bacteria	70		
		80	Lucas *et al.* 1986
		30	Armstrong *et al.* 1987
		47	Probyn 1987
		50	Verheye-Dua and Lucas 1989
Microcosm bacteria		75	Lucas *et al.* 1987
H-flagellates (2-20 µm)	20		
		90	Lucas *et al.* 1987
Micro-zooplankton (20-200 µm)	240		
			none available
Meso-zooplankton (.2-2 mm)	4600		
Max. all zooplankton		2000	Hutchings *et al.* 1984
Mean copepod and euphausiid		590	Pillar 1986
Mean zooplankton .2-.5 mm		170	Verheye and Hutchings 1988
Mean zooplankton >.2 mm		1400	Verheye and Hutchings 1988

All values given in mgC m^{-3}, except for the predicted and observed meso-zooplankton biomasses, which are given in mgC m^{-2}.

size fraction (Table 7). Bacterial peak values are comparable with maximum values observed both in the field and in microcosm experiments (Table 7), and those for heterotrophic flagellates are lower but of the same order as observations in microcosm experiments. There are at present no field observations in the southern Benguela region to compare with the model micro-zooplankton results. Meso-zooplankton simulations are also difficult to compare with field observations, since most observations have been made on all zooplankton or on categories that cover only part of the size range modelled here. Furthermore, zooplankton standing stocks are usually integrated through the water column, and are estimated per unit of surface area, rather than per unit volume as in the model. To make model units for mesozooplankton comparable with observed values, we have assumed a 20 m deep productive zone, giving a rough estimate of 4600 mgC. m^{-2} (Table 7). Mean and maximum observed values in the Table suggest that the predicted maximum for meso-zooplankton may be on the high side, but is probably of the right order of magnitude.

The timing of the phytoplankton bloom some 7 days after upwelling matches the findings of Brown and Hutchings (1987) in drogue studies in the southern Benguela system and the microcosm experiments of Painting *et al.* (1989), if one allows for their upwelled water having been 2-3 days old at the start of incubations, as seems likely. Coincidence of the bacterial and phytoplankton blooms, and the one-day lag between bacterial and H-flagellate peaks also matches the microcosm studies (Painting *et al.* 1989). The succession of organisms as the model microplankton community develops after upwelling, is very much as predicted by Margalef (1958, 1968) and found in previous models of the plankton community after upwelling (e.g. Moloney *et al.* 1986).

In spite of the grossly simplified assumptions made in this model, it appears to broadly reflect the kind of succession that is likely to occur after upwelling under idealised conditions in the Benguela system. Due to the difficulty of following a parcel of water in an upwelling system, it has not been possible to obtain a complete time series of observations in nature and the simulation model has given an integrated and plausible picture of the sequence of events that is likely to occur in the plankton community, according to the hypotheses upon which it is based. The results are consistent with microcosm experimental results for the first ten days when the 60 litre microcosm should not be too badly affected by wall effects (Painting *et al.* 1989), and it matches the order of magnitude of observed biomasses in the field and in the laboratory (Table 7).

NETWORK ANALYSIS

In order to study the flows of nitrogen in the microplankton community, the model results were analysed further using the techniques of network analysis (Wulff *et al.* 1989). Transfers of nitrogen occur from compartment to compartment through uptake, feeding, excretion, defaecation etc. as the succession unfolds. Thus the food web may be viewed as a network of nitrogen transfers. A bookkeeping routine in the model allowed us to keep track of these transfers and sum them daily, so that the changing food web could be represented quantitatively in matrix form. Methods are reported by Kay *et al.* (1989). The daily nitrogen flows were each analysed by a variety of techniques but only a few are reported here. Details are given in Field *et al.* (1989).

Figure 4 shows the daily summed outflows of nitrogen from several compartments, alongside their biomasses. It can be seen that there is little outflow of nitrogen from the nitrogen pool until the pool reserves decline rapidly after Day 5. Similarly, the flows from bacteria peak on day 8, as the bacterial biomass collapses. Largest flows lag behind the biomass peaks. Thus maximum flows occur when biomasses are changing most rapidly and are furthest from steady-state. Furthermore, this figure demonstrates the large flows from bacteria and micro-zooplankton relative to their respective biomass peaks. This is caused by the rapid metabolism and turnover of the relatively minor biomass of these small organisms. Although meso-zooplankton have a considerable standing stock at their peak, excretory outflows of nitrogen from meso-zooplankton are always relatively small and those from micro-zooplankton (both through excretion and predation) are always larger (Figure 4). For simplicity of viewing, the food web may be condensed into a

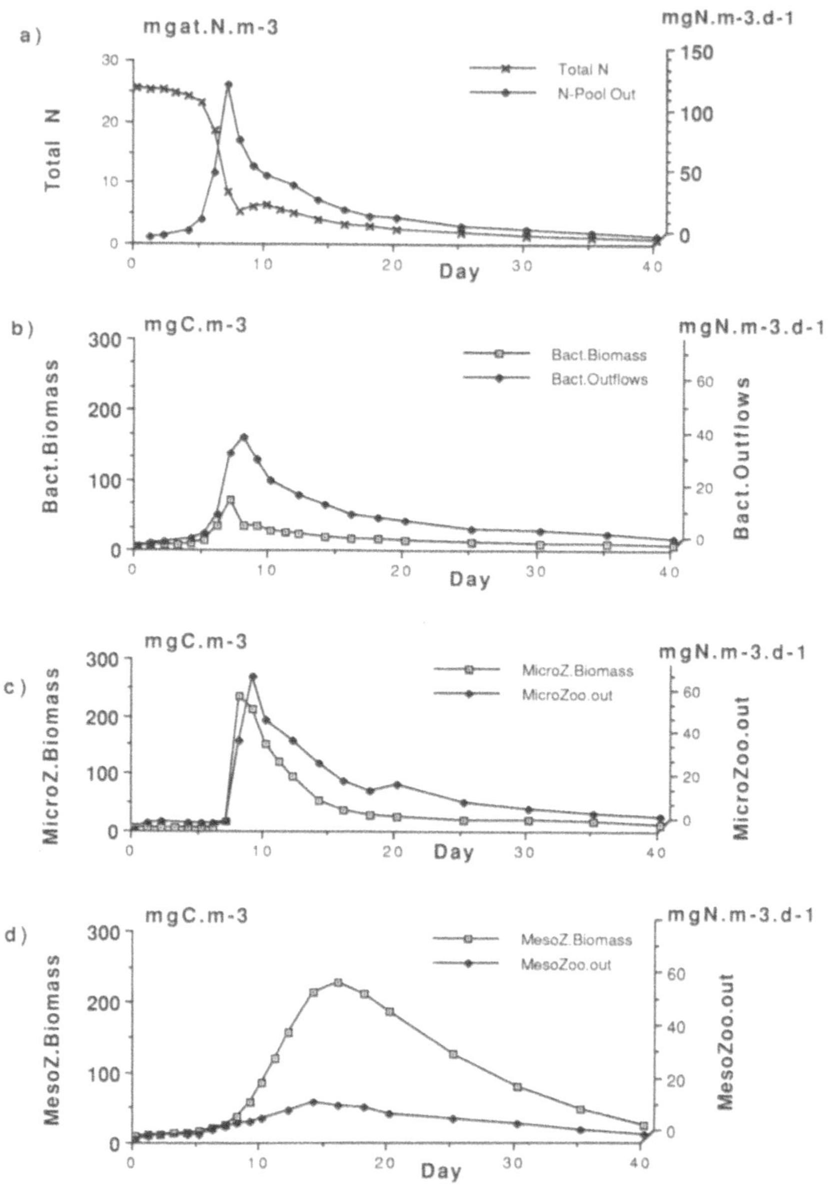

Fig. 4. Network analysis showing examples of summed daily nitrogen outflows from compartments in the model (from Field et al. 1989) along with the corresponding standing stocks. (a) Nitrogen pool outflows ($mgN.m^{-3}.d^{-1}$) with nitrogen concentrations ($mgat\text{-}N.m^{-3}$) in the pool; (b) bacterial outflows ($mgN.\ m^{-3}.d^{-1}$) and standing stocks ($mgC.\ m^{-3}$); (c) microzooplankton outflows ($mgN.\ m^{-3}.d^{-1}$) and standing stocks ($mgC.\ m^{-3}$); (d) mesozooplankton outflows ($mgN.\ m^{-3}.d^{-1}$) and standing stocks ($mgC.\ m^{-3}$)

hypothetical food chain, or the network may be collapsed into a 'Lindeman Spine' (Kay *et al.* 1989, Wulff and Ulanowicz 1989). It is well known that few animals are true herbivores or carnivores; most animals feed at several different trophic levels and indeed the trophic level concept has been strongly criticised (Cousins 1985). Using network analysis techniques, it is possible to compute the amount of nutrition received via pathways of different lengths, and therefore to allocate the intermediate trophic positions of micro-zooplankton, mesozooplankton, etc.

Table 8 shows the average trophic positions of the model components on two days of the simulation. It is noteworthy that the trophic positions of micro- zooplankton and mesozooplankton change with time as the food web changes and nutrition is received along alternative pathways in different proportions.

Table 8. TROPHIC POSITIONS OF COMPONENTS OF THE MODEL PLANKTON SYSTEM AS CALCULATED FROM NETWORK ANALYSIS OF NITROGEN FLOWS. TROPHIC POSITIONS ARE DEFINED RELATIVE TO THE NON-LIVING NITROGEN POOL, WHICH IS ALLOCATED A TROPHIC POSITION OF 1

	Day 8	Day 20
N-Pool	1	1
Nano-plankton	2	2
Net-plankton	2	2
Bacterio-plankton	2	2
H-Flagellates	3	3
Micro-zooplankton	3.02	3.52
Meso-zooplankton	3.01	4.01

Figure 5 depicts two examples of Lindeman Spines of the model microplankton community network based on nitrogen flows. The Lindeman Spine groups model components into 'trophic categories', numbered in Roman numerals. The trophic category indicates the number of steps in the food web to reach that position, starting with non-living matter in category I. Thus primary producers in this model are in trophic category II. Model components are listed below the trophic categories with their proportional activity in each category in parentheses. The upper panel shows the Lindeman Spine on Day 8 at the peak of phytoplankton and bacterial biomass, the lower panel shows Day 20 of the simulation when there was considerably more mesozooplankton biomass than any other component.

On Day 8 there was considerable export of nitrogen from the euphotic zone through net-phytoplankton sinking (14.6 mg $N.m^{-3}.d^{-1}$), but even more through animal faeces (26.7 mg $N.m^{-3}.d^{-1}$). Bacteria are the only heterotrophs in Category II, hence they must be responsible for all the regeneration in that category. Bacteria were at their peak biomass on Day 8, and if the model of bacterial excretion is correct (see Methods), they dominated in nitrogen regeneration, contributing 31.9 mg $N.m^{-3}.d^{-1}$ out of a total of 51.8 mg $N.m^{-3}.d^{-1}$ being regenerated in the model system. The total amount of nitrogen being passed on through grazing and predation is given above the arrows connecting the boxes from left to right, and can be seen to decline rapidly along the Lindeman Spine.

By Day 20 the values of transfers in the Lindeman Spine are smaller, since much of the nitrogen in the model system has sunk out of the euphotic zone. Most of the export is now from sinking faeces of animals in trophic categories III and IV, and considerably more nitrogen is being passed by predation to category IV than on Day 8. Meso-zooplankton comprise almost all of category IV and contribute over one third of all regenerated nitrogen. If bacterial regeneration is considerably over-estimated in this model of excretion, it is possible that meso-zooplankton may contribute up to two thirds of regenerated nitrogen by Day 20.

The simulation model enabled us to explore the consequences of various hypotheses concerning the relationships between size classes of autotrophs and heterotrophs in the plankton

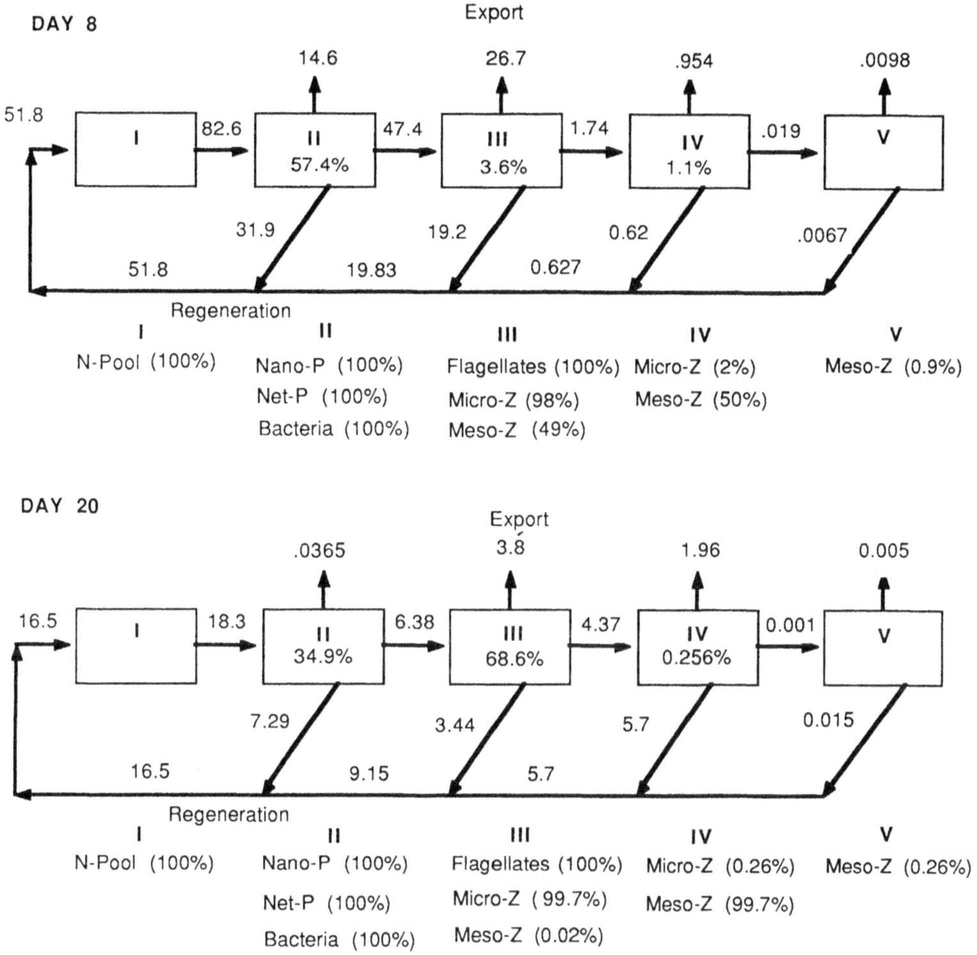

Fig. 5. Network analysis showing Lindeman spines for days 8 and 20 of model output after a simulated upwelling event (from Field et al. 1989). The boxes represent hypothetical trophic categories, with Roman numerals indicating the number of steps removed from non-living matter. The trophic efficiency of grazing transfer through each category along the spine is given as a percentage inside each box. Arrows indicate the flows of nitrogen (mg $N.m^{-3}.d^{-1}$), with regenerated nitrogen recycling below the boxes. Vertical arrows indicate export by sinking out of the surface layer. Model components are listed below the categories, with the percentage of their activity in each trophic category

community. The output from this model was in the form of a time-series of standing stocks, showing a succession of size-classes increasing and decreasing as they interacted. This gives a dynamic and consistent picture of the consequences of size-based hypotheses of trophic interactions which is difficult to achieve either in the field or by laboratory experimentation. The network analyses take this a step further and allow us to explore the flows of nitrogen in the model ecosystem, as compared with the standing stocks output by the simulation model. Thus we have seen that some components that have a small biomass, such as bacteria and micro-zooplankton, may have a much larger role in recycling nitrogen than equivalent biomasses of meso-zooplankton because of their much faster metabolic rates. The network analyses have shown where different components function in the trophic food web, and how these change with time during the succession. Thus we have been able to define an analogue of 'food chain length' for a food web, and we have seen, for example, that meso-zooplankton functioned almost equally at three and four steps up the trophic ladder on day 8 of the simulation, whereas by day 20, over 99% of their activity was four steps along the trophic ladder, or Lindeman Spine i.e. different components of the community dominate nitrogen cycling as the community develops in succession.

CONCLUSIONS

The model reported here is based upon rate parameters that are derived objectively from allometric first principles. The model therefore explores the consequences of the hypothesis that trophic interactions in the planktonic food web and the rates of metabolic processes are largely size-dependent. No tuning was involved, although the model is sensitive to the choice of starting values which are founded on field observations.

The size-based model shows a realistic succession of biomass maxima, with phytoplankton and bacteria peaking together about seven days after the simulated upwelling event. This is consistent with field observations and 60 litre laboratory microcosm experiments with upwelled Benguela Current water (Painting *et al.* 1989). Heterotrophic flagellates and microzooplankton peak on successive days after bacteria, also agreeing with microcosm results.

The succession observed by simulation modelling cannot be completely validated in nature, because due to drogue slippage, mixing and other problems (Brown and Hutchings 1987), it is very difficult to follow a parcel of water in the open ocean for a period of weeks. It is also difficult to validate the model in laboratory microcosm studies, because wall effects are likely to become important after 10-20 days (Painting *et al.* 1989). Nevertheless, the model results are consistent with most observations in the southern Benguela upwelling system and the magnitudes of biomass in model results are of the same order as those observed in the field. The simulation model thus generated a succession of biomass peaks in different size-classes which are the logical consequence of the size-dependency hypothesis in plankton communities (Azam *et al.* 1983). Prior to the development of simulation models such as the one reported here, there was no consistent view as to what occurs in all the different components of the micro- and mesoplankton community after an upwelling event.

Network analyses have enabled us to investigate the flows of nitrogen that are likely to occur among the components of the plankton community after upwelling. It has been shown that small heterotrophs (e.g. bacteria, hetero-flagellates) may have a much larger role in regenerating nitrogen than their biomasses would suggest, because they have relatively fast metabolic rates. It has also been shown that the relative roles of various components change with time after upwelling, with micro-heterotrophs playing a large role soon after upwelling and meso-zooplankton becoming more important in regeneration later. It had not been possible to place the experimental results of nitrogen uptake and release (e.g.

Probyn 1985, 1987) into this time-perspective without the combined use of simulation models and network analyses.

Network analysis of nitrogen flows in the model succession may help resolve the controversy concerning the relative importance of the microbial loop and meso-zooplankton in nitrogen cycling. Although there is some uncertainty concerning the accuracy of the present model for bacterial nitrogen excretion [see equation (5)], even if the contribution of bacteria is grossly over-estimated, our results suggest that the microbial components are likely to be dominant in nitrogen cycling in the euphotic zone within about 12 days of a disturbance (such as an upwelling or mixing event). On the other hand, the results indicate that the meso-zooplankton are likely to become much more important in nitrogen regeneration if surface water remains stratified for over 15 days. The controversy has probably arisen because different workers have sampled in the ocean under different conditions and at varying times after disturbances such as wave-mixing or upwelling events.

REFERENCES

Andrews WH, Hutchings L (1980) Upwelling in the southern Benguela. Prog Oceanogr 9:1-81

Armstrong DA, Mitchell-Innes BA, Verheye-Dua F, Waldron H, Hutchings L (1987) Physical and biological features across an upwelling front in the southern Benguela. In: Payne AIL, Gulland JA, Brink KH (eds) The Benguela and comparable ecosystems. S Afr J Mar Sci 5:171

Azam F, Fenchel T, Field JG, Gray JS, Meyer-Reil L-A, Thingstad F (1983) The ecological role of water-column microbes in the sea. Mar Ecol Prog Ser 10:257-263

Baldock BM, Baker JH, Sleigh MA (1980) Laboratory growth rates of six species of freshwater Gymnamoebia. Oecologia (Berl) 47:156-159

Banse K (1982) Mass-scaled rates of respiration and intrinsic growth in very small invertebrates. Mar Ecol Prog Ser 9:281-297

Banse K (1976) Rates of growth respiration and photosynthesis of unicellular algae as related to cell size - a review. J Phycol 12:135-140

Barthel K (1983) Food uptake and growth efficiency of *Eurytemora affinis* (Copepoda: Calanoida). Mar Biol 74:269-274

Berman T, Holm-Hansen O (1974) Release of photo-assimilated carbon as dissolved organic matter by marine

phytoplankton.Mar Biol 28:305-310
Bienfang PK (1985) Size structure and sinking rates of various microparticulate constituents in oligotrophic Hawaiian waters. Mar Ecol Prog Ser 23:143-151
Bienfang PK, Harrison PJ (1984) Co-variation of sinking rate and cell quota among nutrient replete marine phytoplankton. Mar Ecol Prog Ser 14:297-300
Blueweiss L, Fox H, Kudzma V, Nakashima D, Peters R, Sams S (1978) Relationships between body size and some life history parameters. Oecologia (Berl) 37:257-272
Brody S, Procter RC, Ashworth US (1934) Basal metabolism endogenous nitrogen creatinine and neutral sulphur excretions as functions of body weight Missouri. Agr Exp Sta Res Bull 220:1-40
Brown PC, Field JG (1986) Factors limiting phytoplankton production in a nearshore upwelling area J Plankt Res 8:55-68
Brown PC, Hutchings L (1987) The development and decline of phytoplankton blooms in the southern Benguela upwelling system. I Drogue movements hydrography and bloom development In: Payne AIL, Gulland JA, Brink KH (eds) The Benguela and comparable ecosystems. S Afr J mar Sci 5:357
Bruland KW, Silver MW (1981) Sinking rates of fecal pellets from gelatinous zooplankton (Salps Pteropods Doliolids). Mar Biol 63:295-300
Burns NM, Rosa F (1980) *In situ* measurement of the settling velocity of organic carbon particles and 10 species of phytoplankton. Limnol Oceanogr 25:855-864
Calder WA III (1985) Size and metabolism in natural systems In:Ulanowicz RE, Platt T (eds) Ecosystem theory for Biological Oceanography. Can Bull Fish Aquat Sci 213:65-75
Cammen LM (1980) Ingestion rate: an empirical model for aquatic deposit feeders and detritivores. Oecologia (Berl) 44:303-310
Capriulo GM (1982) Feeding of field collected tintinnid microzooplankton on natural food. Mar Biol 71:73-86
Cousins SH (1980) A trophic continuum model derived from plant structure animal size and a detritus cascade. J theor Biol 82:607-618
Cousins SH (1985) The trophic continuum in marine ecosystems: structure and equations for a predictive model In:Ulanowicz RE, Platt T (eds) Ecosystem Theory for Biological Oceanography. Can Bull Fish Aquat Sci 213:76-93
Dagg MJ (1976) Complete carbon and nitrogen budgets for the carnivorous amphipod *Calliopus laeviusculus* (Krøyer). Int Rev ges Hydrobiol 61:297-357
Dickie LM, Kerr SR, Boudrou PR (1987) Size-dependent processes underlying regularities in ecosystem structure. Ecol Monogr 57:233-250
Dugdale RC, Goering JJ (1967) Uptake of new and regenerated forms of nitrogen in primary productivity. Limnol Oceanogr 12:196-206
Eppley RW, Peterson BJ (1979) Particulate organic matter and planktonic new production in the deep ocean. Nature 282:677-680
Fenchel T (1974) Intrinsic rate of natural increase: the relationship with body size. Oecologia (Berl) 14:317-326
Fenchel T (1980) Suspension feeding in ciliated protozoa:functional response and particle size selection.

Microb Ecol 6:1-11
Fenchel T, Finlay BJ (1983) Respiration rates in heterotrophic free-living protozoa. Microb Ecol 9:99-122
Field JG, Moloney CL, Attwood CG (1989) Network analysis of simulated succession after an upwelling event In: Wulff FV, Field JG, Mann KH (eds) Network analysis in marine ecology:methods and applications Lecture notes on coastal and estuarine studies. Springer-Verlag New York (In press)
Hall DJ, Threlkeld ST, Burns CW, Crowley PH (1976) The size-efficiency hypothesis and the size structure of zooplankton communities. Ann Rev Ecol Syst 7:177-208
Hemmingsen AM (1960) Energy metabolism as related to body size and respiratory surfaces and its evolution. Rep Steno Memorial Hospital and Nordisk Insulin Laboratorium 9:1-110
Hutchings L, Holden C, Mitchell-Innes B (1984) Hydrological and biological shipboard measurements of upwelling off the Cape Peninsula. S Afr J Sci 80:83-89
Ikeda T (1970) Relationship between respiration rate and body size in marine plankton animals as a functioln of the temperature of the habitat. Bull Fac Fish Hokkaido Univ 21:91-112
Ikeda T (1977) Feeding rates of planktonic copepods from a tropical sea. J exp mar Biol Ecol 29:263-277
Ikeda T, Motoda S (1978) Zooplankton production in the Bering Sea calculated from 1956-1970 Oshoro Maru data. Mar Sci Commns 4:329-346
Ivleva IV (1980) The dependence of crustacean respiration rate on body mass and habitat temperature. Int Rev ges Hydrobiol 65:1-47
Kay JJ, Graham L, Ulanowicz RE (1989) A detailed guide to network analysis In: Wulff FV, Field JG, Mann KH (eds) Network analysis in marine ecology: methods and applications. Coastal and estuarine studies. Springer-Verlag New York p 15
Kleiber M (1932) Body size and metabolism. Hilgardia 6:315-353
Kleiber M (1947) Body size and metabolic rate. Physiol Rev 27:511-541
Lampert W (1977) Studies on the carbon balance of *Daphnia pulex* de Geer as related to environmental conditionsII The dependence of carbon assimilation on animal size temperature food concentration and diet species. Arch Hydrobiol Suppl 48:310-355
Lancelot C (1983) Factors affecting phytoplankton extracellular release in the southern Bight of the North Sea. Mar Ecol Prog Ser 12:115-121
Lapidus L, Seinfeld JH (1971) Numerical solution of ordinary differential equations. Academic Press New York
Lucas MI (1986) Decomposition in pelagic marine ecosystems. J Limnol Soc Sth Afr 12:99-122
Lucas MI, Painting SJ, Muir DG (1986) Estimates of carbon flow through bacterioplankton in the S Benguela upwelling region based on 3-H thymidine incorporation and predator-free incubations. GERBAM- 2nd International Colloquium of Marine Bacteriology. CNRS Brest October 1984. IFREMER Actes de Coll 3:375-383
Lucas MI, Probyn TA, Painting SJ (1987) An experimental study of microflagellate herbivory: further evidence for the importance and complexity of microplanktonic interactions.

In: Payne AIL, Gulland JA, Brink KH (eds) The Benguela and comparable ecosystems. S Afr J mar Sci 5:791

MacIsaac JJ, Dugdale RC (1969) The kinetics of nitrate and ammonia uptake by natural populations of marine phytoplankton. Deep-Sea Res 16:45-57

Margalef R (1958) Temporal succession and spatial heterogeneity in phytoplankton. In: Buzzati-Traverso (ed). Perspectives in Marine Biology University of California Press Berkeley p 323

Margalef R (1968) Perspectives in ecological theory. University of Chicago Press Chicago p 1

Miller CA, Landry MR (1984) Ingestion-independent rates of ammonium excretion by the copepod *Calanus pacificus*. Mar Biol 78:265-270

Moloney CL (1988) A size-based model of carbon and nitrogen flows in plankton communities. PhD Thesis University of Cape Town Rondebosch South Africa

Moloney CL, Bergh MO, Field JG, Newell RC (1986)The effect of sedimentation and microbial nitrogen regeneration in a plankton community: a simulation investigation. J Plankt Res 8:427-445

Moloney CL, Field JG (1989) General allometric equations for rates of nutrient uptake ingestion and respiration in plankton organisms. Limnol Oceanogr 34:1290-1299

Moloney CL, Field JG (1990) The size-based dynamics of plankton food webs. I. Description of a simulation model of carbon and nitrogen flows. J Plankt Res (submitted)

Paffenhöfer GA (1971) Grazing and ingestion rates of nauplii copepodids and adults of the marine planktonic copepod *Calanus helgolandicus*. Mar Biol 11:286-298

Painting SJ, Lucas MI, Muir DG (1989) Fluctuations in heterotrophic bacterial community structure and production in response to development and decay of phytoplankton in a microcosm. Mar Ecol Prog Ser 53:129-141

Peters RH (1983)The ecological implications of body size. Cambridge University Press Cambridge

Pillar SC (1986) Temporal and spatial variations in copepod and euphausiid biomass off the southern and south-western coasts of South Africa in 1977/78. S Afr J mar Sci 4:219-230

Platt T (1985) Structure of the marine ecosystem:its allometric basis. In: Ulanowicz RE, Platt T (eds) Ecosystem Theory for Biological Oceanography. Can Bull Fish Aquat Sci 213:55

Platt T, Denman K (1978) The structure of pelagic marine ecosystems. Rapp P-v Reun Cons perm int Explor Mer 173:60-65

Platt T, Silvert W (1981) Ecology physiology allometry and dimensionality. J theor Biol 18:564-573

Probyn TA (1985) Nitrogen uptake by size-fractionated phytoplankton populations in the southern Benguela upwelling system. Mar Ecol Prog Ser 22:249-258

Probyn TA (1987) Ammonium regeneration by microplankton in an upwelling environment. Mar Ecol Prog Ser 37:53-64

Probyn TA, Lucas MI (1987) Ammonium and phosphorus flux through the microplankton community in Agulhas Bank waters. In: Payne AIL, Gulland JA, Brink KH (eds) The Benguela and comparable ecosystems. S Afr J mar Sci 5:209-221

Ross RM (1982) Energetics of *Euphausia pacifica*. I. Effects of body carbon and nitrogen and temperature on measured and predicted production. Mar Biol 68:1-13

Shannon LV (1985) The Benguela ecosystem Part I. Evolution of the Benguela physical features and processes. Oceanogr Mar Biol Ann Rev 23:105-182
Sheldon RW, Prakash A, Sutcliffe WH Jr (1972) The size-distribution of particles in the ocean. Limnol Oceanogr 17:327-340
Sheldon RW, Sutcliffe WH Jr, Paranjape MA (1977) Structure of pelagic food chains and relationship between plankton and fish production. J Fish Res Bd Can 34:2344-2353
Silvert W, Platt T (1980) Dynamic energy flow model of the particle size distribution in pelagic ecosystems. In: Kerfoot W C (ed) Evolution and ecology of zooplankton communities. Univ Press of New England Dartmouth NH p 754
Syrett PJ (1981) Nitrogen metabolism of microalgae. In:Platt T (ed) Physiological bases of phytoplankton ecology. Can Bull Fish Aquat Sci 210:182
Taylor WD, Shuter BJ (1981) Body size genome size and intrinsic rate of increase in ciliated protozoa. Am Nat 118:160-172
Tempest DW, Neijsel OM (1978) Eco-physiological aspects of microbial growth in aerobic nutrient-limited environments. Adv Microb Ecol 2:105-153
Terry KL (1982) Nitrogen uptake and assimilation in *Thalassiosira weissflogii* and *Pageodactylum tricornutum* :interactions with photosynthesis and with the uptake of other ions. Mar Biol 69:21-30
Verheye HM, Hutchings L (1988) Horizontal and vertical distribution of zooplankton biomass in the southern Benguela May 1983. S Afr J mar Sci 6:255-266
Verheye-Dua F, Lucas MI (1988) The southern Benguela frontal region: I. Hydrology phytoplankton and bacterioplankton. Mar Ecol Prog Ser 47:271-280
Wulff FV, Field JG (1983) The importance of different trophic pathways in a nearshore benthic community under upwelling and downwelling conditions. Mar Ecol Prog Ser 12:217-228
Wulff FV, Field JG, Mann KH (eds) (1989) Network analysis in marine ecology: methods and applications. Coastal and estuarine studies. Springer-Verlag New York
Wulff FV, Ulanowicz RE (1989) A comparative anatomy of the Baltic Sea and Chesapeake Bay ecosystems. In: Wulff FV, Field JG ,Mann KH (eds) Network analysis in marine ecology: methods and applications. Coastal and estuarine studies Springer-Verlag New York p 232

STATUS AND FUTURE NEEDS IN PROTOZOAN ECOLOGY

Lawrence R Pomeroy
Department of Zoology and Institute of Ecology
University of Georgia
Athens, Georgia 30602
USA

ESTABLISHING A POINT OF VIEW

Those of us who have been developing the concept of a marine microbial food web have believed for many years that Protozoa are major movers of energy and nutrients. Indeed, Vernadskii (1926) repeatedly referred to the special significance of microorganisms, including specific references to Protozoa as movers of materials on a global, biogeochemical scale. However, until recently this remained more a matter of opinion than established fact. The conferees brought together an impressive body of facts regarding both the magnitude and the diversity of protozoan activity. That protozoa are a dominant force in the ocean's metabolism is no longer in doubt. A comprehensive understanding of protozoan metabolism in the ocean is a prerequisite to a full understanding of and to quantitatively model cycles of carbon and other biologically essential elements on a global scale. During this conference we defined the limits of our knowledge of protozoan ecology and highlighted a number of pressing needs, both for basic understanding of the biosphere and its function and for answering questions of immediate practical importance.

When we consider 'protozoan ecology' we are potentially considering organisms and their interactions on several hierarchical levels, from the level of the individual organism to that of the ecosystem or biosphere. This is a range of five or six hierarchical levels, depending on how one chooses to slice them (Figure 1). Two recent publications (Allen and Starr 1982, O'Neill *et al.* 1986) have pointed out the need to recognise the existence of hierarchy in biological systems and to structure our concepts and our research accordingly. The ecological hierarchy is nested in a larger one extending back

NATO ASI Series, Vol. G 25
Protozoa and Their Role in Marine Processes
Edited by P. C. Reid et al.

through biology, chemistry, and physics. Allen and Starr (1982) distinguish small-number systems (e.g. planets in our solar system), large-number systems (e.g. chemical reactions involving large numbers of atoms) and middle-number systems. Ecosystems and communities are middle-number systems, too complex for simple calculus but not rich enough in replication for the simplest statistical interpretations. Because they are systems containing a relatively small number of complex interactions, they are inherently difficult to analyse.

BIOSPHERE	MANY	WEAK	HEUERISTIC
ECOSYSTEMS	VARIABLES	INFERENCE	MODELS
COMMUNITIES			
SPECIES			
DEMES			
ORGANISMS			
ORGAN SYSTEMS			
ORGANS			
TISSUES	FEW	STRONG	PREDICTIVE
CELLS	VARIABLES	INFERENCE	MODELS

Fig. 1. Living systems of the planet decomposed into a series of hierarchical levels of increasing complexity of interactions

Several characteristics of hierarchical systems and middle-number systems are important for our understanding of ecology. Hierarchies define potential interactions. Strong interactions occur on any given hierarchical level and weaker interactions occur between levels, becoming weaker with increasing hierarchical distance. Thus, in asking ecological questions we must be careful to ask questions about a specific level and seek answers largely through potential interactions within that level. One session at this conference asked: "What caused a *Chrysochromulina* bloom in the Kattegat in the spring of 1988?". This is a question about ecosystem function: subtle interactions among the many species making up the plankton in the Kattegat. Because it is a middle-number system, the potential interactions are too many for a simple mathematical analysis. Even if we knew all of the appropriate factors involved in interspecies competition and predation in that plankton community, modelling it would be at least as challenging (and fruitless) as modelling weather on a similar geographic scale. Instead of beating our heads against this stone wall of intractable interactions, we need to decompose the original question into manageable segments. This will not merely define our ignorance precisely, but will also set us on a road toward answering the original question some day, if not now.

Hierarchy theory tells us why it is usually impossible to do experiments with whole ecosystems, even if we could afford financially to try it. It tells us why most ecosystem models have more heuristic than predictive value. We cannot expect to accomplish what a physicist or astronomer can do with a small-number system or what a chemist can do with a large-number system. We must recognise the inherent limits in the analysis of a middle-number system, do what is appropriate to such systems, and not judge, or allow our results to be judged, on the basis of criteria appropriate to small- or large-number systems. In this regard, much of ecology is like geology and requires similar treatment (Pomeroy *et al.* 1988, Kitts 1977). While such an analysis of a scientific discipline may seem esoteric, it is really very practical, because it permits us to ask questions we can answer, avoid ones we cannot, and know when we do indeed have the best answer we can expect to achieve.

When this conference is viewed in the light of hierarchy theory, one level is conspicuous by its absence. If this meeting was representative of people studying protozoan ecology, and probably it was, then few people are concerned primarily with processes at the level of species population interactions: optimal foraging, predator-prey relationships, competition, co-evolution. While some of these topics were mentioned in passing, none was seriously explored. Moreover, a perusal of four recent textbooks on population ecology reveals only one example based on protozoans, the classical work of Gause (1934). So it appears that there is an empty niche in protozoology waiting to be filled. Some of Fenchel's work shows that population processes are significant (Fenchel 1982), and Riemann's (1985) experiments with enclosures suggest that there are some complex predator-prey interactions in marine systems.

Historically, marine ecologists have approached the explanation of plankton bloom events as a release from nutrient limitation. This may be correct as far as it goes, but blooms must have causes and consequences at the level of population interactions as well. Since many species compete for nutrients, understanding such interactions will be challenging. Indeed, a reversion to simple model systems, but not quite as simple as Gause's, may be one fruitful approach.

THE IMPACT OF BIOTECHNOLOGY

Biotechnology promises to have a revolutionary impact on many aspects of our lives. While it may have more drastic effects in other areas, it is clear that it will markedly enhance our research capability in marine microbial ecology. The effects are being seen first with respect to bacteria, because they are comparatively easy subjects for genetic hybridisation and because their lack of morphology makes conventional systematics difficult. Potentially, DNA hybridisation permits us to identify and enumerate automatically every individual of every species of bacteria in a sample of water or sediment by combining hybridisation with flow cytometry (Porter 1988). Since these procedures are in practice laborious, substantial progress probably awaits further simplification or automation of the technology of DNA hybridisation (Holben and Tiedje 1988).

In terms of practical applications, more radical possibilities abound. Protozoans as well as bacteria, possibly of marine origin, may be genetically modified to perform tasks of waste and pollution treatment more effectively. The safety of the release of such genetically modified organisms is currently subject to debate. However, there is little doubt that we are going to see such releases, and whether or not they result in problems greater than those they were intended to solve, they will provide work for many marine microbial ecologists!

During this conference we saw several applications of biotechnology to problems of protozoology. **Tony Soldo** explained killer effects in *Paramecium* as a naturally-occurring DNA hybridisation phenomenon involving a change in host extranuclear DNA. **Johan Wikner** used a plasmid as a marker of bacterial minicells for feeding studies. So biotechnology is already in the protozoology tool kit as well as in the tool kits of some of the microorganisms themselves.

ENERGY FLUX THROUGH THE MICROBIAL FOOD WEB

Both at this conference and in the recent literature there has been discussion of the significance of the microbial food web (often called the microbial loop) with respect to its modulation of the flow of energy between primary producers and metazoan consumers. Larry Pomeroy (1974) speculated that the microbial food web was an energy sink, and Pomeroy and Wiebe (1988) offered further reasons why this might be so. Many others disagree. In general, those who have approached this question through modelling have seen the microbial food web as a potential energy link, while empirical experiments have implicated it as an energy sink. There are two significant issues here: efficiency and the number of trophic transfers. Because modellers tend to condense information, their representation of the microbial food web may include fewer trophic transfers than a typical community in nature. At this conference we were informed of a range of different types of trophic transfers: ciliates eating other ciliates, ciliates eating phytoplankton and salvaging the chloroplasts, myxotrophic, defecating dinoflagellates, and even radiolarians that eat copepods. All of these transfers are potentially included in empirical mesocosm experiments, such as those of Ducklow *et al.*(1986) and Azam *et al.* (1984) which suggest that the microbial food web dissipates >90% of primary production. Moreover, most extant models were written using early estimates of bacterial assimilation efficiency of the order of 50%. A number of recent studies suggest that bacterial assimilation efficiency is often significantly less than 50% when the bacteria are utilising the relatively refractory or nutrient poor substrates that are commonly available to them (Newell and Linley 1984, Newell *et al.* 1981). Discussions of the assimilation efficiency of protozoans at this conference suggest that we are in the early, optimistic stage. More work on protozoan efficiency, especially under the less-than-ideal conditions protozoans often encounter in nature, is needed (e.g. Lucas *et al.* 1987). Moreover, we need a system-level tracer experiment, not confined in a bag. Is there some naturally occurring or bomb-generated radionuclide we can follow? Will stable isotopic ratios help us enumerate the number of trophic transfers in a food web?

FALLING SNOW, FALLING WHALES AND FALLEN FLUFF

Several conferees brought out the importance of non-living particles in the sea to both the structure and function of pelagic microbial communities. Particulate matter in the sea tends to aggregate into what may often become macroscopic particles (sea snow). When some of this reaches the bottom and is further processed in the nepheloid layer, it becomes more homogeneous in appearance, both macroscopically and microscopically and is called fluff. While there is some experimental evidence that these materials are biologically dynamic, with a relatively short turnover time, there is much that we do not know about the production and fate of these organic aggregates in the sea. Probably many of the particles formed in the upper mixed layer either are utilised there or are caught in the thermocline and processed there. While there are long-standing physical explanations for the oxygen minimum layers of the oceans, it is interesting that they tend to follow the isopycnal of the thermocline across ocean basins. Lande and Wood (1987) modelled the fate of particles sinking through the upper mixed layer of the ocean, showing that particles heavy enough to fall through the thermocline do so quickly, while others that reach the thermocline are trapped and utilised there. The few large, heavy particles passing through the thermocline (including living and senescent phytoplankton) are the principal flux of energy to the abyssal benthos. However, Cho and Azam (1988) showed that there is significant utilisation of falling particulate matter as it passes through the ocean's interior. This implies that further processes not yet fully defined occur as particles sink. Is there mediation by metazooplankton?

Even less is known about fluff, the material of the bottom nepheloid layer. Is it the remains of aggregates that have fallen from the upper mixed layer, somewhat re-worked, or it is produced *in situ* from fallen diatoms (Smetacek 1984)? Fluff is sometimes, but not always, heavily populated by bacteria, suggesting that like snow it is dynamic, transitory material that is utilised by the epibenthic microbial food web and probably by metazoans as well. **Carol Turley** suggested that there is competition between macro- and microconsumers for such transitory material. **Paul Kemp** indicated that in shallow water 80% of the production is bacterial biomass in fine sediments and 50% in sand whereas **Karin Lochte** said that a majority of the abyssal biomass is made of foraminifera. This suggests the presence of an active epibenthic microbial food web everywhere.

With recent discoveries of falling diatom blooms and microbially active fluff on the abyssal sea bed, we have tended to discount the old theory of Anton Bruun that the abyssal benthos is fed by falling whales. However, **Andy Gooday** suggested that the frequency of falling whales might be as great as one per $mile^2$ but gave no time factor for this. Even if this were the rate per century, it would have a significant impact on benthic trophic dynamics. **Diane Stoecker** pointed out the diversity of habitats around a dead whale on the bottom, including an anaerobic oil community under the whale. **Alan Longhurst** warned that in modelling the benthos we should recognise the high spatial and biological diversity implied by whales falling on a plain of fluff and macroscopic foraminifera.

CONCEPTS OF THE MICROBIAL FOOD WEB

Many participants at this meeting contributed to the concept that the microbial food web is diverse, complex, and varied through time and space. **Barry Sherr** said the principal bacterivores are flagellates <55 µm, responsible for 80% of bacterivory. Spermivory was said by **Helena Galvao** to be dominated by 5-10 µm flagellates. Cryptomonads eat algae and sperm. However, small ciliates also eat bacteria as well as phytoplankton (Sherr *et al.* 1986). **Evelyn Lessard, Victor Smetacek**, and **Gotram Uhlig** cited the significance of heterotrophic dinoflagellates, in addition to ciliates. In the Kiel Bight, heterotrophic dinoflagellates have the same biomass as ciliates, while in McMurdo Sound dinoflagellates have 10 times the biomass of ciliates. In Chesapeake Bay, ciliate biomass is relatively constant, while dinoflagellates are seasonal. The heterotrophic abilities of foraminifera, acantharia, and radiolaria were described by **David Caron, John Lee**, and **Jelle Bijma**. **David Caron**'s collections by mid-ocean diving showed differences in prey selection by the three groups. Relatively large organisms may be captured. **Jelle Bijma** showed the capture of copepods by foraminifera. This sort of detailed observational material reinforces the suggestion that the microbial food web is complex and extensive, with many potential energy transfers and much potential switching among prey organisms.

A discussion of estimates of grazing rates from dilution techniques by **Dian Gifford, Madhu Paranjape** and **Suzanne Ström** raised more questions than were answered. **Suzanne Ström** found grazing intensity to be inversely related to light intensity, both with ciliates in culture and with natural communities at Pacific Station P. All investigators found production to be in excess of grazing intensity, raising questions about how grazing control operates. **Fereidoun Rassoulzadegan** queried how ciliates and flagellates survive between patches of algae and bacteria. He estimated that cilates must survive through two doublings between one feeding bout and the next. He also noted that flagellates were consuming about 50% of bacterial production, which again raises a question about grazing control. He asked if we might be overestimating bacterial production. The probable answer is yes, in view of the results with new and more specific methods for measuring incorporation of tritiated thymidine into nucleic acids (Findlay *et al.* 1984, Wicks and Robarts 1988, Robarts *et al.* 1986).

Because it relates to the link-or-sink question, consumption of protozoans by copepods was a subject of wide interest. Comparing Pacific Station P and Terrebonne Bay, Louisiana, **Dian Gifford** showed that copepods consume what is available, switching between algae and protozoans. She also pointed out differences in the preferences of copepod species. **Diane Stoecker** pointed out that nauplii eat small ciliates while adult copepods eat large ones, as well as detritus particles and their own nauplii. **Milos Legner** said that copepods were as sloppy in their feeding on protozoa as they were known to be on phytoplankton, ingesting only about 30% of captured biomass. While copepods tend to be opportunistic switchers, **Torbjørn Dale** pointed out that there are some very specific trophic links. Some species of fish larvae eat only specific protozoan species (e.g. Lasker 1978, Lasker and Zweifel 1978). This discussion showed that there is energy transfer from the microbial food web to metazoans, but that it is highly variable in space and time. **Gerry Capriulo** pointed out that there are typically about 34 amoebae per litre, but little is known about their feeding or selectivity. Information on the predators of pelagic protozoa is also poor. Attempts to extract generalisations from our limited knowledge of this complex situation were unsuccessful. Here was another instance in which we were doing mental acrobatics across several hierarchical levels, asking questions spanning the range from local species populations to the regional ecosystem. While the organism-level observations were instructive and necessary, they could not be extrapolated into ecosystem-level rate processes.

Having established through empirical studies that a substantial fraction of energy fixed through photosynthesis in the sea is utilised by microorganisms, we must now develop a more detailed understanding of how this is possible. How do bacteria and protozoans compete with metazoans, especially for particulate food sources? Azam and Cho (1987) have made a major contribution to the conceptualisation of microbial trophic dynamics in the ocean. Reviewing and elaborating on this concept at this conference **Farooq Azam** stressed the need for physical contact by bacteria with particulate substrates, because they have coupled hydrolysis and uptake systems on their cell surfaces. He said we do not see bacteria attached to phytoplankton, because they enzymatically cut away some surface polysaccarides and move off, presumably to avoid anti-microbial substances secreted by the phytoplankton. According to this concept, there is a 'phycosphere' of microbial activity surrounding each phytoplankter. Bacteria are obtaining dissolved materials from the phytoplankton at concentrations far above those in bulk seawater, and protozoans are attracted to the phycospheres, where they will feed. Unfortunately, if such structure exists in seawater, we destroy it by our sampling methods. Even in controlled laboratory situations, it is as yet impossible to prepare a system in which we can directly observe the undisturbed distribution of microorganisms, floating or sinking, in three dimensions.

While the concept developed by Azam and Cho helps us to postulate how bacteria may efficiently receive organic material from healthy, living phytoplankton, empirical observations suggest that the larger part of bacterial substrates are from other sources, such as zooplankton (including protozoan) excretions and sloppy feeding (Eppley *et al.* 1981). **Victor Smetacek** pointed out that phytoplankton are typically bloom-and-bust organisms, and because they cannot reproduce fast enough to utilise phytoplankton blooms effectively, copepods engage in superfluous feeding, converting the phytoplankton to faecal pellets. Some 99% of those pellets are utilised by microorganisms within the upper mixed layer (Bathmann *et al.* 1987). The result is the production of more protozoans to feed the copepods and to extend the effective length of the bloom for the copepods, thereby improving their fitness.

In bloom situations, the processes **Victor Smetacek** described lead to another process which increases the complexity of the microbial food web. Bacteria produce adhesive extracellular polysaccharides which cause them to adhere to one another (presumably for the purpose of doing some *ad hoc* genetic engineering) and to everything particulate in the water. This appears to be the origin of organic aggregates (Biddanda 1988). These grow to a size visible to the eye (sea snow) and sink to the thermocline or the bottom, sweeping up materials as they go (Figure 2). As they sink, they are being utilised by the bacteria within. Protozoans are attracted to these particles and enter them to feed on bacteria (Silver and Alldredge 1981). Most of this process of conversion of sea snow to protozoans occurs within the mixed layer, and it is another source of protozoan food for copepods.

Nutrients diffusing out of aggregates contribute to their concentration in bulk seawater, and potentially this supports the minibacteria that are capable of utilising nanomolar concentrations of substrates. The morphology of the free-living heterotrophic bacteria in the sea suggests that they are adapted to utilising very low nutrient concentrations. They may survive from day to day on the nanomolar concentrations of amino acids and peptides in seawater, dividing only when something good happens. For example, a passing copepod too large to utilise them as food excretes concentrated nutrients, both organic and inorganic.

Lehman (1986) has shown that phytoplankton are prepared to

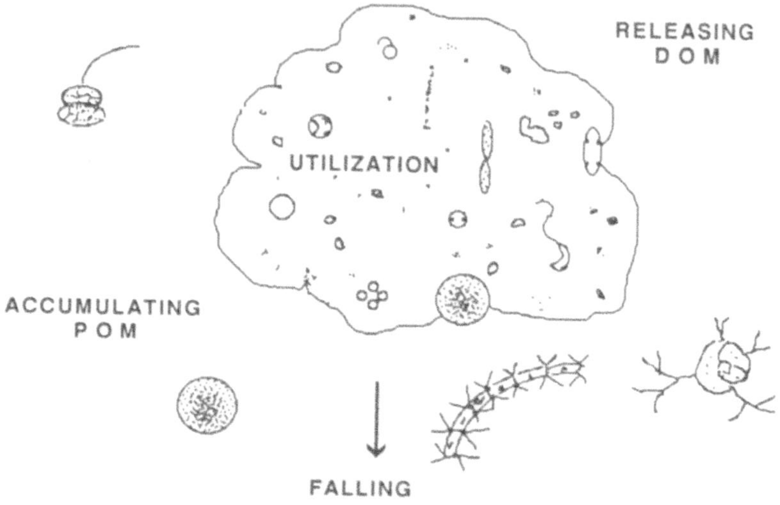

Fig. 2. Events occurring as a macroscopic organic aggregate falls through the ocean

quickly take up one or two days' supply of inorganic nutrients during such an event (Lehman and Scavia 1986a, b). Perhaps the free-living bacteria can do the same with organic substrates. Because experimental approaches are difficult, Jackson (1987) has explored the potential for uptake of local concentrations of dissolved materials by bacteria in terms of the small-scale physics of diffusion, and indeed there is a range of conditions in which bacteria may be expected to utilise temporary concentrations of substrates, either passively or through active orientation as suggested by Azam and Cho. Protozoans may be involved in these processes as happy feeders at sites of bacterial concentration or as contributors to small concentrations of substrates, particularly inside organic aggregates. Further experimental tests of these hypotheses are needed.

We can postulate a number of processes in the upper mixed layer of the ocean that bring about a series of changes of dissolved or finely divided particulate matter into larger particulate matter. While much remains to be done in terms of verifying and quantifying these processes, we are beginning to understand the microbial food web and trophic transfers from it to metazoans. But our understanding of processes in the ocean's interior is more rudimentary. Cho and Azam (1988) have shown that utilisation of particulate matter by bacteria occurs as the particles are sinking through the sub-thermocline water. But **Karin Lochte** pointed out that surface bacterial populations cannot function well at low temperature and high pressure in the ocean's interior. Are abyssal bacteria transported upward to meet the sinking particles? Several potential mechanisms by which this might be accomplished were identified. These included vertically migrating organisms and bubbles from siphonophores and other possible sources. Of course, physical processes such as double diffusion, by which water is slowly advected upwards, also bring upward with them bacteria that are too small to sink. However, it may not be necessary to invoke upward transport of bacteria. Those bacteria that have been near the ocean's surface for many generations may still have the genetic capacity to deal with low temperature and high pressure. However, they may require a rather long induction period if they are suddenly carried down with a falling particle. It is my observation that in cold water induction is longer, sometimes days longer, so we need to know if bacteria can adapt to changing temperature and pressure before a falling particle has carried them right through the ocean's interior to the bottom. To answer this, we need to know more not only about the inherent abilities of bacteria but also more about the sinking rates of particles. It is possible to adapt some geochemical technology to this end. Moore and Dymond (1988) used a ^{210}Pb budget to determine fallout rates of particulate

Table 1. THE ORGANIC AGGREGATE COMMUNITY IN SEAWATER

Passive Aggregation	Active Biota
Broken or nutrient-starved phytoplankton	Phytoplankton
Zooplankton, including protozoan, faeces	Heterotrophic bacteria
Appendicularian houses	
Dead zooplankton	Ciliates
Atmospheric fallout: clay, organic, soils, pollen, ash, soot.	Amoebae

organic matter. Some short-lived radioisotopes which are falling from the atmosphere (e. g. ^{7}Be, ^{32}P) might also help us measure rates of sinking of specific particle sizes or types.

DESIGNER FOOD WEBS?

With biotechnologists already delivering designer organisms, dare we speak about designing food webs? Indeed, we already have. Although the matter was not discussed in those terms during the conference, steps have been taken to utilise our present limited knowledge of microbial food webs to enhance fish productivity. Stockner and colleagues have demonstrated a causal connection between microbial food web structure and productivity in oligotrophic lakes of British Columbia (Stockner and Porter, 1988). The structure is in turn related to the concentration of inorganic plant nutrients. Thus it may require only some empirical trials to permit us to adjust the lake food webs to their most productive configuration.

Farther away, and still in the land of science fiction, is the ability to control or modify marine plankton blooms, which are usually a boon to fish productivity and occasionally a toxic disaster. The biotechnologists may provide additional impetus to get on with this, either by designing organisms that require some ecological expertise in their introduction to the world beyond culture flasks or by designing organisms that run amok.

FUTURE NEEDS

The need that clearly emerged from this NATO conference was to integrate protozoan ecology into global studies of marine food webs. Doing this involves several elements, currently in varied stages of development. Knowledge of the natural history of protozoans is basic to all further ecological studies, and this is more advanced for some protozoan groups than others. If the conference reports are representative, our knowledge of ciliates and the testate Sarcodina in the ocean is more advanced than is knowledge of the smallest flagellates and the naked Sarcodina. Knowledge of estuarine and coastal communities is more advanced than that of oceanic ones, although there is substantial information about all regions. It is unfortunate that as science has become more complex and specialised there has been a tendency to view natural history as a hobby of 18th and 19th century gentlemen and certainly not something to be supported financially by our governmental granting agencies. For the cellular and molecular biologists, natural history has indeed become incidental intelligence that is of little direct utility. For ecologists however, natural history is bricks and mortar from which concepts and models are made. It is the first draft of our description of the biosphere and one we dare not omit from our conceptual development of ecology. At the hierarchical level of the ecosystem, we absolutely must know who lives where, who eats whom, who dominates or replaces whom in space and time. This sort of knowledge of the planet is incomplete for all groups of living organisms, but it is especially limited for some of the protozoa. In general, there is less knowledge of natural history of microorganisms than of metazoans. Gathering this information is costly and requires financial support just as much as any scientific endeavour. Ecologists and oceanographers, whether or not they themselves wish to pursue the study of natural history, have to convey to the grantors of support the message that natural history, by whatever name might be more fashionable, is a necessary part of the continuing scientific effort. Of course, it should be done with modern technology and rigour, and the connection to its use by 'consumer' ecologists should be a close one.

One challenge in understanding the relations of protozoans to one another and to other organisms in the sea is their small size. Observations of protozoans *in situ*, except for the largest of them, are currently impossible. As **Farooq Azam** has pointed out, most of our current concepts assume a randomly distributed mixture of microorganisms in seawater, although there are compelling reasons why microorganisms may be distributed in highly clumped, negative binomial distributions.

If they are, this has a profound influence on community function. The act of collection disrupts and indeed may randomise the distribution of organisms in samples. Observing microorganisms in their truly natural state remains a challenge and an important one. New developments in holography and fibre optic systems offer hope for direct observation, but the problem is challenging. The sort of visualisation of behaviour that Strickler and collaborators have developed for zooplankton (Strickler 1982, Price *et al.* 1983) must be extrapolated downward three or four orders of magnitude to sub-micrometer organisms. Currently, the challenge is to visualise in any possible way the spatial relations of microorganisms in natural water. Ultimately, we must be able to do this in the interior of the ocean and on its abyssal bottom. These environments are still relatively poorly known, in spite of recent advances described by several of the participants at this conference.

The post-Lindeman focus of ecosystem ecologists on the flux of energy led us to the realisation that microorganisms, because of the relation of metabolic rate to body size, are major movers of energy and materials. However, our knowledge of the energetics of microorganisms remains limited. In the case of bacteria, we have come to realise that conditions in culture diverge so greatly from those in the natural world that we can learn little about bacterial production or efficiency in nature from work with laboratory cultures. Even the use of simple, defined substrates in experiments with natural communities probably led to erroneous estimates of bacterial efficiency. Our knowledge of protozoa still emerges largely from the laboratory, and probably it suffers from the same limitations we discovered with laboratory studies of bacteria. The challenge for the future is to further test our laboratory-derived knowledge in the ocean.

REFERENCES

Allen TBH, Starr TB (1982) Hierarchy: Perspectives for Ecological Complexity. Univ. Chicago Press, Chicago

Azam F, Cho BC (1987) Bacterial utilisation of organic matter in the sea. Symposium 41, Soc Gen Microbiol p 261

Azam F, Cowles T, Banse K, Osborne J, Harrison P J (1984) Free-living pelagic bacterioplankton: Sink or link in a marine foodweb? Eos 65: 926
Bathmann UV, Noji TT, Voss M, Peinert R (1987) Copepod faecal pellets: abundance, sedimentation and content at a permanent station in the Norwegian Sea in May/June 1986. Mar Ecol Prog Ser 38: 45-51
Biddanda BA (1988) Microbial aggregation and degradation of phytoplankton-derived detritus in seawater. I. Microbial succession. Mar Ecol Prog Ser 42: 79-88
Cho BC Azam F (1988.) Major role of bacteria in biogeochemical fluxes in the ocean's interior. Nature 332:441-443
Ducklow HW, Purdie DA, Williams PJ leB, Davies JM (1986) Bacterioplankton: A sink for carbon in a coastal marine plankton community. Science 232: 865-867
Eppley RW, Horrigan SG, Fuhrman JA, Brooks ER, Price CC, Sellner K (1981) Origins of dissolved organic matter in southern California coastal waters: experiments on the role of zooplankton. Mar Ecol Prog Ser 6: 149-159
Fenchel T (1982) Ecology of heterotrophic microflagellates. IV. Quantitative occurrence and importance as consumers of bacteria. Mar Ecol Prog Ser 9: 35-42
Findlay SEG, Meyer JL, Edwards RT (1984) Measuring bacterial production via rate of incorporation of ^{3}H thymidine into DNA. J Microbiol Methods 2:57-72
Gause GF (1934) The Struggle for Existence. Williams and Wilkins Baltimore
Holben WE, Tiedje JM (1988) Application of nucleic acid hybridisation in microbial ecology. Ecology 69: 561-568
Jackson GA (1987) Physical and chemical properties of aquatic environments. Symposium 41, Soc Gen Microbiol p 213
Kitts DB (1977) The Structure of Geology. Southern Methodist Univ Press, Dallas
Lande R, Wood AM (1987) Suspension times of particles in the upper ocean. Deep-Sea Res 34: 61-72
Lasker R (1978) The relation between oceanographic conditions and larval anchovy food in the California Current: Identification of factors contributing to recruitment failure. Rapp P-v Reun Cons int Explor Mer 173: 212-230
Lasker R, Zweifel JR (1978) Growth and survival of first-feeding northern anchovy larvae (Engraulis mordax) in patches containing different proportions of large and small prey. In: JH Steele (ed) Spatial Patterns in Plankton Communities p 329
Lehman JT (1986) Microscale nutrient patchiness in plankton communities. ASLO Abstracts of Ann Meeting p 46
Lehman JT, Scavia D (1982a) Microscale patchiness of nutrients in plankton communities. Science 216: 729-730
Lehman JT, Scavia D (1982b) Microscale nutrient patches produced by zooplankton. Proc Nat Acad Sci US 79: 5001-5005
Lucas MI (1987) An experimental study of microflagellate bacterivory: Further evidence for the importance and complexity of microplanktonic interactions. In: Payne RIL Gulland JR Brink KH (eds) The Benguela and Comparable Ecosystems. S Afr J Sci 5: 791-808
Moore WS, Dymond J (1988) Correlation of ^{210}Pb removal with organic carbon fluxes in the Pacific Ocean. Nature 331:

339-341.
Newell RC, Linley EAS (1984) Significance of microheterotrophs in the decomposition of phytoplankton: Estimates of carbon and nitrogen flow based on the biomass of plankton communities. Mar Ecol Prog Ser 16: 105-119
Newell RC, Lucas MI, Linley AES (1981) Rate of degradation and efficiency of conversion of phytoplankton debris by marine microorganisms. Mar Ecol Prog Ser 6: 123-136
O'Neill RV, DeAngelis DL, Waide JB, Allen TFH (1986) A Hierarchical Concept of Ecosystems. Princeton Univ Press, Princeton
Pomeroy LR (1974) The ocean's food web, a changing paradigm. BioScience 24: 499-504
Pomeroy LR, Wiebe WJ (1988) Energetics of microbial food webs. Hydrobiologia 159: 7-18
Pomeroy LR, Hargrove EC, Alberts JJ, (1988) The ecosystem perspective. In: Pomeroy LR, Alberts JJ (eds) Concepts of Ecosystem Ecology. Springer-Verlag New York p 1
Porter KG (1988) Cell sorting techniques for microbial ecology. Ecology 69: 558-560
Price HJ, Paffenhofer G-A, Strickler JR (1983) Modes of cell capture in calanoid copepods. Limnol Oceanogr 28: 116-123
Riemann B (1985) Potential importance of fish predation and zooplankton grazing on natural populations of freshwater bacteria. Appl Environ Microbiol 50: 187-193
Roberts RD, Wicks RJ, Sephton LM (1986) Spatial and temporal variations in bacterial macromolecule labelling with [methyl-^{3}H] thymidine in a hypertrophic lake. Appl Environ Microbiol 52: 1368-1373
Sherr EB, Sherr BF, Fallon RD, Newell SY (1986) Small aloricate ciliates as a major component of the marine meterotrophic nanoplankton. Limnol Oceanogr 31: 177-183
Silver MW Aldredge AL (1981) Bathypelagic marine snow: deep-sea algal and detrital community. J Mar Res 39: 501-530
Smetacek V (1984) The supply of food to the benthos. In: Fasham, M J R (ed) Flows of energy and materials in marine ecosystems: Theory and Practice. Plenum, New York, p 517
Stockner JG, Porter KG (1988) Microbial food webs in freshwater planktonic ecosystems. In: Carpenter SR (ed) Complex Interactions in Lake Communities, Springer-Verlag, New York, p 80
Strickler JR (1982) Calanoid copopods, feeding currents, and the role of gravity. Science 218: 158-160
Vernadskii V I (1926) Biosfera. Leningrad
Wicks RJ, Robarts RD (1988) The extraction and purification of DNA labelled with {methyl-3H}thymidine in aquatic bacterial production studies. J Plankton Res (in press)

LIST OF PARTICIPANTS AT NATO ASI

Dr Necdet Alpaslan, Dokuz Eylul University, Engineering and Architecture Facility, Department of Environmental Engineering, Bornova 35100, Izmir, Turkey

Dr Tenshi Ayukai, Laboratory of Oceanography, Faculty of Agriculture, Tohoku University, Sendai, Miyagi 980, Japan

Dr Farooq Azam, Scripps Institute of Oceanography, UC San Diego, La Jolla, CA 92093, USA

Dr Sylvie Becquevort, Universite Libre de Bruxelles, Groupe de Microbiologie des Milieux Aquatiques, Plaine Batiment A bd du Tromphe, 1040 Ixelles, Belgium

Prof Tom Berman, Kinneret Limnological Laboratory, PO Box 345, Tiberias 14-201, Israel

Dr Catherine Bernard, Station Zoologique, BP 28, 06230 Villefranche sur Mer, France

Ms Ulrike-Gabriele Berninger, Freshwater Biological Association, The Ferry House, Ambleside, Cumbria, LA22 0LP, UK

Dr Jelle Bijma, Geol and Paleontological Institute, Sigwartstrasse 10, 7700 Tuebingen, FRG

Dr Peter Koefoed Bjornsen, Marine Biology Laboratory, University of Copenhagen, Strandpromenaden, DK-3000, Helsingor, Denmark

Dr Knut Borsheim, University of Bergen, Department of Microbiology and Plant Physiology, Allegt 70, N-5007, Bergen, Norway

Dr Peter Burkill, Plymouth Marine Laboratory, Prospect Place, West Hoe, Plymouth, Devon, PL1 3DH, UK

Dr Gerard Capriulo, State University of New York, Division of Natural Sciences, Purchase, New York 10577, USA

Dr David Caron, Woods Hole Oceanographic Institution, Water Street, Woods Hole, MA 02543, USA

Ms Marina Carstens, Institut für Hydrobiologie und Fischereiwissenschaft, Abt Biologische Oceanographie, Zeiseweg 9, D-2000 Hamburg 50, FRG

Dr Andrew Cowling, Culture Collection of Algae and Protozoa, Freshwater Biological Association, The Ferry House, Ambleside, Cumbria LA22 0LP, UK

Dr Torbjørn Dale, Sogndal College, PB 39, N-5801 Sogndal, Norway

Dr John Dolan, University of Maryland, MEES Department of Zoology, College Park, MD 20742, USA

Dr Hugh Ducklow, Horn Point Laboratory, Centre for Environmental and Estuarine Studies, University of Maryland, PO Box 775, Cambridge, Maryland 21613, USA

Dr Malte Elbrächter, Biologische Anstalt Helgoland, Litoralstation List/Sylt, Hafenstrasse 43, D-2882 List/Sylt, FRG

Dr Mike Fasham, Institute of Oceanographic Sciences, Deacon Laboratory, Brook Road, Wormley, Godalming, Surrey GU8 5UB, UK

Prof John Field, Zoology Department, University of Cape Town, Rondebosch 7700, RSA

Dr Gregory Gaines, University of Maryland, Department of Zoology, 2001 N Adams Street 903, Arlington, VA 22201, USA

Miss Helena Galvao, Abt Marine Mikrobiologie, Institut für Meereskunde, Düsternbrooker Weg 20, D-2300 Kiel 1, FRG

Dr Dian Gifford, Louisiana Universities Marine Consortium, Chauvin, Louisiana 70344, USA

Dr Guy Gilron, University of Guelph, College of Biological Science, Department of Zoology, Guelph, Ontario, Canada

Dr Juan Miguel Gonzalez Grau, Departamento Microbiologia, Facultad de Ciencias, Universidad del Pais Vasco, Apto 664, 48080 Bilbao, Spain

Dr Andrew Gooday, Institute of Oceanographic Sciences, Deacon Laboratory, Brook Road, Wormley, Godalming, Surrey GU8 5UB, UK

Dr Gabriel Gorsky, Station Zoologique, 06230 Villefranche sur Mer, France

Mr Hans Dieter Gortz, Zoology Institut, University of Munster, Schlossplatz 5 D4400 Munster, FRG

Mr Rolf Gradinger, Institut für Meereskunde Kiel, Düsternbrooker Weg 20, D-2300 Kiel, FRG

Dr John Green, Plymouth Marine Laboratory, Citadel Hill, Plymouth, Devon PL1 2PB, UK

Dr Dean Jacobson, Department of Oceanography, University of British Columbia, 6270 University Blvd, Vancouver BC, V6T 1W5, Canada

Dr Ian Jenkinson, Agence de Conseil et de Recherche Océanographicques, Lavergne, F-19320, La Roche Canillac, France

Dr Malcolm Jones, Polytechnic South West, Drake Circus, Plymouth, Devon, UK

Mr Klaus Jürgens, Deutsche Gesellschaft für Limnologie, Institut für Seenforschung und Fischereiwesen, D-7994 Langenargen, FRG

Dr Michael Karydis, University of the Aegean, 12-14 Kavetzou Street, 81100 Mytilini, Greece

Dr Paul Kemp, Brookhaven National Laboratory, Oceanographic Sciences Division, Upton, NY 11973, USA

Dr Johanna Laybourn-Parry, University of Lancaster, Department of Biological Sciences, Lancaster, LA1 4YQ, UK

Dr Michele Laval-Peuto, Laboratoire de Biologie Animale et Cytologie, Universite de Nice, Campus Valrose, 06034 Nice Cedex, France

Dr Barry Leadbeater, University of Birmingham, Department of Plant Biology, PO Box 363, Birmingham B15 2TT, UK

Dr Ray Leakey, British Antarctic Survey, High Cross, Madingley Road, Cambridge CB3 0ET, UK

Prof John Lee, City College of New York, Department of Biology, Convent Avenue at 138 Street, New York 10031, USA

Dr Milos Legner, University of Toronto, Erindale College, Mississauga, Ontario L5L 1C6, Canada

Dr Juha-Markku Leppanen, Finnish Institute of Marine Research, PO Box 33, SF-00931 Helsinki, Finland

Dr Evelyn Lessard, University of Maryland, Horn Point Environmental Laboratories, PO Box 775, Cambridge , MD 21613, USA

Dr Karin Lochte, Alfred-Wegener-Institut für Polar- und Meeresforschung, Am Handelshafen 12, D-2850 Bremerhaven, FRG

Dr Alan Longhurst, Biological Oceanography Division, Bedford Institute of Oceanography, PO Box 1006, Dartmouth, Nova Scotia B2Y 4A2, Canada

Dr Denis Lynn, University of Guelph, Guelph, Ontario N1G 2W1, Canada

Dr Masachika Maeda, Ocean Research Institute, University of Tokyo, Minamidai, Nakano, Tokyo 164, Japan

Dr Harvey Marchant, Antarctic Division, Channel Highway, Kingston, Tasmania 7050, Australia

Ms Ingrid Martinussen, Department of Microbiology and Plant Physiology, University of Bergen, Allegt 70, N-5007 Bergen, Norway

Dr George McManus, Institute of Ecosystem Studies, Box AB, Millbrook, New York 12545, USA

Dr David Montagnes, Department of Zoology, University of Guelph, Guelph, Ontario N1G 2W1, Canada

Dr Eva Nöthig, Alfred-Wegener-Institut für Polar- und Meeresforschung, Am Handelshafen 12, D-2850 Bremerhaven, FRG

Dr Madhu Paranjape, NW Atlantic Fisheries Centre, PO Box 5667, St John's, Newfoundland A1C 5X1, Canada

Prof Larry Pomeroy, University of Georgia, Institute of Ecology, Athens, GA 30602, Georgia, USA

Dr Julian Priddle, British Antarctic Survey, High Cross, Madingley Road, Cambridge, CB3 OET, UK

Dr Fereidoun Rassoulzadegan, Station Zoologique, Ecologie du Microzooplankton, BP 28, 06230 Villefranche sur Mer, France

Dr Chris Reid, Plymouth Marine Laborarory, Citadel Hill, Plymouth, Devon, PL1 2PB, UK

Prof Werner Reisser, Pflanzenphysiologisches Institut, Untere Karspuele 2, D-3400 Göttingen, FRG

Mr Robert Sanders, University of Georgia, Department of Zoology, Athens, GA 30602, USA

Dr Barry Sherr, College of Oceanography, Oregon State University, Oceanography Admin Bldg 104, Corvallis, OR 97331, USA

Dr Evelyn Sherr, College of Oceanography, Oregon State University, Oceanography Admin Bldg 104, Corvallis, OR 97331, USA

Dr Mary Jo Sibbald, Simon Frazer University, Burnaby, British Columbia, V5A 156, Canada

Dr Michael Sieracki, Virginia Institute of Marine Science, College of William and Mary, Gloucester Point, Virginia 23062, USA

Prof Michael Sleigh, University of Southampton, Department of Biology, Medical and Biological Sciences Building, Bassett Crescent East, Southampton S09 3TU, UK

Prof Gene Small, University of Maryland, Department of Zoology, College Park Campus, Maryland 20742, USA

Prof Victor Smetacek, Alfred-Wegener-Institut für Polar- und Meeresforschung, Am Handelshafen 12, D-2850 Bremerhaven, FRG

Mr James Sniezek, University of Maryland, Department of Zoology, College Park, Maryland 20742, USA

Prof Anthony Soldo, Veterans Administration Medical Center Research Service (151), 1201 NW 16 Street, Miami, FL 33125, USA

Dr Tony Stebbing, Plymouth Marine Laboratory, West Hoe, Plymouth, Devon PL1 3DH, UK

Dr Diane Stoecker, Woods Hole Oceanographic Institution, Biology Department, Woods Hole, MA 02543, USA

Miss Suzanne Ström, University of Washington, School of Oceanography WB-10, Seattle, WA 98195, USA

Dr Kozo Takahashi, Woods Hole Oceanographic Institution, Woods Hole, MA 05243, USA

Mr Arnold Taylor, Plymouth Marine Laboratory, West Hoe, Prospect Place, Plymouth, Devon PL1 3DH, UK

Dr Benno Herman Ter Kuile, Hebrew University of Jerusalem, The Interuniversity Institute of Eilat, PO Box 469, Eilat 88103, Israel

Dr Carol Turley, Plymouth Marine Laboratory, Citadel Hill, The Hoe, Plymouth, Devon PL1 2PB, UK

Dr Gotram Uhlig, Biologische Anstalt Helgoland, Meeresstation, D-2192 Helgoland, FRG

Dr Peter Verity, Skidaway Institute of Oceanography, PO Box 13687, Savannah, Georgia 31416, USA

Mr Johan Wikner, Umeå University, Department of Microbiology, 90187 Umea, Sweden

SUBJECT INDEX

NATO ASI Series G

NATO ASI Series G

Vol. 22: **Evolutionary Biogeography of the Marine Algae of the North Atlantic.** Edited by D. J. Garbary and G. R. South. 439 pages. 1990.

Vol. 23: **Metal Speciation in the Environment.** Edited by J. A. C. Broekaert, Ş. Güçer, and F. Adams. 655 pages. 1990.

Vol. 24: **Population Biology of Passerine Birds. An Integrated Approach.** Edited by J. Blondel, A. Gosler, J.-D. Lebreton, and R. McCleery. 513 pages. 1990.

Vol. 25: **Protozoa and Their Role in Marine Processes.** Edited by P. C. Reid, C. M. Turley, and P. H. Burkill. 516 pages. 1991.

GPSR Compliance

The European Union's (EU) General Product Safety Regulation (GPSR) is a set of rules that requires consumer products to be safe and our obligations to ensure this.

If you have any concerns about our products, you can contact us on ProductSafety@springernature.com

In case Publisher is established outside the EU, the EU authorized representative is:

Springer Nature Customer Service Center GmbH
Europaplatz 3
69115 Heidelberg, Germany

Batch number: 10761697

Printed by Printforce, the Netherlands